Partially Ordered Systems

Partially Ordered Systems
Editorial Board: L. Lam • D. Langevin

Solitons in Liquid Crystals
Lui Lam and Jacques Prost, Editors

Bond-Orientational Order in Condensed Matter Systems
Katherine J. Strandburg, Editor

Diffraction Optics of Complex-Structured Periodic Media
V.A. Belyakov

Fluctuational Effects in the Dynamics of Liquid Crystals
E.I. Kats and V.V. Lebedev

Nuclear Magnetic Resonance of Liquid Crystals
Ronald Y. Dong

Electrooptic Effects in Liquid Crystal Materials
L.M. Blinov and V.G. Chigrinov

Liquid Crystalline and Mesomorphic Polymers
Valery P. Shibaev and Lui Lam, Editors

Micelles, Membranes, Microemulsions, and Monolayers
William M. Gelbart, Avinoam Ben-Shaul, and Didier Roux, Editors

William M. Gelbart   Avinoam Ben-Shaul
Didier Roux   Editors

Micelles, Membranes, Microemulsions, and Monolayers

With 220 Illustrations

Springer-Verlag

New York Berlin Heidelberg London Paris
Tokyo Hong Kong Barcelona Budapest

William M. Gelbart
Department of Chemistry
and Biochemistry
University of California
405 Hilgard Avenue
Los Angeles, CA 90024-1569
USA

Avinoam Ben-Shaul
Department of Physical
Chemistry
Hebrew University
91904 Jerusalem
Israel

Didier Roux
CRPP
Domaine Universitaire
33405 Talence Cedex
France

Library of Congress Cataloging-in-Publication Data
Micelles, membranes, microemulsions, and monolayers / [edited by] William M.
  Gelbart, Avinoam Ben-Shaul, Didier Roux.
     p.   cm. — (Partially ordered systems)
  Includes bibliographical references and index.
  ISBN-13:978-1-4613-8391-8
  1. Surface active agents.   2. Micelles.   3. Emulsions.
  I. Gelbart, W. (William)   II. Ben-Shaul, A. (Avinoam)   III. Roux, D. (Didier)
  IV. Series.
  TP994.M53   1994
  541.3′3 — dc20   94-15496

Printed on acid-free paper.

©1994 Springer-Verlag New York, Inc.
Softcover reprint of the hardcover 1st edition 1994

Production managed by Hal Henglein; manufacturing supervised by Vincent Scelta.
Camera-ready copy prepared from the editors' TeX files.

9 8 7 6 5 4 3 2 1

ISBN-13:978-1-4613-8391-8   e-ISBN-13:978-1-4613-8389-5
DOI: 10.1007/978-1-4613-8389-5

Preface

Over the course of the past one to two decades, the study of surfactant solutions has been profoundly transformed by a dramatic infusion of new ideas and techniques from Chemistry, Physics, and Materials Science. This renaissance in fundamental research activity has been sparked largely by the many connections and analogies that have been established between micellar phases and microemulsions on the one hand and polymer systems and interfacial films on the other. Consequently, many otherwise intractable conceptual and technical problems arising in the self-assembly area have now become feasible to study because of theoretical and experimental breakthroughs in the general field of complex fluids. For example, the theory of critical phenomena and polymer structure/dynamics (including scaling and renormalization group ideas) has been especially useful, as have new high-resolution scattering and magnetic resonance spectroscopies.

The purpose of this book is to develop a systematic account of the exciting developments referred to above. Part of our effort is devoted to providing a general introduction to the broad range of phenomena involved and part to offering a critical consensus of what is presently understood and what is not. While the book consists of twelve chapters by different authors, we have specifically edited them so that they reinforce one another in content, format, and notation. Almost every page of each chapter contains an explicit reference to related sections of other chapters. A single subject index at the back of the book refers to all chapters simultaneously. In the remainder of this Preface we make some general remarks about the physical systems and problems involved, and about how they are treated in the present monograph.

First, though, a few words about the evolution of this particular volume. For several years, the editors had been encouraged to write a book that would go beyond the usual collections of disjointed articles and compendia of conference proceedings, etc. These latter types of volumes, after all, do not provide the "uninitiated" but interested reader with a sufficiently incisive or critical introduction to the field. Often, in explaining our work to colleagues and visiting researchers in different areas of physical chemistry and condensed matter science, we have been asked for instructive references where they might follow up on these discussions. In response, we came up with the idea of an edited volume, where the chapters would be written by different experts in the various subfields but where each contribution would be revised to complement and dovetail with all of the others, i.e., each author would have to agree beforehand to have their essay substantively edited with this end in mind. While we are confident that none of them has regrets in having assigned us these rights, we are also sure that they will never forgive us for having taken so long before we buckled down and completed this time-consuming job. Since many chapters were contributed for the first time as early as 1987 and 1988, and since several of them were written by colleagues from France, we thought it appropriate that the volume should appear in 1989, in time to celebrate the bicentenary of the French Revolution. When 1993 came along and we still hadn't found sufficient time for revising and transcribing all the texts, we decided to commemorate instead the notorious Terror of 1794. The authors have graciously managed to continue to speak to us over the past few years and to humor us in our belief that the long delays involved might indeed make the subject material more timeless, even if less timely. In most cases, references have been updated and paragaraphs added at the appropriate places to follow up on the earlier work described and to apprise the reader of some most recent developments.

A surfactant, or amphiphile—"loving" ("philo") "both" ("amphi")—molecule is made of two parts that have opposing natures: one is water soluble (hydrophilic) and the other is oil soluble (hydrophobic). The hydrophilic and hydrophobic parts of the surfactant are linked together by a chemical bond and consequently cannot phase separate as they would if the two parts were free. When such molecules are put in water they prefer the (water/air) surface, where the hydrophilic "heads" and hydrophobic "tails" lie, respectively, in and out of the water. Indeed, for a small amount of surfactant, they practically all lie at the interface. As a consequence, the (liquid-vapor) surface tension decreases as the concentration of surfactant increases. Then, at a certain concentration (the Critical Micelle Concentration, or "CMC"), the surface tension levels off and remains nearly constant. Careful study of this phenomenon confirms that the added surfactant molecules no longer go preferentially to the surface but rather go into solution in the bulk of the aqueous phase. There the molecules organize as small aggregates—*micelles*—which are often globular in shape, the tails comprising the interior and the heads coating the surface.

In both the very low concentration regime, where most of the added molecules are at the surface, and in the higher concentration case, where aggregates are formed, the physical phenomenon responsible for such behavior is referred to as the *hydrophobic effect* and is due to a subtle balance between intermolecular energies and entropies. (This is the same hydrophobic effect that underlies the (classic) immiscibility of oil and water.) For concentrations of surfactant that greatly exceed the CMC, there is a negligible number of molecules that sit at the surface or remain as monomers in solution. The aggregated molecules, on the other hand, reveal themselves in a large variety of structures. Indeed, upon increasing further the concentration of surfactant, long-range ordered phases may appear, such as lamellar or hexagonal states: these phases are commonly classified under the name of *lyotropic liquid crystals*. Furthermore, adding oil to micellized surfactant solutions leads to many different structures and phases of *microemulsion*.

The industrial applications of amphiphilic molecules were recognized very early. The cleaning power of soaps is probably one of the oldest and best-known exploitations. But the practical uses of amphiphilic species have increased dramatically in recent times, including a variety of important applications in the pharmaceutical, cosmetics, and oil industries. Perhaps most spectacular of all will be the use of surfactant layers (as Langmuir-Blodgett films) in the fabrication of new optical and electronic devices. With compelling impetus from these developments, fundamental scientists have also recognized the importance of studying surfactants in solution. A large amount of work was devoted in the 1940s and 50s to understanding the phase behavior and structural properties of many amphiphilic systems. In addition, the discovery that biological membranes in living cells are integrally composed of lipids, which are basically "just" surfactant molecules, has inspired many studies of bilayers (*lamellae*) as simple models for cell membranes. In the 1960s, a remarkable series of x-ray studies (which even now remain up to date) established the similarity of behavior between lipids and "simple" amphiphiles and identified the different types of structures that can be found in both biological and "ordinary" surfactant systems.

As already remarked, in the last decade or two, the field of surfactants in solution has undergone profound changes. A good part of this rebirth is due to the development of new understanding in the area of thermotropic ("neat") liquid crystals. Indeed, at the time in the early 1970s when the modern era of liquid crystals was launched, lyotropic systems were actively being investigated as a special class of these aligned fluids. The systematic studies of micellar phase diagrams led to important discoveries such as biaxial nematics (which to this day have not been discovered for thermotropic systems), and in-plane defects in what had long been considered as regular ("classical") lamellae. For obvious reasons, the main interest of the liquid crystal community was focused on long-range ordered phases, with scant attention paid to the isotropic solutions composed of small globular ("spherical") micelles. The oil crisis in the 1970s, and the possibility of using microemulsions for enhanced oil recovery, had (among many other consequences) the effect of attracting the interest of an increasing number of scientists to these fascinating *disordered* phases.

The many problematic definitions of microemulsions in the early literature were a natural reflection of the scientific community's ignorance of what they really were. Now that much more is known, having a precise definition seems to be of less importance! We simply recall that for certain types of surfactants, or combination of surfactants, it is possible to mix water and oil in all proportions with only a small amount of surfactant (just a few percent in favorable cases). The *thermodynamically stable* phase obtained is liquid, isotropic, optically transparent, and is conveniently called microemulsion. A great deal of work has been devoted to understanding the structure and stability of such phases, and only now can a coherent description of this state be given.

From all the work that has been done on the states of surfactants in solution, there emerges a fundamental new concept: these phases are often better described as phases of *surfaces* than as phases of *particles*. The ability of surfactants in solution to aggregate can in fact be seen as the ability to create surfaces in the bulk of the solution. This is true not only for the lyotropic liquid crystal phases but even more so for the isotropic fluids such as microemulsions. The statistical physics of fluctuating surfaces is a field that is currently in initial but fast-moving stages of development: it is too early to describe the full behavior of amphiphilic systems exclusively in terms of phases of surfaces—but it is clearly the direction that will be followed in the next few years. In pursuing this course, we will continue to benefit from describing these phases by means of what is known and well established in other fields, notably those of simple fluids, liquid crystals, polymers, and quasi-two-dimensional systems.

In this spirit, we shall emphasize throughout this monograph both what *can* be learned about amphiphilic systems via analogies with these other areas and—still more importantly—what can *not*. We proceed, then, by dividing the characteristic features of amphiphilic systems into two classes. On the one hand, the properties of surfactant solutions correspond to well understood behaviors of other systems *but with a different range of physical parameters*. This is largely the case, say, for the existence of uniaxial nematic phases (Chapter 3), the isotropic phase of interacting microemulsion droplets that can be considered as an example of colloidal suspension (Chapters 7 and 9), and the critical behavior of micellar and microemulsion systems (Chapter 11). But, on the other hand, entirely new concepts have also to be developed. This is most notably the case for micellar growth in dilute surfactant solutions (Chapters 1 and 2), size/alignment coupling in micellar liquid crystals (Chapters 1 and 3), curvature frustration in systems of parallel films (Chapter 4), fluctuations in dilute lamellar phases (Chapters 5 and 6), and bicontinuous states of microemulsions (Chapters 7-9). Also, in systems of adsorbed surfactant monolayers, dramatically low interfacial tensions can occur (see Chapter 10), and the nature of the special phase transitions that arise is due to the inter- (half 2-D/half 3-D) dimensionality of the interfacial film (Chapter 12). In these latter instances, we are confronted by wholly new phenomena for which it is no longer sufficient to make simple analogies with "ordinary" liquid crystal or polymer fluids.

We stress that the fundamental conceptual differences between micellar solutions and "ordinary" colloidal suspensions do not arise only as special cases involving extreme circumstances. Rather, they are often unavoidable, dominating the observable properties under virtually all conditions. Even in a dilute ("ideal solution") phase of micellar aggregates, for example, new phenomena appear because the basic "particles" involved—*micelles*—*do not maintain their integrity as the concentration or temperature, say, is varied.* Instead, the equilibrium position of the exchange of molecules between aggregates is shifted. That is, not only do we have more "particles" as we add surfactant, but we also see a change in their average size. The extent to which there are increases in the *number* of aggregates, versus changes in their *size*, depends on the details of the surfactant and aqueous solvent involved. At higher concentrations, where interactions between the aggregates become important, the onset of long-range orientational and positional order can be shown to enhance *further* the size of the micelles. Clearly, we are dealing with a situation in which—unlike the ordinary colloidal suspension—the experimentalist can only control the total number of added *molecules*: the number of *aggregates* and hence *the distribution of sizes* will be determined by the statistical thermodynamics of exchange and the degree of long-range order.

After a long period during which the physics of amphiphilic systems was considered as essentially descriptive, important breakthroughs in both experimental and theoretical studies have made possible precise, quantitative accounts of many classes of these systems. On the experimental side, it has become possible to probe directly the intramolecular structure and dynamics of surfactants in aggregates. Selective deuteration and relaxation spectroscopies, developed in the context of nuclear magnetic resonance techniques, have been especially fruitful. Similarly, the overall shapes and sizes of micelles—and the polymorphism and defect characteristics of their ordered phases—have been incisively investigated by a concerted combination of static and dynamic light scattering, synchrotron x-ray diffraction, and small angle neutron spectra. Novel fluorescence and electron microscopies have also been developed and applied. On the theoretical side, physicists and chemists have turned to amphiphilic systems from the more "standard" areas of simple fluids, liquid crystals, polymers, and thin films, bringing with them the powerful conceptual techniques of many-body perturbation theories, continuum and scaling approaches, fluctuation and critical phenomena, symmetry analyses, and the statistical mechanics of model hamiltonians. In this process, several classic problems and phenomena—including ultra-low interfacial tensions and adsorbed monolayer phase transitions—have finally been put on a firm conceptual footing.

There are basically two prevailing and complementary levels of phenomenological description of aggregates in solution. The first is couched in terms of individual molecules and tries to deduce the structure and phase behavior of their aggregates from the hamiltonian and free energy of such molecules in an aqueous solvent. While *in principle* such an approach could proceed from detailed interaction potentials between each surfactant and water molecule, followed by explicit molecular dynamics or Monte Carlo simulations, *in practice* we must await several new generations of computer power before it will be feasible to carry out *a priori* calculations of this kind with enough particles and for sufficiently long times. For example, to describe the spontaneous formation of just a single micelle at low concentrations (10^{-5}M, say), one needs to consider at least 10^6 water molecules and calculate for as long as billions of picoseconds; furthermore, interaction potentials between all of the relevant surfactant and aqueous solution species must be known to far better than current accuracy. Accordingly, comprehensive approaches that start from the individual molecules have necessarily been of the phenomenological kind in which spins on a lattice are introduced, for example, with many-site interactions being used to keep surfactant at the oil-water interface with its preferred curvature. By contrast, the second level of description starts from the outset with already-formed aggregates of specified shapes. Here the focus is on the *interface* between hydrophobic and hydrophilic regions, rather than on any single molecular species. A free energy is associated directly with this interface, featuring separately its *bending (curvature) elasticity* and its *topological entropy*. The power of this continuum approach lies in its natural ease in predicting the coupling between self-assembled (*meso* -scopic) structures on the one hand, and phase transition (*long* -range) behavior on the other, even as it gives up the possibility of describing structure and order on a *single* molecule level.

The aim of the book is to present in a series of twelve chapters a carefully chosen set of problems on which sufficient progress has been made to provide agreed-upon starting points for further work in fundamental amphiphilic science. We intend this monograph specifically to serve as an introduction for both young and established researchers who are interested in moving into this "new" (renewed) and challenging field. As mentioned earlier, much effort has been devoted to cross-referencing the discussion, with the intention of emphasizing the several common concepts that underlie the broad range of physical effects covered. By emphasizing these central themes we hope to focus further attention on the coupling of self-assembly to thermodynamic and long-range ordering variables, the topology of defects, fluctuations, and structure of "inter"-dimensional (i.e., quasi 2D and 3D) systems, and phases of surfaces. All of these ideas play a crucial role in understanding the basic physical properties of amphiphilic systems, and point up the dramatic contrasts with "simple" fluids, "ordinary" liquid crystals and colloidal suspensions, and "conventional" solid-state systems.

Contents

7 The Structure of Microemulsions: Experiments
Loïc Auvray **347**

8 Lattice Theories of Microemulsions
Gerhard Gompper and Michael Schick **395**

1

Statistical Thermodynamics of Amphiphile Self-Assembly: Structure and Phase Transitions in Micellar Solutions

Avinoam Ben-Shaul[1]
William M. Gelbart[2]

1.1 Introduction

The main purpose of this chapter is to present a comprehensive, statistical-thermodynamic framework for treating the sizes and shapes of micellar aggregates in aqueous surfactant solutions. In this first section, after a brief historical perspective on experimental studies and theoretical pictures of micellar structure, surface roughness and other fluctuation effects are discussed in the context of both phenomenological and computer simulation studies. In Sec. 1.2 we present the basic self-assembly theory for dilute systems; the aim is to proceed systematically from a formulation of overall partition functions for the aqueous surfactant solution to a phenomenological discussion of effective chemical potentials. By deriving this latter language, we make contact with the highly useful "law of mass action" approach which has been pursued by most workers in their analyses of micellization and size/shape effects in dilute solution. We also feature there the role of "dimensionality of growth", i.e., the basic differences between the concentration dependence of equilibrium sizes in the case of sphere ("zero-dimensional"= 0D)-, rod (1D)-, and disk (2D)-like aggregates. Following a brief discussion of the structures and relative stabilities of vesicles and mixed micelles, we close Sec. 1.2 with some remarks on the much-neglected

---

[1]Department of Physical Chemistry and The Fritz Haber Research Center for Molecular Dynamics, The Hebrew University, Jerusalem 91904, Israel

[2]Department of Chemistry and Biochemistry, The University of California, Los Angeles, California 90024, USA

and highly problematic question of "rotation/translation" contributions to micellar partition functions (and hence to the effective surfactant chemical potentials).

Section 1.3 treats the problem of separating the effects of amphiphilic "heads" and "tails" in statistical thermodynamic descriptions of micellar aggregates. The prevailing ideas concerning predictions of preferred shapes are discussed critically. We also outline there the nature of chain packing in micelles and the contribution of chain degrees of freedom to the free energies of surfactant molecules in aggregates of different curvature; in this context, a microscopic basis is provided for the bending elasticities of amphiphilic mono- and bi-layers. In Sec. 1.4 we introduce the reader to recently developed ideas concerning the way in which interactions between micelles can lead first to the enhancement of size in isotropic solutions and then to the onset of orientational and positional long-range order in nematic, hexagonal and lamellar states of concentrated surfactant-water systems.

The concept of a *micelle* as an aggregate of surfactant molecules arose in the early part of this century when McBain [1] in 1913 first sought to understand the anomalous concentration dependence of many physical properties of aqueous soap solutions. In the words of Debye [2]: "Soap solutions exhibit even lower osmotic activity than would be predicted if one assumed that soap existed in solution as simple undissociated molecules. They also conduct the electric current far better than would be expected from the observed osmotic effects." Figure 1.1 represents these effects and correlates their breaks in slope with sudden rise of turbidity and with leveling off in surface tension reduction, i.e., with onset of micelle formation in bulk solution. To continue in the words of Debye, writing in 1949: "Since 1913, investigators have shown considerable interest in the determination of the size and shape of the micelle." As an introduction to the subject of this chapter, in fact, it is useful to proceed further with a look at the work of Debye.

The experiments which Debye reported in his 1949 paper [2] involved light scattering studies of a series of n-alkyl trimethyl ammonium bromides $[C_nH_{2n+1}N(CH_3)_3^+Br^-]$. From the plots of turbidity (τ) vs. surfactant concentration (c) it was concluded that, above a *critical micelle concentration* (CMC) aggregates appear as the dominant species: "The curves resemble typical τ vs. c plots for polymer solutions." For these latter systems Debye had already shown that a plot of turbidity vs. concentration would yield the molecular weight M as its slope at the origin ($c \to 0$). Hence the "Debye equation" for molecular weights of micellar aggregates [3]

$$\frac{K(c - CMC)}{\tau} = \frac{1}{M} + 2A_2(c - CMC) \qquad (1.1)$$

where K is the "optical" constant (determined by the incident wavelength

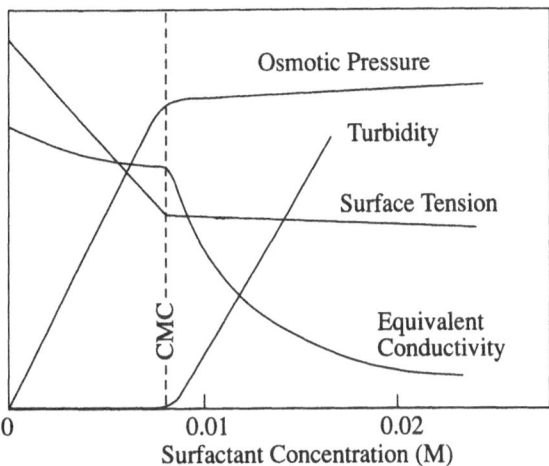

FIGURE 1.1. Effect of micellization on bulk properties of surfactant solutions. Note, for example, that the osmotic pressure is proportional to the total concentration of *particles* (monomers plus micelles). Above the CMC the added surfactants form micelles, and the increase in total particle concentration is small. On the other hand, the turbidity is proportional to the concentration of micelles. The CMC value and the concentration scale correspond to an aqueous solution of SDS (sodium dodecyl sulfate) (after Refs. 3 and 33).

and indices of refraction of the solvent and solution) and A_2 is the second (osmotic pressure) virial coefficient. Figure 1.2, reproduced from the original work of Debye [2], shows the first light scattering estimates of micellar size. "The heights of the vertical lines drawn at the critical concentration represent the reciprocal molecular weights of the micelles" [2]. Note how micellar size increases with carbon number.

Debye also suggested an additional technique for determining micellar size. He observed in particular that "in the presence of relatively high salt concentrations the micelles of the longer-chain detergents are large enough to cause measurable disymmetry in the intensity of the scattered light" [4]. On the basis of such measurements (on n-hexadecyl trimethyl ammonium bromide in 0.2M KBr) it was concluded that the micelles underwent a transition from spherical to rod-like aggregates upon increase in surfactant concentration above the CMC. This idea has been refined and pursued vigorously by many workers during the past thirty-five years. Ikeda *et al.* [5], for example, have measured the angle dependence of the

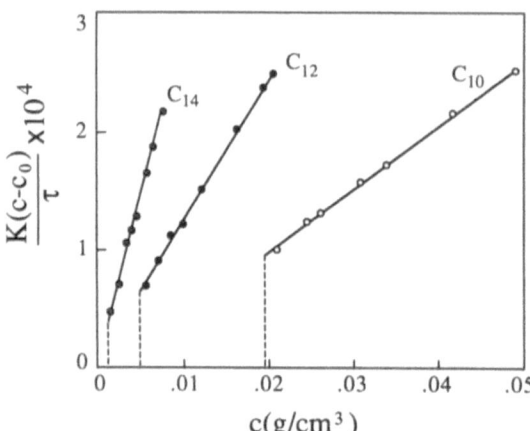

FIGURE 1.2. Reciprocal specific turbidities vs. surfactant concentration for solutions of n-alkyl trimethyl/ammonium bromides in water [2].

turbidity as a function of surfactant concentration in high-salt (0.8M NaCl) aqueous solutions of dilute sodium dodecyl sulfate (SDS). Figure 1.3 shows some of their results for $T = 35°C$. The intercept of each curve gives the reciprocal weight-average molecular weight (assuming $A_2 = 0$ in (1.1)), and the slope gives the radius of gyration. Thus, for the high concentration curve, $c = 1.10 \times 10^{-2} g/cm^3$ (≈ 100 CMC), one finds an aggregation number of ≈ 1000 molecules/micelle (about six times the size at the CMC). The average micellar length, if the micelles are rod-like, is found to be ≈ 600Å, as compared to their diameter which is estimated to be ≈ 40Å. Better estimates of the size can be obtained by fitting the data to (1.1) with A_2 calculated assuming that the micelles are rigid rods [5].

An entirely independent technique for determining the size of micelles involves *dynamic* light scattering in which the homodyne autocorrelation function [6] of the quasielastic intensity is measured. Application of this approach to surfactant solutions was developed primarily by Benedek and coworkers [7] in the U.S. and by Corti and Degiorgio [8] in Italy. Approximately what one does is extract the mean translational diffusion coefficient D from the autocorrelated intensity and then use the Stokes-Einstein relation to infer the hydrodynamic radius $R_H = kT/6\pi\eta D$, where k is Boltzmann's constant and η is the solvent viscosity. (See Sec. 2.2.1 for further discussion of this light scattering approach.) In practice the spread in micellar sizes leads to a nonexponential homodyne signal, and a cumulant analysis is necessary to deduce the first several moments (and hence the average, variance, skewness) of the distribution.

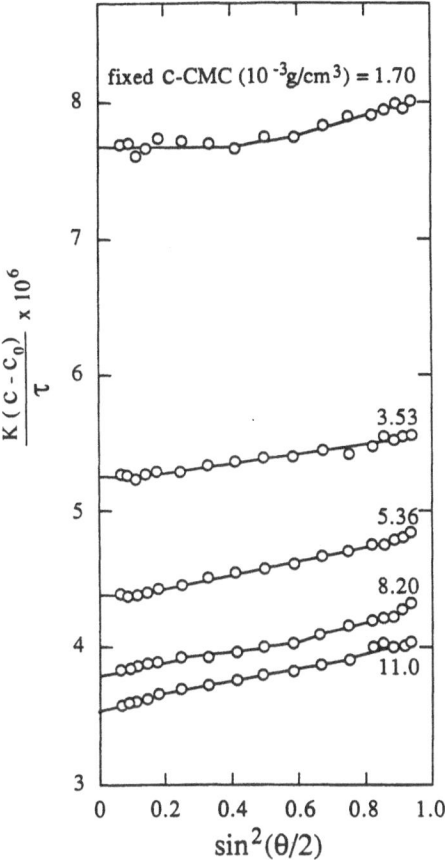

FIGURE 1.3. Angle dependence of light scattering intensity from sodium dodecyl sulfate solutions in 0.80M NaCl at 35°C [5].

Another complication arises because the measured diffusion constant contains contributions from the interactions between micelles. More explicitly, D varies with concentration c according to

$$D = D_0(1 + k_D c) \qquad (1.2)$$

where k_D is a simple functional of the inter-aggregate potential [9]. A similar relation holds for the "apparent" ("intrinsic", i.e., single-micelle) molecular-weights which contribute to the *static* intensity—see (1.1). It follows that the increase of M and R_H with concentration need not imply a growth of micelles, but rather "only" the effect of interactions. The relative importance of the two effects depends on the composition (i.e., surfactant, added salt, etc.) and temperature of the solution in question.

In the late 1970's and early 1980's there was considerable controversy surrounding the interpretation of several quasielastic light scattering studies of micellar size in ionic amphiphilic systems—see [5,7] and the more recent overviews given by these same workers [10]. It is now agreed that a high concentration of added salt is necessary to suppress (screen) the interaggregate (electrostatic) potential contributions to the diffusion coefficient; only at large ionic strength can one deduce safely that the mean micellar size is truly increasing with surfactant concentration. We note that polydispersity and interaction contributions complicate similarly the inference of micellar size from viscosity [11] and dynamic Kerr (electric field-induced birefringence) [12] measurements.

Valuable information about the structure of micelles in dilute solution is also provided by small-angle neutron scattering data [13]. In these experiments one is probing heterogeneities on length scales d of order $10 - 1000\text{Å}$, corresponding to scattering vectors with magnitudes $q = 2\pi/d$ in the range 0.006 to 0.6Å^{-1}. At this spatial resolution the different micellar regions and the solvent can be regarded as continuous media, with the scattering amplitude for the nth aggregate given by

$$A_n(q) = \int_{V_n} dr e^{i\mathbf{q}\cdot\mathbf{r}}[\rho_n(\mathbf{r}) - \rho_s]. \tag{1.3}$$

Here $[\rho_n(\mathbf{r}) - \rho_s] \propto [\sum_{i \varepsilon n} b_i \delta(\mathbf{r} - \mathbf{r}_i) - \rho_s]$ is the excess (with respect to solvent's) density of scattering length in the nth micelle, b_i being the nuclear scattering length for the ith nucleus in n (located at $\mathbf{r}_i$) [13] and V_n is the nth aggregate's volume. The static ("elastic") intensity measures the thermal average of the square of this amplitude, after it has been summed (with appropriate phase factors) over all aggregates:

$$I(q) \propto \left\langle |A(q)|^2 \right\rangle \tag{1.4}$$

with

$$A(q) = \sum_{n=1}^{N} A_n(q) e^{i\mathbf{q}\cdot\mathbf{R}_n} \tag{1.5}$$

where $\mathbf{R}_n$ is the center of mass position of the nth aggregate, and $\langle \cdots \rangle$ denotes the thermal average of $\cdots$.

For globular micelles, weak correlations between the positions of aggregates and their orientations and sizes allow the scattering cross section to be approximated by [14]

$$I(q) \propto \left[\left\langle |A_1(q)|^2 \right\rangle + \langle A_1(q) \rangle^2 \left(S(q) - 1 \right) \right]. \tag{1.6}$$

Here, $S(q)$ is the usual *structure factor* describing the interferences between the centers of mass of different micelles, i.e., the Fourier transform of the pair correlation function. $\langle A_1(q) \rangle$, the single particle *form factor*, is the Fourier transform of the *average* distribution $\rho(\mathbf{r}) - \rho_s$ of excess scattering length within an aggregate, the brackets denoting specifically an average over all sizes. Similarly, $\langle |A_1(q)|^2 \rangle$ is the transform of the average "Patterson function" $\int d\mathbf{r} [\rho(\mathbf{r}) - \rho_s][\rho(\mathbf{r} + \mathbf{R}) - \rho_s]$ for an aggregate [15], containing therefore the effects of nonspherical shape and size polydispersity. Note then that, even in the absence of interparticle interferences (i.e., for situations where only the first term in (1.6) is important) the interpretation of $I(q)$ is problematic. Hayter [16] has shown, for example, that for any solution of *monodisperse ellipsoids* there corresponds a *polydisperse sphere* system which leads to the same form of $\langle |A_1(q)|^2 \rangle$. At higher concentrations, where $S(q) \neq 1$, it becomes only more difficult to infer the micellar sizes and shapes.

Cabane *et al.* [17] have emphasized the importance of large q measurements for determining the fluctuations in micellar size and shape. They argue in particular that for scattering data extending out only to some maximum wavenumber q_m, details in structure corresponding to distances smaller than π/q_m will not be resolved. Using "contrast variation tricks", however, it is possible to measure separately the average radius (R_c) of the hydrocarbon core of a micelle and that (R_p) of the *whole* aggregate (core plus layer). "Contrast" between density profiles is achieved via H/D isotopic substitution which shifts the scattering lengths (b) of methylenes from -0.083 (CH_2) to $+1.999$ (CD_2) and of methyls from -0.457 (CH_3) to $+2.666$ (CD_3); similarly, for $H_2O \rightarrow D_2O, b = -0.168 \rightarrow +1.92$. It is thus possible to "label" a surfactant molecule in many different ways (e.g., degrees of alkyl deuteration); then, by using alternately H_2O or D_2O as a solvent, one can generate several micellar solutions and hence as many relations between the number densities $n_i(r)$ of methyls, methylenes, polar heads, and counterions in and around the micelle of a given proposed structure. More explicitly, one models the density averaged scattering length distribution $\rho(r)$ as a linear combination of nuclear scattering lengths $b_i : \rho(r) = n_1(r)b_1 + n_2(r)b_2 + \ldots$. Then selective deuteration ("contrast") provides the required number of linear relations for determining the spatial distribution of the various functional groups (e.g., methyl, methylene, etc.) in the aggregate. This procedure requires, of course, that such isotopic substitutions do not disturb the micellar structure [18].

In virtually all cases, the elastic neutron scattering techniques described above have been restricted to studies of essentially *globular* (i.e., approximately spherical) micelles [19]. Small q data have been exploited to obtain much information about the average structural features of these surfactant solutions, such as the average aggregation number N and the average charge Z of a micelle. In extracting such information, the structure factor, $S(q)$ in (1.6), has been estimated from approximate analytical theories of interferences between charged hard spheres in water [20]. Since differences between $\langle |A_1(q)|^2 \rangle$ and $\langle A_1(q) \rangle^2$ are small at low q, the micellar model is defined crudely by only two parameters, N and Z, i.e., fluctuation effects involving size/shape and charge are suppressed. $\langle A_1(q) \rangle^2$ and

$S(q)$—and hence $I(q)$— are calculated for each choice of average values, N and Z, and correlated iteratively via successive comparisons between computed and observed (low q) scattering distributions. As already stressed above, *large q* data are required to deduce meaningful information about *fluctuations* from the average micelle. Many workers have discussed how these latter experiments, as well as $q \to 0$ "contrast variation" studies, can provide details on the nature of size and shape distributions for globular ("almost spherical") aggregates and of the structures of interfacial regions and the conformational statistics of alkyl chains in these micelles. Again, however, scattering analyses of this kind are no longer possible in the case of anisotropic and polydisperse (e.g., long rod-like) aggregates at higher concentrations (see Chap. 2).

At high amphiphile concentrations the aggregates organize into positionally ordered phases. In this chapter we focus our interest on *finite* rod-like micelles in *positionally disordered* solutions, but also comment briefly on their evolution in columnar phases (see Sec. 1.4.3). To complement our present introduction, we remark as well on the classic (i.e., essentially infinite aggregate) lamellar and hexagonal phases of surfactant/water systems. (The lamellar states are treated comprehensively in Chaps. 4, 5 and 6.) Here, the basic structural determinations were provided more than thirty years ago by Luzzati, Mustacchi, Husson and Skoulios [21]. They used X-rays to measure organizational changes in solutions of fatty acids in water as a function of temperature and concentration, the Bragg diffraction pattern for each long-range translationally ordered phase being analyzed in terms of its characteristic reflection indices [22]. For uniformly spaced reflections $(1/d, 2/d, \ldots \text{Å}^{-1})$, for example, the aggregation state corresponds to stacked *bilayers* (lamellae) with spacing (one-dimensional lattice constant) d. Knowing d and the overall volume fraction of amphiphile in water one can deduce the lamellar thickness D. The sharp Bragg reflections typically observed in these systems provide direct evidence for the minor role played by fluctuations in the lamellar phase. In other words, bilayer packing of the surfactant molecules is the overwhelmingly preferred curvature mode under high concentration (and/or high salt, added alcohol) conditions. (See, however, the discussion of fluctuation effects in Chaps. 5 and 6.) Similarly, the interaxis spacing d' and diameter D' corresponding to *cylindrical* aggregates in *hexagonal* states can be inferred for each of these soap systems at higher water content. A typical result is shown in Fig. 1.4. Considerable experimental [23,24] and theoretical [25] work has been done recently to further document and explain data of this kind. But many unanswered questions remain concerning the nature of defects in those systems [26–28], the mechanism of their phase transition [26,27], and the microscopic details of the chain packing and head-group organization in them (see Sec. 1.3).

We have already noted that most experimental investigations have been concerned with small globular micelles just above the CMC. Similarly, a considerable theoretical effort has been devoted to describing head-group

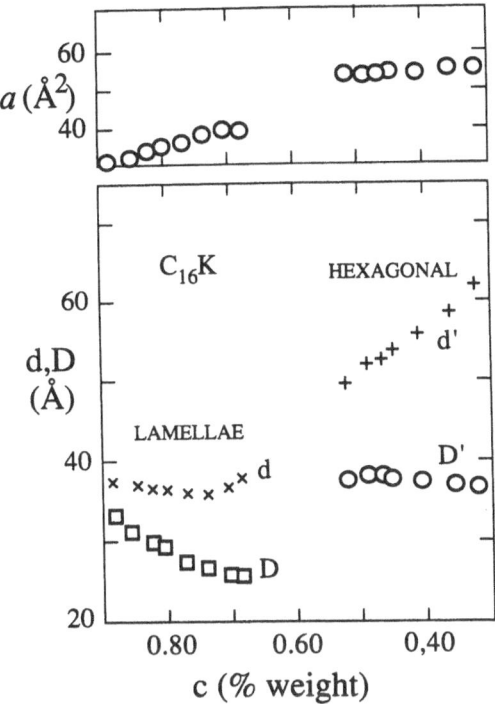

FIGURE 1.4. Distance between sheets (d) and cylinders (d') and their thicknesses (D and D') in lamellar and hexagonal phases of $CH_3-(C_2)_14COO^-K^+$ (palmitic acid salt), as a function of concentration. These data can be used to calculate the area, a, per surfactant head-group in each of the two phases as a function of concentration, as shown in the upper panel [21d].

organization in the interfacial region, and "tail" conformations in the hydrophobic core, of these "minimum" ("almost spherical") aggregates. Gruen [29] has presented a critical synopsis of the available experimental findings. His analysis supports the "standard" picture according to which [30–33]:

- on average, almost all of the hydrocarbon chain of each amphiphile lies within the micellar core.

- hydrophilic species (i.e., head-groups and aqueous solution) are nearly completely excluded form the core.

- the amphiphilic (alkyl) tails fill the core at a nearly uniform, approximately liquid n-alkane, density, the semi-flexible chains showing a high degree of conformational disorder.

In essence, this approach asserts that a useful, first-order, picture of micellar structure follows from a division of the amphiphilic volume into *core* and *interfacial* regions, each described by relatively simple geometries. On a more molecular scale, it argues that many basic features of micellar organization (e.g., chain statistics, solubilization properties, etc.) can be accounted for without giving up the notion of a well-defined aggregate. In particular, as we shall see in Sec.1.3 below, the assumption of a dry, liquid-like core with monomer (CH_2 segment) density $\rho_{CH_2} \approx \rho_{CH_2 (liquid\ alkane)}$ can be shown to yield bond orientational order profiles and labeled segment distributions in close agreement with experiment. The probability of a terminal segment (methyl group) sitting at the surface of a spherical micelle, for example, is significant, even though its most probable location is halfway to the center. Indeed, *all* segments have a non-negligible probability of lying at the surface, because most of the volume of a small globular aggregate is associated with its outer shell. It is this simple geometric fact which reconciles a dry hydrophobic core with large interfacial contact between chain segments and water: there is no need to insist on significant penetration by water. (For alternative models of micelle structure, see, e.g., [34,35].)

Recently, several machine simulation attempts have been made to test the above picture of micellar structure. The earliest efforts, Monte Carlo (MC) calculations by Pratt *et al.* [36], do not treat water or ionic surfactants on a truly microscopic level, introducing instead a phenomenological account of short chains on a lattice. Similarly, the molecular dynamics (MD) calculations of Haile and O'Connell [37] do not directly test the nature of equilibrium aggregates, because they essentially *impose* a given shape by constraining the amphiphile head-groups to undergo translations and small amplitude oscillations within a spherical shell. Aqueous solvent does not appear explicitly in their theory. Instead, the hydrophobic effect involving water and methylenes is incorporated by a short-ranged repulsive shell displaced slightly from the head-group sphere. Nevertheless, one can obtain useful information from these sorts of simulations about the segment density distributions and bond conformational statistics. Indeed, by following the molecular dynamics of chains interacting in this geometry via realistic potentials, one generates an essentially exact solution to the problem of constrained surfactants. In this context Haile and O'Connell present a detailed comparison between their results and those obtained in earlier, mean-field theories which also assume a dry spherical core with head-groups confined near the surface. Good agreement is found, for reasons which we shall explain in Sec. 1.3.3, where we treat in some detail the mean-field approach to chain statistics.

In order to address on a microscopic level the still more fundamental questions of micellization, it is necessary to consider machine simulations which do not impose at the outset any particular form for the equilibrium aggregates. Indeed, the micelle must be observed to form spontaneously from our having simply put surfactant molecules into water at high enough concentration. First steps in this direction have been taken by Jönsson, Edholm and Teleman [38], and by Watanabe, Ferrario and Klein [39], who studied the formation of sodium octanoate micelles in aqueous solution via molecular dynamics simulations. Here, the water is described by the point-charge effective pair potential which is known to give a good account of bulk liquid water properties. Watanabe *et al.* introduce site-site interactions and conformational energies involving the alkyl chains, and Lennard-Jones combining rules for all atoms and pseudo-atoms (e.g., CH_2, CH_3) as well as charges on the head-group oxygens and carbons and, of course, on the sodium ions. Long range Coulomb interactions between charges are calculated via Ewald summation. (Jönsson *et al.* introduce somewhat different charges, and simply truncate the electrostatic forces beyond 10Å.) Periodic boundary conditions are used with a cubic box having side length 34.2Å containing 15 surfactant molecules ("monomers") and 1068 waters. Note that this situation necessarily implies a mole fraction of over 1%, i.e., a concentration of surfactants that is orders of magnitude larger than the CMC and that is already large enough for ordered phases to form! An investigation of micelles in *dilute* solution would require not only many (i.e., 10^2 times) more water molecules, but also that trajectories be followed for much (i.e., 10^3 times) longer times in order to follow the *spontaneous* appearance of the aggregates. Both improvements are well beyond the power of present molecular dynamics computations. Nevertheless, it is not unlikely that these goals will be achieved in the foreseeable future as indicated by the most recent studies of Smit *et al.* [40]. These authors performed molecular dynamics simulations using a parallel algorithm (employing 100 transputers) on a model system of nearly 40000 particles representing surfactant, water and oil molecules. The model particles are drastically simplified versions of the real molecules, yet the system features many of the essential properties of real surfactant solutions, including spontaneous micellization in the water phase.

The simulation in [39] begins with a micelle of 15 octanoates with head-groups constrained to the surface of a sphere whose radius is chosen to give a core with density equal to the liquid alkane value. After the solvent is prepared by equilibrating 1331 water molecules (corresponding to a pure water density of 1g/cc), the micelle is introduced at the box's center and all overlapping waters removed. Finally, sodium counterions are substituted for an equivalent number (15) of waters chosen randomly on a shell of radius of about 6Å larger than that of the micellar core. The whole system is then equilibrated for 50ps with constrained head-groups, after which time these constraints are released and the full molecular dynamics simulation is begun ($T = 300K$, with a time step of 2.5 fs.). Analysis is done on 250 ps trajectories, with most of the data obtained during the last 50 ps of calculation. Because the simulation cannot be carried out for longer times,

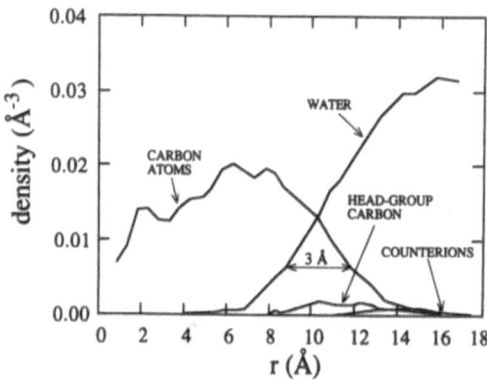

FIGURE 1.5. Density of carbon atoms, water molecules, surfactant head-groups, and counterions, as a function of the distance from the center of the micelle. The results were obtained by molecular dynamics simulations for micelles composed of 15 sodium octanoate molecules in water [39].

it is *assumed*—as already alluded to above—that the *prepared micelle* is stable. That is, in reality, a prepared aggregate can only lose its integrity on much longer time scales (e.g., micro- to milli-seconds). So the molecular dynamics computation falls short of being able to demonstrate the spontaneous formation and persistence of micelles with specific size and shape. Furthermore, many of the *structural* results mentioned below are quite sensitive to the uncertainties in molecular interaction parameters used in the calculations.

In spite of these numerous limitations, it is of interest to consider some of the reported MD results. In particular, the mean radius, derived from the positions of the head-group carboxylate oxygens, is found to be somewhat (15%) larger than the fully stretched (all-trans) chain length with a shape non-sphericity characterized by average principal inertial moment ratios of 1.7:1.5:1. Figure 1.5 shows the total carbon atom density profile, measured from the center of mass, as well as the density profiles for solvent water, head-group carbons, and sodium counterions. Note that the sodiums do not penetrate the core with water, but rather remain in the outer region with the carboxylate head groups. More significantly, *the simulation profiles serve to confirm the existence of a dry hydrophobic core.* (The "hole" at the micellar center is an artifact of the choice of system (box) size which leads to a small negative pressure.) The interfacial region in which water and alkyl segments are mixed is approximately 4Å wide, compared to a micellar radius of roughly 12Å. Furthermore, the calculated bond conformational statistics and chain segment distributions are found to agree closely with the mean-field theory results to be discussed in Sec 1.3.3 and with the earlier model simulations by Pratt et al. [36] and Haile and O'Connell [37]. Recently, Karaborni and O'Connell [41a] have used molecular dynamics computations to examine the effects of alkyl chain length and head-group characteristics on internal micellar structure and chain packing. They confirm the existence of a well-defined, dry, hydrocarbon interior and of a significant degree of conformational disorder. Similar conclusions were reached by Wendoloski et al. [41b]; but again, we stress that these simulations—like those in [37–39]—begin with an already-assembled aggregate and, furthermore, treat the aqueous solvent only implicitly, via a set of effective "boundary forces."

Our interest in the above ideas concerning micellar structure lies primarily in their extension to larger aggregates having distinctly anisotropic shapes. As a rule it is found that average aggregation numbers begin to increase significantly at high enough concentrations (e.g., one to two orders of magnitude) above the CMC. The fact that at still higher concentrations one observes first *nematic* (rod- and disk-) liquid crystalline states and then hexagonal and lamellar *positionally ordered* phases, suggests further that the large aggregates in isotropic solution are themselves rod-like or disk-like. This conclusion is also consistent with the results described above, in our brief introduction to light scattering studies of isotropic phases, as well as in other chapters in this volume. Indeed, the very existence of large rod-like micelles in relatively dilute (i.e., ideal, as far as *inter*-aggregate forces are concerned) solution implies a strong self-assembly preference for cylindrical geometry. After all, if the hydrophobic effect could be satisfied equally well by spherical, rod-like, and disk-like aggregates, then the *minimum* (i.e., globular, spherical) micelle would be favored overwhelmingly. This is because the entropy of dispersion demands that—all other things being equal—the surfactants should organize themselves into the largest number of aggregates. But "all other things" are generally *not* equal, i.e., the *intra*-micellar ("self") free energy is distinctly lower for one particular geometry. Accordingly, one sees cylindrical aggregates predominating in dilute solutions of sodium decyl sulphate and other familiar surfactants, or essentially planar (bilayer) vesicles appearing in the case of phospholipid amphiphiles, and so on (see Chaps. 2 and 3). The fact that amphiphiles self-assemble into structures characterized by well-defined geometries provides additional support for the "standard" picture of the aggregates, as representing a well-defined hydrophobic region surrounded by a head-group mantle. Clearly, overall shape and surface roughness fluctuations do take place, but the basic structures are preserved.

As argued in the following sections, large aggregation numbers—commonly observed at high concentrations of surfactant and/or salt—can only be reconciled with *extended* structures. By extended aggregates we

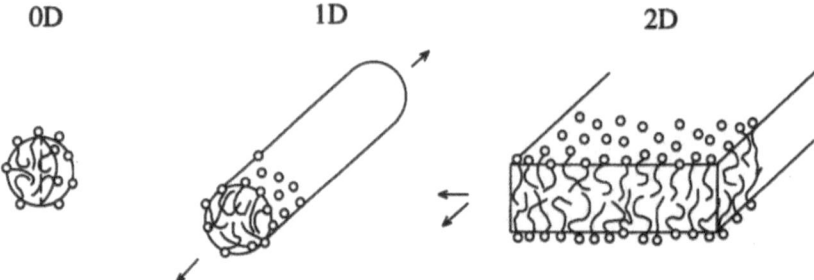

FIGURE 1.6. Schematic illustration of three "basic" aggregation geometries, describing aggregates that can grow along two dimensions (the bilayer), one dimension (the cylinder) and zero dimensions (i.e., no growth possibility, as for the spherical micelles).

mean those derived from minimum micelles via growth in one (rod) or two (disk) dimensions: see Fig. 1.6. Unlike the case of nucleation clusters [42–45] or microemulsion droplets (see Chaps. 7–9), the radius of a globular micelle can never significantly exceed the length of a fully stretched molecule; hence its "molecular weight" is restricted to small values. Accordingly, in the following sections we adopt as a basic premise the fact that big aggregation numbers are associated with a preference for cylindrical or bilayer packing of amphiphilic molecules. We explore the consequences of this premise on the equilibrium size distribution of aggregates and its dependence on concentration, and its consequences for phase transitions to orientationally and positionally long-range ordered states. This "first order" approach is all the more reasonable in light of the extreme difficulty of carrying out more detailed calculations on systems which necessarily involve so many degrees of freedom and such poorly understood interactions. The very few attempts to predict the spontaneous aggregation structures from basic statistical-thermodynamic principles have been necessarily limited to drastically simplified models (see, e.g., [46,47]). Thus we continue to characterize micelles by regular geometric structures, insofar as:

- a wide range of exotic self-assembly phenomena can be systematically explained within this description; and

- there still does not exist any compelling experimental or theoretical evidence which demands an alternative point of view.

1.2   Amphiphile Self-Assembly in Dilute Solutions

An equilibrium solution of amphiphiles in water corresponds (above the CMC) to a system of aggregates (micelles), generally of different sizes and possibly also different shapes, coexisting with a nearly constant concentration of monomers. Unlike in ordinary solutions, the solute particles (i.e., the micelles) in amphiphile solutions can respond to variations in thermodynamic parameters (such as total concentration, temperature or ionic strength) by changing their size and shape distributions. This behavior resembles that of a system governed by multiple chemical equilibria. The preferred aggregation geometry (e.g., spherical micelles, cylindrical aggregates or planar bilayers) and the equilibrium size distribution are determined by the molecular characteristics of the amphiphiles, as well as by the total concentration and other thermodynamic variables. In concentrated solutions, the sizes and shapes of the aggregates are also influenced by inter-aggregate forces, resulting in a rich and complex phase behavior, as discussed in Sec. 1.4 and in more detail in other chapters in this volume.

The discussion in this section is concerned with self-assembly and growth in the dilute solution regime. In Sec. 1.2.1 we introduce the basic statistical-thermodynamic concepts required to describe these phenomena, and in Sec. 1.2.2 we derive the formal expressions for the equilibrium size distribution. We then proceed to discuss the passage from monomers to aggregates (Sec.1.2.3), and to analyze the factors governing micellar growth with special emphasis on the role of the aggregate's dimensionality (Sec. 1.2.4). In Sec. 1.2.5 we briefly consider mixed aggregates and, finally, in Sec. 1.2.6 we comment on the contribution of external rotational and translational degrees of freedom to the micellar chemical potential, an issue which is totally overlooked in the phenomenological treatments of surfactant aggregation, and which still awaits a satisfactory statistical-thermodynamic resolution.

Before turning to the technical discussion, a few remarks should be made regarding the approach adopted here and its relation to alternative treatments. Throughout this chapter, rigorous statistical-thermodynamic derivations and concepts are interwoven with phenomenological descriptions. For instance, we employ elementary statistical thermodynamics to express the standard chemical potentials (which determine the equilibrium size distribution) in terms of the micellar partition functions. But then, realizing how difficult it is to treat rigorously the interactions determining the partition functions, we turn to phenomenological ('semi-empirical') considerations which, albeit approximate, provide important qualitative insights into the mechanisms of amphiphile association.

Many authors [48–55] have presented statistical-thermodynamic theories of micellar solutions, involving different degrees of rigor concerning, for example, the treatment of the solvent, the role of intra- and inter-micelle interactions, or the effects of translational and rotational motions. Unfortunately, in many cases the ultimate results of such theories are formal expressions, e.g., for the partition functions, which are not very useful unless drastic approximations are invoked to simplify them. Other authors have formulated the micellization phenomena in terms of classical thermodynamics, treating micelles as small thermodynamic systems [56–59] ('micro-phases') and (some) analyzing in detail their mechanical properties [60,61]. The prevailing treatments of amphiphile self-assembly and micellar growth are those which combine basic classical thermodynamics with simple phenomenological models for the various contributions to amphiphile-amphiphile and amphiphile-water interactions [30–33,62–67]. One of our aims in the following discussion is to cast these notions into statistical thermodynamic terms, thus explicitly demonstrating the assumptions and approximations involved in the phenomenological approaches and thereby assessing their validity.

1.2.1 UNDERLYING STATISTICAL THERMODYNAMICS

Consider a solution of N amphiphiles and N_w water molecules in volume V and at temperature T. At any given instant N_s of the N amphiphiles are incorporated in $n_s = N_s/s$ aggregates of size $s = 1, 2, \ldots$, with $\sum N_s = N$. $N_1 = n_1$ is, of course, the number of free monomers. Because of monomer association, micellar dissociation and exchange between aggregates, and similar dynamical processes involving groups of monomers or even micelles, the size distribution $\{n_s\}$ is a dynamical quantity, undergoing statistical fluctuations. Thus, the partition function of the system is a sum over all possible distributions of the amphiphiles: $Q_{tot}(N, N_w, V, T) = \sum_{\{n_s\}} Q_{tot}(\{n_s\}, N_w, V, T)$. (The subscript "$tot$" indicates 'total,' i.e., amphiphiles plus water.) However, the fluctuations around the average distribution $\{\bar{n}_s\}$ or, equivalently, the most probable distribution $\{n_s^*\}$, are negligible. Hence, using the usual maximum term method, we can safely replace $\ln Q_{tot}(N, N_w, V, T)$ by $\ln Q_{tot}(\{n_s^*\}, N_w, V, T)$ and express the Helmholtz free energy of the system as

$$A_{tot} = -kT \ln Q_{tot}(\{n_s^*\}, N_w, V, T) \qquad (1.7)$$

The equilibrium size distribution corresponds to the set $\{n_s\} = \{n_s^*\}$ for which $\ln Q_{tot}(\{n_s\}, N_w, V, T)$ is maximal, so that A_{tot} is minimal. To derive $\{n_s^*\}$ we need an explicit expression for Q_{tot}. As a first step we separate A_{tot} into solvent and solute contributions. Assuming that the thermodynamic properties of the water are not affected by the presence of the micelles, one can write $Q_{tot}(\{n_s\}, N_w, V, T) = Q(\{n_s\}, V, T)Q_w$ and, correspondingly, $A_{tot} = A + A_w$. Here, $Q(\{n_s\}, V, T)$ is the partition function of the system of aggregates and monomers, dispersed in the continuous background of the solvent. The free energy of interaction between the micelles and the aqueous solution is included in Q in a approximate fashion, as explained in Sec. 1.3. Q_w and A_w are the solvent contributions.

In dilute solutions, interaction effects between aggregates are negligible, and the free energy of a polydisperse system of $\{n_s\}$ aggregates is given, in analogy to ideal gas mixtures, by

$$
\begin{aligned}
A &= -kT \ln Q(\{n_s\}, V, T) = -kT \ln \prod_s \frac{q_s^{n_s}}{n_s!} \\
&= kT \sum_s n_s(-\ln q_s + \ln n_s - 1)
\end{aligned}
\tag{1.8}
$$

where q_s is the partition function of an s-mer in solution. q_s includes contributions from the internal degrees of freedom of the aggregate, its overall translational and rotational motions, and approximately (via a potential of mean force) interaction effects with the solvent, as discussed in more detail in Secs. 1.2.6 and 1.3.1. It should be noted that the decomposition of Q and A into contributions from aggregates of different *sizes* does not imply that all s-aggregates are necessarily of identical *shape*. The possible existence of different micellar shapes, just like the possibility of small shape fluctuations, are included in q_s (in analogy to the case of a molecule fluctuating between several isomeric forms). Yet if one chooses to classify the aggregates by both size and shape, then Eq. (1.8) still holds, with s specifying both characteristics.

The derivation of $\{n_s^*\}$ from (1.8) is straightforward. However, we defer the derivation to the next section, after introducing several quantities, relationships and definitions which will prove useful later on. To this end we will temporarily regard (1.8) as the free energy of an ideal mixture of a given *fixed* composition $\{n_s\}$.

The chemical potential of species s is given by

$$
\begin{aligned}
\mu_s &= \left(\frac{\partial A_{tot}}{\partial n_s}\right)_{T,V,n_{s'\neq s},N_w} = \left(\frac{\partial A}{\partial n_s}\right)_{T,V,n_{s'\neq s}} \\
&= -kT \ln q_s + kT \ln n_s \\
&= -kT \ln(q_s/V) + kT \ln \rho_s
\end{aligned}
\tag{1.9}
$$

with $\rho_s = n_s/V$ denoting the number density of s-aggregates. Combining (1.8) and (1.9), we find

$$
\begin{aligned}
A &= \sum_s n_s \mu_s - kT \sum_s n_s \\
&= G - \Pi V
\end{aligned}
\tag{1.10}
$$

Here $G = \sum n_s \mu_s$ is the Gibbs free energy of the system of aggregates, not including the solvent's contribution. (The total Gibbs free energy of the solution is given by $G_{tot} = A_{tot} + PV = \sum n_s \mu_s + N_w \mu_w$. P is the external pressure; $P = -(\partial A_{tot}/\partial V)$ with the derivative evaluated at constant composition.) Π is the osmotic pressure which, in the dilute solution limit, reduces to the ideal gas form $\Pi = kT(\sum \rho_s)$. It is easily verified that $\mu_s = (\partial G_{tot}/\partial n_s)_{T,P,N_{s'\neq s},N_w} = (\partial G/\partial n_s)_{T,\Pi,n_{s'\neq s}}$.

The chemical potential is conveniently expressed as

$$\mu_s = \mu_s^{o,\rho} + kT \ln \rho_s \qquad (1.11)$$

with, cf. (1.9),

$$\mu_s^{o,\rho} = -kT \ln(q_s/V) = -kT \ln q_s^o \qquad (1.12)$$

denoting the standard chemical potential "on the number density scale." Since the only volume dependent factor in q_s is the translational partition function, which is proportional to V, $q_s^o = q_s/V$ and consequently $\mu_s^{o,\rho}$ are functions of T only (see Secs. 1.2.6 and 1.3.1). While the number densities ρ_s are most convenient for statistical thermodynamic formulations, the predominant concentration scale in the phenomenological theories, as well as in the experimental literature on amphiphile self-assembly [30–33], involves the mole fractions, X_s, defined via

$$X_s = \frac{N_s}{N_{tot}} = \frac{sn_s}{N_{tot}}, \qquad (1.13)$$

with $N_{tot} = N_w + N = N_w + \sum_s N_s$ denoting the total number of molecules, i.e., amphiphiles and water, in solution. Note that X_s is the mole fraction of amphiphiles incorporated into s-aggregates, as distinguished, say, from the mole fraction of s-aggregates $y_s = n_s/\sum n_s = (N_s/s)/\sum(N_s/s); \sum y_s = 1$. Note also that the X_s do *not* sum up to 1, but instead

$$\sum_s X_s = X , \qquad (1.14)$$

with $X = N/N_{tot}$ denoting the total mole fraction of amphiphiles in solution. In the dilute solution limit which is of interest here, $X = 1 - X_w \ll 1$, and $X_w \simeq 1$. X_s and ρ_s are related by

$$X_s = \frac{sn_s/V}{N_{tot}/V} = \frac{s\rho_s}{\rho_{tot}} \simeq \frac{s\rho_s}{\rho_w} \qquad (1.15)$$

with $\rho_w = N_w/V$ and $\rho_{tot} = \rho_w + \sum s\rho_s$.

From (1.11), (1.12) and (1.15) we find

$$\mu_s = \mu_s^o + kT \ln(X_s/s) \qquad (1.16)$$

where μ_s^o (which we use as a shorthand notation for $\mu_s^{o,x}$) is the standard chemical potential "on the mole fraction scale",

$$\mu_s^o = \mu_s^{o,\rho} + kT \ln \rho_{tot} = -kT \ln(q_s^o/\rho_{tot}). \qquad (1.17)$$

In the dilute solution regime $\rho_{tot} \simeq \rho_w$ and $\mu_s^o(= \mu_s^{o,x})$ is independent of the amphiphile concentration X. In principle, however, since ρ_{tot} depends on X, so does μ_s^o (unlike $\mu_s^{o,\rho}$). Notwithstanding this reservation, in the following discussion we shall generally adopt the mole fraction scale, primarily in order to comply with the more familiar, phenomenological, treatments.

Using (1.13) and (1.16), the Gibbs free energy of the system (not including the solvent) can be expressed as

$$
\begin{aligned}
G &= \sum_s n_s \mu_s = \sum_s N_s \tilde{\mu}_s \\
&= N_{tot} \left[\sum_s X_s \tilde{\mu}_s^o + kT \sum_s (X_s/s) \ln(X_s/s) \right] \\
&\equiv G^o(\{X_s\}) - T\, S_d(\{X_s\}).
\end{aligned}
\tag{1.18}
$$

The quantity

$$
\tilde{\mu}_s \equiv \mu_s/s
\tag{1.19}
$$

will be referred to as the average chemical potential per amphiphile in an s-aggregate. Note in particular that $\tilde{\mu}_1 = \mu_1$ is the monomer's chemical potential. In analogy to (1.19) we define $\tilde{\mu}_s^o = \mu_s^o/s$.

The second term in (1.18), $S_d(\{X_s\})$, is an entropic contribution accounting for the 'dispersity' (or 'mixing') of the micellar size distribution. In particular, S_d is maximal when no aggregation takes place (i.e., when $X = X_1$ and $X_s = 0$ for all $s \geq 2$) demonstrating that micellization, like any aggregation process, is entropically unfavorable. More generally S_d is lowered by, and thus tends to oppose, any process which results in a smaller number (or lower polydispersity) of solute particles in the system. Thus, micellar *growth* is also an entropically unfavorable process. [It should be noted that $S_d(\{X_s\})$ involves only the so-called "mixing entropy" of the *solutes*. Solute-solvent mixing is not accounted for by S_d because G does not include the solvent contribution. Solute-solvent mixing is included in $G_{tot} = G + G_w = G + \sum N_w \mu_w$. The usual, ideal solution, mixing entropy results when we write $\mu_w = \mu_w^o + kT \ln X_w$, thus adding to (1.18) the solvent term $N_{tot} \sum X_w \mu_w^o + kT \sum X_w \ln X_w$ with $X_w = N_w/N_{tot}$.]

1.2.2 The Equilibrium Size Distribution

Expressed in terms of the mole fractions, X_s, the Helmholtz free energy A is given by (cf. (1.10), (1.13) and (1.16)),

$$
A = N_{tot} \sum_s (X_s/s) \left[\mu_s^o + kT \ln(X_s/s) - kT \right]
\tag{1.20}
$$

As noted in the previous section the most probable distribution $\{X_s^*\}$ is the equilibrium size distribution, and $A(N, V, T, \{X_s^*\})$ is the free energy of the self-assembling amphiphilic system. The most probable distribution corresponds to the set of X_s which minimizes A, subject to the conservation constraint $\sum X_s = X$. The conditional minimization can be carried out using the Lagrange multipliers method which, in our case, amounts to solving

$$\frac{\partial}{\partial X_s}\left(\frac{A}{N_{tot}} - \mu X\right) = 0 \quad \text{(all } s) \tag{1.21}$$

with A given by (1.20). Here μ is the Lagrange multiplier conjugate to the constraint $X = \sum X_s$, with this later condition determining the numerical value of μ. The use of A/N_{tot} rather than A in (1.21) ensures that μ is intensive. This follows from the fact that A and $\mu X N_{tot} = \mu N$ must have the same dimensions and the same N-dependence, and that $A \sim N$. We shall soon see that μ is simply the amphiphile's chemical potential.

From (1.21), (1.20) and (1.14) we find that for all s

$$X_s^* = s \cdot \exp\left[-s\left(\tilde{\mu}_s^o - \mu\right)/kT\right] \tag{1.22}$$

The numerical value of μ can be determined by substituting (1.22) into $\sum\{X_s^*\} = X$ and solving the resulting equation for μ. Clearly, since the $\tilde{\mu}_s^o$'s depend on T only, it follows that $\mu = \mu(X,T)$. From (1.16), (1.19) and (1.22) we obtain

$$\begin{aligned} \mu &= \tilde{\mu}_s = \tilde{\mu}_s^o + (kT/s)\ln(X_s/s) \quad \text{(all } s) \\ &= \mu_1 \end{aligned} \tag{1.23}$$

where it should be understood that here, and hereafter, all X_s stand for their most probable values X_s^*.

The last result reveals that μ is nothing else but the chemical potential of the amphiphiles in the solution which, at equilibrium, must be the same everywhere in the system. This includes free monomers, for which $\mu = \tilde{\mu}_1$, as well as amphiphiles incorporated in a micelle of (any) size s, for which $\mu = \tilde{\mu}_s$. Note also that, consistent with the general thermodynamic relation $G = N\mu$, the equality $\mu_s = s\tilde{\mu}_s = s\mu$ indicates that μ_s may be regarded as the Gibbs free energy of an s-aggregate. Of course, these are just different but equivalent interpretations of the equality

$$G = \sum_s n_s \mu_s = \mu \sum_s s n_s = N\mu \tag{1.24}$$

Another obvious and common interpretation of (1.23) derives from the notion that these equalities are the conditions for chemical equilibria in a system in which all the chemical reactions

$$sA_1 = A_s \quad \text{(all } s) , \tag{1.25}$$

as well as linear combinations of these reactions (see below), take place simultaneously. In the present context A_s stands for an s-aggregate and

A_1 for a monomer. Of course one could start as well from (1.25) which implies $\mu_s = s\mu_1$ as the description of dynamical chemical (association-dissociation) equilibrium in the self-assembling solution. Then using (1.16) for μ_s we obtain the law of mass action [30–33,62–64]

$$\frac{(X_s/s)}{X_1^s} = \exp\left[s\left(\mu_1^o - \tilde{\mu}_s^o\right)/kT\right] \equiv K_s \tag{1.26}$$

Here X_s/s corresponds to the concentration of "product", and X_1 to that of "reactant" (raised to the power s, the "stoichiometric coefficient" in (1.25)); finally, $\mu_s^o - s\mu_1^o$ is the *standard free energy change* for the association "reaction." This "equilibrium quotient" result is entirely equivalent to (1.22), since $\mu = \mu_1 = \mu_1^o + kT\ln X_1$. Note, however, that this equivalence does not imply that the set of reactions (1.25) is the actual self-assembly mechanism in the system. (In fact the simultaneous association of s molecules to form an aggregate is a highly unlikely kinetic event.) The multiple chemical equilibria could similarly be described by other sets of independent (but coupled) reactions, e.g., the step-wise association processes

$$A_{s-1} + A_1 = A_s \tag{1.27}$$

for which $\mu_s = \mu_{s-1} + \mu_1$, consistent with (1.23). The equilibrium constants L_s corresponding to (1.27) are simply related to the K_s's, via

$$L_s = \frac{X_s/s}{X_1 X_{s-1}/(s-1)} = \frac{K_s}{K_{s-1}} \tag{1.28}$$

Clearly, this relation, as well as any other expression of the law of mass action corresponding to a set of independent chemical reactions such as (1.25) or (1.27), implies the same equilibrium values of X_s as those given by (1.22).

1.2.3 FROM MONOMERS TO AGGREGATES—THE CMC

The sudden appearance of micelles at a certain monomer concentration (the CMC) and the (near) constancy of the free monomer concentration upon further increase in the total amphiphile concentration (see Fig. 1.1), resemble a phase transition—in which the micelles play the role of a condensed phase and the free monomers the vapor (or the solute). This cooperative behavior indicates that even the smallest micelles formed contain a fairly large number of molecules, typically several dozens. The existence of some minimal aggregation number, m, can be explained by microscopic molecular packing considerations, as discussed in Sec. 1.3. In this section we shall only consider the thermodynamic implications of this fact.

As noted with regard to (1.18), the association of monomers into aggregates involves entropy loss. Thus, spontaneous aggregation must be associated with an "enthalpic" gain. More explicitly, from (1.22) or (1.26) we see that unless $\tilde{\mu}_s^o < \mu_1^o$ (i.e., $K_s > 1$) for at least some $s > 1$, X_s will be much smaller than X_1 for all X and T. If only micelles of sizes $s \geq m$ appear, then clearly $\tilde{\mu}_s^o > \mu_1^o$ for all aggregates in the "gap" $2 \leq s \leq m$, whereas $\tilde{\mu}_s^o < \mu_1^o$ for $s \geq m$. The s-dependence of $\tilde{\mu}_s^o$ for $s \geq m$ determines the size distribution of the micelles as will be discussed in more detail in the next section. For our present goal, which is to characterize the CMC regime, it is both convenient and sufficient to assume for the moment that micelles of only one size, m, are formed in the system. Thus, the solution contains either free monomers at concentration X_1, or m-mers at concentration X_m, with $X_1 + X_m = X$. In terms of the $\tilde{\mu}_s^o$'s this scheme amounts to setting $\tilde{\mu}_s^o = \infty$ for all s except $s = 1$ and m; it corresponds, approximately, to a system of amphiphiles whose overwhelmingly preferred aggregation geometry is spherical (Secs. 1.2.4 and 1.3.2).

The onset of micelle formation and the corresponding saturation of the free amphiphile concentration takes place over a narrow (yet finite) range of the total concentration X. Thus, any X or X_1 in this range may be taken as an operational definition of the CMC. Generally, we can specify the CMC as the (monomer or total) concentration corresponding to a solution in which a certain fraction σ of the amphiphiles are micellized. (It is common [30] to use $\sigma \approx 0.01 - 0.1$; yet, larger values, e.g., $\sigma = 1/2$ are also common and for mathematical purposes are often more convenient.) Let us denote by $\hat{X} = $ CMC, $\hat{X}_1 = (1 - \sigma)\hat{X}$ and $\hat{X}_m = \sigma\hat{X}$, the critical values of the total, free and micellized amphiphile concentrations, respectively. Then from (1.26) we get

$$
\begin{aligned}
\ln(CMC) &= -\left(\frac{m}{m-1}\right)(\mu_1^o - \tilde{\mu}_m^o)/kT + \frac{1}{m-1}\ln\left[\frac{\sigma}{m(1-\sigma)^m}\right] \\
&\approx -(\mu_1^o - \tilde{\mu}_m^o)/kT
\end{aligned}
\tag{1.29}
$$

In passing to the second equality we have taken into account that, typically, $m \simeq 20 - 100$ [30–33]. Neglecting the second term in the first equality is justified (for all reasonable σ) since generally $(\mu_1^o - \tilde{\mu}_m^o)/kT \geq 10$. Choosing σ to be small implies that the free monomer concentration and the CMC are the same, i.e.,

$$
\hat{X}_1 \approx \text{CMC} \approx \exp\left[-(\mu_1^o - \tilde{\mu}_m^o)/kT\right]
\tag{1.30}
$$

To demonstrate the sharpness of the transition from free monomers to micelles at the CMC we first use (1.26) to write

$$
X_m = X - X_1 = \hat{X}_m\left(X_1/\hat{X}_1\right)^m
\tag{1.31}
$$

Using $\hat{X}_1/\hat{X}_m = (1-\sigma)/\sigma$ we find

$$
\frac{\partial X_1}{\partial X} = \left[1 + \alpha\left(\frac{X_1}{\hat{X}_1}\right)^{m-1}\right]^{-1}
\tag{1.32}
$$

where $\alpha = m\sigma/(1-\sigma)$ is of order 1. This very simple relation shows clearly how the fact that $m \gg 1$ is reflected in the monomer concentration. Below the CMC, i.e., as long as $(X_1/\hat{X}_1) < 1$, the second term in the denominator on the r.h.s. (right hand side) of (1.32) is negligible, so that $\partial X_1/\partial X \approx 1$ and $X_1 \approx X$, i.e., all the amphiphiles are free monomers. On the other hand, once X_1 exceeds, even slightly, its CMC value $\hat{X}_1$, we find $\partial X_1/\partial X \sim (\hat{X}_1/X_1)^{m-1} \approx 0$, and hence $X_1 \approx \hat{X}_1 = $ constant. As a numerical example consider, say, a surfactant (not very different from SDS) for which CMC $\approx \hat{X}_1 = 10^{-5}$ and $m = 50$ (suppose $\sigma \approx 0.1$ so that $\hat{X}_1 \approx 10\hat{X}_m$). It follows from (1.31) that when $X_1 = 0.9\hat{X}_1$ for instance, $X_m \approx 10^{-4}\hat{X}_1$ is still negligible. On the other hand, even at very high concentrations, e.g., $X = 1000$ CMC $= 1000\hat{X}_1$, the monomer concentration has only increased to $X_1 \approx 1.15\hat{X}_1$.

Most of the free energy gain associated with amphiphile packing in a micelle reflects the preference of the amphiphile's hydrocarbon chain for the hydrophobic environment inside the micellar core, as compared to the aqueous environment of the free amphiphile. The CMC or, equivalently, the saturation concentration (solubility limit) of free monomers, decreases exponentially with this preference which is measured by $\Delta\tilde{\mu}^o = \mu_1^o - \tilde{\mu}_m^o$: see (1.30). For most amphiphiles $\Delta\tilde{\mu}^o$ increases nearly linearly with the amphiphile chain length [30].

It should be noted that relative to $\hat{X}_1$, the sharpness of the monomer-micelle transition depends entirely on m. However, the absolute width, i.e., the concentration range characterizing the transition, depends on $\hat{X}_1$ as well. The lower is the CMC, the sharper is the transition. The dependence on m reflects the *cooperativity* of the transition. As m increases, the transition resembles more closely a real, first order phase transition, in which the micelles correspond to a condensed phase, and the monomers are the vapor. Several treatments of micelle formation are based on this analogy [57–59]. Such treatments can account for the behavior at the CMC and for monomer-aggregate coexistence. On the other hand, since the micelles are regarded as macroscopic phases, the size dependence of their thermodynamic properties is necessarily ignored (corresponding to $S_d \equiv 0$ in (1.18)). Consequently, this "microphase" view of the micelle is, for instance, inadequate for discussing micellar growth.

Although the above technical discussion has been limited for convenience of illustration to a monodisperse micellar system, all the qualitative conclusions remain valid when larger micelles are also possible. We shall comment briefly on this point in the next section.

1.2.4 MICELLAR GROWTH

Micellar aggregates appear in various shapes and sizes, depending on the molecular nature of the constituent amphiphiles (Sec. 1.3) as well as on the total concentration and other thermodynamic parameters. There is, however, one basic structural feature commonly shared by all aggregates: the width (diameter) of their hydrophobic core is always on the order of the amphiphile's molecular length. More precisely, since holes within the hydrophobic region are energetically intolerable, the core diameter cannot exceed $2l$ where l is the length of the fully stretched hydrocarbon tail, cf. Fig. 1.6. Consequently, at least one linear dimension of the hydrophobic core in all aggregates is of order l. This simple notion is essential for understanding micellar growth, and suggests a classification of the aggregates into three general categories as follows:

1. Globular aggregates—in which all three linear dimensions are of order l. A spherical micelle of radius $R \leq l$ can serve as a prototype of this class.

2. Rod-like aggregates—in which two dimensions, those perpendicular to the long (rod) axis, are of order l. A cylindrical micelle of radius $R \leq l$ and length $h > 2l$ is the typical example of this class. Such micelles are often described as spherocylinders, i.e., as cylinders capped by hemispheres at their two ends (Fig. 1.6). As we shall see below, end effects play an essential role in micellar growth; however, the structural details of the ends are (qualitatively) irrelevant.

3. Bilayers—in which only one dimension is of order l, namely, the thickness, which cannot exceed $2l$. An oblate, disk-like aggregate of thickness $w \approx 2l$ and diameter $h > 2l$ can serve as a representative of this class. Again, a more specific geometry (e.g., a disk-shaped body surrounded by a semi-toroidal rim) may be assumed for these aggregates; however, as with rod-like micelles, their growth characteristics are governed by their dimensionality and not by structural details.

Quite generally, micellar growth is driven by the tendency to reduce unfavorable end ("edge" or "surface") effects. In particular, we shall see that above a certain concentration disk-like micelles tend to grow, actually to undergo a phase transition, to infinitely large sheet-like aggregates. Another possibility for bilayers to overcome the excess rim energy is to close upon themselves and hence to form spherical vesicles [30–33]. (Similarly, rod-like micelles can close into tori [66].) Vesicles serve as model systems for biological membranes and are interesting entities in their own right. However, unlike spherical, cylindrical and disk-like aggregates whose growth characteristics are dictated by the dimensionality and end effects which we want to emphasize here, other factors control the size behavior of vesicles, primarily curvature elasticity (Sec. 1.3.3). Thus, except briefly in Sec. 1.2.5, we shall not discuss vesicles in this chapter.

1.2.4.1   The phenomenological approach [30–33]

As noted already in Secs. 1.2.1 and 1.2.3, micellar growth, like any other aggregation process in dilute solution, reduces the translational ("mixing")

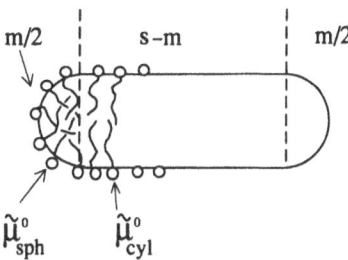

FIGURE 1.7. Schematic illustration of a rod-like micelle, described as a simple sphero-cylinder.

entropy of the system. For growth to occur, the *internal* free energy per molecule within the aggregate must decrease with size; that is, $\tilde{\mu}_s^o$ must decrease as s increases. Consider for example a spherocylindrical rod-like micelle of s amphiphiles, of which $m/2$ comprise each of the two hemispherical ends and $s-m$ constitute the cylindrical body, cf. Fig. 1.7. Clearly, such micelles tend to grow if the molecules feel more "comfortable" in a cylindrical rather than in a spherical environment. This is more quantitatively expressed by the inequality $\tilde{\mu}_{cyl}^o < \tilde{\mu}_{sph}^o$, with $\tilde{\mu}_{cyl}^o$ and $\tilde{\mu}_{sph}^o$ denoting the average free energy (standard chemical potential) of the molecules packed in cylindrical and spherical environments, respectively. We similarly define $\tilde{\mu}_{bil}^o$ as the free energy per amphiphile in a planar bilayer.

The $\tilde{\mu}_g^o$ ($g = sph, cyl, bil$) play a central role in the phenomenological approach to micellar growth which we follow largely in this section [30–33,62–66]. In particular, this approach provides a simple way to analyze the dependence of micellar sizes on the competition between the tendency of molecules to organize into larger aggregates (when $\tilde{\mu}_{sph}^o > \tilde{\mu}_{cyl}^o$ or $\tilde{\mu}_{cyl}^o > \tilde{\mu}_{bil}^o$) and the accompanying loss in translational entropy. Note that larger aggregates also correspond to lower (average) curvature of the hydrocarbon-water interface. The $\tilde{\mu}_g^o$ depend on the molecular characteristics of the amphiphiles, and can be calculated from molecular models or inferred from experimental data (see Sec. 1.3). In the present discussion we shall be mainly concerned with their relative magnitudes.

A key assumption in the phenomenological approach is that μ_s^o of an arbitrary aggregate can be expressed as a superposition of the $\tilde{\mu}_g^o$'s corresponding to its different microenvironments. That is,

$$\mu_s^o = \sum_g s_g \tilde{\mu}_g^o \tag{1.33}$$

with s_g denoting the number of molecules in microenvironment g. For example, for the spherocylindrical micelle containing $s - m$ molecules in the cylindrical body and $m/2$ molecules in each hemispherical end (Fig. 1.7), we have

$$\begin{aligned} \mu_s^o &= s\tilde{\mu}_s^o = (s-m)\tilde{\mu}_{cyl}^o + m\tilde{\mu}_{sph}^o \\ &= s\tilde{\mu}_{cyl}^o + m\left(\tilde{\mu}_{sph}^o - \tilde{\mu}_{cyl}^o\right) \ (s \geq m) \end{aligned} \tag{1.34}$$

Similarly, for a disk-like micelle, μ_s^o can be expressed as a linear combination of $\tilde{\mu}_{bil}^o$ and $\tilde{\mu}_{rim}^o$ corresponding, respectively, to the $s - k$ molecules in the central part, and the k molecules in the semi-toroidal rim (which for large s becomes semi-cylindrical [68], i.e., $\tilde{\mu}_{rim}^o = \tilde{\mu}_{cyl}^o$, and $k \sim s^{1/2}$). More generally, for a (large) d-dimensional aggregate, we can write

$$\mu_s^o = s\tilde{\mu}_s^o = s\tilde{\mu}_\infty^o + kT\,\delta s^{(d-1)/d} \tag{1.35}$$

Here $\tilde{\mu}_\infty^o = \mu_{s\to\infty}^o$ denotes the asymptotic standard chemical potential, i.e., the free energy per molecule in the main body of the aggregate and δkT is a measure of the excess edge ("surface") free energy. Note that both $\tilde{\mu}_\infty^o$ and δ depend on the aggregate's dimensionality (or, more precisely, on its geometry), e.g., for spherocylinders $(d = 1), \tilde{\mu}_\infty^o = \tilde{\mu}_{cyl}^o$ and $\delta = m(\tilde{\mu}_{sph}^o - \tilde{\mu}_{cyl}^o)/kT$. A relation similar to (1.35), known as the "capillarity approximation", is used in nucleation theory for the standard chemical potential of a nucleation cluster [42–45]. In nucleation problems, one is mainly concerned with $d = 3$, in which case $\tilde{\mu}_\infty^o$ is identified as the "bulk" chemical potential corresponding to the interior of the cluster, while the second term in (1.35) accounts for the surface free energy, i.e., $\delta s^{2/3} \sim 4\pi R^2 \gamma$, where R is the cluster's radius and γ is the surface tension. (Note that, in general, for a d-dimensional cluster of radius R, the "volume" varies as $R^d \sim s$ and the "surface" as $R^{d-1} \sim s^{(d-1)/d}$.)

1.2.4.2   Dimensionality and micellar growth

Since at least one linear dimension of a micelle is restricted to be of order l, micellar growth is limited to either $d = 1$ (as in rod-like micelles) or $d = 2$ (planar, or disk-like aggregates). Based on the expressions derived for X_s in Sec. 1.2.2 and the simple relation (1.35), one can show that micellar growth in $d = 1$ and $d = 2$ (or, more generally, $d > 1$) are markedly and qualitatively different from each other. In particular, we shall see that for $d = 2$ a first-order phase transition from finite micelles (or even directly from monomers) to infinite aggregates takes place at some low amphiphile concentration, $\bar{X}$. On the other hand, for $d = 1$, the average micellar size varies continuously with X and remains finite at all concentrations.

Let us first derive the condition for coexistence between a saturated (albeit dilute) solution containing monomers and finite micelles, and an infinite aggregate which, thermodynamically, constitutes a macroscopic condensed phase. According to the capillarity approximation and the phenomenological approach described above, the amphiphile chemical potential (i.e., the free energy per molecule) in the infinite aggregate is $\tilde{\mu} = \tilde{\mu}_\infty^o$.

If these aggregates coexist in equilibrium with the monomers and finite micelles, then $\tilde{\mu}_\infty^o = \mu_1 = \tilde{\mu}_s$ and hence

$$
\begin{aligned}
\tilde{\mu}_\infty^o &= \mu_1^o + kT \ln \bar{X}_1 \\
&= \tilde{\mu}_s^o + (kT/s) \ln(\bar{X}_s/s) \quad (s \geq m)
\end{aligned}
\tag{1.36}
$$

with $\bar{X}_1$ and $\bar{X}_s/s$ denoting the concentrations of monomers and s-micelles at saturation. For further reference we rewrite the first equality in (1.36) as

$$
\bar{X}_1 = \exp\left[-\left(\tilde{\mu}_1^o - \mu_\infty^o\right)/kT\right] .
\tag{1.37}
$$

For the second equality we use (1.35) to obtain

$$
\bar{X}_s = s \cdot \exp\left[-\delta s^{(d-1)/d}\right] \quad (s \geq m)
\tag{1.38}
$$

Now the difference between $d = 1$ and $d = 2$ (or any $d > 1$) is apparent. The sum

$$
\bar{X} = \bar{X}_1 + \sum_{s=m}^{\infty} \bar{X}_s
\tag{1.39}
$$

converges to a finite total concentration $\bar{X}$ for $d = 2$, but *diverges* for $d = 1$. Since (1.36) cannot be satisfied for $d = 1$, the conclusion is that an infinitely long, one-dimensional, aggregate cannot coexist with an ideal solution of finite, rod-like, micelles. Conversely, as we shall see below, for $d = 1$ the condition $X = X_1 + \sum X_s$ can be satisfied for all X, with the appropriate X_s for finite rod-like micelles in dilute solutions. (Recall, however, that the discussion here is limited to the dilute solution regime, where interaction effects between micelles are negligible; see Sec. 1.4.)

Consider now the $d = 2$ case (the conclusions apply to all $d > 1$). Below saturation, i.e., when $X < \bar{X}$, the solution contains only monomers or finite micelles, because $\tilde{\mu}_s < \tilde{\mu}_\infty^o$ for $X_s < \bar{X}_s$; see (1.36). When $X > \bar{X}$ a saturated dilute solution of monomers and micelles (with $X_s = \bar{X}_s$) coexists with infinite two dimensional bilayer lamellae containing all the other $(X - \bar{X})$ amphiphiles: adding more amphiphiles just leads to more bilayers, without affecting the micellar size distribution $\{\bar{X}_s\}$. Thus, $\bar{X}$ marks the onset of a first-order phase transition. Note, finally, that for disk-like micelles the surface term in (1.35), hence in (1.38), varies as $s^{1/2}$ because we have assumed large s (e.g., $s > 100$); for smaller sizes the dependence on s is more complicated [68]. In practice, these details are irrelevant because for systems of bilayer-forming amphiphiles (for which $\tilde{\mu}_{bil}^o < \tilde{\mu}_{sph}^o, \tilde{\mu}_{cyl}^o$) $\bar{X}$ is generally negligible compared to the total amphiphile concentrations of interest. (E.g., for long chain phospholipids $\delta > 5$ and $m > 50$, implying extremely small $\bar{X}_s$ for all $s > m$. Furthermore, for these molecules $\tilde{\mu}_1^o - \mu_\infty^o > 20$, implying negligible $\bar{X}_1$; see (1.37)).

The different qualitative behaviors of $d = 1$ and $d > 1$ aggregates reflect inverse "hierarchies" of the "surface free energy" and the translational entropy contributions to $\tilde{\mu}_s$. Although this important point may already be clear from the discussion above, let us consider it from a more explicit approach, in the spirit of the analysis of Landau and Lifshitz [69]. To simplify the discussion, we again suppress the effects of polydispersity, assuming that all micelles are of one size s, hence $X_{s'} = \delta_{ss'}X$. (We also ignore the monomers, which are irrelevant for micellar growth; see below.) Thus, using (1.23) and (1.35), we have

$$
\begin{aligned}
\tilde{\mu}_s &= \tilde{\mu}_s^o + (kT/s)\ln(X/s) \\
&= \tilde{\mu}_\infty^o + kT\,\delta s^{-1/d} + (kT/s)\ln(X/s)\,.
\end{aligned}
\tag{1.40}
$$

Hence,

$$
\frac{\partial \tilde{\mu}_s}{\partial s} = \frac{kT}{s^2}\left[-bs^{(d-1)/d} + \ln\left(\frac{s}{eX}\right)\right]
\tag{1.41}
$$

with $b = \delta/d$.

Consider first the $d > 1$ case. For large enough s, say $s > \bar{s}$, the first term in the square brackets, corresponding to the decrease in surface free energy of the system (per molecule) is larger than the second term, $\ln(s/eX)$, which accounts for the loss of translational entropy as s increases. Hence, beyond $\bar{s}$, $\tilde{\mu}_s$ falls off without bound ($\partial \tilde{\mu}_s/\partial s < 0$), reflecting the tendency of the amphiphiles to form infinite aggregates. (For $d = 2$ and typical values such as $b \approx 5$ and $X \approx 10^{-3}$ one finds that $\tilde{\mu}_s$ starts decreasing already above $\bar{s} \approx 3$.) On the other hand, for $d = 1$ the surface term is constant, hence $\partial \tilde{\mu}_s/\partial s = kT/s^2[-b+\ln(s/eX)]$ vanishes (corresponding to the state of minimum free energy) for a *finite* aggregation number, $s = X\exp(\delta+1)$. In other words, no matter how large the growth parameter (the "surface free energy"), or the total concentration X, the entropic contribution is always sufficient to keep the micelles finite. This is a striking manifestation of the "theorem" concerning "the impossibility of the existence of phases in a one dimensional system" [69].

We now turn to consider the polydispersity of micellar sizes. There is little to say about disk-like ($d = 2$) micelles, whose size distribution reaches its saturation value, as given by the $\bar{X}_s$ of (1.38), already at low concentrations. On the other hand, rod-like micelles are characterized by a broad and X-dependent size distribution, as discussed in the next section. The main results obtained there are well known [31–33,53,63–68], but the presentation and emphasis are somewhat different.

1.2.4.3 From short to long micelles—the "sphere-to-rod transition" [31, 64–66]

We consider a solution of amphiphiles preferentially aggregating into rod-like micelles, which for the sake of concreteness will be treated as spherocylinders; see Fig. 1.7. Thus, we express $\tilde{\mu}_s^o$ as in (1.34) or equivalently (1.35) with $d = 1$. Upon substitution into either of the general expressions (1.26) or (1.22), X_s can be rewritten in the simple form

$$
X_s = se^{-(\alpha s+\delta)} \quad (s \geq m)
\tag{1.42}
$$

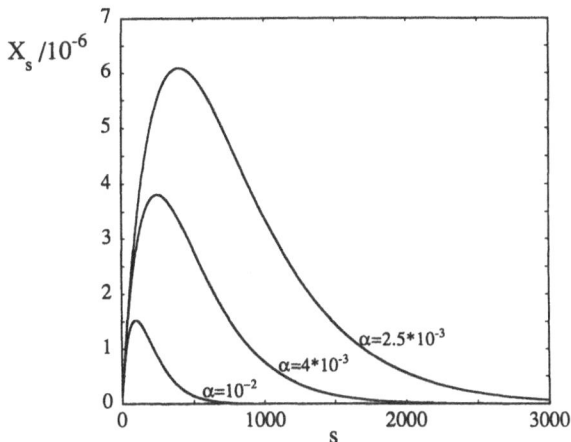

FIGURE 1.8. Size distribution of rod-like micelles in dilute solution, calculated using Eq. (1.42) for $\delta = 15, m = 50$, and three values of α. The total concentrations corresponding to $\alpha = 10^{-2}, 4 \times 10^{-3}$ and 2.5×10^{-3} are $X = 3.77 \times 10^{-4}, 2.54 \times 10^{-3}$ and 6.57×10^{-3}, respectively.

with $\delta = m(\tilde{\mu}_m^o - \tilde{\mu}_\infty^o)/kT > 0$ denoting the rods' growth parameter. (Note that $\tilde{\mu}_m^o = \tilde{\mu}_{sph}^o$ and $\tilde{\mu}_\infty^o = \tilde{\mu}_{cyl}^o$.) The quantity α can be expressed in several different but equivalent forms; e.g.,

$$
\begin{aligned}
\alpha &= 1/s^* = (\tilde{\mu}_\infty^o - \mu)/kT \\
&= \ln(\bar{X}_1/X_1) = (1/s)\,\ln(\bar{X}_s/X_s)
\end{aligned}
\tag{1.43}
$$

The first equality identifies $1/\alpha$ as s^*, the most probable aggregation number, as follows directly from $\partial X_s/\partial s = 0$ and (1.42). In the second equality μ is the chemical potential of amphiphiles in solution, cf. (1.22) and (1.23). In the third and fourth equalities, α is related to the limiting ("saturation") concentrations of monomers, $\bar{X}_1$, and s-micelles, $\bar{X}_s$, respectively, cf. (1.36). From the discussion in the previous section we know that a solution of rod-like micelles ($d = 1$) is never saturated. Thus, as X increases, all X_s approach $\bar{X}_s$ very closely but never quite reach these values. Correspondingly α is positive and approaches zero as X increases. A typical progression of X_s-distributions, upon increasing X, is shown in Fig. 1.8. Given the very simple form of X_s in (1.42), one can easily calculate any desired characteristic of the size distribution, such as the averages $\langle s \rangle$, $\langle s^2 \rangle$, etc. The monomer contribution to these quantities can be included explicitly and exactly. However, for a number of reasons we prefer to completely ignore the monomer terms in the discussion of micellar growth. First, this simplifies all derivations. Second, we are explicitly interested in concentrations well above the CMC where the monomer concentration is both negligible and essentially constant, i.e., $X \gg X_1 \approx \bar{X}_1 = constant$. To see this more clearly, we note that in this concentration regime $\hat{X}_1 < X_1 < \bar{X}_1$, with $\hat{X}_1$ denoting the monomer concentration at the CMC as given by (1.30), and $\bar{X}_1$ the limiting concentration (1.37). (From (1.30) and (1.37) we find $\hat{X}_1/\bar{X}_1 \sim \exp(-\delta/m)$ with δ/m ranging typically between 0.1 and 0.5.) We shall see below that at high concentrations $s^* \approx \langle s \rangle \approx 1/\alpha \gg 1$, hence $X_1/\bar{X}_1 = \exp(-\alpha) \approx \exp(-1/s^*) \approx 1$. Finally, we note that although X_s is often expressed in terms of X_1 (see e.g., (1.26)), the monomers have no influence on the micellar size distribution. This is because the growth of spherical into cylindrical micelles, which is sometimes called "the sphere-to-rod transition", is governed completely by the two parameters δ and α; δ measuring the difference in free energy between spheres and cylinders and α reflecting the total concentration. None of these parameters depend on monomer properties!

The kth moment of X_s is defined by

$$
\begin{aligned}
M_k &= \sum_{s \geq m} s^k X_s = e^{-\delta} \sum_{s \geq m} s^{k+1} e^{-\alpha s} \\
&= e^{-\delta} \left(-\frac{\partial}{\partial \alpha} \right)^{k+1} \sum_{s \geq m} e^{-\alpha s} \\
&\approx e^{-\delta}(k+1)!/\alpha^{k+2}
\end{aligned} \tag{1.44}
$$

where in the second equality we have used (1.42). In the passage to the last equality we have used the approximation $\sum_{s \geq m} \exp(-\alpha s) = \exp(-\alpha m)/(1 - \exp(-\alpha)) \approx 1/\alpha$, corresponding to the assumption $\alpha m = m/s^* \ll 1$. Based on this last assumption, we shall replace below the summations $s \geq m$ by $s \geq 0$.

The zeroth moment is simply the *total* amphiphile concentration, $M_0 = \sum X_s = X - X_1 \approx X = \exp(-\delta)/\alpha^2$. From M_1 we obtain the familiar result for the *weight averaged* size [31,32,63,64,66,70]

$$
\begin{aligned}
\langle s \rangle_w &= \sum_s s X_s / \sum_s X_s = 2/\alpha = 2s^* \\
&= 2(Xe^\delta)^{1/2}
\end{aligned} \tag{1.45}
$$

Similarly, for the *width* of the micellar size distribution we obtain

$$
\begin{aligned}
\sigma_s &= \left[\langle s^2 \rangle_w - \langle s \rangle_w^2 \right]^{1/2} = \langle s \rangle_w / \sqrt{2} \\
&= (Xe^\delta/2)^{1/2}
\end{aligned} \tag{1.46}
$$

indicating significant size polydispersity, as reflected in Fig. 1.8.

Weight averages are calculated with respect to the "weight distribution", $P_w(s) = X_s/X$, expressing the fraction of molecules incorporated in micelles of size s. Similarly, *number* averages are calculated using the "number distribution", $P_n(s) = n_s/\sum n_s = (X_s/s)/\sum(X_s/s)$, corresponding to the fraction of micelles of size s. Thus, the number averaged size is defined and given by

$$
\begin{aligned}
\langle s \rangle_n &= \sum_s s(X_s/s)/\sum_s (X_s/s) = M_0/M_{-1} \\
&= 1/\alpha = \langle s \rangle_w /2 = s^*
\end{aligned}
\tag{1.47}
$$

Combining (1.42) and (1.47), we find

$$
P_n(s) = \frac{1}{\langle s \rangle_n} \exp\left[-s/\langle s \rangle_n\right]
\tag{1.48}
$$

with $\sum P_n(s) \approx \int P_n(s)ds = 1$.

The analogous expression for $P_w(s) = X_s/X$ reads

$$
P_w(s) = \frac{4s}{\langle s \rangle_w^2} \exp\left[-2s/\langle s \rangle_w\right]
\tag{1.49}
$$

which is another form of (1.42), appropriate for $\langle s \rangle_w \gg m$.

The basic characteristics of 1D micellar growth in dilute solution have been confirmed by many experiments. For instance, light scattering measurements [7,10] of the average size (hydronamic radius, $\bar{R}_H$) of ionic (e.g., SDS) micelles corroborate (1.45). In particular, from plots of $\bar{R}_H$ vs. T it has been possible to deduce the growth parameter δ for different values of the total salt (NaCl) concentration in the solution. The Coulomb repulsion between surfactant head-groups is partially screened by the added counterions. Thus, the added salt favors more strongly the cylindrical over the spherical packing environment (see Sec. 1.3.2), implying larger $\delta = m(\tilde{\mu}_{sph}^o - \tilde{\mu}_{cyl}^o)/kT$ and larger micellar sizes. The above experiments enable quantitative evaluation of δ, which for SDS is ≈ 20.

Deviations from the simple rod growth model described in this section are obviously expected at high surfactant concentrations, where aggregation numbers are large and inter-micelle forces are no longer negligible (see Sec. 1.4). However, deviations are also expected in dilute solutions whenever μ_s^o is not exactly a linear function of s. This, for example, may be the result of curvature effects on the electrostatic interactions in ionic surfactant solution [71], the addition of cosurfactants, or the flexibility of large "worm-like" micelles [78–81]. These issues as well as a more detailed account of experimental results are discussed by Porte in the next chapter (see also Sec. 1.2.5 below).

1.2.5   OTHER AGGREGATES

The discussion in Sec. 1.2.4 has been focused on the role of dimensionality in micellar growth. Accordingly, we have considered only three basic aggregation geometries: spheres, cylinders and planar bilayers, and their "combinations"—e.g., as spherocylinders or disk-like micelles with semitoroidal rims. Furthermore, it has been assumed that all aggregates in solution have the same geometry. Clearly, this is an approximation, and the possibility that aggregates of different shapes, e.g., rod-like and disk-like, can co-exist in solution must be taken into account.

To examine the above possibility, let us consider two aggregation geometries g and g'. (The symbol g has been previously used, e.g., in (1.33), to denote a "basic" geometry such as *sphere* or *cylinder*. Hereafter, we shall also use it for "complex" geometries such as *spherocylinders*.) From (1.22) or (1.26) one expects that if the free energy per molecule in the two geometries is similar (yet not strictly equal), namely $\tilde{\mu}^o_{s,g} \approx \tilde{\mu}^o_{s,g'}$, then $X_{s,g} \approx X_{s,g'}$, i.e., the solution will contain similar numbers of s,g and s,g'-aggregates. This conclusion is valid for small s only; *for large s there will always be a single dominant geometry.* This can be understood quite generally using (1.35) from which it follows that for large s, $\Delta\mu^o_s = \mu^o_{s,g'} - \mu^o_{s,g} \approx s[\tilde{\mu}^o_{\infty,g'} - \tilde{\mu}^o_{\infty,g}] \equiv s\Delta\tilde{\mu}^o_\infty$, where we have neglected the surface term in μ^o_s. (Recall that $\tilde{\mu}^o_{\infty,g}$ is the chemical potential in the main part of the aggregate, e.g., for $g = rod$, $\tilde{\mu}^o_{\infty,g} = \tilde{\mu}^o_{cyl}$ and for $g' = disk$, $\tilde{\mu}^o_{\infty,g'} = \tilde{\mu}^o_{bil}$.) Now, suppose for concreteness that $\Delta\tilde{\mu}^o_\infty = \tilde{\mu}_{\infty,g'} - \tilde{\mu}^o_{\infty,g} > 0$, i.e., g is the (asymptotically) more stable geometry. Clearly, even if $\Delta\tilde{\mu}^o_\infty$ is not large compared to kT, this ("monomeric") difference is largely magnified when comparing large aggregates, for which $\Delta\mu^o_s = s\Delta\tilde{\mu}^o_\infty \gg kT$, implying $X_{s,g'}/X_{s,g} = \exp[-\Delta\mu^o_s/kT] \ll 1$.

Free energy differences corresponding to amphiphile packing in different aggregation geometries, $\Delta\tilde{\mu}^o_\infty$, are typically a few tenths of kT [30–33]. Thus, if e.g., $\Delta\tilde{\mu}^o_\infty/kT \approx 0.1$ then already for $s \approx 100$ one expects a single geometry to prevail. However, for these low aggregation numbers one also has to consider the contribution of the surface terms (cf. (1.35)) which have been ignored in the above analysis. Theoretical analyses indicate that in some systems, at low concentrations, when the micelles are typically small, the solution contains different micellar shapes, e.g., oblate (or disk-like) and prolate (or rod-like) particles [68]. It is also possible that a cross-over from dominance by one geometry to another will take place at some range of concentration and hence of micellar sizes [55,68,82]. However, since these behaviors concern small aggregates they are very difficult for experimental measurements to probe.

The (possible) appearance of micellar aggregates in different geometries may also be regarded as shape fluctuations of these complex particles. Thus, not surprisingly, such fluctuations are expected to be more pronounced when the aggregates are small. Shape fluctuations of different kinds are also observed for large micelles. Of particular interest in this context are the "giant" rod-like micelles which, similar to polymers, show a high degree of flexibility, and are often described as *"worm-like"* micelles. For these aggregates the form (1.34), for the standard chemical potential of a (rigid) rod-like aggregate, should be extended to include contributions accounting for the conformational flexibility of the micelle [66]: see Sec. 2.2.3.

Semi-flexible rod-like micelles (even relatively short or somewhat rigid ones) can bend so that their ends fuse, thus forming a ring (torus) shaped aggregate. The bending free energy associated with the formation of the ring (see Sec. 1.3.4) provides a positive, hence unfavorable, contribution to μ_s^o. Furthermore, ring closure involves a loss of conformational entropy. On the other hand, the ring closure relieves the unfavorable free energy price corresponding to the two ends, cf. (1.34). Since this last term is the driving force for the growth of rod-like aggregates, it is clear that the relative abundance of rings and rods will be determined by the difference between the bending and the edge free energies [66].

A more interesting case where "bending and fusion" relieves unfavorable edge free energy corresponds to the formation of *vesicles*, i.e., bilayers which close upon themselves to form a nearly spherical bubble [30–33,53,83–87]. Recall from Sec. 1.2.4 that large planar disk-like micelles are unstable. Thus vesicle formation not only relieves the excess free energy associated with the rim of a disk-like micelle, but also allows for the possibility of finite aggregates composed of (non-planar) bilayers. As usual, the size distribution of vesicles in dilute solution is determined by the variation of μ_s^o with s, which correlates with the vesicular radius R. In the simplest approximation it is assumed, based on simple molecular packing constraints (see Sec. 1.3.2), that the amphiphiles cannot organize into vesicles of radius R (size s) smaller than some critical value R_c (size s_c), and that $\tilde{\mu}_s^o$ is constant for $s \geq s_c$. Then, clearly, entropy considerations imply that most vesicles will be of radius R_c. (From (1.22) or (1.26) we see that if $\tilde{\mu}_s^o$ is constant then X_s/s decreases exponentially with s.) A rather different approximation corresponds to assuming that $\tilde{\mu}_s^o \sim k/R^2$ with k denoting the bilayer's bending constant (Sec. 1.3.4). Then, assuming that the average area per molecule is the same for all vesicles, it follows that $s \sim R^2$ and hence $\tilde{\mu}_s^o \sim 1/s$ or $\mu_s^o = constant$. This model yields $X_s = As[X_1 \exp(\beta \mu_1^o)]^s = As \exp(s\beta\mu)$, with $A = \exp(-\beta\mu_s^o) = constant$ (cf. (1.22) and (1.26)). This distribution is considerably broader than the one corresponding to $\tilde{\mu}_s^o = constant$. It can be improved by including the (logarithmic) dependence of the elastic constant on vesicle size, which also improves the agreement with experiment [83].

Finally, we note that so far we have only considered "pure", that is, single component, aggregates. Generally speaking, the formation and growth of *mixed aggregates* are governed by the same principles prevailing in solutions of pure aggregates [88–90]. However, additional complexities (and possibilities) arise, because of the new thermodynamic degree of freedom corresponding to amphiphile composition. Consider, for example, a solution of two amphiphiles A and B, each of which separately prefers the formation of aggregates of different geometry. Suppose, for example, that A prefers a cylindrical environment whereas B has a slight preference for a spherical geometry. Because of the natural thermodynamic tendency for mixing, the two amphiphiles will tend to form mixed aggregates (although the formation of separate A and B aggregates is also a possibility). Most likely, the mixed aggregates formed in this system will be rod-like micelles in which the B molecules will tend to concentrate in the spherical ends, while the A's prefer the cylindrical body. In other words, the molecular compositions in the two regions of the micelle can be different. (This corresponds to minimizing the packing free energy; see Sec. 1.3.) Furthermore, the overall A/B ratio in a micelle may depend on its size. These qualitative notions can be cast in standard statistical thermodynamic terms, and although the mathematical procedure is more involved than in Sec. 1.2.4, the derivations are rather straightforward. For example, in the case of a mixed system of rod-like micelles it has been shown that $\langle s \rangle_w \sim X^{2/5}$ as compared to the $\langle s \rangle_w \sim X^{1/2}$ behavior for pure aggregates [88]. Surfactant segregation ("repartitioning") into different regions of a micelle has been observed in several experimental studies [91–93]. However, the variation of $\langle s \rangle_w$ according to $X^{2/5}$ still awaits experimental verification.

Another example of an interesting and extensively studied class of systems corresponds to binary mixtures of phospholipids (e.g., lecithin) and 'detergents' (e.g., bile salts) and ternary systems including cholesterol as well. Here, the stable aggregation geometry as well as the micellar size depends on overall composition. For example, below a certain bile salt/lecithin ratio a dilute solution of these molecules consists of stable mixed vesicles. However, above this ratio the vesicular form becomes unstable and the solution (becomes transparent and) contains mixed *micelles* instead. It is believed that these micelles are composed of a disk-like (bilayer) body containing mainly lecithin, which is surrounded by a bile-salt rim [92].

There are many other examples of multi-component amphiphilic systems exhibiting diverse structures in dilute solutions. Even richer polymorphism is found in concentrated solutions, as is amply demonstrated in Sec. 1.4 and in some of the other chapters in this volume.

1.2.6   THE AGGREGATE'S PARTITION FUNCTION: ROTATION-TRANSLATION CONTRIBUTIONS

The basic premise of the phenomenological approach to micellar growth described in the previous sections is that the standard chemical potential of an s-aggregate, μ_s^o, is a weighted sum of contributions from the different regions of the aggregate, as expressed generally in (1.33) or, for simple geometries by (1.34) and (1.35). In the standard treatments of micellar growth, such expressions are usually postulated, without attempting to justify or assess their validity. This may be related to the common interpretation of μ_s^o as "the internal free energy of the aggregate", in analogy to the free energy of a (stationary) macroscopic system. Eq. (1.33) corresponds to extending this analogy to macroscopic systems composed of several subsystems g.

For *finite* aggregates some caution is required in identifying the degrees of freedom contributing to the free energy represented by μ_s^o. To this end we now return to the basic statistical thermodynamic relation (1.12) (or (1.17))

$$\mu_s^o = -kT \ln(q_s/V) \qquad (1.50)$$

and examine to what extent (1.33) is consistent with the general expression for the aggregate's partition function. (Note that μ_s^o in (1.50) is in fact $\mu_s^{o,p}$ which differs from μ_s^o of (1.17) by an additive constant.) As stressed at the very beginning of Sec. 1.2.1, a rigorous calculation of q_s is hopeless, because micellar aggregates are such complex systems. Nevertheless, one could expect that the more modest goal of evaluating the s dependence of q_s is, perhaps, more feasible. We shall see, however, that even this goal involves some highly nontrivial difficulties.

Let $f = t+r+c$ denote the number of degrees of freedom (d.f.) of a single amphiphile. For a monomer in solution, $t = 3$ is the number of translational (center of mass) degrees of freedom, $r = 3$ is the number of overall rotations of the molecule, and c is the number of internal or conformational degrees of freedom. Consider, for example, a simple amphiphile of the form $\mathbf{P}$–$(CH_2)_{n-1}$–CH_3, with $\mathbf{P}$ denoting the polar head group, which for simplicity will be assumed to be structureless. It is well known that the flexibility of these molecules is due, almost exclusively, to internal rotations around C–C bonds (see Sec. 1.3.3). All other modes, such as C–H and C–C stretches and C–C–C, C–C–H or H–C–H bends can be treated as frozen. (In fact, all we need to assume in order to ignore these modes in the present discussion is to establish that they are not affected by the state of aggregation of the amphiphiles.) Thus, the number of relevant internal degrees of freedom is $c = n - 1$, corresponding to the number of internal rotations around the backbone (non-terminal) C–C bonds.

In a micelle, due to the high packing density and the strong interactions between neighboring molecules, all f degrees of freedom of a given molecule are strongly coupled to each other, as well as to those of the other molecules. The translational motions are strongly hindered, and are more adequately described as vibrations, with occasional small jumps of the center of mass (or head-group) of the molecule. Without specifying the exact character of these motions, we shall assume that they can be treated classically. (This assumption may not be valid, for example, for the crystalline states of lipid bilayers.) The overall molecular rotations are even more strongly hindered. In fact, in the aggregate an overall rotation of the amphiphile, with a given (frozen) conformation, is obviously impossible; any rotation must involve conformational changes of both the rotating chain itself and its neighbors. Thus, in the aggregate, it is more appropriate to describe the conformational state of a molecule by $c + r$ (rather than by c) degrees of freedom; c numbers specify the bond (e.g., trans/gauche) sequence $\mathbf{b} = b_1 \cdots b_c$ of the chain (with $\{b_i\}$ the successive dihedral angles), and $r = 3$

numbers specify the overall orientation of the chain, Ω, with respect to some fixed coordinate system. Here, and in Sec. 1.3, we shall refer to $\alpha \equiv b, \Omega$ as the conformational state of the molecule. For concreteness let us assume that the corresponding $r + c$ d.f. are non-classical.

Based on the division of the $s \cdot f$ d.f. of the aggregate into 3s (hindered) classical translations and $(f - 3)s$ conformational d.f., the aggregate's partition function is given by

$$q_s = \frac{1}{h^{3s} s!} \int dp^{3s} \int dr^{3s} \sum_{\alpha^s} \exp\left[-\beta H\left(p^{3s}, r^{3s}, \alpha^s\right)\right] \tag{1.51}$$

where h is Planck's constant and $\beta = 1/kT$. r^{3s} and p^{3s} refer to the 3s position coordinates and 3s conjugate momenta of the s molecules, and α^s to their chain conformations. The Hamiltonian can be separated into kinetic and potential energy contributions

$$H = \sum_{i=1}^{3s} p_i^2/2m + W\left(r^{3s}, \alpha^s\right) \tag{1.52}$$

The potential energy W includes all the intermolecular interaction potentials, the internal (conformational) chain energies, and the interactions between the aggregated molecules and the surrounding solvent. With H given by (1.52) the integration over momenta in (1.15) is immediate, yielding

$$q_s = \frac{Z_s}{\lambda^{3s}} \tag{1.53}$$

with $\lambda = (h^2/2\pi mkT)^{1/2}$ denoting the de Broglie wavelength of a surfactant molecule, m being the amphiphile's mass. Z_s is the aggregate's configurational integral

$$Z_s = \frac{1}{s!} \int dr^{3s} \sum_{\alpha^s} \exp\left[-\beta W\left(r^{3s}, \alpha^s\right)\right] \tag{1.54}$$

Substituting (1.53) into (1.50) we can express the standard chemical potential as a sum of momentum and configurational terms

$$\begin{aligned} \mu_s^o &= \mu_s^{o,m} + \mu_s^{o,c} \\ &= skT \ln \lambda^3 - kT \ln(Z_s/V) \end{aligned} \tag{1.55}$$

Recall from (1.26) that the micellar size distribution X_s is governed by the s-dependence of $\mu_s^o - s\mu_1^o = s(\tilde{\mu}_s^o - \mu_1^o)$ or, equally, by $\tilde{\mu}_s^o - \tilde{\mu}_r^o$ with r denoting an arbitrary size. Since $\tilde{\mu}_s^{o,m} = kT \ln \lambda^3$ is a constant, it is clear that the *momentum* term in (1.55) is of no consequence for the micellar size distribution [53]. More explicitly, from (1.26) and (1.55), we see that X_s,

$$X_s = sX_1^s \frac{(Z_s/V)}{(Z_1/V)^s} \tag{1.56}$$

depends only on the configurational factors in the aggregate's partition function. Note also that the momentum terms in μ_s^o satisfy trivially the phenomenological expression $\mu_s^o = \sum s_g \tilde{\mu}_g^o$, cf. (1.33). (This follows immediately from the fact that we can define $kT \ln \lambda^3 \equiv \tilde{\mu}_g^{o,m} = constant$, hence $\tilde{\mu}_s^{o,m} = skT \ln \lambda^3 = \sum s_g \tilde{\mu}_g^{o,m}$.) Thus, not only X_s but also the validity of (1.33) depends only on the configurational part of μ_s^o.

For a given α^S, the potential energy $W(r^{3S}, \alpha^S)$ can be expressed as a function of $3(s-1)$ relative position coordinates $r_i - r_j$, or in terms of $3(s-1)$ (independent) coordinates measured with respect to the aggregate's center of mass. Accordingly, the remaining three coordinates may be chosen to represent a particular ("reference") particle, or the center of mass. Integrating over these coordinates in (1.54) yields a factor of V, implying that Z_s/V is volume independent. Note further that the factor V extracted from Z_s is the configurational integral associated with the translational motion of the aggregate as a whole. Accordingly, Z_s/V can be interpreted as the configurational integral of an *immobile* aggregate. More precisely, this corresponds to a non-translating aggregate because there is no restriction on its overall rotational motions. For an aggregate with well defined geometry, g, one can always separate three (Euler) angles, specifying its overall orientation in space [42–45]. Since W_s is independent of these angles, an additional factor of $8\pi^2$, corresponding to the integral over the three angles, can be extracted from Z_s. Thus, $Z_s^o = Z_s/8\pi^2 V$ or, see (1.54),

$$Z_s^o = \frac{1}{8\pi^2 V s!} \int dr^{3S} \sum_{\alpha^S} \exp\left[-\beta W\left(r^{3S}, \alpha^S; g\right)\right] \qquad (1.57)$$

can be interpreted as the configurational integral of a completely stationary (i.e., non-translating and non-rotating) aggregate. The symbol g in $W(r^{3S}, \alpha^S; g)$ is meant to emphasize that the integral in (1.57) includes only those configurations corresponding to a stationary aggregate of a well-defined geometry g.

From the discussion above, it follows that $\mu_s^{o,c} = -kT \ln Z^o$ is the (configuration part of the Helmholtz) free energy of a stationary aggregate. In the limit of a macroscopic aggregate ($s \to \infty$) of a single basic geometry g, $\mu_s^{o,c}$ becomes an extensive thermodynamic property and hence $\mu_s^{o,c}/s \to \tilde{\mu}_g^{o,c} = constant$. Similarly, for an aggregate comprising *several* "macroscopic regions" g we expect $\mu_s^{o,c} \to \sum s_g \tilde{\mu}_g^{o,c}$, as suggested by the phenomenological representation (1.33). Thus, for large aggregates (1.33) is certainly an adequate expression for μ_s^o. Whether (1.33) is equally appropriate for small micelles is not as obvious, and requires detailed analysis of Z_s.

An exact calculation of Z_s remains beyond our scope. But even the more qualitative question regarding the s dependence of Z_s for finite aggregates is highly non-trivial. For example, it can be shown that the transformation of dr^{3S} from laboratory to center-of-mass coordinates involves a Jacobian containing a factor s^3 [44]. Additional s^k factors arise from the separation of the overall rotational motion (see below). These factors in Z_s suggest that μ_s^o contains, in addition to terms varying linearly with s, also terms proportional to $\ln s$. The $\ln s$ terms become negligible as $s \to \infty$, but for finite s their inclusion in μ_s^o can significantly affect the calculation of X_s [54]. It should be stressed, however, that it is not clear that the s^k factors resulting from coordinate and angle transformations account for the entire s-dependence of Z_s. Very similar questions arise in nucleation theory [42–45], where considerable controversy still exists regarding the most consistent way to separate the external motions of a cluster from the internal motions of its constituent particles.

In (1.53), based on (1.52) and (1.51), we have expressed q_s as a product of configurational and momentum factors. But this is not the only reasonable representation of q_s. Another one corresponds to a factorization of q_s into a product of translational, rotational and internal partition functions,

$$q_s = q_s^T q_s^R q_s^I . \tag{1.58}$$

This form has been adopted by a number of authors in their analyses of amphiphile self-assembly [48–52,54]. We shall briefly consider (1.58) below, because it clearly demonstrates the (possible) appearance of s^k factors in q_s. Eq. (1.58) represents a common factorization of molecular partition functions; namely, for diatomic or polyatomic molecules, it is always possible to separate out the center-of-mass translation [94]. Neglecting vibration-rotation coupling, one can also separate a rotational partition function, so that q_s^I corresponds to a vibrational partition function. Similarly, it can be shown that (1.58) is valid for any aggregate with well-defined center-of-mass and moments of inertia (i.e., well-defined shape and mass distribution) [42,43]. Accordingly, we can write

$$\mu_s^o = \mu_s^{o,TR} + \mu_s^{o,I} \tag{1.59}$$

with

$$\mu_s^{o,TR} = -kT \ln \left(q_s^T q_s^R / V \right) \tag{1.60}$$

and

$$\mu_s^{o,I} = -kT \ln q_s^I \tag{1.61}$$

The translation-rotation partition function $q_s^T q_s^R = q_s^{TR}$ involves six degrees of freedom; three corresponding to center of mass translation and three to the overall rotations of the aggregate when regarded as a rigid body. Thus, the internal partition function involves $3s - 6$ internal position and $3s - 6$ internal momentum coordinates, defined relative to an aggregate-fixed system of coordinates. Formally, q_s^I can be expressed as

$$q_s^I = \frac{1}{h^{3s-6} s!} \int dp^{3s-6} \int dr^{3s-6} \sum_{\alpha^s} \exp\left[-\beta H_I \left(p^{3s-6}, r^{3s-6}, \alpha^{3s} \right) \right]$$

$$\tag{1.62}$$

The s-dependence of this quantity is extremely complicated and generally unknown except in highly idealized cases. As noted already with respect to Z_s there are, for example, s-dependent factors associated with the Jacobian of the transformation from the laboratory to the aggregate-fixed coordinate system. (Also, the $1/s!$ factor can be modified, as "part of it" should be absorbed into the symmetry numbers of the rotational partition function.)

On the other hand, the translational-rotational factors in (1.58) are simple. The translational partition function is $q_s^T = V/\Lambda_s^3$, with $\Lambda_s = \lambda/s^{1/2} = (h^2/2\pi smkT)^{1/2}$ denoting the deBroglie wavelength of the s-mer. The rotational partition function is given by $q_s^R = 8\pi^2/(\Lambda_{rot,s})^3$ with $\Lambda_{rot,s} = (h^2/2\pi I_s kT)^{1/2}$, I_s being the geometric mean of the aggregate's three moments of inertia, e.g., for rod-like micelles $I_s \sim s^{7/2}$. Thus for rods, say, $\mu_s^{o,TR}$ includes a $-5\ln s$ term ($5 = 3/2 + 7/2$). Now, if we assume that the separation $\mu_s^o = \sum s_g \tilde{\mu}_g^o$ should be applied to the internal free energy part only ($\mu_s^{o,I}$) in (1.59), then obviously μ_s^o will contain $\ln s$ contributions. E.g., for rods, it is easy to see that, instead of $\mu_s^o = s\tilde{\mu}_\infty^o + \delta kT$ (set $d = 1$ in (1.35)) we get $\mu_s^o = s\tilde{\mu}_\infty^o + (\delta - 5\ln s)kT$. Using this expression for μ_s^o, one finds a slower than $X^{1/2}$ increase of $\langle s \rangle$ with total concentration [54], which can be attributed to the fact that the translation-rotation entropy decreases with $\langle s \rangle$ and thus resists growth.

In the classical limit, when all the $3s$ degrees of freedom corresponding to molecular translations are classical, q_s is given generally by (1.51). In this limit the two representations (1.53) and (1.58) are equivalent in principle, provided the aggregate structure is consistently defined. (This sets constraints on coordinate integrations in the calculation of partition functions.) For dimers ($s = 2$) held together by a harmonic potential, this can be rigorously shown, provided the dimer is "tight", so that the vibrational amplitude is small compared to the distance between the two particles. The generalization to larger s-clusters, even simple ones like a linear array of "atoms" bound by harmonic restoring forces, is highly non-trivial. Thus, although we recognize that (in the classical limit) the two schemes (1.53) and (1.58) are equivalent, we cannot evaluate the exact s-dependence of q_s, q_s^I or Z_s. Clearly, both $\mu_s^{o,c} = -kT \ln Z_s^o$ and $\mu_s^{o,I} = -kT \ln q_s^I$ represent "internal free energies." Since we are equally ignorant about their exact s-dependence (except in one case—see below), the phenomenological expression may be arbitrarily applied to either $\mu_s^{o,c}$ or $\mu_s^{o,I}$. The only difference is that in the second case $\mu_s^o = \mu_s^{o,I} + \mu_s^{o,TR}$ will contain the $\ln s$ terms mentioned above. In the asymptotic ($s \to \infty$) limit, the difference disappears, but for finite aggregates it may affect the calculated size distribution.

It should be noted that there is one limiting, albeit somewhat hypothetical regime, where (1.58) is exact, whereas (1.53) is totally inadequate. This is the case of "rigid" (or solid) aggregates, i.e., aggregates in which all the intermolecular distances are fixed. Of course, in this case only the external rotations and translations (i.e., 6 degrees of freedom) can be treated as classical. (Hence (1.51) is not applicable.) In this case the internal free energy $-kT \ln q^I$ is simply the internal energy of the aggregate, which can be expressed as a sum of contributions from different microenvironments, i.e., $\sum s_g \bar{\epsilon}_g$, with $\bar{\epsilon}_g$ denoting the energy per molecule in environment g. As a simple (idealized) example, suppose that a rod-like micelle is a rigid linear string of s beads, held to each other by nearest neighbor attractive potentials. Then, of course, the internal energy is exactly $2\epsilon_1 + (s - 2)\epsilon_2$, with ϵ_1 and ϵ_2 corresponding to the energies of the terminal and non-terminal beads, respectively.

General discussions of rotation/translation vs. internal contributions to the micelle size distributions are found in the literature as early as 1957, when Hoeve and Benson [48] considered a factorization of q_s, which is virtually identical to (1.58). They calculated q_s^o assuming a liquid-drop model of the micellar interior, in conjunction with an estimate of the surface free energy. Then they determined the size distribution along the lines prescribed in Sec. 1.2.4. This approach has been extended from the canonical to the constant pressure ensemble by Aranow [49]. A detailed statistical-thermodynamic analysis, discussing various representations, and considering (qualitatively) the connection to the phenomenological descriptions has been given by Wulf [50]. Poland and Scheraga have also explicitly separated out the "external" (translation and rotation) contributions from q_s, writing them as proportional to s^n, with $n = 3/2$ (from translation) $+5/2$ (from rotation) in the case of spheres [51]. (Note that the mean moment of inertia $I_s \sim (I_A I_B I_C)^{1/3}$ scales as $s^{5/2}$ for spherical aggregates.) These $s^{5/2}$ factors in q_s give rise to $\ln s$ terms in μ_s^o which are partially cancelled in the Poland-Scheraga theory by contributions of the same form from $\ln q_s^I$. This partial cancellation between "external" and "internal" contributions has been considered most systematically by Nagarajan and Ruckenstein [52], with a rather different set of approximations invoked to treat the micellar interior. But, clearly, the subtle questions raised in this section require a great deal of further consideration before any firm conclusions can be reached concerning the relevant s-dependence of standard chemical potentials in the phenomenological treatment of self-assembly.

1.3 Molecular Organization of Aggregates

In Sec. 1.2 we have elaborated upon the central role of μ_s^o in amphiphile self-assembly and micellar growth. In the phenomenological approach (see (1.33)) μ_s^o is expressed as a superposition of the $\tilde{\mu}_g^o$'s, with $\tilde{\mu}_g^o$ representing the free energy per molecule in a microenvironment characterized by a single geometry g. So far, however, we have not considered at all the microscopic aspects of these quantities, namely, their relation to molecular parameters and intermolecular interactions. This is our goal in the present section. We shall describe several theoretical models for the structure and thermodynamics of amphiphilic aggregates, and examine their underlying assumptions from a statistical thermodynamic point of view.

The importance of molecular theories of amphiphilic aggregates extends beyond their implications for micellar growth. There are numerous systems in chemistry, biology and other fields where the *internal structure* and properties of these systems are of utmost importance. To mention one important issue, the conformational statistics of the phospholipid molecules constituting biological membranes play a crucial role in determining the fluidity, elasticity and stability of these systems. Thus, Sec. 1.3.3 is devoted to a discussion of chain packing and conformational statistics of amphiphiles in various microenvironments. Curvature elasticity of amphiphilic films is treated explicitly in Sec. 1.3.4. Beforehand, we consider in Sec. 1.3.1 the separation of head-group and chain (or, surface vs. bulk) contributions to the free energy. Simple yet instructive and useful models for the structure and relative stability of amphiphilic aggregates are the subject matter of Sec. 1.3.2.

1.3.1   SEPARATION OF HEAD AND TAIL CONTRIBUTIONS

In this section we focus on the statistical-thermodynamic properties of amphiphiles packed in a microenvironment of well-defined geometry g. Specifically, the thermodynamic system of interest is an (immobile) aggregate, or portion thereof, containing s amphiphiles which (neglecting end effects) can be treated as equivalent particles. A large section of a long cylindrical micelle, a sheet of a planar bilayer, a spherical micelle or a spherical vesicle are typical examples.

Let $F(s, g, T)$ denote the configurational free energy of the system, with $f = F/s$ the free energy per molecule. We are interested in the *macroscopic* $(s \to \infty)$ limit, where end effects are negligible and $f = f(g, T)$ is strictly intensive. (The $s \to \infty$ limit is clearly an idealization for small micelles or vesicles.) In this limit $f(g, T)$ is identical to the configurational part of $\tilde{\mu}_g^o$, as discussed in Sec. 1.2.6, i.e.,

$$f(g, T) \equiv \tilde{\mu}_g^{o,c} \tag{1.63}$$

Recall also that when $s \to \infty$ the various standard chemical potentials, e.g., $\tilde{\mu}_g^{o,c}, \tilde{\mu}_g^o, \mu_{s,g}^o/s$ or $\mu_s^{o,I}/s$, differ from each other by uninteresting additive constants. For notational convenience we shall use $f(g, T)$ rather than any of the above μ^o's.

As usual, F is related to a configurational integral Z,

$$F(s, g, T) = -kT \ln Z(s, g, T) \tag{1.64}$$

with

$$Z(s, g, T) = C \int dr^{3S} \sum_{\alpha^s} \exp\left[-\beta W\left(r^{3S}, \alpha^s; g\right)\right] . \qquad (1.65)$$

As in (1.54) W is the potential energy of the aggregate assuming that the surrounding solvent is a continuous medium, so that W is in fact a potential of mean force [94]. Note that $Z(s, g, T)$ corresponds to the $s \to \infty$ limit of Z_s° in (1.57), which is the configurational integral of a stationary aggregate. Accordingly, we can identify C as $1/8\pi^2 V s!$. ($1/s!$ corrects for the overcounting of configurations implied by the (unrestricted) integration over the position coordinates of the (identical) molecules. The $1/8\pi^2 V$ factor cancels out after integrating over the center of mass coordinates and rotation angles of the aggregate.)

As emphasized in the previous section, the calculation of Z is practically impossible because of the great complexity of amphiphilic aggregates. The most detailed approaches to date involve computer simulations, mainly molecular dynamics studies of the kind already discussed in Sec. 1.1 [36–41,95–105]. However, as stressed there, even these studies involve many difficulties and must employ drastic simplifying assumptions, as well as rely on uncertain intermolecular potentials. Some interesting results derived from computer simulations have been described in Sec. 1.1, and others will be mentioned in Sec. 1.3.3 below. However, most of the following discussion will be devoted to simple, but general, phenomenological models (Sec. 1.3.2) and mean-field theories (Sec. 1.3.3) where the "standard picture of micelles" outlined in Sec. 1.1 serves as a basic premise.

According to the "standard picture" every aggregate is composed of two well-defined regimes:

1. A hydrophobic core containing the hydrocarbon tails, packed such that the monomer (chain segment) density is uniform and liquid-like.

2. A surface region containing the polar head-groups, surrounded by solvent molecules (including counter ions in the case of charged amphiphiles), see Fig. 1.6. The two regimes are separated by a well-defined interface. The area per head-group, a, at the hydrocarbon-water interface and the principal curvatures $\mathbf{c} = c_1, c_2$ of this interface fully specify the geometry of the aggregate, i.e., $g \equiv a, \mathbf{c}$.

In the above model the head-groups always see the same (rather smooth) hydrocarbon surface, regardless of the many-chain configuration α^s within the hydrophobic core. Thus, associating $r_1, \cdots, r_s$ in $W(r^{3S}, \alpha^s; g)$ with the head-group positions relative to the interface, we can write $W(r^{3S}, \alpha^s; g) = W_h(r^{3S}; g) + W_t(\alpha^s; r^{3S}, g)$. W_h accounts for all the contributions to W from the head-group (surface) region. That is, $W_h = W_{hh} + W_{hs} + W_{ht} + W_{st}$ includes, respectively, head-group/head-group, head-group/solvent, head-group/hydrocarbon (tail) and tail/solvent interactions. A simple approximation for W_{st}, which includes the interactions between the solvent and those segments of the tails residing at the interface, would be $W_{st} = \gamma A$, with γ denoting the (effective) hydrocarbon-water interfacial tension and $A = sa$ being the total surface area. The geometry $(g = a, \mathbf{c})$ dependence of W_h enters mainly through its dependence on a, the $\mathbf{c}$ dependence being weaker.

The tail term W_t accounts for all the interactions between chain segments inside the hydrophobic core. In Sec 1.3.3 we show that W_t, which determines the probabilities $P(\alpha^s)$ of the various chain configurations α^s, depends strongly on the aggregation geometry, g. On the other hand, the dependence of W_t on head-group positions, r^{3s}, is weak. Because of head-tail connectivity the statistical weights of the α^s are correlated with r^{3s}. Imagine, however, that the head-groups were clamped at some fixed lattice points corresponding to their equilibrium positions at the hydrocarbon water-interface. Then r^{3s} is fully specified by the geometry $g = a, \mathbf{c}$ of the interface, implying $W_t = W_t(\alpha^s; g)$. Similarly, averaging over r^{3s} would also yield $W_t = W_t(\alpha^s; g)$. Since only small deviations from this behavior are expected when the head-groups are fluctuating around the equilibrium positions, we can safely write

$$W(r^{3s}, \alpha^s; g) = W_h(r^{3s}; g) + W_t(\alpha^s; g) \qquad (1.66)$$

It follows from this decomposition that (cf. 1.65)

$$Z(s, g, T) = Z_h(s, g, T) Z_t(s, g, T) \qquad (1.67)$$

with

$$Z_h(s, g, T) = C \int dr^{3s} \exp\left[-\beta W_h\left(r^{3s}; g\right)\right] \qquad (1.68)$$

$$Z_t(s, g, T) = \sum_{\alpha^s} \exp\left[-\beta W_t(\alpha^s; g)\right] \qquad (1.69)$$

From (1.67) we see that the aggregate's free energy is a sum of head and tail contributions, $F = F_h + F_t$ with $F_h = -kT \ln Z_h$ and $F_t = -kT \ln Z_t$. Similarly, $f = F/s$ is also a sum of head and tail terms

$$f(g) = f_h(g) + f_t(g) \qquad (1.70)$$

with both $f_h = F_h/s$ and $f_t = F_t/s$ depending on the aggregation geometry g.

In the next section we describe a phenomenological theory of $f(g)$. The term "phenomenological" used here refers to a theory which, based on some basic physical principles and assumptions, postulates a certain functional form for the free energy ($f(g)$ in our case). In other words, phenomenological theories circumvent the difficulties associated with direct calculation of partition functions. Quite often in complex systems the evaluation of partition functions involves so many approximations and assumptions that more physical insights are gained from phenomenological models. Nevertheless, in Sec 1.3.3 we shall describe a mean-field theory for $f_t(g)$ which corresponds to an approximate evaluation of Z_t and which provides considerable insight into the microscopics of chain packing statistics in hydrophobic environments.

1.3.2   HYDROCARBON DROPLET MODELS

Two phenomenological models for $f(g)$ have largely dominated the discussion of amphiphile packing and micellar size and shape in recent years. These models, one due to Tanford [30] and the other to Israelachvili, Mitchell and Ninham (IMN) [31] share many common features but differ markedly in some important aspects. Consistent with the "standard picture of the micelle", both theories treat the hydrophobic core as a liquid hydrocarbon droplet. More specifically, it is assumed that the hydrocarbon tails in the hydrophobic regions of the aggregates behave like the corresponding chains in the bulk liquid alkane. Following this assumption, both models assume that the tail term in (1.70) is a constant, independent of the micellar geometry, i.e.,

$$f_t(g) = f_t = constant . \qquad (1.71)$$

Furthermore, in both the Tanford and the IMN models, the dependence of $f_h(g) = f_h(a, \mathbf{c})$ on the area per head-group a is evaluated in terms of the "opposing forces" (see Sec. 1.3.2.2 below). The difference between the two approaches is manifested in the different criteria for determining the optimal aggregation geometry, i.e., the one corresponding to minimal $f(g)$.

In this section we present, in a common language, the basic concepts underlying these different phenomenological models, and comment critically on the various approximations involved. We shall mainly follow the IMN approach which provides a simple and elegant ("first order") scheme for determining the optimal geometry. The relation to Tanford's model will be mentioned at the appropriate junctures.

1.3.2.1   Geometric packing constraints and surface-volume relations

Let l denote the length of a fully extended amphiphile chain. l scales linearly with the number of chain segments, n; e.g., for simple alkyl tails, $-(CH_2)_{n-1}-CH_3$, the length of an all-trans chain is [30] $l(\text{Å}) \approx 1.5 + 1.27n$. The volume per chain, v, as measured in the bulk liquid alkane, also scales with n, e.g., for alkyl chains [30] $v(\text{Å}^3) \approx 27.5 + 27n$. Since there are no holes in the hydrophobic core, its smallest dimension must be smaller than $2l$. For instance, in a spherical micelle the radius R cannot exceed the length of a fully stretched chain, i.e., $R \leq l$. The same constraint applies to the radius R of (the circular cross section of) a cylindrical aggregate. Similarly, in planar bilayers $R \leq l$, where $R \equiv D/2$ is the half-thickness of the bilayer. In the following discussion we shall continue to consider $g =$ *sphere, cylinder* and *bilayer*, the three "basic" geometries.

For a spherical micelle of radius R which contains s amphiphiles, the total volume of the hydrophobic core is $V = (4/3)\pi R^3 = sv$, and its total surface area is $A = 4\pi R^2 = sa$, implying $a = 3v/R$. Similar considerations yield $a = 2v/R$ and $a = v/R$ for cylinders and planar bilayers, respectively. From the packing constraint $R \leq l$ it now follows that the areas per head-group in the three basic geometries must satisfy

$$a \geq a_{\min}(g) = i_g \left(\frac{v}{l}\right) \tag{1.72}$$

with $i_g = 1, 2$ and 3 for spheres, cylinders and bilayers, respectively. (Note $i_g = 3 - d$ with d denoting the dimensionality of the aggregate, cf. Sec. 1.2.4.)

As a numerical example illustrating the significance of (1.72) consider, say, a 12-carbon tail such as in SDS. Setting $n = 12$ in the above expressions for the volume per chain and all-trans length, we find for this molecule $v \approx 350\text{Å}^3$ and $l \approx 16.7\text{Å}$. (According to Tanford [30], the first CH_2 group of the tail resides mainly outside the hydrophobic core, so that $v \approx 320\text{Å}^3$ and $l \approx 15.5\text{Å}$, but we ignore these subtleties here.) These values imply that the minimal area per SDS head-group (see (1.72)) in a spherical micelle is $a \approx 63\text{Å}^2$, and the corresponding lower limits for cylinders and bilayers are 42Å^2 and 21Å^2, respectively. The last value, 21Å^2, is indeed close to the common estimate of the minimal cross sectional area of (all-trans) chains in bilayers and monolayers (see Chap. 12).

This example has demonstrated that SDS chains, for example, can be packed into different types of aggregates, provided the area per molecule, a, complies with (1.72). We now turn to consider which of these geometries is the *optimal* one.

1.3.2.2   The opposing forces

In the phenomenological models f is expressed as an explicit function of a, whereas its dependence on curvature $\mathbf{c}$ centers only as a correction. Thus, for the present discussion we shall assume $f(g) = f(a, \mathbf{c}) = f(a)$. (We treat curvature elasticity of amphiphile *films*, i.e., the $\mathbf{c}$ dependence of f, in Sec. 1.3.4.) Both IMN and Tanford express $f(a)$ as

$$\begin{aligned} f(a) &= f_h(a) + f_t \\ &= \gamma a + \frac{\kappa}{a} + f_t \end{aligned} \tag{1.73}$$

with f_t representing the constant tail contribution, as assumed in (1.71).

The first two terms in the second equality correspond to one of the simplest representations of f_h in terms of the "opposing forces" [30–33]. The first term, γa, is the interfacial free energy associated with hydrocarbon-water contact, with γ denoting the effective surface tension. (Recall that

γa is related to the W_{st} term of W_h, as described in Sec. 1.3.1.) Since γa increases with a, this term tends to reduce the area per head group, and in this sense corresponds to an attractive interaction between head-groups. On the other hand, as a decreases, head-group–head-group repulsion increases due to electrostatic or excluded volume interactions, or both. The second term in (1.73), κ/a, is the leading term of the repulsive potential (per molecule) in a system of ionic or zwitterionic head-groups. The constant κ can be estimated from simple models, or empirically (together with γ) from experimental data [30–33,53]. Many authors have studied the electrostatic free energy in the interfacial region in a more quantitative fashion [30,33,55,106–111] based either on solving the Poisson-Boltzmann equation [55,106–108] or via computer simulations [111]. For the qualitative discussion here it suffices to represent these effects by the simple form κ/a. If the repulsion is predominantly steric (excluded area), other representations may be more adequate than κ/a, e.g., a van der Waals-like repulsion $\ln[1/(a - \hat{a})]$, with $\hat{a}$ denoting the excluded area per head-group [112].

Let a_h denote the area per head-group which minimizes $f_h(a)$. From (1.73) we find that $a_h = \sqrt{(\kappa/\gamma)}$. The parameters κ and (to a lesser extent) γ depend on the nature of the head-groups and the properties of the surrounding solution (e.g., the ionic strength). Thus a_h is also determined by the properties of the head-group and the solution, but is independent of the amphiphile tail. Substituting $\kappa = \gamma a_h^2$ into (1.73), we get

$$f(a) = 2\gamma a_h + \frac{\gamma}{a}(a - a_h)^2 + f_t . \qquad (1.74)$$

The second term here accounts for the excess free energy associated with packing the amphiphile with an area per head-group a different from the optimal area a_h. The common estimates of γ range between $(25–50)$ erg/cm^2 which, at room temperature, correspond to $\approx (0.05 - 0.1)kT/\text{Å}^2$. Thus, for example, if $a_h = 30\text{Å}^2, a = 40\text{Å}^2$ and $\gamma = 0.05kT/\text{Å}^2$ we find that the excess free energy is $\approx 0.1kT$ per molecule.

1.3.2.3   The optimal aggregation geometry

The physical content of the assumption (1.71) is that the chains can adjust conformationally so as to comply with any packing geometry, *at no free energy cost*. The accuracy of this approximation is discussed in more detail below (cf. Sec. 1.3.3). Subject to this approximation, $a = a_h$ minimizes not only $f_h(a)$, but $f(a) = f_h(a) + f_t$ as well. Thus, in the IMN model a_h represents the *optimal* area per head-group.

The next, and very crucial, step in the IMN model is to assume that in all aggregation geometries the area per amphiphile head-group is always $a = a_h$, thus ensuring that $f(a)$ is *always* at a minimum $(= 2\gamma a_h + f_t)$. But the condition $a = a_h$ can be satisfied by more than one aggregation geometry, and is thus not sufficient to determine the optimal geometry. For example, if for a given amphiphile $a_h \geq 3(v/l)$, then according to (1.72) the

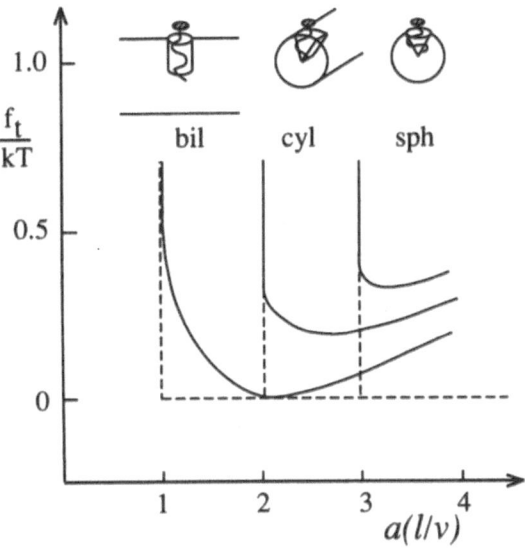

FIGURE 1.9. The packing (conformational) free energy of amphiphile tails in the three basic aggregation geometries, as a function of the area per head-group. The dashed lines, indicating a constant value for f_t for all values of a which are consistent with the molecular packing constraints, correspond to the hydrocarbon-droplet assumption (see text). The full lines represent typical results obtained from more detailed theories of chain packing in the different microenvironments [115, 116, 120-122].

molecules can be packed into spheres, cylinders or planar bilayers at the same free energy cost: $f(a = a_h)$. According to IMN the amphiphiles will preferentially organize in spherical micelles, because for a given total concentration, this arrangement maximizes the "mixing" (disperson) entropy of the system—as discussed in Sec. 1.2.2. If, however, $3(v/l) > a_h > 2(v/l)$, spherical micelles are unstable because they require $a \geq 3(v/l) > a_h$. In this case, for similar reasons as above (i.e., maximum dispersion entropy), cylindrical micelles are the preferred micellar geometry. Finally, if $2(v/l) > a_h \geq (v/l)$, a planar bilayer is the only stable aggregation geometry (among the three basic shapes considered). These conclusions are summarized schematically in Fig. 1.9. Figure 1.9 displays also the dependence of f on a as derived from mean field theories which take into account the variation of f_t with a, i.e., without assuming $f_t(a) = constant$ (see Sec. 1.3.3.) The differences between the minima of the $f(a)$ curves corresponding to the different aggregation geometries are typically of the order of $(0.1 - 0.5)kT$, i.e., of the same order of magnitude implied by (1.74) for $a - a_h \approx 10\text{Å}^2$.

We conclude this section with a comment on Tanford's interpretation [30] of the (hydrocarbon droplet) assumption $f_t = constant$, which differs diametrically from that of IMN. He assumes that the chains preserve the same degree of conformational freedom in all the aggregates. In particular, he assumes that, as in bulk hydrocarbon liquids, the mean end-to-end chain length, b, obeys $b \approx 0.75l$ in *all* aggregation geometries. Combining this requirement with simple geometric considerations (of the type leading to (1.72)), one can calculate the values of a corresponding to any aggregation geometry (spheres, oblate or prolate micelles, etc.). Tanford then uses (1.73) to calculate $f(a)$ for each possible geometry, and the optimal one is chosen to be the one corresponding to the minimal value of $f(a)$. In general, the average a corresponding to the stable geometry is different from a_h, i.e., a_h does not play any significant role in Tanford's model.

1.3.3   CHAIN PACKING STATISTICS

All the information about chain organization in an aggregate of geometry g (or portion thereof) is contained in the many-chain distribution function

$$P(\alpha_1, \alpha_2, \ldots, \alpha_s) = \frac{1}{Z_t} \exp\left[-\beta W_t(\alpha^s, g)\right] \tag{1.75}$$

which gives the probability of finding the s chains in configuration $\alpha_1, \alpha_2 \cdots \alpha_s = \alpha^s$. $Z_t = Z_t(s, g, T)$ is the configurational integral (1.69). Thermodynamic functions can also be expressed in terms of $P(\alpha^s)$. In particular, the configurational free energy is given by

$$\begin{aligned} F_t &= E_t - TS_t = -kT \ln Z_t \\ &= \sum_{\alpha^s} P(\alpha^s) W_t(\alpha^s) + kT \sum_{\alpha^s} P(\alpha^s) \ln P(\alpha^s) \end{aligned} \tag{1.76}$$

with $E_t = \sum_{\alpha^s} P(\alpha^s) W_t(\alpha^s) = kT^2(\partial \ln Z_t/\partial T)$ and $S_t = -k \sum P(\alpha^s) \ln P(\alpha^s) = k \ln Z_t + kT(\partial \ln Z_t/\partial T)$ denoting the system's energy and entropy, respectively.

Owing to the great complexity of amphiphilic aggregates, computer simulations appear to be the only reasonable approach for calculating many-chain properties related to $P(\alpha^s)$. However, these studies encounter various kinds of technical difficulties, as noted already in previous sections. On the other hand, it should be realized that most of the experimental measurements of chain conformational statistics in micelles and bilayers involve single chain properties such as bond order parameter profiles or segment spatial distributions. These properties are determined by the singlet probability distribution function (*pdf*) of chain conformations $P(\alpha)$, which gives the probability of finding a given chain in conformation α, regardless of (i.e., averaged over) the conformations of all other chains. The singlet probability of finding chain 1, for instance, in conformation $\alpha_1 = \alpha$

is given by

$$P(\alpha) = \sum_{\alpha_2, \cdots, \alpha_S} P(\alpha, \alpha_2, \cdots \alpha_S) \tag{1.77}$$

In addition to conformational properties, $P(\alpha)$ can be used to calculate, in the mean-field approximation, any thermodynamic function of interest (see below).

The calculation of $P(\alpha)$ and derivable conformational and thermodynamic properties is the central objective of the mean-field (or "single-chain") theories of chain packing in micelles and bilayers. Several authors, beginning with Marcelja's study of lipid chain packing and phase transitions in bilayers [113], have formulated such theories, differing from each other in various respects (e.g., lattice vs. non-lattice models) but also sharing some common important features (e.g., the assumption of uniform liquid-like hydrophobic core) [113–128]. The common and discrepant aspects of the various approaches have been critically analyzed elsewhere [121].

In the discussion below, we follow closely the approach described in [120–126]. Our goal is to describe the basic conformational and thermodynamic characteristics of chains packed in microenvironments of different geometries. Our treatment is based largely on the application of a simple explicit expression which we shall derive for $P(\alpha)$. Underlying our derivation are simple geometric packing considerations which will be discussed first.

1.3.3.1   Chain packing constraints

Figure 1.10 describes schematically a portion of the hydrophobic core of an aggregate containing s chains, originating from an interface of total area A. The local geometry $g = a, c_1, c_2$ is specified by the average area per molecule $a = A/s$ and the local curvatures $c_1 = 1/R_1$ and $c_2 = 1/R_2$, with R_1, R_2 denoting the principal radii of curvature. Let $a(x)$ denote the average cross sectional area available per chain at (normal) distance x from the interface. Using $c > 0$ for convex interfacial curvatures (viewed from the hydrophobic core), we have

$$a(x) = a \left[1 - (c_1 + c_2)x + c_1 c_2 x^2 \right] \tag{1.78}$$

with $a = a(0)$. Thus, for planar interfaces ($c_1 = c_2 = 0$) such as in lipid bilayers or surfactant monolayers $a(x) = a = constant$. In spherical micelles, where $c_1 = c_2 = 1/R$, the area diminishes rapidly towards the center: $a(x) = a(1 - x/R)^2$. For chains packed in cylindrical geometries ($c_1 = 1/R, c_2 = 0$) the area decreases linearly with the distance from the interface: $a(x) = a(1 - x/R)$. On the other hand, $a(x)$ *increases* with x for chains packed in *concave* geometries such as the inner leaflet of a vesicle, the surface of a water-in-oil microemulsion droplet, or the cylinders constituting reversed hexagonal phases.

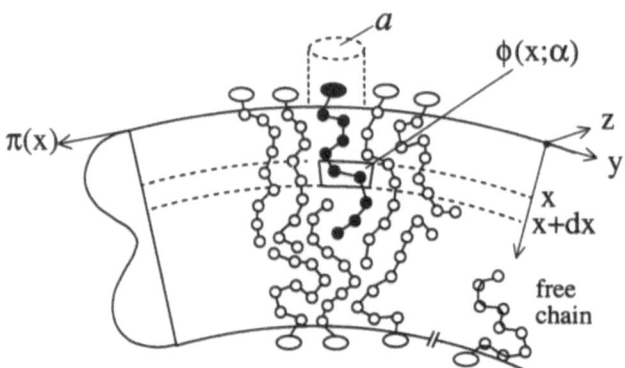

FIGURE 1.10. Schematic illustration of the origin of chain packing (uniform monomer density) constraints in the hydrophobic core. $\phi(x;\alpha)dx$ is the volume occupied by (or the number of monomers of) the 'central' chain within the layer $x, x+dx$ when the chain is in conformation α. $\pi(x)$ is the lateral pressure profile (see text). Also illustrated is a "free chain", i.e., one with no neighbors around and which is thus not subjected to packing constraints.

Let $\phi(x;\alpha)dx$ denote the volume taken up by a chain in conformation α, in the shell $x, x+dx$ of the hydrophobic core; see Fig. 1.10. More precisely, in our calculations we identify $\phi(x;\alpha)dx/\nu$ with the number of (centers of) segments of an α-chain within $x, x+dx$, where ν is the volume per chain segment in the bulk liquid hydrocarbon phase. Note that ϕ has dimensions of *area*. (For alkyl chains $\nu = \nu(CH_2) \approx 27Å^3$, except for the terminal CH_3 group which can be counted as two segments: $\nu(CH_3) \approx 2\nu(CH_2)$ [30].) Except for lipid chains in the crystalline ("gel") bilayer phase, which is not of interest here, $\rho_l = 1/\nu$ represents an upper limit to the monomer density in the hydrophobic regions of amphiphilic assemblies.

The local monomer density at distance x from the interface is $\rho(x) = \rho_l \langle\phi(x)\rangle/a(x)$ where $\langle\phi(x)\rangle = \sum P(\alpha)\phi(x;\alpha)$. Since $\rho(x) \leq \rho_l = 1/\nu$ we get

$$\langle\phi(x)\rangle = \sum_{\alpha} P(\alpha)\phi(x;\alpha) \leq a(x) \qquad (1.79)$$

with the equality holding when the monomer density is uniform and liquid-like throughout the hydrophobic core. Note, however, that the condition of uniform density $\rho(x) = \rho = constant$ is weaker than $\rho(x) = \rho_l$. ρ must satisfy $\rho = \rho_l(v/\hat{v})$, where $v = \int \langle\phi(x)\rangle dx = \int \phi(x;\alpha)dx$ is the chain's total volume, and $\hat{v} \equiv \int a(x)dx$ is the *available* volume per chain.

Following the discussion in Sec. 1.1, we shall assume that the condition of uniform monomer density applies to "compact" aggregates such as micelles and lipid bilayers. An example of a "non-compact" system is a surfactant monolayer adsorbed at a water-air or water-oil interface [124]. We shall often refer to monolayers throughout this section, but our interest will be focussed on compact aggregates. We therefore rewrite (1.79) for the important case of uniform monomer density,

$$\langle \phi(x) \rangle = \sum_{\alpha} P(\alpha)\phi(x; \alpha) = a(x) \tag{1.80}$$

which represents a key mathematical constraint on the singlet *pdf* $P(\alpha)$.

The extension of (1.80) to mixed aggregates is straightforward, e.g., for binary aggregates containing chains of types A and B, in proportions X_A and $X_B = 1 - X_A$, respectively, we have

$$X_A \sum_{\alpha} P_A(\alpha)\phi_A(x; \alpha) + X_B \sum_{\beta} P_B(\beta)\phi_B(x; \beta) = a(x) , \tag{1.81}$$

with α and β representing the conformations of chains A and B, respectively. This condition is also applicable to hydrophobic cores which are packed by segments of chains originating from different interfaces, such as in lipid bilayers [122].

1.3.3.2 The singlet distribution

The potential energy $W_t(\alpha_1 \cdots, \alpha_s)$ in (1.75) and (1.76) is a sum of single (internal) chain energies $\epsilon(\alpha_i)$ and an intermolecular interaction potential $U(\alpha_1 \cdots, \alpha_s)$. The internal energies $\epsilon(\alpha)$ are easily calculated, e.g., in the rotational isomeric state model [129] (see below) and their contribution to $P(\alpha^s)$ and Z_t is simple. On the other hand, the effects of U are complicated, and their analysis requires a number of approximations.

Our first approximation is to separate u into two terms $U(\alpha^s) = U_{hc}(\alpha^s) + U_{att}$. The first term, $U_{hc}(\alpha^s) = \sum U_{hc}(\alpha_i, \alpha_j)$, is a sum of short range, or more specifically hard-core, repulsive interchain potentials. Thus, $u_{hc}(\alpha^s) = \infty$ for all configurations where (segments belonging to any) two chains or more overlap in space; otherwise, that is, for non-overlapping configurations, $U_{hc}(\alpha^s) = 0$. The second term, U_{att}, incorporates the long-range attractive potentials, and is assumed to be independent of $\alpha_1, \cdots, \alpha_s$ for chains packed at uniform monomer density. In other words, $U_{att} = su$ is regarded as a uniform attractive background, with u denoting the average "cohesive" energy per chain: for simplicity, we shall set $u = 0$. According to the above separation of $U(\alpha^s)$, the configurational properties are determined by the hard-core (excluded volume) repulsive potential $U_{hc}(\alpha^s)$. This approximation is inspired by a similar representation of U which underlies many theories of simple liquids whose structure is known to be dominated by the short range potential [130].

From the discussion above it follows that

$$W_t(\alpha^s) = \begin{cases} \sum_i \epsilon(\alpha_i), & \text{non-overlapping } \alpha^s \\ \infty, & \text{overlapping } \alpha^s \end{cases} \qquad (1.82)$$

Thus, $P(\alpha^s) = 0$ for forbidden configurations, whereas for the allowed configurations $P(\alpha^s) \propto \exp[-\beta \sum \epsilon(\alpha_i)]$. Using the definition of E_t in (1.76) and of $P(\alpha)$ in (1.77), we find that (1.82) implies $E_t = \sum_i \sum_\alpha P(\alpha_i)\epsilon(\alpha_i) = s \sum_\alpha P(\alpha)\epsilon(\alpha) = s\langle\epsilon\rangle$. Thus, E_t becomes a simple sum of single chain energies. Clearly, (1.82) does not greatly simplify the calculation of Z_t, since exact counting of the allowed configurations remains an impossible task. However, it allows the derivation of a simple expression for $P(\alpha)$, as outlined next by two approaches.

(i) Mean-field approach

In this approach $P(\alpha^s)$ is expressed as a product of singlet pdf's

$$P(\alpha_1, \cdots, \alpha_s) = P(\alpha_1) \cdots P(\alpha_s) \qquad (1.83)$$

thus neglecting interchain correlations. Substituting (1.83) into (1.76) and using $E_t/s = \sum_\alpha P(\alpha)\epsilon(\alpha) = \langle\epsilon\rangle$ we find that the free energy per chain, $f_t = F_t/s$, is a simple functional of $P(\alpha)$,

$$f_t = \sum_\alpha P(\alpha)\epsilon(\alpha) + kT \sum_\alpha P(\alpha) \ln P(\alpha) \qquad (1.84)$$

with $\langle\epsilon\rangle = \sum P(\alpha)\epsilon(\alpha)$ and $s_t = -k \sum P(\alpha) \ln P(\alpha)$ representing the energy and entropy contributions to f_t.

Eq. (1.84) is the exact free energy per chain only when (1.83) is valid, i.e., when the chains are independent, as would be the case if the chains are far apart from each other. Otherwise, (1.84) is an approximation to F/s even if $P(\alpha)$ is the exact singlet pdf. In general, $P(\alpha)$ is not known, but then f_t can be used as a variational free energy for deriving the best approximation for $P(\alpha)$: the desired pdf is the one which minimizes f_t subject to whichever constraints $P(\alpha)$ must fulfill.

Consider first the limit of non-interacting (or "free") amphiphiles corresponding to a system in which $a \gg a_f$, where a_f is the effective cross sectional area of a single, isolated, chain; see Fig. 1.10. In this case, $P(\alpha)$ is not subject to any constraint, except the normalization condition $\sum P(\alpha) = 1$. It can be shown that in this case f_t is minimized by [120]

$$P_f(\alpha) = \frac{1}{y_f} \exp[-\beta\epsilon(\alpha)], \qquad (1.85)$$

where $y_f = \sum_\alpha \exp[-\beta\epsilon(\alpha)]$ is the partition function of the free chain. Note that the conformational freedom of the free chain is limited by the existence of the hydrocarbon-water interface, which restricts the chain to its hydrophobic side. Clearly, the number of allowed chain conformations, and consequently y_f, depend on the curvature of the interface, e.g., y_f is smaller for micelles (because of their convex interface), as compared to planar bilayers or inverted micelles. For micelles, bilayers and other compact aggregates, the free-chain is a hypothetical limiting case. In monolayers it represents the real chain in the limit of low surface density [124].

In the free chain limit $\langle \phi_f(x) \rangle \ll a(x)$, and (1.79) is trivially satisfied. This packing constraint becomes relevant only when a (recall that $a = a(x = 0)$) reduces to a value such that $\langle \phi_f(x) \rangle$ exceeds $a(x)$ for some x. (This may be taken as an operational definition of a_f.) At smaller a's excluded volume interactions become important, and the chains must be stretched, thus "sacrificing" conformational freedom, in order to satisfy the packing constraint. The packing constraints (1.79) for monolayers and (1.80) for compact liquid-like hydrophobic cores are the mathematical expression of (1.82) in the mean-field approach.

The minimization of f_t (see (1.84)) subject to (1.80) yields

$$P(\alpha) = \frac{1}{y} \exp \left[-\beta \epsilon(\alpha) - \beta \int \pi(x) \phi(x; \alpha) dx \right] \qquad (1.86)$$

with

$$y = \sum_\alpha \exp \left[-\beta \epsilon(\alpha) - \beta \int \pi(x) \phi(x; \alpha) dx \right] \qquad (1.87)$$

Here $\pi(x)$ is a continuous set of Lagrange multipliers conjugate to the uniform packing constraint $\langle \pi(x) \rangle = a(x)$. $\pi(x)$ can be interpreted as the *lateral pressure* or (negative) stress profile; see Fig. 1.10. Indeed, $P(\alpha)$ is a generalized *isothermal-isobaric* probability distribution and y is the corresponding partition function. The integral $\int \pi(x) \phi(x; \alpha) dx$ is a generalized "PV" term, representing *the "mean-field" experienced by a "central" chain in conformation α due to the presence of close-by neighbor chains.* Consistent with this interpretation, we note that the free chain limit, (1.85), corresponds to $\pi(x) \equiv 0$. The integral $\int \pi(x) dx = \pi_t$ is the lateral pressure required to squeeze the chains so as to satisfy the packing constraint (1.80).

The numerical values of the $\pi(x)$ are determined by the packing constraints; namely, substituting (1.86) and (1.87) into (1.80), one gets, for every value of x

$$\sum_\alpha [\phi(x; \alpha) - a(x)] \exp \left[-\beta \epsilon(\alpha) - \beta \int \pi(x') \phi(x'; \alpha) dx' \right] = 0 . \qquad (1.88)$$

These "self-consistency" equations can be solved numerically for the $\pi(x)$, provided of course that the $\phi(x; \alpha)$ are known. A convenient solution procedure is to divide the hydrophobic core into L layers of thickness Δx. Correspondingly the integral in (1.88) (and similarly in (1.86) and (1.87)) transforms into a layer sum, and hence (1.88) becomes a set of L coupled nonlinear equations, for $i = 1, \ldots, L$:

$$\sum_{\alpha} [\phi_i(\alpha) - a_i] \exp \left[-\beta \epsilon(\alpha) - \beta \sum_{j=1}^{L} \pi_j \phi_j(\alpha) \right] = 0 \qquad (1.89)$$

with the correspondence $a(x)\Delta x \to a_i$, $\phi(x;\alpha)\Delta x \to \phi_i(\alpha)$ and $\pi(x) \to \pi_i$. (Note that both a_i and ϕ_i now have the dimensions of *volume*.) Clearly, the π_i and hence $P(\alpha)$ depend on the aggregation geometry, which in (1.89) is fully accounted for by the a_i.

The $\phi_i(\alpha)$ depend on the nature of the hydrophobic tails. For amphiphiles with simple alkyl tails, $P-(CH_2)_{n-1}-CH_3$, the rotational-isomeric-state (RIS) scheme [129] provides an excellent model (P denotes the polar head). In this model each of the $n-1$ internal bonds has three possible ("isomeric") states, t, g^+, g^- (t = trans, g = gauche) corresponding to dihedral angles of $0°, +120°$ and $-120°$, respectively. The dihedral angle of bond C_k-C_{k+1} is the angle between the planes defined by C_{k-1} C_k C_{k+1}, and C_k C_{k+1} C_{k+2}. All C C C bond angles are fixed (at $112°$) and similarly the C–C bond length (at 1.53Å). The energy of a $g^{\pm}$ bond is $\epsilon_g \approx 500\text{cal/mole}$ relative to $\epsilon_t \equiv 0$. Thus the internal chain energy is $\epsilon(\alpha) = n_g(\alpha)\epsilon_g$, where $n_g(\alpha)$ is the number of $g^{\pm}$ bonds in conformation α. A certain fraction of the 3^{n-1} possible bond sequences, $\{b\}$, are self-intersecting (non-self-avoiding) and are discarded from the calculation [122–126].

A bond sequence, b, is specified by the $n-1$ rotational isomeric states of the (non-terminal) C–C bonds, e.g., $b = t, t, \ldots, t$ ("all trans"), $b = t, t, \ldots, g^-$, etc. Each bond sequence specifies the coordinates of all chain segments relative to any triad of segments, say $P-C_1-C_2$, but not the over-all positional orientation of the chain. We thus characterize an arbitrary conformation α by $\alpha = b, \Omega, \delta$ with Ω denoting the three Euler angles specifying the overall orientation of the chain, and δ denoting the position of the head group P within a small interval around the interface. In most calculations so far, all 3^{n-1} bond sequences b have been generated, and (arbitrarily) sampled with 36 Ω, δ values for each b, thereby obtaining a total of $36 \times 3^{3n-1} \alpha$'s. This number is considerably reduced after discarding those α which violate either the interface inpenetrability or the chain's excluded volume conditions. That is, a chain is restrained from either passing through the interface (into the aqueous region) or intersecting itself. For each α generated, one calculates $\phi_i(\alpha), \epsilon(\alpha) = \epsilon(b)$, and any conformational property of interest, such as bond order parameters or segment distributions in space. The $\phi_i(\alpha)$ are substituted into (1.89) to solve for the π_i. Then, $P(\alpha) = (1/y)\exp[-\beta\epsilon(\alpha) - \beta\sum\pi_i\phi_i(\alpha)]$ is known and can be used to calculate any conformational or thermodynamic property of interest.

The lateral pressure profile in a planar bilayer composed of $-(CH_2)_{11}-CH_3$ chains modeled by the above RIS scheme is shown in Fig. 1.11a. Also shown is the density profile of a free chain $\langle \phi(x) \rangle_f$, as compared to the constant density profile in a compact bilayer, $\langle \phi(x) \rangle = a$. Note that

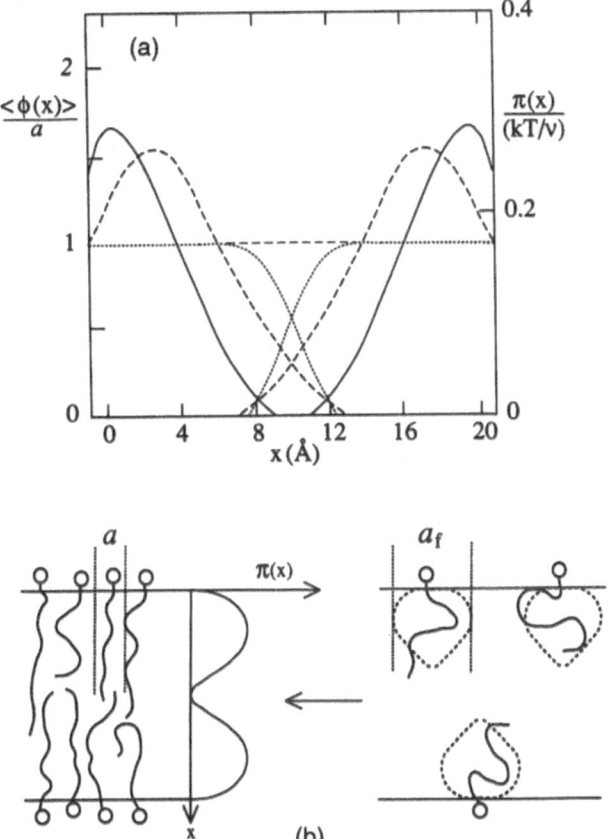

FIGURE 1.11. (a) Normalized density profiles ($\langle\phi(x)\rangle_f/a$, dashed curves) of *free* chains originating from either one of the two interfaces of a planar bilayer composed of twelve-carbon chains, as a function of the distance from one interface. The dotted curves represent the density profiles in each of the two monolayers when the chains are packed with an area per head-group $a = 33\text{Å}^2$. Their sum is the constant monomer density, as described by the horizontal dashed line. The solid lines describe the (normalized) lateral pressure experienced by the chains when packed at the above value of a [124, 125]. (b) Schematic qualitative illustration of the origin of the results shown in (a).

the shape of $\pi(x)$ is qualitatively similar to the chain distortion profile $\langle\phi(x)\rangle_f - a$. Similarly, it has been shown that $\pi(x)$ increases in magnitude as a decreases, consistent with increased chain stretching [124]. A pictorial explanation of these notions is provided in Fig 1.11b.

The singlet *pdf* resulting from the minimization of f_t subject to the *inequality* constraint (1.79) is *also* given by (1.86). As expected, it can be shown that for those values of x for which $\sum P(\alpha)\phi(x,\alpha) = \langle\phi(x)\rangle < a(x)$, the packing constraint is irrelevant, and $\pi(x) \equiv 0$. On the other hand, for those x where $\langle\phi(x)\rangle = a(x)$ we get $\pi(x) > 0$. [Note, however, that since, for all α, $\int \phi(x;\alpha)dx = v$ is the constant chain volume, $P(\alpha)$ and hence also $s_t = -k\sum P(\alpha)\ln P(\alpha)$ remain the same if $\pi(x) \rightarrow \pi(x)+$ *constant*. The choice $\pi(x) > 0$ is consistent with the interpretation of π as a pressure. Further support for this choice is given in the alternative derivation of $P(\alpha)$, outlined in section (ii) below.]

Substituting (1.86) into (1.84) and using $\langle\phi(x)\rangle = a(x)$, we get

$$f_t = -kT\ln y - \int \pi(x)a(x)dx \qquad (1.90)$$

$$= -kT\ln z \qquad (1.91)$$

with the second equality (1.91) serving as the definition of z. Since the integral on the right hand side of (1.90) is a "PV" term, and since f_t is a Helmholtz free energy, the function

$$\mu_t = -kT\ln y \qquad (1.92)$$

is the Gibbs potential per chain, i.e., it is the (configurational part of the tail's) chemical potential. Note also that (1.92) is the familiar relationship between the Gibbs free energy (here μ_t) and the isothermal-isobaric partition function (here y). Consistent with these interpretations, it can be shown (see below) that for a planar layer, where $a(x) = a$,

$$\pi_t = \int \pi(x)dx = -\partial f_t/\partial a \qquad (1.93)$$

is the total lateral pressure acting on the chains.

Recall that f_t, as given by (1.84), is an approximation to the system's free energy even if $P(\alpha)$ is the exact singlet *pdf*. On the other hand, it is not clear if, and to what extent, Eq. (1.86) for $P(\alpha)$ is an approximation, even though it was derived by the minimization of (1.84). In fact, as will be demonstrated below, conformational properties calculated using (1.86) are in excellent agreement with results from many-molecule computer simulations. An explanation for the high accuracy of (1.86) is provided by an alternative derivation of $P(\alpha)$ which does not invoke explicitly a mean-field approximation. We next sketch briefly this derivation.

(ii) Derivation of $P(\alpha)$ from Z_t

From the definitions (1.75) and (1.77) of $P(\alpha^s)$, and from the assumption (1.82) that $W_t(\alpha^s)$ is dominated by excluded volume interactions, it follows that

$$P(\alpha) = \frac{Z_t(s-1, \tilde{g}(\alpha), T)}{Z_t(s, g, T)} \exp\left[-\beta\epsilon(\alpha)\right] \qquad (1.94)$$

Here, $\tilde{g}(\alpha)$ is the geometry of the volume available to the $s-1$ chains surrounding a (central) chain in conformation α. Apart from the Boltzmann factor associated with the self energy $\epsilon(\alpha)$, $Z_t(s-1, \tilde{g}(\alpha), T)$ is the statistical weight of conformation α. The shape of the "hole" prescribed by the α-chain uniquely specifies its area profile $\phi(x; \alpha)$. The geometry, g, of the s-aggregate is characterized by $A(x) = sA(x)$, cf. (1.78). Similarly, the geometry of the $(s-1)$-aggregate, $\tilde{g}(\alpha)$, will be characterized by $A(x) - \phi(x; \alpha)$.

Noting that $\ln Z_t$ is extensive (i.e., of order s), expanding $\ln Z_t(s-1, \tilde{g}(\alpha), T) = \ln Z_t(s-1, \{A(x) - \phi(x; \alpha)\}, T)$ about $A(x)$ and neglecting terms of order $1/s$, we get [120]

$$\ln Z_t(s-1, \{A(x) - \phi(x; \alpha)\}, T) - \ln Z_t(s, \{A(x)\}, T)$$
$$= -\frac{\partial \ln Z_t}{\partial s} - \int dx \left(\frac{\delta \ln Z_t}{\delta A(x)}\right) \phi(x; \alpha)$$
$$= \beta\mu_t - \beta \int dx \pi(x)\phi(x; \alpha) . \qquad (1.95)$$

Here, $\mu_t = -kT\partial(\ln Z_t/\partial s)_{A(x),T} = (\partial F_t/\partial s)_{A(x),T}$ is the chain's chemical potential, and the functional derivative

$$\pi(x) = kT\frac{\delta \ln Z_t}{\delta A(x)} = -\frac{\delta F_t}{\delta A(x)} = -\frac{\delta f_t}{\delta a(x)} \qquad (1.96)$$

is the lateral pressure profile. Substituting (1.95) into (1.94) we regain $P(\alpha)$ as given by (1.86), with a rigorous thermodynamic interpretation of $\pi(x)$ as the lateral pressure and of $\beta\mu_t = -kT \ln y$ as the Gibbs free energy per molecule (cf. (1.92)). Consistent with this interpretation, one can derive additional useful relations, e.g.,

$$\frac{\delta\mu_t}{\delta\pi(x)} = kT\frac{\delta \ln y}{\delta\pi(x)} = a(x) \qquad (1.97)$$

which, using (1.87), is recognized as the packing constraint (1.80) in its more explicit form (1.88). Note also that the integral (over x) of (1.96) describes $-\partial f_t/\partial a$ for the case $\partial a(x)/\partial a = 1$, thus leading to (1.93).

The key step in the above derivation is the assumption that $\tilde{g}(\alpha)$ is uniquely specified by $A(x) - \phi(x; \alpha)$. This implies equal statistical weights for all conformations with the same $\{\phi(x; \alpha)\}$ regardless of their exact shape. (The internal energy $\epsilon(\alpha)$ is treated exactly.) A direct way to test

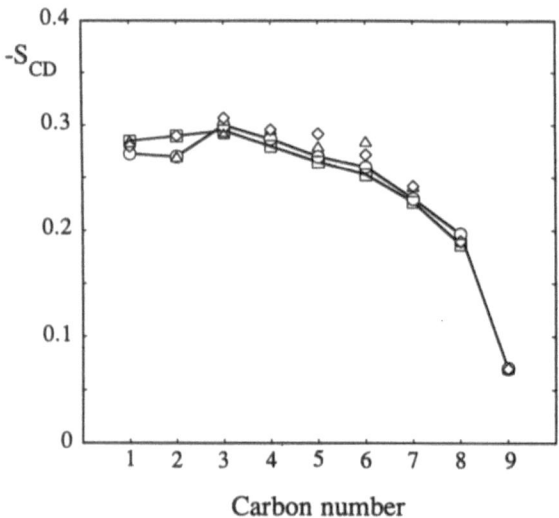

FIGURE 1.12. C–D bond order parameter profiles of C_9 chains packed in a planar bilayer—the area per head-group is $a = 25\text{Å}^2$ [120]. The triangles represent the experimental results of Seelig and Niederberger [131], the squares correspond to the molecular dynamics simulations of van der Ploeg and Berendsen [95] and the diamonds describe the results of Gruen's mean-field theory [116]. The circles describe the results obtained [120] by the mean field theory described in the text.

the accuracy of this prediction and hence of (1.86) is by comparison with computer simulations.

Figure 1.12 shows four C–H orientational bond order parameter profiles, S_k, of -$(CH_2)_8$–CH_3 chains, packed in a planar bilayer with an area per head-group $a = 25.5\text{Å}^2$. The order parameter of C_k–H bond is defined by

$$S_k = \sum_{\alpha} P(\alpha) P_2(\cos \theta_k(\alpha)) \qquad (1.98)$$

where $P_2(x) = (3x^2 - 1)/2$ is the second-order Legendre polynomial. $\theta_k(\alpha)$ is the angle between bond C_k–H of a chain in conformation α and the "director", i.e., the normal to the hydrocarbon-water interface. In the limit that all chains are fully stretched (i.e., all-trans) and are normal to the interface, $S_k = -0.5$, since the C–H bonds are then parallel to the interface. On the other hand, for a random distribution of bond orientations, $S_k = 0$. The C–H bond order parameters are simply related to $\tilde{S}_k$, the *skeletal* (C–C bond) order parameters, via $\tilde{S}_k = -2S_k$. $\tilde{S}_k$ is defined as in (1.98) but with θ_k denoting the angle between the director and the vector connecting carbons C_{k-1} and C_{k+1}.

One of the curves in Fig. 1.12 represents the experimental data of Seelig and Niederberger, obtained from magnetic resonance measurements of the quadrupole splittings of selectively deuterated C–H (i.e., C–D) bonds [131]. These results were accurately reproduced through many-chain molecular dynamics simulations by van der Ploeg and Berendsen [95], who employed realistic intra- and inter-chain interaction potentials. Their results, corresponding to $a = 25.5\text{Å}^2$ are also shown in Fig. 1.12. The value of a is the only input parameter which has been used to calculate the third S_k profile, as outlined above. More explicitly, this value of a was used to calculate $P(a)$ of (1.86) by solving (the discretized) self-consistency equations (1.89) for C_9 chains modeled by the RIS scheme [120]. The agreement between the MD results and the (much simpler) calculations based on (1.86) is obviously excellent. Similar agreement is found with respect to other conformational properties, as well as for average chain energies $\langle \epsilon \rangle$, and entropies $s_t = -k \sum P(a) \ln P(a)$. The fourth S_k profile shown in Fig. 1.12 was obtained by Gruen [116], based on an assumed form for $P(a)$ which is essentially identical to (1.86) (although Gruen's method of calculation was different).

1.3.3.3   Chain conformational characteristics

Using $P(a)$ one can calculate any single chain property of interest. Furthermore, with the aid of mean-field expressions such as (1.84), it is possible to calculate and analyze various thermodynamic functions, for instance, elastic moduli of amphiphilic films, as described in the next section. In this section we briefly discuss some of the basic characteristics of chain packing in planar and curved aggregates.

Some basic qualitative characteristics of chain conformational statistics in micelles and bilayers can be inferred from Fig. 1.11 which illustrates the conformational distortions experienced by free chains upon close-packing among neighboring chains. On the average, a free chain is a "random-coil" with a globular ("turnip-like") shape. The confinement of the chain to one (the hydrophobic) side of the interface includes a certain degree of chain alignment, affecting mainly the first few chain segments connected to the interface. This "residual" ordering is clearly reflected in the order parameter profile of the free $-(CH_2)_{11}-CH_3$ chain anchored to the flat interface of a planar bilayer, as shown in Fig. 1.13. The confining effect of the interface is somewhat more stringent in convex aggregates (such as spherical micelles), implying slightly larger S_k values of the free chains, as confirmed by detailed calculations [121]. Similarly, the conformational free energy, f_t, of a free chain in a micelle is expected to be higher than in a bilayer, because of the greater loss of conformational freedom (see Fig. 1.9). However, in general, the interfacial restrictions on chain conformational freedom are

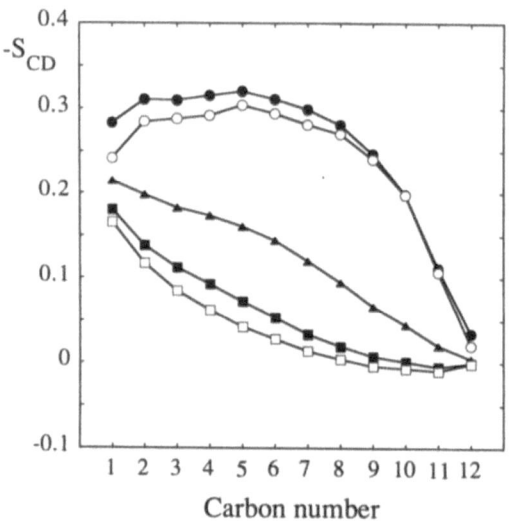

FIGURE 1.13. C–D bond order parameter profiles for C_{12} chains in a planar bilayer. The circles correspond to chains packed with area per head-group $a = 25Å^2$; the solid circles are for chains with gauche energy $\epsilon_g = 500$ cal/more, whereas the open circles are for "athermal" chains, i.e., $\epsilon_g \equiv 0$. All other data are for $\epsilon_g = 500$ cal/more. The solid triangles represent the results for $a = 32Å^2$ and the solid squares for $a = 40Å^2$. The open diamonds are the results for a free chain. (Calculated by D. Hornreich and by I. Szleifer.)

considerably less important than those implied by the packing constraints $\langle \phi(x) \rangle = a(x)$.

As the area per chain in a bilayer decreases, the chains are further squeezed and stretched compared to their free state, resulting in increased values of S_k, cf. Fig. 1.13. Recall from Sec. 1.3.2 that $a = v/R$ in a bilayer, where v is the chain's volume and R is the bilayer's half-thickness. The minimal area per chain, a_{min}, corresponds to fully extended, all-trans, chains so that $a_{min} = v/l$, where l is the chain length. (For alkyl chains, $a_{min} \approx 21Å^2$.) The initial plateau, followed by a rather rapid fall, characterizing S_k in bilayers, can be explained as follows. If the chains were infinitely long we would obviously expect $S_k = constant$, since $a(x) = constant$. However, since the chains are finite, at some point (say, $\bar{x}$) down the imaginary cylinder implied by $a(x) = a$, they begin to terminate with some distribution of chain ends. Beyond $\bar{x}$, increased conformational freedom (larger effective a) is available to the "dangling ends" of the chains which have not yet terminated, resulting in lower S_k values for the large k [120–122].

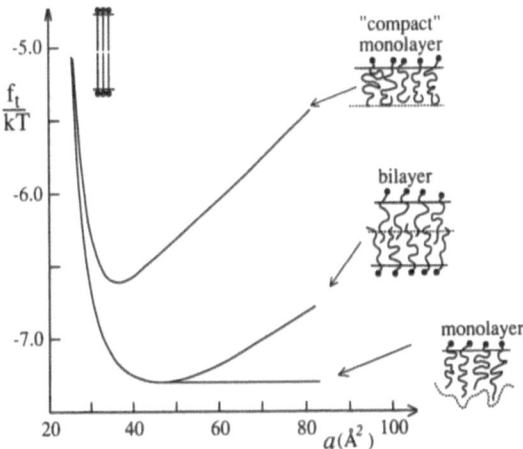

FIGURE 1.14. The conformational free energy of C_{12} chains as a function of the area per molecule, for chains packed in a planar bilayer, monolayer and a "compact" monolayer [124].

One of the S_k curves in Fig. 1.13 corresponds to "athermal" chains, i.e., chains characterized by $\epsilon(\alpha)/kT = 0$ for all α (or $\epsilon_g \equiv 0$). This curve is very similar to the one corresponding to chains with nonzero ϵ_g, demonstrating that *the internal chain energy plays a secondary role in determining chain conformational properties*, as compared to the packing constraints $\langle \phi(x) \rangle = a$ (except at very low temperatures) [122].

In Fig. 1.13 we also see that when $a = a_f \approx 45\text{Å}^2$ the real chain behaves as if it were free, indicating that for this area per head-group the packing constraint $\langle a(x) \rangle = a$ is satisfied easily, with a negligible extent of chain distortion. Indeed, the calculations also show that for this area per chain $\pi(x) \approx 0$ for all x. Furthermore, since the free chain corresponds to maximal conformational freedom, and hence minimal free energy, we expect that f_t is minimal at $a = a_f$. This behavior is quantitatively confirmed in Fig. 1.14, which shows f_t vs. a for C_{12} chains packed in a planar bilayer. Recall that $f_t = \epsilon_t - T s_t$, where $\epsilon_t = \langle \epsilon(\alpha) \rangle$ is the average conformational energy and s_t is the entropy per chain. At the minimal area per chain, $a = a_{\min}$, the chain is in the "pure" all-trans state, so that both $\epsilon_t = 0$ and $s_t = 0$. As a increases and more conformations are allowed, ϵ_t increases, and so does s_t. It is clearly seen, however, that s_t is the dominant term, since f_t decreases.

Figure 1.14 shows also f_t for the C_{12} chains in a single monolayer, for which $\langle\phi(x)\rangle \leq a$, cf. (1.79). The behavior of f_t is very similar for monolayers and bilayers for $a \leq a_f$, indicating that a bilayer is to a very good approximation a superpositon of two interdigitating monolayers. In monolayers, $f_t = constant$ for $a \geq a_f$, because in this regime $\langle\phi(x)\rangle \leq a$ becomes an irrelevant constraint, and thus the chain is free [124]. The slow increase of f_t for bilayers with $a > a_f$ is not due to chain packing but, rather, to the fact that at very large a's the chains may "bump into" the opposite interface, thus losing conformational freedom. Note, however, that typical areas per chain in bilayers are generally well below $a_f \approx 45\text{Å}^2$. The "compact" monolayer, as illustrated in Fig. 1.14, is characterized by a step-function density profile, corresponding to $\langle\phi(x)\rangle = a$ for all $x \leq v/a$. This is a hypothetical limiting case (corresponding, approximately, to chains in contact with a very poor oil solvent) demonstrating the constraining effect of the "opposite wall." The conformational freedom available to the "dangling chain ends" in a simple monolayer, or in the interdigitating monolayers comprising a bilayer, is severely limited in the compact monolayer, resulting in a steep increase of f_t with a which sets in at relatively small areas per head-group.

The schematic description in Fig. 1.11b suggests that the free chain is not significantly distorted when packed in a spherical (or cylindrical) micelle because, qualitatively, the turnip-like shape of the free chain is not very different from the conical (or wedge-like) average shape in the micelle. Indeed, calculated conformational properties such as S_k or segment spatial distributions appear similar to those of the corresponding free chains. In line with this behavior, f_t varies very weakly with a in spherical and cylindrical micelles. This is because in those geometries $a \geq 3a_{min}$ and $2a_{min}$, respectively; see (1.72). This generally implies $a \geq a_f$ where f_t varies weakly with a, as we have seen (for bilayers) in Fig. 1.14. The weak dependence of f_t increases from bilayers to cylinders to spheres, because of the increasingly stronger confining effect of the interface.

The different f_t values for the different aggregation geometries—and their dependencies on area per molecule a—suggest the inclusion of the chain contribution in the analyses of micellar growth, using geometry dependent f_t's in μ_s^o, cf. (1.63) and (1.70). Recently, this has been taken into account by Blankschtein and coworkers [112], revealing that in some cases the inclusion of f_t is crucial.

1.3.4 ELASTIC PROPERTIES

The dynamics and thermodynamics of shape variations of amphiphilic films, such as the bending or stretching of monolayers and bilayers, are governed by their elastic properties [132–146]. The central role of curvature (bending) elasticity in the thermodynamics of microemulsions and lamellar phases is discussed in several chapters of this volume. Another large class of systems whose elastic characteristics are of great importance are the biological membranes. The shape fluctuations (flickering) of red blood cells is perhaps the most familiar such phenomenon [142,143].

The elasticity of a membrane is characterized by its elastic moduli, which measure the free energy changes, $\delta f = f(g_{eq} + \delta g) - f(g_{eq})$, involved with small deviations from the membrane equilibrium geometry. Corresponding to our specification of a film geometry by three independent variables $g = a, c_1, c_2$, one needs three independent elastic constants, associated with three "normal modes of deformation", in order to calculate δf for an arbitrary δg.

For small changes δa in the film area at constant curvature, the deformation free energy per unit area (to order $(\delta a)^2$) is given by [132]

$$\frac{\delta f}{a} = \frac{1}{2a}\left(\frac{\partial^2 f}{\partial a^2}\right)_{c_1,c_2}(\delta a)^2 = \frac{1}{2}k_s\left(\frac{\delta a}{a}\right)^2 \tag{1.99}$$

with $a = a_{eq}$ denoting the equilibrium area per molecule, i.e., $\partial f/\partial a|_{a=a_{eq}} = 0$. The second equality provides the definition of the stretching modulus, $k_s = a(\partial^2 f/\partial a^2)_{c_1,c_2}$, in terms of the second derivative of f with respect to a, evaluated at $a = a_{eq}$.

Following Helfrich, the *curvature* dependence of f is commonly expressed in the form [132,133]

$$\frac{f}{a} = \frac{f_o}{a} + \frac{1}{2}k(c_1 + c_2 - c_o)^2 + \bar{k}c_1c_2 \tag{1.100}$$

where k and $\bar{k}$ are the curvature elasticity moduli, and the constant c_o is called the spontaneous curvature. $f_o = f(a, c_1 = 0, c_2 = 0) - (k/2)ac_o^2$ is the free energy per molecule in a planar film. The constant k is called the *splay* modulus and is associated with changes in the *mean* curvature, $(c_1 + c_2)/2$. $\bar{k}$ is the *saddle-splay* modulus and is associated with changes in the *Gaussian* curvature c_1c_2. If $-2k < \bar{k} < 0$, the film is characterized by a stable equilibrium curvature, corresponding to a minimum of f, at $c_1 = c_2 = c_{eq} = [k/(2k + \bar{k})]c_o$. For symmetric films such as a simple bilayer, $c_{eq} = c_o = 0$. From (1.100) we can obtain the deformation free energy, $\delta f = f - f_{eq} = f(c_1, c_2) - f(c_1 = c_2 = c_{eq})$, in the form

$$\frac{\delta f}{a} = \frac{1}{2}k(\delta c_1 + \delta c_2)^2 + \bar{k}(\delta c_1)(\delta c_2) \tag{1.101}$$

with $\delta c_i = c_i - c_{eq}$.

The constants $k, \bar{k}$ and c_o can be related to (first and second) derivatives of f with respect to c_1 and c_2 or linear combinations thereof [132–135]. A convenient representation involves $c_+ = c_1 + c_2$ and $c_- = c_1 - c_2$ in terms of which we find (using 1.100)) [125],

$$\frac{1}{a}\frac{\partial f}{\partial c_+} = -\frac{1}{2}kc_o \tag{1.102}$$

$$\frac{1}{a}\frac{\partial^2 f}{\partial c_+^2} = k + \frac{\bar{k}}{2} \qquad (1.103)$$

$$\frac{1}{a}\frac{\partial^2 f}{\partial c_-^2} = \frac{\bar{k}}{2} \qquad (1.104)$$

with all derivatives evaluated at the planar geometry $c_1 = c_2 = 0, (c_+ = c_- = 0)$.

Various experimental techniques are available for the measurement of the elastic constants. On the other hand very few theoretical calculations of these constants have been reported to date. Some of the existing treatments [50,107,108] consider only the head-group (surface) contributions to f, i.e., they assume $\delta f = \delta f_h(g)$ (cf. (1.70)), thus ignoring the important tail contributions $\delta f_t(g)$. Other theories [125,144–146] focus on the tail term. Although in many cases certain elastic constants may be dominated by either the tail or the head contribution, a unified theoretical approach which takes both contributions into account is certainly desirable.

Let us first briefly consider the stretching constant k_s, as defined in (1.99). Using the simple representation of f_h in terms of the opposing forces (cf. (1.73) or (1.74)) and using (1.93) for f_t, we find

$$k_s = 2\gamma(a_h/a)^2 - a(\partial \pi_t/\partial a) \qquad (1.105)$$

Here, $a = a_{eq}$ is the equilibrium area per molecule corresponding to the minimum of $f = f_h + t_t$, as determined by $\partial f/\partial a = -\pi_h - \pi_t = \gamma[1 - (a_h/a)^2] - \pi_t = 0$. The equilibrium area per chain depends on the ion concentration in the aqueous solution (which affects mainly π_h and a_h) and temperature (which affects both π_h and π_t). If one assumes, as in the phenomenological models described in Sec. 1.3.2, that a_{eq} is fully determined by the balance between the opposing forces operating in the head-group region (so that $\partial f_t/\partial a = \pi_t = 0$) then $a_{eq} = a_h$ and $k_s = 2\gamma$, which according to common estimates implies $k_s \approx (0.1 - 0.2)kT/\text{Å}^2 \approx (50 - 100)erg/cm^2$. However, from Fig. 1.14 we note that around typical values of a_{eq}, say $\approx 25\text{Å}^2$, f_t decreases steeply with a, so that $\pi_t = -\partial f_t/\partial a$ indicates a strong chain repulsion. Furthermore, in this regime $\partial \pi_t/\partial a < 0$. As a consequence of this strong chain repulsion, $a_{eq} > a_h$, implying a smaller contribution from the first (head-group) term to k_s, and a non-zero one from the second (tail) term in (1.105). Since the representation of the head-group contribution to k_s in terms of the opposing forces is highly approximate, any estimate of k_s based on (1.74) is rather speculative. More reliable treatments of f_h, e.g., approaches based on solving the electrostatic equations governing the head-group region [107,108], or computer simulations, should be combined with detailed chain packing calculations of f_t in order to get better values for k_s, as well as for the curvature elastic constants considered below.

Experimental measurements of the bending modulus k of amphiphilic bilayers reveal strong dependence of this elastic modulus on the chain length of the constituent molecules. In mixed layers k varies markedly with the molecular composition. One finding of particular interest is that the addition of short-chain cosurfactants to a bilayer or monolayer of long chain amphiphiles can lead to a dramatic lowering of k [137–139]. These results support the notion that the curvature elasticity of many amphiphilic layers is predominantly determined by the (relatively thick) hydrophobic region. The tail contribution to k can be calculated based on the mean-field theory of chain packing described earlier in this section [125]. Note that since $f = f_h + f_t$, the elastic constants can also be expressed as sums of head and tail terms. We shall conclude this section describing some important conclusions provided by this approach. We emphasize again, however, that a more complete analysis requires consideration of the head-group (interfacial) region as well.

One, conceptually simple, but numerically awkward procedure for evaluating the curvature elastic constants, is based on calculating f_t for several values of c_1, c_2 around c_{eq}, and determining k and $\bar{k}$ using (1.100) or, equivalently, (1.101). For instance, to determine k for a symmetric bilayer ($c_{eq} = 0$) we can calculate f_t for cylindrical deformations, i.e., $c_1 = \delta c_1 \neq 0$ and $c_2 = 0$, in which case $f = \delta f = (1/2)kc_1^2$. $\bar{k}$ can then be evaluated from a calculation of f for spherical deformations ($c_1 = c_2 \neq 0$) using the previously determined k. Since δf is typically very small, this type of calculation involves considerable numerical effort.

An alternative, more elegant as well as more efficient, approach is based on substituting our explicit expression for f_t, namely (1.90) with y from (1.87), into the definitions (1.102)-(1.104) of the phenomenological constants. Following some algebra it can be shown that this procedure yields [125] (see also [144])

$$kc_o = \int x\pi(x)dx \qquad (1.106)$$

$$k = -\int x \left[\frac{\partial \pi(x)}{\partial c_+} \right]_{c_1,c_2=0,0} dx + h_b \left[\pi(x); a(c) \right] \qquad (1.107)$$

$$\bar{k} = -\int x^2 \pi(x)dx \qquad (1.108)$$

with the integrals extending over the layer's thickness. h_b is a "relaxation function", discussed below. Equations (1.106) and (1.108) were originally derived by Helfrich, based on purely mechanical considerations [133].

The (negative) lateral stress profile $\pi(x)$ and its derivatives appearing in (1.106)-(1.108) are evaluated at the planar ($c_1 = c_2 = 0$) geometry. The second term in (1.107) is zero for monolayers whereas for bilayers, besides being a functional of $\pi(x)$, it also depends on the mode of deformation. More specifically, it depends on the fashion in which the area per molecule in the "inner" (I) and "outer" (O) leaflets of the bilayer (or, equivalently, the "mole fractions" X_I and $X_O = 1 - X_I$ of molecules in the two leaflets) vary with curvature. In the case of "blocked-exchange" [135] which corresponds to $X_I = X_O = const$, molecules cannot diffuse laterally (or flip from one monolayer to another) and $h_b \equiv 0$. This represents a rather stiff layer in which no relaxation can take place in the course of bending. In this case k is high, largely because bending at constant X_O implies large deviations of head-group areas a_I and a_O from the equilibrium (undeformed) value a_{eq}. A more realistic scenario is one for which, in the course of bending, molecules diffuse laterally so as to ensure $a_O = a_I = a_{eq}$. Both the curvature derivative of $\pi(x)$ in (1.107) and the second term in this equation, h_b, can be expressed and evaluated in terms of moments of the stress profile and density correlation functions in the *planar* layer. In these equations, which we skip writing here, the extent of relaxation (e.g., blocked vs. free exchange) is accounted for by h_b [125].

Amphiphilic molecules in bilayers and monolayers are usually packed with areas per chain ($a_{eq} \leq 30\text{Å}^2$) which are considerably smaller than the optimal area ($a_t \approx 40\text{Å}$) corresponding to a minimum of f_t (see Fig. 1.14). In other words, the chains are rather strongly stretched. For these low values of a_{eq}, the packing free energy increases rapidly with chain length, n, and (for a given n) decreases rapidly as the area per chain a increases. Since curvature deformations involve stretching and squeezing of chains, one expects that k increases strongly with n and with $1/a$ (for $a < a_{eq}$). In fact, a simple "compressional" model [125,144] suggests that k scales with n and a, roughly according to $k \sim n^3/a^5$.

These qualitative trends are borne out by the more detailed calculations based on the mean-field theory of chain packing and Eqs. (1.106)-(1.108). The strong dependence on a is demonstrated in Fig. 1.15, for -$(CH_2)_{n-1}$- CH_3 chains ($n = 6 \rightarrow 12$) packed in a bilayer. In the calculation it is assumed that the area per molecule, at both interfaces, remains constant during the bending deformation. Detailed calculations reveal that the variation of k with a is even stronger than that predicted by the compressional model: $k \sim a^{-7}$ as compared to $k \sim a^{-5}$.

Finally, Fig. 1.16 shows how k varies with chain length for $n = 8 \rightarrow 16$ in a single component bilayer and how it varies with the fraction of short (C_8) chains in a mixed bilayer of C_8/C_{16} chains. Both curves correspond to bending at constant area per molecule ($a = 31.6\text{Å}^2$). The strong increase of k with n is apparent ($k \sim n^{3.2}$). Even more dramatic is the lowering of k upon addition of short chains (C_8) to a bilayer composed of long chains. We note that when the proportion of short chains reaches $X_s \approx 1/3$, k has already dropped to its value for a pure C_8 bilayer. A qualitative explanation of this behavior is easily given as follows. At small distances from the hydrocarbon-water interface the short and the long chains experience similar packing constraints (similar area per chain). However, beyond the region where the short chains terminate, the area per long chain increases significantly (e.g., from $\approx 30\text{Å}^2$ to $\approx 40\text{Å}^2$ when $X_s \approx 0.3$). Clearly then, the tails of the longer chains resemble free chains, the bending of which

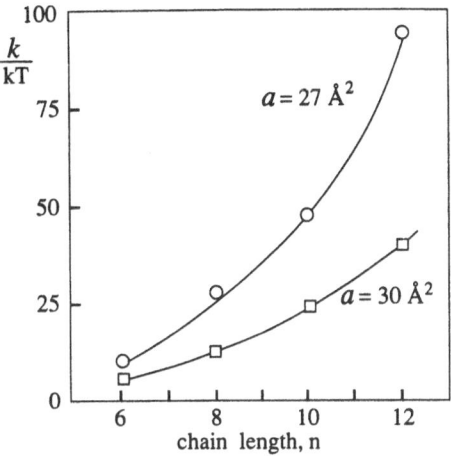

FIGURE 1.15. Bending elastic constants of bilayers as a function of the chain length of the constituent molecules. Note the strong increase of k with chain length and with increasing packing density, $1/a$ [125].

involves negligible free energy cost. Thus, the short chains can be regarded as "spacers" between the long chains. The shallow minimum in the k vs. X_{short} curve implies that the free energy gain of the long chain resulting from the introduction of spacers is larger than the bending free energy difference between long and short chains packed, separately, in pure bilayers. As noted previously, the membrane softening effect of short chain surfactants in monolayers and bilayers has been verified by various recent experiments [137–139]. Some solutes, e.g., cholesterol, have the opposite (i.e., stiffening) effect on membrane elasticity. This effect can also be explained on the basis of packing arguments. (Qualitatively, the chains neighboring the bulky cholesterol molecule are forced to stretch out, thus effectively reducing their cross sectional area and contributing to an increase in k).

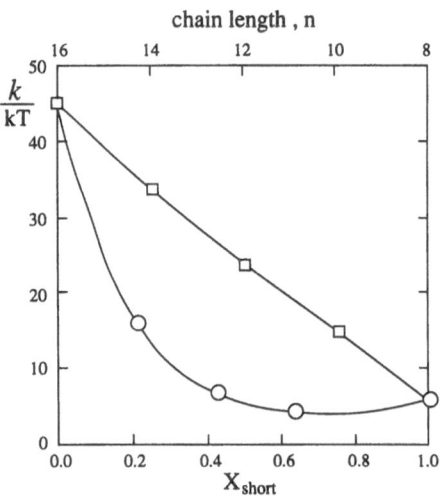

FIGURE 1.16. Bending elastic constants of a mixed C_{16}/C_8 bilayer as a function of the short chain mole-fraction (circles, lower abscissa). Also shown (squares, upper abscissa) are the bending constants of a single component bilayer as a function of chain length. All data, for both the mixed and the pure bilayers, are for chains packed with an average area per head-group of $a = 31.6 Å^2$. In these calculations the bending takes place at constant a [125].

1.4   Interaction Effects in Dilute Solution and Beyond

Throughout this entire chapter we have assumed that micellar self-assembly proceeds without being influenced by interactions between the aggregates. This assumption came into play at the very outset of our discussion of the underlying statistical thermodynamics. Specifically, in passing from Eq. (1.7) to (1.8), we wrote the total partition function as a product of those for the individual aggregates. This led to the familiar ideal solution form for the Gibbs and Helmholtz free energies (1.18) and (1.20) and to all of our subsequent results for micellar size distributions, dependence of average aggregation number on concentration, etc. In this concluding section we outline briefly a general method for including explicitly the effects of interactions on these various properties. Since intermicellar forces are the driving force for phase transitions in more concentrated solutions, we also comment on the problems inherent in treating these changes of state.

1.4.1 ISOTROPIC SOLUTION EFFECTS

We have already described how interactions between aggregates can contribute to static and dynamic light scattering measurements and hence to *apparent* molecular weights and diffusion coefficients. (See in particular Mukerjee's early treatment of integrated intensities and the problem of separating intermicellar potential corrections from size distribution effects [63].) In all of these cases, however, the information deduced from experiment is compared against ideal-solution estimates of average aggregation number, concentration dependence and polydispersity. At sufficiently high amphiphile number density this theoretical approach must break down because the total partition function can no longer be written as a product of those for the individual aggregates comprising the micellized solution. This leads to a chemical potential μ_s which—in addition to the entropy of mixing $[-kT \ln(X_s/s)]$ and the standard contribution (μ_s^o)—contains *activity* corrections of the kind familiar from the classic theory of dilute but *nonideal* solutions [147]. In the case of hard-particle (excluded volume, or "steric") interactions, we have shown [148] how the leading terms of this kind can affect the micellar size distribution.

Specifically, instead of (1.16), we now write

$$\beta\mu_s = \beta\mu_s^o + \ln(X_s/s) + \chi_s \qquad (1.109)$$

with $\chi_s = \ln\gamma_s$ accounting for the effects of inter-aggregate interactions on the chemical potential of an s-aggregate; γ_s is the activity coefficient. In the second virial approximation

$$\chi_s = \rho_{tot} \sum_{s'} B_2(s, s') \frac{X_{s'}}{s'} \qquad (1.110)$$

where ρ_{tot} ($\approx \rho_w$ for dilute solutions) is the total number density of the solution (recall from Sec. 1.2.1 that $\rho_{tot}X_s/s = \rho_s$ is simply the number density of s-aggregates). $B_2(s, s')$ is the second virial coefficient associated with aggregates s and s'. Taking the intermicellar interactions to be of the hard core type, the second virial coefficient reduces to (one half) the pair excluded volume $v_{ss'}$ associated with aggregates of sizes s and s':

$$2B_2(s, s') = -\frac{1}{V} \int d\tau_s \int d\tau_{s'} \left[e^{-\beta u_{ss'}(\tau_s, \tau_{s'})} - 1 \right] \equiv v_{ss'} \qquad (1.111)$$

with the identity holding for the hard-core repulsive potential

$$u_{ss'} = \begin{cases} \infty & \text{if } s \text{ and } s' \text{ overlap} \\ 0 & \text{otherwise} \end{cases} \qquad (1.112)$$

The integrations in (1.111) are carried out over the translational and orientational coordinates of the two micelles.

Recall that the equilibrium size distribution in ideal solution is determined by the s-dependence of μ_s^o, as has been discussed in detail in
Sec. 1.2.4. In particular we have seen that for rod-like micelles the linear
s-dependence $\beta\mu_s^o = s\beta\tilde{\mu}_{cyl}^o + \delta_{rod}$ (cf. (1.34)) implies a size distribution
which decreases exponentially with s (cf. (1.42)) with its average size con-
trolled by the "growth parameter" δ_{rod} (cf. (1.45)) which is the energy
cost associated with "capping" a cylinder. (In this section we shall use δ_{rod}
and δ_{disk} for the edge free energy δ in rod- and disk-like aggregates.) In
the analysis below we shall show that the *interaction* term χ_s in (1.110)
is also a linear function of s, just as in the *intramicellar* term μ_s^o, i.e.,
$\chi_s \approx s\chi_{rod} + \bar{\chi}_{rod}$. Thus, the effective growth parameter in the presence
of inter-micelle interactions is $\delta_{rod} + \bar{\chi}_{rod}$. Furthermore, since $\bar{\chi}_{rod}$ turns
out to be positive and to be an increasing function of the overall surfactant
concentration, the effect of interactions is to make the micelles bigger, and
this size enhancement becomes more important as the solution is concentrated [148].

The effect described immediately above can be illustrated most simply
by neglecting for a moment the fact that the solution of micelles is polydisperse. That is, we imagine that all of the surfactant molecules are incorporated into aggregates of a single size, $s : X_{s'} = X_s\delta_{ss'} = X\delta_{ss'}$. The χ_s
in (1.110) becomes

$$\chi_s = \rho_{tot}Xv_{ss}/s \qquad (1.113)$$

where we have used the fact that X_s is equal to X, the *total* mole fraction
of surfactant, when only one size (s) of aggregates is present. Then the
s-dependence of χ_s is determined by that of the pair excluded-volume v_{ss}.
For long spherocylindrical rods, we have that [149]

$$v_{ss} \approx (\pi/2)DL_s^2 + 2\pi D^2 L_s \qquad (1.114)$$

where L_s and D describe the rod length and diameter, respectively. Recall
that D is of order twice the length of a surfactant molecule, independent
of the aggregation number s. L_s , on the other hand, is proportional to s,
i.e., $L_s/s \equiv \lambda = $ constant. It then follows from (1.113) and (1.114) that

$$\chi_s^{rod} \approx a_1\phi s + a_2\phi \qquad (1.115)$$

Here $\phi \equiv Nv/V$ is the volume fraction of surfactant in the solution, with N
and v denoting the number and volume of individual surfactant molecules,
respectively. (We have used the fact that $\rho_{tot} \equiv N_{tot}/V$ and $X \equiv N/N_{tot}$,
with $N_{tot} \equiv N + N_w$ the total number of surfactant and water molecules
in solution, cf. Sec. 1.2.1.) The constants a_1 and a_2 are given by $a_1 \simeq a_2 \simeq$
$D\lambda^2/v$, where all dimensionless factors of order unity have been dropped.

The fact that the interaction term χ_s^{rod} is positive and has the same form
as the intramicellar contribution μ_s^o suggests immediately, as mentioned

earlier, that it will enhance the average size of aggregates. More explicitly, using (1.115) we can define an "effective" standard chemical potential $\tilde{\mu}_s^o$ which, like μ_s^o, varies linearly with s:

$$
\begin{aligned}
\mu_{s,rod}^{o,eff} &= \mu_s^o + kT\chi_s^{rod} \\
&= s(\tilde{\mu}_{cyl}^o + kTa_1\phi) + (\delta_{rod} + a_2\phi)kT \\
&\equiv s\tilde{\mu}_{cyl}^{o,eff} + \delta_{rod}^{eff}
\end{aligned}
\tag{1.116}
$$

Now, using (1.109), we can write the rod's chemical potential in the ideal solution form

$$
\mu_{s,rod} = \mu_{s,rod}^{o,eff} + kT\ln(X_s/s)
\tag{1.117}
$$

with the effects of inter-micelle interactions incorporated into the effective standard chemical potential $\mu_{s,rod}^{o,eff}$. Recall from Sec. 1.2.4 that in the ideal solution the average aggregation number is an exponentially increasing function of the growth parameter δ_{rod}. A similar behavior is expected in the presence of interactions but with δ_{rod} replaced by $\delta_{rod}^{eff} = \delta_{rod} + a_2\phi$. Since δ_{rod}^{eff} increases (linearly) with the total surfactant concentration ϕ, the average rod size increases with ϕ faster than in dilute (ideal) solution.

A similar result is found for disks. For this geometry, we have [149]

$$
v_{ss} \approx 2\left(\frac{\pi}{4}\right)^2 H_s^3
\tag{1.118}
$$

for large disks, i.e., ones whose diameter H_s greatly exceeds the thickness D. Here D (like D in the case of rods) is necessarily of order twice a molecular length, independent of aggregation number. H_s, on the other hand, depends on micellar size as $(as)^{1/2}$, where a is—apart from numerical factors of order unity—the area per molecule in the bilayer portion of the disk. This follows from the fact that, for large disks, $H_s^2 \approx as$ is the total surface area of the aggregate. Using (1.118) in (1.110) leads immediately to

$$
\chi_s^{disk} \simeq a_3\phi s^{1/2} ,
\tag{1.119}
$$

where $a_3 \simeq a^{3/2}/v$ is of order unity (as are the factors 2 and $(\pi/4)^2$, etc., which have been dropped altogether). We note that, again (as in the case of rods), the interaction contribution χ_s is found to have the same s-dependence as the intra-micellar term μ_s^o. Specifically, we have from Eq. (1.35) that

$$
\beta\mu_{s,disk}^o = s\beta\tilde{\mu}_{bil}^o + \delta_{disk}s^{1/2}
\tag{1.120}
$$

Accordingly, the micellar size enhancement due to interaggregate excluded volume forces appears to hold for disks as well as rods.

The above arguments are intentionally crude, invoking as they do the drastic approximations of monodispersity and neglect of third- and higher-order virial corrections. As a result, we were able to describe the phenomenon of interaction-induced growth by means of transparent dimensional and scaling arguments. It turns out, however, that one can remove

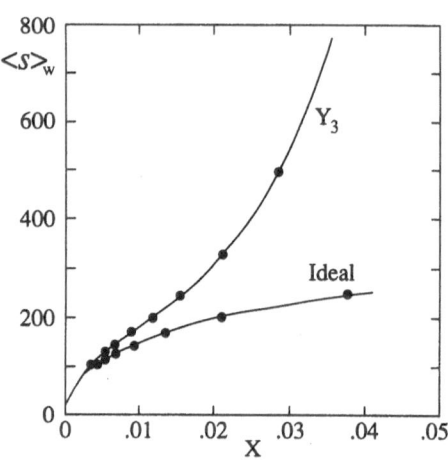

FIGURE 1.17. The weight averaged aggregation number of rod-like micelles as a function of total surfactant concentration, calculated with ("Y_3") and without ("Ideal") the contribution of intermicellar interactions [148b].

these approximations and still obtain essentially the same result. In particular, we have allowed for all sizes of rod-like micelles and replaced the second virial correction by an interaction contribution which sums the effects of excluded-volume to all orders in the density [148b]. For this latter purpose we use the "y-expansion", three terms of which have been shown to provide an accurate representation of hard-particle fluid thermodynamics throughout the compressed gas and liquid density range [150]. Figure 1.17 here shows typical results for the (weight) average aggregation numbers, $\langle s \rangle_w$, without and with intermicellar interaction contributions. (For this particular plot $\delta_{rod} = 13.2$, $\mu_1^o - \tilde{\mu}_{cyl}^o = 10kT$, the spherocylinders diameter is $D = 24\text{Å}$, the minimal micellar size is $m = 20$ and $\rho_{tot} = 0.033\text{Å}^{-3}$.) Note that $\langle s \rangle_w$ is enhanced by interactions by as much as a factor of three, but that these effects become significant only at concentrations which are orders of magnitude larger than the CMC (CMC $\approx 8 \times 10^{-5}$ in the case of the system in Fig. 1.17).

1.4.2 ISOTROPIC-TO-NEMATIC TRANSITION

In Chap. 3 Boden discusses the special nature of orientationally ordered—nematic—phases of rod-like micelles. These states arise when the isotropic solution of micellar rods is sufficiently concentrated: the self-assembling rods, interacting through excluded volume forces and getting bigger with increasing concentration, are no longer able to keep out of each other's way without undergoing a spontaneous alignment. The statistical mechanics of this isotropic-to-nematic transition of hard rods was first treated by Onsager [149]. He showed explicitly how the packing entropy of hard rod fluids could be enhanced by long-range alignment and how at high enough concentrations this *interaction* effect would dominate the *single-particle* orientational entropy loss. We can use Onsager's formulation in our present discussion, in an effort to explore the possibility that *self-assembling* rods might have their sizes affected—not only by the interaction effects described above for *isotropic* solutions, but also—by the onset of *long-range orientational (nematic) order*.

Again, to illuminate the phenomenon most clearly, we consider the case of a single size of rod-like micelles and determine how this size responds first to concentration increase and then to the transition from isotropic-to-nematic states. At the second-virial level of approximation, we generalize the interaction term χ_s from Eqs. (1.109) and (1.110) to

$$\chi_s = \rho_s \int d\Omega \int d\Omega' f_s(\Omega) f_s(\Omega') v_{ss}(\Omega, \Omega') \qquad (1.121)$$

Here $\rho_s = \rho_{tot} X_s/s$ is the number density of s-micelles ($\rho_s = \rho_{tot} X/s$ in the monodisperse solution approximation). $f_s(\Omega)$ is the fraction of s-micelles having orientation Ω ($f_s = 1/4\pi$ in the isotropic phase) and $v_{ss}(\Omega, \Omega')$ is the pair excluded-volume associated with two s-rods having orientations Ω and Ω'. Eq. (1.121) describes the *packing* entropy contribution to the free energy of interacting hard rods in the presence of long-range alignment. There is, however, another way in which we must generalize the chemical potential μ_s: we must add a term which accounts for the *single*-particle ("ideal solution") entropy loss due to long-range orientational order. This term takes the familiar form

$$\sigma_s = \int d\Omega f_s(\Omega) \ln f_s(\Omega) \qquad (1.122)$$

Using the exact result for $v_{ss}(\Omega, \Omega')$ appropriate to spherocylinders, and taking Onsager's single-parameter choice for $f_s(\Omega) [\sim \cosh(\alpha_s \cos \theta)]$, it is straightforward to show again that χ_s varies linearly with s [151,152]:

$$\chi_s \simeq s\chi + \bar{\chi} \qquad (1.123)$$

with

$$\left\{ \begin{array}{c} \chi \\ \bar{\chi} \end{array} \right\} \simeq \left\{ \begin{array}{c} \phi/m\alpha^{1/2} \\ \phi \end{array} \right\} \text{ in the nematic phase} \qquad (1.124a)$$

and

$$\left\{ \begin{array}{c} \chi \\ \bar{\chi} \end{array} \right\} \simeq \left\{ \begin{array}{c} \phi/m \\ \phi \end{array} \right\} \text{ in the isotropic phase} \qquad (1.124b)$$

Here ϕ is the total surfactant volume fraction, and m is the minimal micellar size. $\alpha = \alpha_s$ is the orientational distribution parameter, whose numerical value is determined by minimization of the free energy and by the requirements of equal pressures and chemical potentials in the coexisting phases (see below). In deriving Eqs. (1.124) we have explicitly used the fact that the rods are large ($s \gg 1$) in both phases and highly aligned ($\alpha \gg 1$) in the nematic, and have again dropped all numerical factors of order unity. Similarly, the same approximations can be invoked to write

$$\sigma_s \simeq \begin{cases} \ln \alpha & \text{in the nematic phase} \\ 0 & \text{in the isotropic phase} \end{cases} \qquad (1.125)$$

Note that the usual orientational order parameter $\eta = \langle P_2(\cos\theta) \rangle$ is related to α via $\eta \approx 1 - 3/\alpha$.

Using the above forms for χ_s and σ_s in the μ_s's for the isotropic (I) and nematic (N) phases, it can be shown that s increases significantly upon alignment [151]. Specifically, minimizing $\mu_s(N)$ with respect to α and s and $\mu_s(I)$ with respect to s, and equating the resulting μ_s's, leads to the estimate that at the transition $\alpha_{transition} \approx 17.5$, and the ratio between rod sizes in the nematic and isotropic phases is

$$\left(\frac{s_N}{s_I} \right)_{transition} \approx 1.36 \qquad (1.126)$$

independent of self-assembly parameters.

Eq. (1.126) demonstrates the growth of micellar rods at the I-N transition. That this effect is due primarily to the orientational "entropy of mixing" term σ_s is clear as soon as one observes (from $\partial\mu_s(N)/\partial\alpha = 0$ that $\alpha \sim s^{1/2}$ and hence from (1.125)) that

$$\frac{1}{s}\sigma_s \sim \frac{\ln s}{s} . \qquad (1.127)$$

Thus on a per-*molecule* basis, the single-particle orientational entropy loss *decreases* with aggregation number, i.e., the corresponding free energy is minimized by reorganization of the system into a smaller number of larger rods. When the nematic phase is properly allowed to have a higher volume fraction of surfactant than the coexisting isotropic phase, the ratio $(s_N/s_I)_{transition}$ is found to be even larger (because of the $\bar{\chi}$ term in χ_s which scales as ϕ—see (1.124)). Finally, as outlined immediately below, one can abandon the assumption of monodispersity and include explicitly the contributions of all sizes of micellar rods. Again one finds that the nematic rods are larger than those in the coexisting isotropic phase because less orientational entropy is lost upon alignment when the aggregates rearrange into fewer, larger, rods.

Consider, then, a more rigorous and unifying formulation of the several effects of intermicellar interactions on size distributions in isotropic and nematic solution [152]. It incorporates size (s) and orientational (Ω) degrees of freedom on an equal footing, and reduces naturally to the limits of ideal micellar solution on the one hand, and aligned fluids of non-self-assembling rods on the other. More explicitly, extending Onsager's expression for the Helmholtz free energy of a monodisperse rod mixture [149] to a polydisperse system, we write

$$\beta A = \sum_s n_s \int d\Omega f_s(\Omega) \beta a_s(\Omega) \tag{1.128}$$

with

$$\beta a_s(\Omega) = \beta \mu_s^{o,\rho} + \ln \left[\frac{4\pi n_s f_s(\Omega)}{V} \right] - 1 + \sum_{s'} \left(\frac{n_{s'}}{V} \right) \int d\Omega' f_{s'}(\Omega') v_{ss'}(\Omega, \Omega') . \tag{1.129}$$

Here n_s is the number of s-rods in solution and $n_s(\Omega)d\Omega \equiv n_s f_s(\Omega)d\Omega$ is the number of s-rods having orientation between Ω and $\Omega+d\Omega$, i.e., $f_s(\Omega)$ is the orientational distribution of s-rods. Thus, $a_s(\Omega)$ is the contribution of one s, Ω-rod to A. $v_{ss'}(\Omega, \Omega')$ is the pair excluded volume associated with s, Ω and s', Ω' rods. Accordingly, the last term in (1.129) represents the second virial contribution of inter-aggregate forces to the free energy, as discussed several paragraphs earlier. The first three terms in (1.129) describe the ideal solution effects associated with the (size-angle) distribution $n_s(\Omega) = n_s f_s(\Omega)$. $\mu_s^{o,\rho}$ is the standard chemical potential "on the number density scale" as discussed in Sec. 1.2.1 (cf. Eqs. (1.9)-(1.12) and recall $\rho_s = n_s/V = (N_s/sV)$, etc.). This may be seen more clearly as follows: since all s-rods, regardless of their orientation, must have the same chemical potential, $\mu_s = (\partial A/\partial n_s)_{T,V,n_{s'}} = \mu_s(\Omega) = \partial A/\partial n_s(\Omega))_{T,V,n_{s'}(\Omega')}$. Thus, using (1.128) and (1.129) we get

$$\beta \mu_s = \beta \mu_s^{o,\rho} + \ln \left[\frac{4\pi n_s f_s(\Omega)}{V} \right] + \sum_{s'} \left(\frac{n_{s'}}{V} \right) \int d\Omega' f_{s'}(\Omega') v_{ss'} . \tag{1.130}$$

In the dilute solution limit all $n_s/V \to 0$ and the last term can be neglected. Also, in this limit all $f_s(\Omega) \to 1/4\pi$ and hence $\beta \mu_s \to \beta \mu_s^{o,\rho} + \ln \rho_s$, ($\rho_s = n_s/V$), as expected.

When the rods are "ordinary" molecules, rather than self-assembled micellar aggregates, then one can in principle impose monodispersity ($n_{s'} = n_s \delta_{ss'} = n\delta_{ss'}$) on the colloidal suspension, without any approximation. Furthermore, to find $f_s(\Omega)$ one need not minimize the free energy with respect to size, since s is fixed once and for all by the particular choice of macromolecule (rigid rod). Instead, the single variational condition $\delta(\beta A/n)/\delta f - \lambda = 0$ leads directly to the familiar Onsager integral equation for the orientational distribution function:

$$f(\Omega) = \frac{1}{4\pi} \exp \left[\lambda - \beta \mu^* - \rho \int d\Omega' f(\Omega') v(\Omega, \Omega') \right] . \tag{1.131}$$

Here we have dropped the (fixed) s subscripts on f, μ^* and v; $\beta\mu^* \equiv \beta\mu^{o,\rho} + \ln(n/V)$ is the ideal solution approximation to the chemical potential. Note that λ is the Lagrange multiplier conjugate to the normalization constraint $1 - \int d\Omega f(\Omega) \equiv C[f] = 0$, i.e., λ in (1.131) is chosen so as to assure that f is normalized. Eq. (1.131) for f has been solved numerically by Lasher [153] in the case of hard spherocylinders, using the exact pair excluded volume $v(\Omega, \Omega')$ derived by Onsager. For $\rho < \rho^\dagger$, only the isotropic solution $f(\Omega) = 1/4\pi$ is found, whereas for $\rho > \rho^\dagger$ a nematic $f(\Omega) \neq 1/4\pi$ also appears. The actual phase transition (coexistence) occurs for values of the density, ρ_I and ρ_N, for which the chemical potentials $\beta\mu = \beta A/n + \Pi/\rho$ for the two phases are equal. For sufficiently *long* rods (i.e., $L \gg D$), the densities $\rho^\dagger, \rho_I$ and ρ_N—in units of $1/L^2 D$—are all on the order of unity. Equivalently, the *volume* fractions $\phi = D^2 L\rho = (D/L)(L^2 D\rho)$ are as small as D/L: *the longer the rods the lower the concentration at which long-range orientational ordering appears.* It is this latter fact, incidentally, which provided the original motivation for Onsager's use of the virial expansion.

Now, what about the original, more general, case in which the rods are not only polydisperse but correspond to self-assembled micellar aggregates comprised of surfactant molecules? We return to the full free energy expressions given by (1.128) and (1.129) and consider the minimization of A with respect to the "joint" size-orientation distribution $\{n_s(\Omega)\}$, i.e.,

$$\frac{\delta(\beta A)}{\delta n_s(\Omega)} = 0 . \tag{1.132}$$

The minimization must be carried out subject to the (mass conservation) condition

$$\sum_s \int d\Omega s n_s(\Omega) = N \tag{1.133}$$

where N is the total number of amphiphiles. In this connection we note that $sn_s(\Omega) = sn_s f_s(\Omega)$ is the number of amphiphiles incorporated in s, Ω-aggregates, which is related to the size (mole-fraction) distribution $\{X_s\}$ of Sec. 1.2 via

$$n_s(\Omega) = n_s f_s(\Omega) = N_{tot}\left(\frac{X_s}{s}\right) f_s(\Omega) \tag{1.134}$$

where $N_{tot} = N + N_w$ is the total number of (amphiphile + water) molecules in solution.

Solving (1.132) using (1.128), (1.129) and (1.133) gives

$$\frac{4\pi}{V} n_s(\Omega) = \exp\left[-\beta\mu_s^{o,\rho} + \lambda s - \sum_{s'} \int d\Omega' n_{s'}(\Omega') v_{ss'}(\Omega, \Omega')\right] \tag{1.135}$$

The Lagrange multiplier λ can be determined from the $s = 1$ (monomer) or from the $s = m$ (spherical micelle) realizations of (1.135). Explicitly, for $s = 1$ (ignoring monomer-monomer and monomer-micelle interactions), we have $v_{1s'} \equiv 0$ and $n_1(\Omega)/V = (n_1/V)f_1(\Omega) \equiv \rho_1/4\pi$ so that $\rho_1 = \exp(-\beta\mu_1^{o,p} + \lambda)$ or

$$\lambda = \beta\mu_1^{o,p} + \ln\rho_1 = \beta\mu_1^o + \ln X_1 = \beta\mu_1 = \beta\mu \qquad (1.136)$$

where we have used $\rho_1 = X_1\rho_{tot}$ and $\mu_1^{o,p} + kT\ln\rho_{tot} = \mu_1^o$ (cf. (1.17)). That is, the Lagrange multiplier turns out to be the chemical potential (in units of kT) of the single surfactant molecule μ_1 which is equal to μ—the surfactant chemical potential—since all surfactants, monomeric or aggregated, must have the same chemical potential. As discussed already in Sec. 1.2.2, this follows from the molecular exchange equilibrium which defines the self-assembly phenomenon: Instead of a constraint on the number of micelles of any particular size, we insist only that the total number of surfactant molecules is conserved (see Eq. (1.133)).

Eq. (1.135) is a nonlinear integral equation for $n_s(\Omega)$ which must be solved numerically. Armed with $n_s(\Omega)$ we can then evaluate βA from (1.128—9) and calculate its derivatives

$$\Pi = -\left(\frac{\partial A}{\partial V}\right)_{T,N} \quad \text{and} \quad \beta\mu = \left(\frac{\partial\beta A}{\partial N}\right)_{T,V}. \qquad (1.137)$$

The isotropic-nematic coexistence follows from finding the pair of densities (ρ_I and ρ_N) which allow for simultaneous equality of the Lagrange multipliers $\lambda(= \beta\mu)$ and osmotic pressures Π for the two phases. Figure 1.18 shows the results obtained by McMullen et al. [152] for the size distributions of rod-shaped micelles (with $m = 20$ and $\delta_{rod} = 33$) in coexisting isotropic (I) and nematic (N) solutions. $X_N = 5.2 \times 10^{-3}$, the total surfactant molefraction in the nematic, is significantly larger than that in the isotropic phases: $X_I = 4.5 \times 10^{-3}$. But this density difference is not the cause of the micelles being larger in the nematic: rather, the enhanced size is due to the coupling between self-assembly and long-range orientational order described above, immediately following (1.127).

The above theory, while systematic in its treatment of intermicellar interactions to all orders, has been restricted so far to the case of excluded-volume repulsions. McMullen et al. [154] have considered the inclusion of *electrostatic* forces between rod-like micelles, but the size enhancement and isotropic-nematic transitions are found to be driven by interactions in a way which is qualitatively the same as that in the hard-particle case. Similarly, incorporation of the effects of rod *flexibility* have been discussed by Odijk [155], but again micellar growth and alignment are shown to proceed with essentially the same characteristics. In the limit of sufficiently *long* rod-like aggregates, isotropic micellar systems are shown to behave very much like semi-dilute solutions of flexible polymer chains. This latter situation is discussed by Porte in Chap. 2.

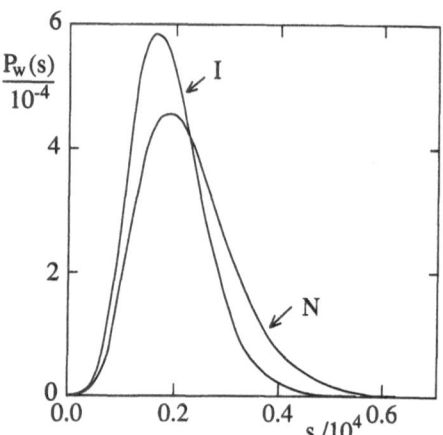

FIGURE 1.18. Size distributions of rod-shaped micelles (with growth parameter $\delta_{rod} = 33$) in the coexisting isotropic (I) and nematic (N) phases. The total surfactant concentrations in the two coexisting phases are $X_I = 4.5 \times 10^{-3}$ and $X_N = 5.2 \times 10^{-3}$. The dashed curve describes the size distribution in the isotropic phase for $X = X_N = 5.2 \times 10^{-3}$, demonstrating that the increased size in the nematic is not simply due to the higher surfactant concentration there but, rather, to the coupling between the self-assembly and long-range orientational order [152].

The one important case of intermicellar interactions which has been virtually untouched, in the context of their effect on aggregate size distributions, is that of *attractions*. The basic difficulty here is that dispersional forces between objects whose sizes are not small compared to their separation distances are quite varied and complicated. Even for relatively simple situations such as parallel planes or rods of infinite extent, the distance dependence of the attraction energy changes markedly from one length scale regime to another [156]. For *finite* rods, with *arbitrary* relative orientations, no tractable results are available. Consequently, it has not been possible to formulate a theory for the interaction contribution χ_s in the case of dispersional forces, especially with polydispersity included. It has been argued [148b] that these effects should act counter to those for the excluded-volume interactions, i.e., in response to dispersional attractions between the micelles they will reorganize themselves into a *larger* number of *smaller* aggregates. But a more rigorous theoretical account of this possibility is clearly needed, in particular to explain the behavior reported by Boden *et al.* (see Chap. 3, particularly Fig. 3.17), according to which the average micellar size is seen to decrease with concentration at high enough surfactant weight fraction (≥ 0.5). This experimental observation is especially intriguing insofar as the size decrease sets in at the *same* concentration along each of many isotherms, *regardless* of whether the system is in its isotropic, nematic, or smectic state at that point (again, see Fig. 3.17).

It must be emphasized that even in the dilute solution limit there are still other factors determining micellar size distributions which are poorly understood. In particular, it is only very recently that theoretical attention has been focussed on the extent to which *viscous forces* due to flow can affect the average aggregation number [157–159]. Two kinds of velocity gradient have been considered: elongational and shear. In the former case the primary effect of flow is shown [157, 159] to be due to the stretching deformation of the rod-like micelles: the axial component of the viscous forces gives rise to a tension throughout the rod (maximum at its center) which imposes an upper bound on the length of rod which can "survive" in flow. This maximum length is found in particular to decrease strongly with velocity gradient (as the inverse one-half power). In the case of *shear* flow, on the other hand, the axial (tension) component of the viscous forces goes to zero (rather than a maximum, as in elongational flow) for aligned rods, and the average micellar size is predicted to increase (linearly) with velocity gradient [158]. Furthermore, this effect is shown to become important only when the flow rate becomes of the order of the rotational diffusion coefficient of the micellar rods. While these predictions are qualitatively consistent with preliminary experimental measurements, it is clear that much further work needs to be done before the basic phenomena involved are well-understood.

1.4.3 NEMATIC-TO-HEXAGONAL TRANSITION

In generalizing the theory of the preceding sections to the case of *positionally* ordered phases, such as the hexagonal state of micellar solutions, we will be brief, since hardly any theoretical calculations have yet been performed [160]. Also, as we shall explain, only a few experiments have been reported which probe directly the *size* of cylindrical aggregates in hexagonal phases. The key point which we wish to stress is that the prevailing, classical, view of the hexagonal state is that it consists of infinite rod-like micelles whose cross-sections form a two-dimensional triangular lattice. Thus, for example, the *only* micellar curvature involved is the cylindrical one (see, e.g., the discussion in Sec. 1.1). In the point of view which we propound below, however, the aggregates are *finite*, characterized by a broad distribution of sizes which depends sensitively on overall concentration. Accordingly, higher curvature than cylindrical is involved: each finite rod is "capped" by half-spherical "ends" in the way described at length throughout earlier sections. We shall argue in particular that the number of such high-curvature ends, and hence the average size of micelles, is determined by a competition between the surfactant's preference for spherical over cylindrical geometry and the colloidal suspension's need to pack its aggregates as efficiently as possible.

We illustrate the above point first with a specific example. The nonionic system "$C_{12}E_6$" (i.e., dodecyl-hexapolyethyleneoxide) in water is known to show a strong preference for satisfying the hydrophobic effect via formation of maximum-curvature aggregates [161]. That is, at and for a large range of concentration above the CMC, the micelles are found to be globular, essentially spherical. Unlike the systems treated in Sec. 1.2.4, the aggregates do not "grow" into increasingly long rods, say, as surfactant is added to the solution. Recall that this growth was driven there by $\delta_{rod} \propto (\tilde{\mu}^o_{sph} - \tilde{\mu}^o_{cyl}) >$ 0, i.e., by a molecule being better off in the cylindrical body than in the half-spherical cap; the only reason the rods remain finite (thereby retaining their higher free energy ends) is because of the entropy of dispersion which they gain in this way. When $\delta_{rod} < 0$, on the other hand, the spherical aggregates are favored by both *intramicellar free energy* considerations, i.e., preference for maximum curvature in satisfying the hydrophobic effect, *and* by *dispersion entropy*.

When the isotropic phase of these globular micelles becomes too concentrated, however, *interaggregate* interactions begin to contribute significantly to the free energy. These latter terms increase with concentration because of the inefficient packing of spheres, as explained below. Bagdassarian *et al.* [27] have shown, in fact, that a first-order transition to an aligned solution of the cylindrical aggregates allows for a decrease in this interaction energy which more than compensates for the higher "self"-energy of the micelles (which have been forced to give up their preferred, spherical, curvature). This is because aligned rods pack more efficiently than spheres, for any given volume fraction of surfactant.

What do we mean by "packing efficiency", and what is its relevance to interaction free energy? We note first that monodisperse *spheres* can be packed to a volume fraction of no more than $\pi/\sqrt{18}$ or 0.74, and that this is possible only if hexagonal close-packed long-range order is present. For *random* configurations, the maximum packing fraction is more than 10% lower [162]. When the colloidal particles are infinite *cylinders*, on the other hand, they can be packed to a volume fraction as high as $\pi/\sqrt{12}$, or 0.91, and this limit corresponds to hexagonal ordering of the long axes. Finally, for infinite *sheets*, there need not be any "wasted" space, and the volume fraction at closest packing is 1.

Consider now a micellar solution of essentially monodisperse spherical aggregates. At volume fractions of roughly 50%, say, the isotropic phase of these particles has practically run out of packing entropy because the micelles are virtually touching one another. Alternatively, if the surfactant is partially ionized, the free energy is high at these concentrations because of the strong repulsive interactions associated with the micellar surfaces being so close to one another. Now imagine that this same volume fraction of material is organized as aligned cylinders. Then the interaction free energy is significantly lower because of the enhanced packing entropy and decreased repulsive energy (greater average distance between micelles). At sufficiently high concentrations this free energy lowering can more than compensate for the self-energy ($\tilde{\mu}^o_{cyl} > \tilde{\mu}^o_{sph}$) and dispersion entropy contributions, and the system undergoes a discontinuous change in both micellar structure *and* long-range order.

We can follow the above scenario further, into the hexagonal phase itself. Specifically, recalling that spherical curvature is locally preferred over cylindrical, we consider the possibility that the rods in the ordered phase will be "as short as possible" in order to retain the maximum amount of high curvature, i.e., in order to have as many spherical end-caps as possible. Then as the concentration is raised, the rods will lengthen so that fewer ends are present and the interaction free energy is minimized. More explicitly, since longer rods mean (for any given volume fraction of surfactant) a greater distance between the hexagonally packed columns of rods, both the inter-aggregate packing entropy and the inter-aggregate repulsion energy will be optimized by reorganization of the system into a smaller number of longer rods. Ultimately, even cylinders which are essentially infinite no longer allow for an efficient enough packing of the aggregates: a phase transition occurs to a system of stacked, essentially planar, bilayers. (Recall that cylinders cannot be packed to volume fractions greater than 91%.) We say "essentially planar" here because—just as we argued that cylinders in hexagonal phases will "sneak in" higher (than cylindrical) curvature in the form of half-spherical caps—the bilayers in lamellar states will be riddled with transmembrane pores and cracks. These latter defects allow locally for the presence of higher (than planar) curvature in the form of half-cylinders which close on themselves as the half-toroidal "rims" of pores or meander with larger radii of curvature as "lips" of cracks. Once again, at still higher concentrations, even these last remaining traces of preferred curvature will be squeezed out by the dominant effect of inter-aggregate forces which demand a greater average distance between layers.

In Sec. 1.4.4 we shall treat explicitly the suppression of curvature in lamellar states, referred to immediately above, whereas in concluding this present section we comment further on the situation for *hexagonal* phases. Again, in order to illustrate most simply the physical ideas involved, it is useful to continue here in terms of the second virial approximation, with the understanding that more realistic theories will necessarily incorporate the effects of interaction through *all* orders in the density. First we note that the fundamental expression (1.128) must be generalized to allow for the possibility of long-range *positional* order. An additional integration appears in (1.128) and $n_s(\Omega) = n_s f_s(\Omega)$ is replaced by $n_s(\Omega, \mathbf{r}) = n_s f_s(\Omega, \mathbf{r})$, the number of rods having orientation Ω, aggregation number s and position $\mathbf{r}$. Similarly, $n_{s'}(\Omega')v_{ss'}(\Omega, \Omega')$ appearing in (1.129) becomes $\int d\mathbf{r}' n_{s'}(\Omega', \mathbf{r}')[-\Phi_{ss'}]$, where $\Phi_{ss'}(\mathbf{r} - \mathbf{r}', \Omega, \Omega')$ is the Mayer function associated with a pair of s- and s'-rods having orientations Ω and Ω' and positions $\mathbf{r}$ and $\mathbf{r}'$. The resulting free energy is sufficient to describe the coupling between short-range self-assembly and long-range orientational/positional ordering, such as that which occurs at the nematic→hexagonal transition.

Since we are also interested in describing *ribbon-like* and other non-classical *lamellar* phases, we must be careful to allow for the possibility of lower-than-cylindrical-symmetry aggregates. Consider, for example, micelles which are biaxial "slabs" of dimensions D (of order twice the molecular surfactant length), $L' \geq D$ and $L \geq L'$. $L \approx L' \approx D$ implies a globular (spherical, minimum-size, maximum-curvature) aggregate; $L \gg L' \approx D$ a prolate (rod-like) micelle; $L \approx L' \gg D$ an oblate (disk-like) one, and $L \gg L' \gg D$ a ribbon-like one. Let m, b and c denote the number of surfactant molecules residing in the spherical, bilayer and cylindrical positions of an s-aggregate having dimensions D, L' and L. Writing $\mu_s^o = m\tilde{\mu}_{sph}^o + b\tilde{\mu}_{bil}^o + c\tilde{\mu}_{cyl}^o$, and expressing m, b and c as explicit functions of D, L' and L (noting that $s \equiv m + b + c \simeq DL'L/v$), it remains only to substitute this L'- and L-dependent μ_s^o into the generalized free energy, with its contribution involving the L'- and L-dependent Mayer function $\Phi_{ss'}$ for hard (slab-like) particles. Since the orientational order is almost completely saturated at the onset of the nematic→hexagonal transition, it is reasonable to simplify this free energy further by taking the alignment to be perfect. Indeed, this is fully consistent with the recent series of Monte Carlo simulations of hexagonal, smectic and crystalline states performed by Frenkel, Lekkerkerker and Stroobants on parallel rods [163]. Similarly, virtually all of the density-functional theories of these same transitions for *non* self-assembling systems have started from perfectly aligned nematics [164]. In our present situation, where self-assembly is explicitly included, it is convenient to stipulate further that only one aggregation number (s) is allowed, to be determined statistical thermodynamically for each total surfactant concentration of interest. The variational problem associated with the micellar free energy then reduces to a simultaneous minimization with respect to L' and L (or s) and $n_{s'}(\Omega', \mathbf{r}') = \delta_{s,s'}\delta(\Omega')f(\mathbf{r}')$. Here $f(\mathbf{r})$ is a trial function specifying the lattice constants (d_H and $d_{H'}$, say) and amplitude (A) of the density wave describing the distorted hexagonal state. In the nematic phase we have $A \equiv 0$, with $D = L' \ll L$ rods whose length (L) increases with concentration. Upon the onset of hexagonal order ($A \neq 0$) the rods will be still longer and the micellar symmetry $D = L'$ will correspond to undistorted hexagonal long-range order in which $d_H = d_{H'}$. Note that even when spherical curvature is preferred locally over cylindrical (i.e., $\tilde{\mu}_{sph}^o < \tilde{\mu}_{cyl}^o$), increasing concentration will imply longer rods (fewer spherical "ends") because otherwise the interaction terms involve too much excluded volume. At still higher concentrations we expect L' to grow away from D, forming biaxial slabs consistent with a ribbon or "rectangular-like" ($d_H \neq d_{H'}$) phase which can be thought of as a precursor to the lamellar state.

Herzfeld and coworkers [160] have pursued ideas of this kind in their formulation of nematic and columnar (hexagonal) ordering in micellar solutions of aligned rod-like aggregates. Instead of a virial expansion, they use scaled-particle theory to treat the excluded volume interactions between aggregates, and introduce an ingenious separation between these packing entropy effects on the one hand and the entropy losses associated with partial positional order on the other. They demonstrate a coupling between average micellar size and long-range order which is consistent with the scenario sketched above and with the experimental results obtained by Boden *et al.* in their measurements on triphenylpolyoxyethylene in water systems [165] where rod-like aggregates are seen to form nematic and hexagonal states. These exciting developments, while preliminary, are certain to stimulate much further work on the coupling of micellar size/shape and long-range orientational/position order.

1.4.4 INTERACTION-INDUCED SUPPRESSION OF CURVATURE IN LAMELLAR PHASES

In Sec. 1.4.2 we considered the possibility that a surfactant-in-water system with a sufficiently strong preference for maximum (globular) curvature could undergo a transition from an isotropic solution of globular micelles to a nematic state of rod-like aggregates. This phase change results from a competition between intra- and inter-micellar contributions to the free energy. Specifically, the *self*-energy prefers maximum curvature, whereas the *packing* entropy (or, alternatively, the Coulombic repulsion) of the aggregates is optimized by aligned micelles of lower curvature (cylindrical, here). Similarly, we have just discussed how the demands of packing entropy at still higher concentrations lead to a partial *positional* (and enhanced orientational) ordering of the aligned rods. More explicitly, by organizing themselves into hexagonally arrayed columns, the micelles are able to pack in more ways (at a greater average distance from one another) and thereby lower their free energy in a way that (at sufficiently high density) more than compensates for their loss of positional disorder. At the same time they stay as short as they can, to maximize the number of spherical/globular "ends" and hence the amount of higher curvature which they prefer on a local scale. Of course, the fewer such ends, the larger the packing entropy; accordingly, the higher curvature (associated with these ends) is necessarily "squeezed out" of the system as the surfactant concentration is raised in the hexagonal phase. Ultimately, as described in the previous section, the cylindrical aggregates become biaxial, i.e., ribbons with increasingly large planar (zero-curvature bilayer) portions, eventually giving way at still higher concentration to continuous lamellar sheets which stack upon one another.

Even in highly organized bilayers, however, which arise because of how efficiently they can keep out of each other's way and hence raise the pack-

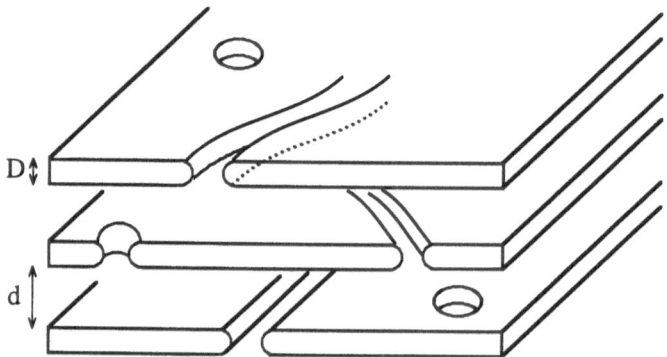

FIGURE 1.19. Schematic illustration of line and pore (curvature) defects in the lamellar phase [168].

ing entropy, curvature defects will persist because of the local free energy lowering associated with their greater ability to satisfy the hydrophobic effect. In the case where *cylindrical* curvature is preferred, for example, it is particularly easy to imagine how defects in the bilayers can both help and hurt, i.e., lower and raise the free energy. Figure 1.19 shows transmembrane "cracks" in neighboring bilayers, and also some "holes." Note that each "crack"—or *line* defect—involves a pair of opposing half-cylindrical "lips." In these lips the standard chemical potential $\tilde{\mu}^o_{lip} \approx \tilde{\mu}^o_{cyl}$ is lower than that ($\tilde{\mu}^o_{bil}$) for molecules in the planar, bilayer portions of the lamellae. We write $\tilde{\mu}^o_{lip} \approx \tilde{\mu}^o_{cyl}$ here because the lip geometry is almost but not quite that of a cylindrical micelle, since the cracks are not quite straight. That is, they bend on length scales (R) large compared to that of the bilayer thickness (molecular dimension: $D = 2l$). Similarly, transmembrane "holes" are shown in Fig. 1.19 in whose curved cylindrical (half-toroidal) lips the standard chemical potential is also lower than that in the planar lamellar portions. These defects, then, allow for a local, self-energy, lowering of the micellar free energy. Their introduction into the bilayer sheets, however, simultaneously results in a *raising* of the free energy because of a greater interaction *between* the lamellae. More explicitly, in order to "create" them, one (or, rather, the Maxwell demon) needs to remove surfactants from the bilayer and accommodate them elsewhere (since we work at constant volume fraction). This is done by *thickening* the bilayer, i.e., packing surfactant in the planar portions at smaller area per molecule, *or* by creating *new* bilayers. Either way, the resulting average distance between lamellae is smaller and the packing entropy is decreased and/or the repulsion energy increased. One expects, then, that at higher concentration, where these inter-lamellar interactions become more important, it will become progressively more difficult to "sneak" cracks and holes into the bilayers. Conversely, at low enough concentration, where inter-lamellar interactions are greatly reduced, these defects will proliferate and thereby mediate a phase transition out of the bilayer state.

The above scenario has been quantified in several ways, one of which we feature here. Note first that, independent of any model for the defects and their statistical mechanics, we can argue for simplicity that defects only increase the *number* of bilayers and not their (half) *thickness* (l), i.e., that the area per molecule $a = v/l$ is constant. The number of molecules per bilayer is then given by

$$N_l = \frac{2A(1 - \phi_d)}{a} ,$$ (1.138)

where A is the area of each lamellar sheet and ϕ_d is the fraction of each bilayer surface covered with defect. N_l decreases with ϕ_d, but—at fixed surfactant volume fraction ϕ—so does the distance between lamellar surfaces:

$$d = \frac{2v(1 - \phi - \phi_d)}{\phi a}$$ (1.139)

This latter relation follows immediately from Eq. (1.138) for N_l and the definition $\phi = [N_l \ (\text{no. of layers}) \ v/AH] = N_l v/(2l + d)A$, where H is the total "height" of the system (dimension normal to the area A) and (no. of layers) $= H/(2l + d)$. d decreasing with ϕ_d reflects the increase in interaction free energy *lowering* associated with the local preference for curvature. Also, cracks contribute to the topological disorder within each lamellar sheet, thereby stabilizing further the presence of defects.

The free energy terms alluded to above are represented most conveniently in dimensionless form, on a per molecule basis:

$$\frac{\beta F v}{V} \equiv f_b + f_d + f_e$$ (1.140)

Here, the subscript b describes the interactions between bilayers and is a decreasing function of the intersurface distance d, where the latter depends only on the area fraction ϕ_d according to (1.139). The second term refers to the (negative) "core" energy of the defects, i.e., an energy lowering proportional to the total length of cracks. Finally, f_e is the entropic contribution arising from the in-plane disorder which depends—like f_d (see below)—on both the total length of cracks *and* on their topology.

To proceed, we partition the surface of a bilayer into a square lattice with lattice constant ξ, and randomly divide the lattice squares into type A and B, with ψ defined as the area fraction of A (see Fig. 1.20). The *equilibrium* value of ψ will follow from free energy minimization for given surfactant

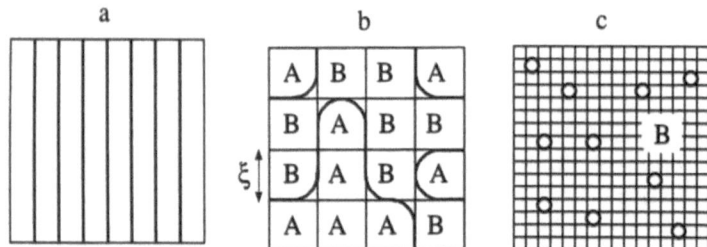

FIGURE 1.20. Stripe (a) → random line (b) → pore (c) progression with increasing surfactant volume fraction (decreasing defect density), according to the line defect model [168].

volume fraction ϕ and similarly for the equilibrium value of ϕ_d. Note that A and B do not refer to different species or phases; rather, they denote different regions of the bilayer, *separated by line defect* (i.e., a water "crack" of width w). This formulation is motivated by the phenomenological theory of microemulsions described in Chap. 9, and especially by the description of planar surfaces in the isotropic L_3 ("sheety" or "sponge") phases of dilute surfactant solutions [166]—see also Sec. 2.4.2. In the present situation, ψ does not describe a composition, but (as in the sponge phase case) is instead *a measure of the topological organization of defect* (sponge interface). Furthermore, in the random mixing approximation, there is a direct relation between the lattice constant ξ and the two variational parameters ψ and ϕ_d; for each pair ψ and ϕ_d, there corresponds a unique ξ. (Note that in both the microemulsion and sponge phase theories, ϕ_d is fixed by the total amount of surfactant present.)

At the interface of an A-square and a B-square, we lay down our strip of line defect; the defect is *transmembrane* (i.e., a channel), but we are now concerned only with the *surface* features of the bilayer. Figure 1.20 shows typical patterns constructed from the model. The defects are pictured as self-avoiding at positions of possible crossings. Also, the model does not permit lines that simply "end": the allowed patterns are closed. Accordingly, the building-block of all defect patterns is the line defect, and the associated energy is described by a single quantity, the defect energy per unit length. Let L be the total defect length per bilayer, as measured by the overall length of A-B interface. Then

$$L = 4N\psi(1 - \psi)\xi \tag{1.141}$$

in the random mixing approximation, with $N = M^2/\xi^2$ the number of squares in the planar $M \times M$ lattice ($M \sim \sqrt{A}$). Since every defect is of width w, $\phi_d = Lw/M^2$ and we have from (1.141) and $N = M^2/\xi^2$ that

$$\xi = 4w\psi(1 - \psi)/\phi_d . \tag{1.142}$$

This result suggests a natural cutoff value for ξ which serves to define a *pore* phase, i.e., as ψ decreases, the defects become loops around isolated A-squares. More explicitly, if solution of the extremization equation gives $\xi < w$, we set $\xi = w$ and (using $\phi_d = 4\psi(1 - \psi)$—see (1.142) with $\xi = w$) treat the resulting free energy as a function of a single variational parameter (ϕ_d or ψ; see (1.142)).

Because the building-block of the defect pattern is a line, it is natural to introduce a local elastic energy per unit length, $-\Delta + \kappa c_1^2$. Here $-\Delta$ is the "core" energy per unit length of the locally *straight* line defect, and c_1 is the single curvature needed to characterize its bending in the bilayer plane. $\kappa(> 0)$ is a harmonic bending constant, having dimensions of energy times length. We are interested in cases where $\Delta > 0$ since $-\Delta$ reflects the fact that defects provide the opportunity for molecules to pack with large head-group areas. Recalling that $L = \phi_d M^2/w$ is the total length of defect in a bilayer, and that $c_1 \neq 0$ only in the bends, the defect energy per sheet can be written as $-\Delta\phi_d M^2/w + [(\kappa c_1^2)$ (number of bends) (length per bend)]. Estimating the number of bends within the random mixing approximation, taking the length per bend to be $\xi/2 = 1/c_1$, and recalling Eq. (1.138) for the number of surfactants per bilayer, the dimensionless defect free energy per molecule becomes

$$f_d = -\frac{\Delta\phi_d\phi}{1 - \phi_d} + \frac{\kappa w^3 \psi(1 - \psi)\phi}{\xi^3(1 - \phi_d)} \tag{1.143}$$

Here, Δ and κ have been rendered dimensionless and are related to their earlier incarnations via $\Delta \equiv \Delta a/2wkT$ and $\kappa \equiv 2\pi\kappa a/w^3 kT$.

The entropy contribution f_e involves the connection between the A-B mixing entropy and that of the defects. Since the A-B squares comprise a fictitious division of the bilayer surface, this scheme is simply a convenient device for arriving at f_e. In this way, we find the familiar-looking result

$$f_e = \left\{ \frac{a\phi}{2\xi^2(1 - \phi_d)} \right\} [\psi \ln \psi + (1 - \psi) \ln(1 - \psi)] \tag{1.144}$$

where it should be noted that the factor in curly brackets is $N\phi/N_l = M^2\phi/N_l\xi^2$.

The remaining contribution to f is the bilayer-bilayer interaction free energy, f_b. Assuming that the only effect of introducing defects is to decrease the intermembrane spacing according to Eq. (1.139), and taking the bilayer-bilayer free energy to be equal to $f_o M^2/d$ (as arises, say, from solving the Poisson-Boltzmann equation in the appropriate limit), we find

$$f_b = \frac{f_o\phi^2}{(1 - \phi_d)(1 - \phi - \phi_d)} , \tag{1.145}$$

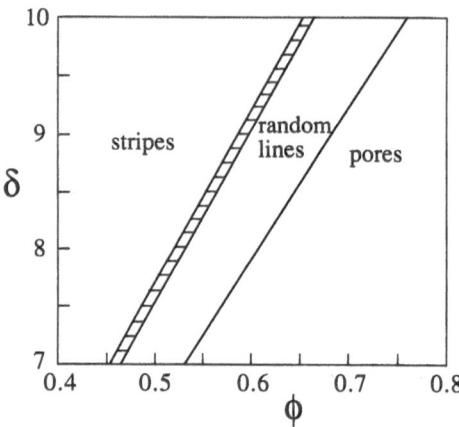

FIGURE 1.21. Phase diagram obtained from the line defect model, for $\kappa = 4$ [168].

with f_o now dimensionless via $f_o \equiv f_o a^2/4vkT$. Note that, at fixed surfactant volume fraction ϕ, the overall effect of increasing ϕ_d (and hence increasing d) is an increase in f_b, i.e., interbilayer interactions do indeed serve to suppress the proliferation of defects, as discussed earlier.

Finally, we consider a *stripe* phase which corresponds to the defect channels being parallel to each other throughout the bilayer surface. For such a pattern we can eliminate the bending energy and topological entropy terms in f and describe this phase by the straight-line defect energy $(-\Delta\phi_d\phi/(1-\phi_d))$ and bilayer interaction (f_b) only.

Recalling Eq. (1.142) for $\xi = 4w\psi(1-\psi)/\phi_d$ we can proceed to minimize the *line*-phase free energy $f_d + f_e + f_b$ with respect to both ϕ_d and ψ, for each of successively larger values of ϕ; the *stripe* phase free energy is varied with ϕ_d only. The resulting thermodynamic potentials, $f_{\text{line}}[\phi_d^*(\phi), \psi^*(\phi); \phi]$ and $f_{\text{stripe}}[\phi_d^*(\phi); \phi]$ as functions of ϕ, yield the possible phase coexistences via common tangent constructions. A typical phase diagram with Δ vs. ϕ is given [168] in Fig. 1.21. The progression of phases can be described qualitatively, as follows. Below a certain value of ϕ, which depends on Δ and κ, the free energy minimum corresponds to a *random line* phase for which $\psi = 1/2$. Above this ϕ-threshold ψ decreases continuously from $1/2$ to give a true pore-like ($\xi = w$) phase at high ϕ. Indeed, at these large surfactant concentrations the bilayers are effectively defect-free. (Note that, for any fixed Δ and κ the defect area fraction ϕ_d decreases steadily with increasing ϕ, because the interbilayer interactions become dominant.)

Because the stripe phase free energy is distinct from that of the random line phase, the associated phase transition is necessarily first-order: coexistence arises between a lamellar phase with ribbon-like defect patterns and another with random channels, the latter with smaller ϕ_d. One can imagine this discontinuous transition upon increase in surface concentration ϕ as one from a phase with high concentration (area fraction) of defects to a more "perfect" bilayer system. On the other hand, the order of the random line to *pore* transition depends on the values of Δ and κ. Increasing the value of κ has the effect of decreasing the slope of the lines separating the stripe-line and line-pore domains in Fig. 1.21: at any given δ, both stripe and random line phases survive to higher surfactant concentration for larger κ.

It should be clear by now that the presence of defects in lamellar phases results from an attempt to satisfy *molecular* packing demands within the constraints of the *aggregate* packing required at moderate to high surfactant volume fractions. In order to make connection with the work of Charvolin and Sadoc (see Chap. 4), we can describe this competition in terms of the frustration between local, molecular, forces and the large-scale liquid crystalline order when the amphiphile molecules are packed in homogeneous, perfect bilayers. Resolution of this frustration involves allowing for some of the molecules to reside in highly curved defect lips which leads to pattern formation on the bilayer face. Similarly, in the geometrical approach of Charvolin and Sadoc, local stresses in opposing *monolayers* lead to a frustration which is also resolved by large scale topological reorganization of the amphiphilic system. The common feature shared by their work and ours is a locally stressed bilayer in which the molecular packing requirements lead to a frustration. Our approach, however, explores directly the statistical thermodynamic consequences of this frustration as a function of surfactant concentration.

Recent experimental work indicates that lamellar phases are indeed pierced by pores or channels. We have seen that as Δ, the core defect energy, becomes larger, the bilayers are more fragmented by channels. Neutron scattering studies [169] on the lamellar phase of the ternary system sodium decyl sulfate (SdS)/decanol/water show that, for constant water fraction, the bilayers have a greater area fraction of defects as the surfactant/alcohol ratio is increased (thereby favoring larger curvature—more positive Δ). Furthermore, as this ratio is increased, the periodicity of the lamellar phase decreases—indeed our model requires that the periodicity (d spacing) decrease with increasing defect density. What is happening here is clear. Both the surfactant and the alcohol have the same tail group, but the former has the larger head: as the amount of surfactant relative to decanol is increased, there is need to accommodate the larger head groups, and this promotes defects of positive curvature. Indeed, it seems that the SdS molecules will be packed preferentially in the defect lips [170]. As is expected, at still higher surfactant/alcohol ratios (at constant water content), these curvatures proliferate to the point where the lamellar phase is disrupted via a phase transition.

Similarly, upon *dilution* of this system at *fixed* SdS/decanol ratio, we predict an increase in defect density with increasing water content. Experimentally it is found that the dilution of this system leads, by a first-order transition, to a phase of infinite ribbons arranged in a 2-dimensional rectangular lattice [171]. Though it is not stated that the bilayer defect area fraction increases with dilution in the *lamellar* phase, it is reasonable to infer that this is the case and that the phase transition is mediated by the increasing defect density, perhaps even through our proposed stripe phase which is suggestive of the experimental ribbon phase. Finally, we point to Helfrich's analysis [133] of the potassium caprylate/decanol/water phase diagram which can be found in the classic review by Ekwall [172]. There it is suggested that two coexisting lamellar phases, which can be connected by a dilution path at fixed surfactant/alcohol ratio, differ in that the more dilute one contains a high concentration of pores.

1.5   Concluding Discussion

In this chapter our goal has been to provide a systematic statistical thermodynamic introduction to micellization phenomena in aqueous surfactant solutions. Most of our discussion has been restricted to the case of *dilute* solutions, where interaction *between* micellar aggregates can be neglected. This regime is of special interest because it allows for the most definitive conclusions to be deduced about the *microscopic structure* of micelles. In particular, high resolution light, X-ray and neutron scattering experiments can be analyzed without the complications of interference (i.e., correlation) contributions involving more than one aggregate. Similarly, in this same dilute-solution limit it becomes possible to formulate a statistical thermodynamic theory without needing to guess the intermicellar forces about which so little is known at present. Nevertheless, it is precisely the interactions *between* aggregates which give rise to the staggeringly rich variety of micellar structures and long-range order symmetries that characterize aqueous surfactant systems at higher concentrations. Accordingly, we have devoted the final fourth of this chapter (Sec. 1.4) to a brief exposition of some recent theoretical approaches to understand those novel phase behaviors in concentrated amphiphilic solutions.

In the introduction (Sec. 1.1), we reviewed some key experiments which led to the contemporary "picture" of micelles and described specifically the strengths and limitations of computer simulations of surfactant aggregates. The idea was to "argue the case" for regular structures providing the basic building blocks for micellized solutions. That is, while acknowledging that there are many interesting features underlying the micellar surface roughness and fluctuations on Å and μsec scales, we insist nevertheless that one must begin with a "zero-order" description in which the aggregates can be described by simple, competing geometries. More explicitly, we argue that it is useful to classify surfactant-in-water systems according to their characteristic mode of organization in dilute solution: just above the CMC the amphiphilic molecules will be found in predominantly one *shape* of aggregate—globular/spherical, cylindrical or bilayer. Again, while recognizing that each preferred shape can be represented by a distribution of *sizes* and *distortion* modes, the fact remains that one *basic geometry* is generally dominant. The aim, then, is to predict the nature of this distribution in dilute solution, e.g., the dependence of average size (aggregation number) on the concentration of surfactant and of added salt, etc. (see Sec. 1.2). Similarly, one wants to be able to derive the ways in which the amphiphilic "heads" and "tails" pack—organize structurally—in each of these micellar geometries (Sec. 1.3).

As stressed in the Introduction and referred to again immediately above, it is premature at this time to look to computer simulation for the kind of definitive help that it has provided in the case of much simpler fluids. That is because, in the case of aqueous surfactant solutions:

1. Too little is known about the microscopic interactions amongst amphiphilic species (especially ionic ones) and water.

2. It is not yet feasible to simulate large enough systems for sufficiently long times to explore the spontaneous formation of micelles at low concentrations. For mole fractions comparable to those in dilute solutions just above the CMC, for example, one would need at least tens of thousands of water molecules and to compute for at least tens of thousands of picoseconds. Consequently, "all" one can do at present with molecular dynamics (MD) computation, say, is to work at much higher concentrations (i.e, mole fractions greater than a tenth) and to start with an already-formed micelle. One then explores the microscopic details of the structure and dynamics of such an aggregate, subject to the specific choice of interaction potentials for amphiphiles and water. (See, however, the initial attempts [40] to study spontaneous micellization via application of massively parallelized computers to idealized surfactant solutions.)

The statistical thermodynamic approach which we have featured in this chapter pursues a very different point of view. For the self-assembly process itself—that is, the spontaneous appearance of micelles and their subsequent "growth" with increasing concentration—we establish the explicit connections between the microscopic partition functions relevant to the problem and the phenomenological standard chemical potentials which abound in the older literature. We then proceed to discuss the molecular bases for different possible dependences of micellar standard chemical potentials on aggregation number. Open questions remaining here are those which involve the role of rotation and translation degrees of freedom (see Sec. 1.2.6) as well as those pertaining to more exotic geometries (Sec. 1.2.5) such as closed bilayers (vesicles) and cylinders (rings). As far as the separation of "heads" and "tails" is concerned, we have emphasized the fact that head-group effects, in particular those arising in the case of *ionic* species, are especially problematic. This is due to the peculiar structuring behaviors associated with water near interfaces and to the dramatically large variation in micellization phenomena attendant upon small changes in head-group, e.g., replacement of Li^+ by Na^+ counterions in the case of alkyl sulphate salts. Accordingly, it is difficult to make any general statistical thermodynamic arguments concerning the role of head-groups in determining micellar structures. (For a most recent exception to this rule, however, see the scaling behavior discussed by Carale and Blankschtein [173] for polyoxyethylene head groups.) In the case of "tails", on the other hand, we have stressed in Secs. 1.3.3 and 1.3.4 how both chain packing statistics and elastic properties can be understood from a microscopic *and yet generic* point of view, because of the universal nature of the packing constraints which govern semi-flexible chains with excluded volume.

Finally, we believe that the most rich and far-reaching open problems in micellar systems involve the coupling between self-assembly and long-range order which becomes dominant at concentrations well above the CMC. These questions have been featured in Sec. 1.4 above, where we treat inter-micellar interactions in the particular contexts of non-ideal isotropic solutions (Sec. 1.41.), nematic states (Sec. 1.4.2), hexagonal phases (Sec. 1.4.3) and more generally in lamellar systems where the naturally preferred curvature has been suppressed but for defects (Sec. 1.4.4). Here all of the thorny problems of partial positional and orientational ordering in neat, thermotropic, liquid crystals become compounded by the fact that neither size nor shape distributions are fixed, but rather couple directly to the interaction strength (concentration) and long-range order (symmetry). Accordingly, new ideas will be needed to provide a systematic theory of the novel phenomena arising in these systems.

Acknowledgments

Over the past ten years we have benefited greatly from countless discussions with theoretical and experimental colleagues, both "here" and "abroad", who share our excitement about this modern era of research on self-assembling systems. Without attempting to list them all, we thank them heartily. We do want, however, to be explicit about the graduate students, postdoctorals and visitors who gamely joined our efforts, in particular (in approximate chronological order): Bill McMullen, Igal Szleifer, Andy Masters, Zhong-Ying Chen, Jean-Louis Viovy, Danny Rorman, Carey Bagdassarian, Didier Roux, Shi-Qing Wang, Diego Kramer, Zhen-Gang Wang, Linda Steenhuizen and Debbie Fattal-Hornreich. For financial support we thank the U.S. National Science Foundation (WMG), the U.S.-Israel Binational Science Foundation (ABS and WMG), the Yeshaya Horowitz Association, and the Fund for Basic Research of the Israeli Academy of Science (ABS).

References

1. J. W. McBain, Trans. Faraday Soc. **9**, 99 (1913).

2. P. Debye, Ann. New York Acad. Sci. **51**, 575 (1949); see also P. Debye, J. Phys. Colloid Chem. **51**, 18 (1947).

3. D. J. Shaw, *Introduction to Colloid and Surface Chemistry* (Butterworths, London, 1970).

4. P. Debye and E. W. Anacker, J. Phys. Colloid Chem. **55**, 644 (1951).

5. S. Hayashi and S. Ikeda, J. Phys. Chem. **84**, 744; see also S. Ikeda, S. Hayashi and T. Imae, J. Phys. Chem. **85**, 106 (1981) and reference contained therein.

6. B. B. Berne and R. Pecora, *Dynamic Light Scattering* (John Wiley, New York, 1976).

7. N. A. Mazer, G. B. Benedek and M. C. Carey, J. Phys. Chem. **80**, 1075 (1976); C. Y. Young, P. J. Missel, N. A. Mazer, G. B. Benedek and M. C. Carey, *ibid.* **82**, 1375 (1978) and **84**, 1044 (1980). See also N. A. Mazer in *Dynamic Laser Scattering: Applications of Photon Correlation Spectroscopy*, R. Pecora, ed. (Plenum, New York, 1981).

8. M. Corti and V. Degiorgio, in *Photon Correlation Spectroscopy and Velocimetry*, H. Z. Cummins and E. R. Pike, eds. (Plenum, New York, 1977), p. 450; Ann. Phys. (Paris) **3**, 303 (1978) and J. Phys. Chem. **85**, 711 (1981).

9. G. K. Batchelor, J. Fluid Mech. **74**, 1 (1976); B. U. Felderhof, J. Phys. **A11**, 929 (1978).

10. See papers by G. B. Benedek, N. A. Mazer, D. F. Nicoli, M. Corti and V. Degiorgio in *Physics of Amphiphiles: Micelles, Vesicles and Microemulsions*, V. Degiorgio and M. Corti, eds. (North Holland, Amsterdam, 1985).

11. See, for example: S. Ozeki and S. Ikeda, J. Colloid Interface Sci. **77**, 219 (1980); H. Hoffmann, G. Platz and W. Ulbricht, J. Phys. Chem. **85**, 1418 (1981); R. Nagarajan, J. Colloid Interface Sci. **90**, 477 (1982) and papers cited therin.

12. D. F. Nicoli, J. G. Elias and D. Eden, J. Phys. Chem. **85**, 2866 (1981) and W. Schorr and H. Hoffmann, *ibid.* **85**, 3160 (1981).

13. See, for example, S. H. Chen, Ann. Rev. Phys. Chem. **37**, 351 (1986) and W. Marshall and S. W. Lovesey, *Theory of Thermal Neutron Scattering* (Oxford, 1971).

14. J. B. Hayter and J. Penfold, Colloid and Polymer Sci. **261**, 1022 (1983).

15. B. Cabane, R. Duplessix and T. Zemb, in *Surfactants in Solution*, B. Lindman and K. L. Mittal, eds. (Plenum, New York, 1984), Vol. 1, p. 373.

16. J. B. Hayter, in *Physics of Amphiphiles: Micelles, Vesicles and Microemulsions*, V. Degiorgio and M. Corti, eds. (North Holland, Amsterdam, 1985).

17. B. Cabane, R. Duplessix and T. Zemb, J. de Physique **46**, 2161 (1985).

18. B. Jacrot, Rep. Prog. Phys. **39**, 911 (1976) and H. B. Sturman and A. J. Miller, J. Appl. Cryst. **11**, 325 (1978).

19. See, for example: D. Bendedouch, S. H. Chen and W. C. Koehler, J. Phys. Chem. **87**, 153 and 2671 (1983).

20. J. B. Hayter and J. Penfold, Mol. Phys. **42**, 109 (1981).

21. V. Luzzati, H. Mustacchi and A. Skoulios, (a) Nature **180**, 600 (1957); (b) Disc. Faraday Soc. **25**, 43 (1958); (c) V. Luzzati, M. Mustacchi, A. Skoulios and F. Husson, Acta Cryst. **13** 660 (1960); (d) F. Husson, H. Mustacchi and V. Luzzati, Acta Cryst. **13**, 668 (1960).

22. V. Luzzati, in *Biological Membranes*, D. Chapman, ed. (Academic Press, New York, 1968), p. 71.

23. See references 21 and 22 and G. L. Kirk, S. M. Gruner and D. L. Stein, Biochemistry **23**, 1093 (1984), and S. M. Gruner, M. W. Tate, G. L. Kirk, P. T. C. So, D. C. Turner, D. T. Keane, C. P. S. Tilcock and P. R. Cullis, Biochemistry **27**, 2853 (1988), and other work cited therein.

24. E. Sternin, B. Fine, M. Bloom, C. P. S. Tilcock, K. F. Wong and P. R. Cullis, Biophys. J. **54**, 689 (1988). See also Ref. 126 below.

25. A. Parsegian, Trans. Faraday Soc. **62**, 848 (1966); D. Mather, J. Colloid Interface Sci. **57**, 240 (1976); and B. Jönsson and H. Wennerström, *ibid.* **80**, 482 (1981).

26. See, for example: G. Lahajnar, S. Zumer, M. Vilfan, R. Blinc and L. W. Reeves, Mol. Cryst. Liq. Cryst. **113**, 85 (1984); P. J. Photinos, L. J. Yu and A. Saupe, *ibid.* **67**, 277 (1981); and G. Chidichimo, L. Coppola, C. LaMesa, G. A. Ranieri and A. Saupe, Chem. Phys. Lett. **145**, 85 (1988).

27. C. Bagdassarian, W. M. Gelbart and A. Ben-Shaul, J. Stat. Phys. **52**, 1307 (1988).

28. M. C. Holmes and J. Charvolin, J. Phys. Chem. **88**, 810 (1984); M. C. Holmes, D. J. Reynolds and N. Boden, J. Phys. Chem. **91**, 5257 (1987); P. Kekicheff, B. Cabane and M. Rawiso, J. de Physique Lett. **45**, 813 (1984); and P. Kekicheff and G. J. T. Tiddy, J. Phys. Chem. **93**, 2520 (1989).

29. D. W. R. Gruen, Prog. Colloid Polymer Sci. **70**, 6 (1985). See also Y. Chevalier and T. Zemb, Rep. Prog. Phys. **53**, 279 (1990).

30. C. Tanford, *The Hydrophobic Effect*, 2nd edition (Wiley, New York, 1980).

31. J. N. Israelachvili, D. J. Mitchell and B. W. Ninham, J. Chem. Soc., Faraday Trans. 2 **72**, 1525 (1976); J. N. Israelachvili, S. Marcelja and R. G. Horn, Quart. Rev. Biophys. **13**, 121 (1980); J. N. Israelachvili, in *Physics of Amphiphiles: Micelles, Vesicles and Microemulsions*, V. Degiorgio and M. Corti, eds. (North Holland, Amsterdam, 1985).

32. J. N. Israelachvili, *Intermolecular and Surface Forces* (Academic, New York, 1985).

33. H. Wennerström and B. Lindman, Physics Reports **52**, 1 (1979).

34. F. M. Menger and D. W. Doll, J. Am. Chem. Soc. **106**, 1109 (1984) and F. M. Menger, Acc. Chem. Res. **12**, 111 (1974). See also F. Menger, Angew. Chem. Int. Ed. Engl. **30**, 1086 (1991), for a revised version of his picture of the micelle.

35. P. Fromherz, Chem. Phys. Lett. **77**, 460 (1981).

36. S. W. Haan and L. R. Pratt, Chem. Phys. Lett. **79**, 436 (1981) and B. Owenson and R. L. Pratt, J. Phys. Chem. **88**, 2905, 6048 (1984).

37. J. M. Haile and J. P. O'connell, J. Phys. Chem. **90**, 1875 (1986).

38. B. Jönsson, O. Edholm and O. Teleman, J. Phys. Chem. **85**, 2259 (1986).

39. K. Watanabe, M. Ferrario and M. L. Klein, J. Phys. Chem. **92**, 819 (1988).

40. B. Smit, P. A. J. Hilbers, K. Esselink, L. A. Rupert, N. M. van Os and N. M. Schlijper, J. Phys. Chem. **95**, 6361 (1991) and Nature **348**, 624 (1990).

41. (a) S. Karaborni and J. P. O'Connell, J. Phys. Chem. **94**, 2624 (1990) and Langmuir **6**, 905 (1990); (b) J. J. Wendoloski, S. J. Kimatian, L. E. Schutt and F. R. Salemme, Science **243**, 636 (1989).

42. F. F. Abraham, *Homogeneous Nucleation Theory* (Academic Press, New York, 1974).

43. J. Lothe and G. M. Pound, J. Chem. Phys. **36**, 2082 (1962), **45**, 630 (1966) and **48**, 1849 (1968).

44. H. Reiss, Adv. Colloid Interface Sci. **7**, 1 (1977) and H. Reiss, J. L. Katz and E. R. Cohen, J. Chem. Phys. **48**, 5553 (1968).

45. H. Reiss, A. Tabazadeh and J. Talbot, J. Chem. Phys. **92**, 1266 (1990).

46. F. H. Stillinger, J. Chem. Phys. **78**, 4654 (1983).

47. K. A. Dawson, B. L. Walker and A. Berera, Physica A **165**, 320 (1990); A. Berera and K. A. Dawson, Phys. Rev. A **42**, 3618 (1990); R. G. Larson, J. Chem. Phys. **91**, 2479 (1989), **89**, 1642 (1988), R. G. Larson, L. E. Scriven and H. T. Davis, J. Chem. Phys. **83**, 2411 (1985); C. M. Care, J. Phys. C **20**, 689 (1987); T. Kawakatsu and K. Kawasaki, J. Colloid Interface Sci. **145**, 413, 420 (1991); D. Chowdhury and D. Stauffer, Phys. Rev. A **44**, 2247 (1991); see also G. Gompper and M. Schick, Chap. 8 and references to earlier work contained therein.

48. C. A. J. Hoeve and G. C. Benson, J. Phys. Chem. **61**, 1149 (1957).

49. R. H. Aranow, J. Phys. Chem. **67**, 556 (1963).

50. A. Wulf, J. Phys. Chem. **82**, 804 (1978).

51. D. C. Poland and H. A. Scheraga, J. Colloid Interface Sci. **21**, 273 (1966); J. Phys. Chem. **69**, 2431 (1965).

52. R. Nagarajan and E. Ruckenstein, J. Colloid Interface Sci. **60**, 221 (1977) and **71**, 580 (1979); see also Langmuir **7**, 2934 (1991) for a recent comprehensive theoretical approach by the same authors.

53. D. J. Mitchell and B. W. Ninham, J. Chem. Soc. Faraday Trans. 2 **77**, 601 (1981).

54. W. E. McMullen, W. M. Gelbart and A. Ben-Shaul, J. Phys. Chem. **88**, 6649 (1984).

55. B. Halle, M. Lundgren and B. Jönsson, J. de Physique **49**, 1235 (1988).

56. T. L. Hill, *Thermodynamics of Small Systems*, Vols. I and II (Benjamin, New York, 1963).

57. D. G. Hall and B. A. Pethica, Chap. 16 in *Nonionic Surfactants*, M. J. Schick, ed. (Dekker, New York, 1967); D. G. Hall, in *Nonionic Surfactants*, M. J. Schick, ed. (Marcel Dekker, New York, 1987), p. 247.

58. D. Stigter, J. Colloid Interface Sci. **41**, 473 (1974).

59. K. Shinoda and E. Hutchinson, J. Phys. Chem. **66**, 577 (1962).

60. S. Ljunggren and J. C. Eriksson, Progr. Colloid and Polymer Sci. **74**, 38 (1987).

61. S. Ljunggren and J. C. Eriksson, J. Chem. Soc. Faraday Trans. 2 **82**, 913 (1986) and **80**, 489 (1984); J. C. Eriksson, S. Ljunggren and U. Henriksson, *ibid.* **81**, 833, 1209 (1985).

62. I. Reich, J. Phys. Chem. **60**, 257 (1956).

63. P. Mukerjee, J. Phys. Chem. **76**, 565 (1972), and review in *Micellization, Solubilization, and Microemulsions*, K. L. Mittal, ed. (Plenum, New York, 1977).

64. P. J. Missell, N. A. Mazer, G. B. Benedek, C. Y. Young and M. C. Carey, J. Phys. Chem. **84**, 1044 (1980); see also G. B. Benedek in Ref. 10.

65. G. Porte and J. Appell, J. Phys. Chem. **85**, 511 (1981).

66. See G. Porte, Chap. 2 in this volume and references therein.

67. J.-M. Chen, T.-M. Su and C. Y. Mou, J. Phys. Chem. **90**, 2418 (1986).

68. W. E. McMullen, A. Ben-Shaul and W. M. Gelbart, J. Colloid Interface Sci. **98**, 523 (1984).

69. L. Landau and E. M. Lifshitz, *Statistical Physics* (Pergamon, Oxford, 1968).

70. D. Blankschtein, G. M. Thurston and G. B. Benedek, Phys. Rev. Lett. **54**, 955 (1985) and J. Chem. Phys. **85**, 7268 (1986); S. Puvvada and D. Blankschtein, J. Phys. Chem. **93**, 7753 (1989).

71. S. A. Safran, P. A. Pincus, M. E. Cates and F. C. MacKintosh, J. de Physique **51**, 502 (1990).

72. G. Porte, J. Appell and Y. Poggi, J. Phys. Chem. **84**, 3105 (1980) and J. Colloid Interface Sci. **87**, 492 (1982).

73. J. Appell and G. Porte, Europhys. Lett. **12**, 185 (1990).

74. S. J. Candau, E. Hirsch, R. Zana, J. de Physique **45**, 1263 (1984) and J. Colloid Interface Sci. **105**, 521 (1985).

75. H. Rehage, H. Hoffmann, J. Phys. Chem. **92**, 4712 (1988) and Faraday Discuss. Chem. Soc. **76**, 363 (1983).

76. G. Warr, L. Magid, E. Caponetti and C. Martin, Langmuir **4**, 813 (1988).

77. M. E. Cates and S. J. Candau, J. Phys.: Condens. Matter **2**, 6869 (1990), and references therein.

78. S. Safran, L. Turkevich and P. Pincus, J. de Physique Lett. **45**, L-69 (1984).

79. T. M. Clausen, P. K. Vinson, J. R. Minter, H. T. Davis, Y. Talmon and W. G. Miller, J. Phys. Chem. **96**, 474 (1992).

80. R. Nagarajan, J. Colloid Interface Sci. **90**, 477 (1982).

81. R. Messager, A. Ott, D. Chatenay, W. Urbach and D. Langevin, Phys. Rev. Lett. **60**, 1410 (1988).

82. M. Borkovec, J. Chem. Phys. **91**, 6268 (1989).

83. W. Helfrich, J. de Physique **47**, 321 (1986); see also L. Peliti and S. Leibler, Phys. Rev. Lett. **54**, 1960 (1985) for further discussion of k renormalization.

84. E. W. Kaler, A. K. Murphy, B. E. Rodriguez and J. A. N. Zasadzinski. Sci. **245**, 1371 (1989).

85. D. Lichtenberg and Y. Barenholtz, in Methods of Biochemical Analyses **33**, 337 (1988); D. Lichtenberg in *Handbook of Biological Membranes*, M. Shinitsiky, ed. (Balaban, Rehovot, 1992).

86. S. A. Safran, P. Pincus and D. Andelman, Science **248**, 354 (1990); see also H. Wennerström and D. M. Anderson in *Statistical Thermodynamics and Differential Geometry of Microstructured Material* (Springer, in press).

87. D. D. Lasic, F. G. Martin, J. M. Neugebauer and J. P. Krathovil, J. Colloid Interface Sci. **133**, 539 (1989).

88. A. Ben-Shaul, D. H. Rorman, G. V. Hartland and W. M. Gelbart, J. Phys. Chem. **90**, 5277 (1988).

89. R. Nagarajan, Langmuir **1**, 331 (1985).

90. M. M. Stecker and G. B. Benedek, J. Phys. Chem. **88**, 6519 (1984).

91. T.-L. Lin, S.-H. Chen, N. E. Gabriel and M. F. Roberts, J. Phys. Chem. **91**, 406 (1987).

92. See, e.g.: (a) P. Schurtenberger, N. A. Mazer and W. Känzig, J. Phys. Chem. **89**, 1042 (1985); (b) S. Almog, T. Kushnir, S. Nir and D. Lichtenberg, Biochemistry **25**, 2597 (1986) (see also Ref. 85); (c) P. Vinson, Y. Talmon and A. Watter, Biophys. J. **56**, 669 (1989) and references cited therein.

93. Y. Hendrikx, J. Charvolin and M. Rawiso, J. Colloid Interface Sci. **100**, 597 (1984) and J. M. Pope and J. W. Doane, J. Chem. Phys. **87**, 3201 (1987).

94. See, e.g., T. L. Hill, *An Introduction to Statistical Thermodynamics* (Addison-Wesley, Reading, 1960); D. A. McQuarrie, *Statistical Mechanics* (Harper and Row, New York, 1973).

95. P. van der Ploeg and H. J. C. Berendsen, Mol. Phys. **49**, 233 (1983); J. Chem. Phys. **76** 3271 (1982); O. Edholm, H. J. C. Berendsen and P. van der Ploeg, Mol. Phys. **2**, 379 (1983).

96. E. Egberts and H. J. C. Berendsen, J. Chem. Phys. **89**, 3718 (1988).

97. G. Cardini, J. P. Bareman and M. Klein, Chem. Phys. Lett. **145**, 493 (1988) and Phys. Rev. Lett. **60**, 2152 (1988).

98. J.-P. Ryckaert, M. L. Klein and I. R. McDonald, Phys. Rev. Lett. **58**, 698 (1987).

99. J. Harris and S. Rice, J. Chem. Phys. **89**, 5898 (1988).

100. J. Hautman and M. L. Klein, J. Chem. Phys. **91**, 4494 (1989) and **93**, 7483 (1990).

101. A. Biswas and B. L. Schurmann, J. Chem. Phys. **95**, 5377 (1991).

102. D. Brown and J. H. R. Clarke, J. Phys. Chem. **92**, 2881 (1988).

103. B. Smit, A. G. Schlijper, L. A. M. Rupert and N. M. van Os, J. Phys. Chem. **94**, 6933 (1990); B. Smit, Phys. Rev. A **37**, 3431 (1988); B. Smit, *Ph.D. Thesis*, Rijksuniversiteit Utrecht, The Netherlands (1990).

104. P. W. Pastor, R. M. Venable and M. Karplus, J. Chem. Phys. **89**, 1112 (1988).

105. M. Milik, A. Kolinski and J. Skolnick, J. Chem. Phys. **93**, 4440 (1990).

106. B. Jönsson and H. Wennerström, J. Phys. Chem. **91**, 3381 (1987).

107. M. Winterhalter and W. Helfrich, J. Phys. Chem. **92**, 6865 (1988) and **96**, 327 (1992) and references therein. See also M. Koslov and M. Winterhalter, J. de Physique **2**, 175 (1992).

108. H. N. W. Lekkerkerker, Physica A **159**, 319 (1989); D. J. Mitchell and B. W. Ninham, Langmuir **5**, 1121 (1989).

109. D. Stigter and K. A. Dill, Langmuir **2**, 791 (1986), **4**, 200 (1986), and K. A. Dill and D. Stigter, Biochemistry **27**, 3446 (1988).

110. S.-H. Chen and R. Rajagopalan, eds., *Micellar Solutions and Microemulsions* (Springer, New York, 1990).

111. B. Jönsson and H. Wennerström, in Ref. 110 and references cited therein.

112. S. Puvvada and D. Blankschtein, J. Chem. Phys. **92**, 3710 (1990) and in the Proceedings of the 8th Symposium on "Surfactants in Solution" (Gainesville, 1990); G. Briganti, S. Puvvada and D. Blankschtein, J. Phys. Chem. **95**, 8989 (1991). A recent extension of their molecular thermodynamic theory to mixed micellar solution has recently been presented by Puvvada and Blankschtein; see, e.g., the Proceedings of the Symposium on "Mixed Surfactant Systems" (Oklahoma, 1991).

113. S. Marcelja, J. Chem. Phys. **60**, 3599 (1974) and Biochem. Biophys. Acta **367**, 165 (1974).

114. B. Lemaire and P. Bothorel, J. Polym. Sci. Polym. Phys. Ed. **20**, 867 (1982) and Macromolecules **13**, 311 (1980).

115. D. W. R. Gruen and E. H. B. de Lacey, in *Surfactants and Solution*, K. L. Mittal and B. Lindman, eds. (Plenum, New York, 1984) Vol. 1, p. 559.

116. D. W. R. Gruen, J. Phys. Chem. **89**, 146, 153 (1985). See also J. Colloid Interface Sci. **84**, 281 (1981).

117. K. A. Dill and P. J. Flory, Proc. Natl. Acad. Sci. USA, **77**, 3115 (1980) and **78**, 676 (1981).

118. K. A. Dill and R. S. Cantor, Macromolecules **17**, 380, 384 (1984) and K. A. Dill, D. W. Koppel, R. S. Cantor, J. D. Dill, D. Bendedouch and S.-H. Chen, Nature **309**, 42 (1984).

119. K. A. Dill, J. Naghizadeh and J. A. Marqusee, Ann. Rev. Phys. Chem. **39**, 425 (1988).

120. A. Ben-Shaul, I. Szleifer and W. M. Gelbart, J. Chem. Phys. **83**, 3597 (1985); see also Ref. 10, p. 404 and Proc. Natl. Acad. Sci. USA **81**, 4601 (1984).

121. For a review, see A. Ben-Shaul and W. M. Gelbart, Ann. Rev. Phys. Chem. **36**, 179 (1985). See also A. Ben-Shaul, I. Szleifer and W. M. Gelbart, in *Surfactants in Solution*, K. L. Mittal, P. Bothorel, eds. (Plenum, New York, 1986), Vol. 4, p. 429.

122. I. Szleifer, A. Ben-Shaul and W. M. Gelbart, J. Chem. Phys. **83**, 3612 (1985), **85**, 5345 (1986) and **86**, 7094 (1987).

123. J. L. Viovy, W. M. Gelbart and A. Ben-Shaul, J. Chem. Phys. **87**, 4114 (1987).

124. I. Szleifer, A. Ben-Shaul and W. M. Gelbart, J. Phys. Chem. **94**, 5081 (1990).

125. I. Szleifer, D. Kramer, A. Ben-Shaul, W. M. Gelbart and S. A. Safran, J. Chem. Phys. **92**, 6800 (1990); I. Szleifer, D. Kramer, A. Ben-Shaul, D. Roux and W. M. Gelbart, Phys. Rev. Lett. **60**, 1966 (1988); A. Ben-Shaul, I. Szleifer and W. M. Gelbart in Ref. 136; I. Szleifer, *Ph.D. Thesis*, Hebrew University, Jerusalem (1988).

126. L. Steenhuizen, D. Kramer and A. Ben-Shaul, J. Phys. Chem. **95**, 7477 (1991).

127. F. A. M. Leermakers, J. M. H. M. Scheutjens and J. Lyklema, Biophys. Chem. **18**, 353 (1983) and F. A. M. Leermakers, *Ph.D. Thesis*, Agricultural University, Wageningen, The Netherlands (1988).

128. J. M. H. M. Scheutjens and G. J. Fleer, J. Phys. Chem. **83**, 1619 (1979).

129. P. J. Flory, *Statistical Mechanics of Chain Molecules* (Wiley-Interscience, New York, 1969).

130. J. P. Hansen and I. R. McDonald, *Theory of Simple Liquids* (Academic Press, San Diego, 1986), 2nd edition.

131. J. Seelig and W. Niederberger, Biochem. **13**, 1585 (1974).

132. W. Helfrich, Z. Naturforsch 693 (1973).

133. W. Helfrich, in *Physics of Defects*, Les Houches, Session XXXV, R. Balian, M. Kleman and J. P. Poirier, eds. (North Holland, Amsterdam, 1981).

134. E. A. Evans and R. Skalak, CRC Crit. Rev. Bioeng. **3**, 181 (1979).

135. A. G. Petrov and I. Bivas, Prog. Surf. Sci. **16**, 389 (1984).

136. J. Meunier, D. Langevin and N. Boccara, eds., *Physics of Amphiphilic Layers* (Springer, Berlin, 1987).

137. H.-P. Duwe and E. Sackmann, Physica A **163**, 410 (1990) and H.-P. Duwe, J. Kaes and E. Sackmann, J. de Physique **51**, 945 (1990).

138. C. R. Safinya, E. B. Sirota, D. Roux and G. S. Smith, Phys. Rev. Lett. **62**, 1134 (1989).

139. J.-M. DiMeglio, in Ref. 136; J.-M. DiMeglio, M. Dvolaitzki and C. Taupin, J. Phys. Chem. **89**, 871 (1985).

140. J. Meunier, J. de Physique Lett. **46**, L-1005 (1985); J. Phys.: Condens. Matter **2**, SA347 (1990).

141. P. G. DeGennes and C. Taupin, J. Phys. Chem. **56**, 2294 (1982).

142. See, e.g., J. F. Lennon and F. Brochard, J. de Physique **36**, 1035 (1975) and I. Bivas, P. Hanusee, P. Bothorel, J. Lalanne and O. Aquerre-Charid, *ibid.* **48**, 855 (1987).

143. K. Zeman, M. Engelhard and E. Sackmann, Eur. Biophys. J. **18**, 203 (1990) and K. Berndl, J. Kaes, R. Lipowsky, V. Seifert and E. Sackmann, Europhys. Lett. **13**, 659 (1990).

144. S. T. Milner and T. A. Witten, J. de Physique **49**, 1951 (1988).

145. Z.-G. Wang and S. A. Safran, J. de Physique **51**, 185 (1990) and J. Chem. Phys. **94**, 679 (1991).

146. R. S. Cantor, Macromolecules **14**, 1186 (1981).

147. W. G. McMillan and J. E. Mayer, J. Chem. Phys. **13**, 276 (1945).

148. (a) A. Ben-Shaul and W. M. Gelbart, J. Phys. Chem. **86**, 311 (1982); (b) W. M. Gelbart, A. Ben-Shaul, W. E. McMullen and A. Masters, J. Phys. Chem. **88**, 861 (1984).

149. L. Onsager, Ann. N.Y. Acad. Sci. **51**, 627 (1949).

150. B. Barboy and W. M. Gelbart, J. Stat. Phys. **22**, 709 (1979) and J. Chem. Phys. **71**, 3053 (1979).

151. W. M. Gelbart, W. E. McMullen and A. Ben-Shaul, J. de Physique **46**, 1137 (1985).

152. W. M. McMullen, W. M. Gelbart and A. Ben-Shaul, J. Chem. Phys. **82**, 5616 (1985).

153. G. Lasher, J. Chem. Phys. **53**, 4141 (1970).

154. W. M. McMullen, W. M. Gelbart and Y. Rosenfeld, J. Chem. Phys. **85**, 1088 (1986).

155. T. Odijk, J. de Physique **48**, 125 (1987).

156. J. Mahanty and B. W. Ninham, *Dispersion Forces* (Academic, New York, 1976).

157. S. Q. Wang, W. M. Gelbart and A. Ben-Shaul, J. Phys. Chem. **94**, 2219 (1990).

158. S. Q. Wang, J. Phys. Chem. **94**, 8381 (1990) and Macromolecules **25**, 1153 (1992); R. Bruinsma, W. M. Gelbart and A. Ben-Shaul, J. Chem. Phys. **96**, 7710 (1992).

159. M. E. Cates and M. S. Turner, Europhys. Lett. **11**, 681 (1990).

160. M. P. Taylor and J. Herzfeld, Langmuir **6**, 911 (1990) and Phys. Rev. A **43**, 1892 (1991).

161. D. J. Mitchell, G. J. T. Tiddy, L. Waring, T. Bostock and M. P. McDonald, J. Chem. Soc. Faraday Trans. 2 **79**, 975 (1983).

162. Y. Song, R. M. Stratt and E. A. Mason, J. Chem. Phys. **88**, 1126 (1988).

163. A. Stroobants, H. N. W. Lekkerkerker and D. Frenkel, Phys. Rev. A **36**, 2929 (1987) and Phys. Rev. Lett. **57**, 1452 (1986).

164. See: M. P. Taylor, R. Hentschke and J. Herzfeld, Phys. Rev. Lett. **62**, 800 (1989); A. M. Somoza and P. Tarazona, *ibid.* **61**, 2566 (1988); A. Paniewierski and R. Holyst, *ibid.* **61**, 2461 (1988); and references contained therein.

165. N. Boden, R. J. Bushby, C. Hardy and F. Sixl, Chem. Phys. Lett. **123**, 359 (1986).

166. M. E. Cates, D. Roux, D. Andelman, S. T. Milner and S. A. Safran, Europhys. Lett. **5**, 733 (1988).

167. See, for example, H. Yamakawa, *Modern Theory of Polymer Solutions* (Harper and Row, New York, 1971).

168. C. K. Bagdassarian, D. Roux, A. Ben-Shaul and W. M. Gelbart, J. Chem. Phys. **94**, 3030 (1991); C. Bagdassarian, *PhD. Thesis*, University of California, Los Angeles (1991).

169. Y. Hendrikx, J. Charvolin, P. Kekicheff and M. Roth, Liq. Cryst. **2**, 677 (1987).

170. Y. Hendrikx, J. Charvolin and M. Rawise, J. Colloid Interface Sci. **100**, 597 (1987).

171. Y. Hendrikx and J. Charvolin, J. de Physique **42**, 1427 (1981).

172. P. Ekwall, in *Advanced Liquid Crystals*, G. H. Braun, ed. (Academic Press, New York, 1975), Vol. 1.

173. T. R. Carale and D. Blankschtein, J. Phys. Chem. **96**, 459 (1992).

2

Micellar Growth, Flexibility and Polymorphism in Dilute Solutions

Gregoire Porte[1]

2.1 Introduction

When explaining to nonexperts what basic concepts are hidden behind the word "micelle," one usually draws on the blackboard a globular object consisting of surfactant molecules with polar heads oriented towards its surface and hydrophobic chains confined inside. As a matter of fact, this simple shape is very often what appears spontaneously when a small amount of surfactant is dissolved in water. In contrast, when turning attention to *concentrated* surfactant aqueous dispersion, a wide variety of different *ordered* phases is encountered [1,2]. They exhibit long range orientational alignment and spatial periodicities that are only compatible with micellar shapes distinctly different from that of the initial (i.e., low concentration) globules. From these observations, a major characteristic of surfactant aggregates appears: their extraordinary ability to accommodate very different shapes. It took quite a long time to realize, however, that this polymorphism also takes place in very dilute dispersions. The basic difficulty with dilute solutions is the challenge to characterize unambiguously the shape of aggregates. The average distance between aggregates is much larger than in the concentrated phases, and long range alignment and periodicities are generally absent. Conversely, this difficulty conceals a simplifying opportunity: since they are weak, the interaggregate interactions have a negligible effect on the thermodynamic stability of a given shape. The observed sequence of morphological transformations is then much easier to analyze and interpret since it is determined by *local, intra-aggregate*, mechanisms only.

In this chapter we shall describe briefly the two basic types of morphologies corresponding, respectively, to unidimensional and bidimensional

---

[1]Groupe de Dynamique des Phases Condensées, Université des Sciences et Techniques du Languedoc, 34095 Montpellier Cedex 05, France

growth of the initial globular micelles. Attention will be focused on very dilute dispersions in order to avoid the complications brought about by the intermicellar interactions occuring at higher concentrations–see Sec. 1.4 and Chap. 3 for a discussion of these latter points.

The case of unidimensional growth has been investigated thoroughly during the past fifteen years and an accurate and consistent picture of the so called "giant micelles" can now be drawn. The next section deals with experimental probes and the conditions of formation of the giant micelles (Sec. 2.2.1), with their structure (Sec. 2.2.2), and with the basic thermodynamics of their growth (Sec. 2.2.3). Sec. 2.3 is focused on their flexibility, with special emphasis on the analogy with high polymers.

The case of bidimensional growth (Sec. 2.4) in dilute solutions has received attention only in more recent years. Many of these results, in particular those obtained on very swollen lamellar phases (L_α) as described in Sec. 2.4.1 and swollen anomalous isotropic phases (L_3) (see Sec. 2.4.2) suggest promising fields for future investigations.

In Sec. 2.5 we introduce the basic concept that is thought to provide a unifying interpretation of the micellar polymorphism, i.e., the bending elasticity of the surfactant film (monolayer). Finally, the connection between micellar polymorphism and the formation of microemulsions is presented briefly in Sec. 2.6.

2.2   Globule to Rod Transformation

2.2.1   SCATTERING PROBES AND CONCENTRATION EFFECTS

The formation of a giant unidimensional micelle results from the condensation of several globules to form a large flexible worm like aggregate. Any physical quantity sensitive either to the molecular weight or to the shape anisometry of the objects in suspension is thus well suited to reveal the morphological transformation. Amongst these, the viscosity is certainly the most sensitive one. An example is given in Fig. 2.1 where the full circles represent the relative viscosities (ratio of the viscosity of the solution to that of the pure aqueous solvent) of suspensions of cetylpyridinium chloride (CPCl) in brine (0.2M NaCl) at different concentrations [3]. These points fall on a straight line with slope ≈ 3, consistent with nearly spherical shape for the globular micelles (a slope of 2.5 is expected for exact spheres). Addition of small amounts of alcohol induces extremely steep increase of the viscosity (full triangles in Fig. 2.1), which is totally incompatible with globular shape for the (mixed) micelles. Other examples of micellar growth, indicated by steep variations of the rheological properties of the solution, are reported elsewhere [4,5,6]. However, in spite of its extreme sensitivity and easy availability, viscometry has not been widely used, mainly because

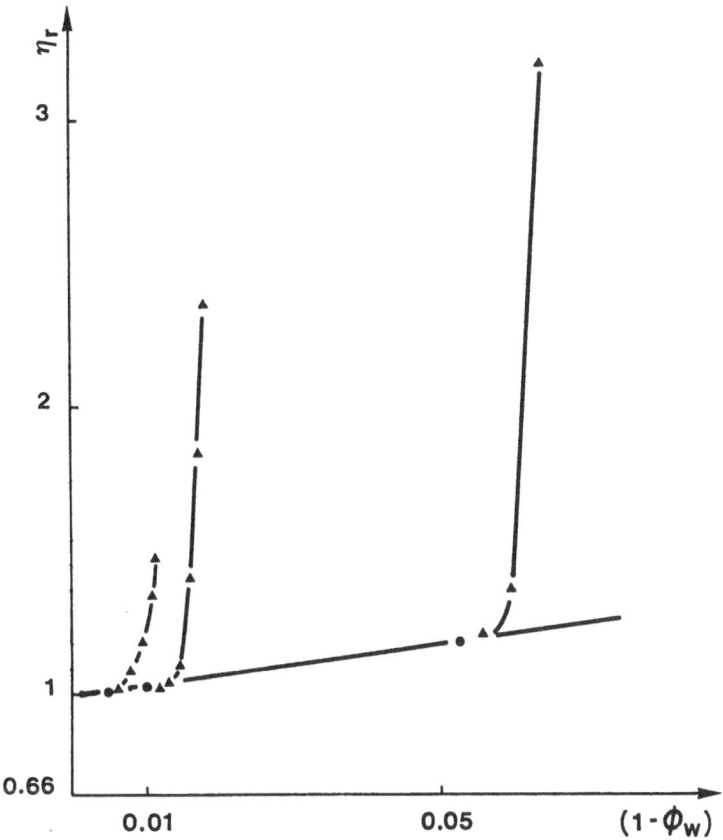

FIGURE 2.1. Relative viscosity η_r of micellar solutions of the system CPCl/hexanol/brine (0.2M NaCl) versus the volume fraction of active matter $(1 - \phi_w)$. Full circles: binary mixtures (CPCl/brine). Full triangles: progressive addition of hexanol. From [3].

of the lack of a reliable theory allowing quantitative data treatment in the case of large objects with complex shapes.

On the other hand, molecular weights of micelles are readily measured by the same methods that are normally used for true macromolecules (see, for example, Eq. (1.1)). Indeed, most existing data revealing micellar growth have been obtained by the use of light scattering.

In the most common geometry for a light scattering experiment [7,8] incident plane polarized monochromatic (wave length λ_0) light illuminates the sample and the component $I(\vec{q})$ of the scattered intensity, polarized perpendicular to the scattering plane, is measured at a distance R_s far from the sample and at an angle θ off the incident beam direction (Fig. 2.2). For

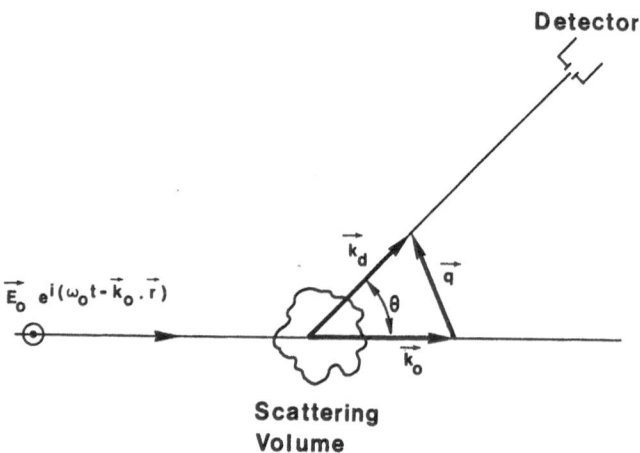

FIGURE 2.2. Geometry for a light scattering experiment.

this geometry, the magnitude q of the scattering wavevector $\vec{q}$ is given by:

$$q = \frac{4\pi n}{\lambda_0} \sin(\theta/2) \tag{2.1}$$

where n is the refractive index of the medium.

The average intensity scattered by the sample is:

$$I(\vec{q}) = \left\langle \int \int a(\vec{r})a^*(\vec{r'}) \exp(i\vec{q} \cdot (\vec{r} - \vec{r'})) d\vec{r} d\vec{r'} \right\rangle \tag{2.2}$$

where $a(\vec{r})$ is the amplitude of the scattered electric field due to polarization at point $\vec{r}$ and the integrations run over the scattering volume V. The angular brackets denote a thermal average over positions.

For a sample isotropic at large scale, $I(\vec{q})$ does not depend on the direction of the wave vector; so $I(\vec{q}) = I(q)$ were q is given by (2.1). If the medium consists of separate, finite-size, particles with *no correlations of their relative orientations*, (2.2) can be factorized as follows:

$$I(q) = K_0 \frac{I_0}{R_s^2} \cdot C \cdot M \cdot P(q) \cdot S(q) \tag{2.3}$$

I_0 is the intensity of the incident radiation and K_0 is

$$K_0 = \frac{2\pi^2 n^2}{\lambda_0^4} \left(\frac{dn}{dc}\right)^2 \tag{2.4}$$

M is the mass of the particle and C is its concentration (mass/volume). The two q-dependent factors $P(q)$ and $S(q)$ are, respectively, the form factor of the particle and the structure factor, related to their average distribution in space. (See also the discussion in Sec. 1.1.)

$P(q)$ represents the *intraparticle* interferences. It is simply equal to the intensity $I_1(a)$ scattered by an isolated particle normalized so that $P(q) = 1$ in the small q limit. Correspondingly it is related to the distribution function $\gamma_1(r)$ of distances *inside the particle* according to:

$$P(q) = \frac{1}{v} \int_v 4\pi r^2 (\gamma_1(r) - 1) \frac{\sin(qr)}{qr} dr \qquad (2.5)$$

Here v is the volume of the micellar particle. $S(q)$, on the other hand, represents the *interparticle* interferences and is defined by:

$$S(q) = 1 + N_p \int_0^\infty 4\pi r^2 (g(r) - 1) \frac{\sin(qr)}{qr} dr \qquad (2.6)$$

where N_p is the number of particles per unit volume and $g(r)$ is the radial distribution function with r the distance of *interparticle* center separation.

In sufficiently dilute solutions, and especially in the presence of added salt that screens the electrostatic repulsions between charged micelles, the interparticle correlations are weak and $S(q) \approx 1$. Furthermore, in light scattering experiments, λ_0 falls in the range of 4 to 8.10^3Å so that the accessible q values (see (2.1)) are small ($q \approx 10^{-4}$ to 5.10^{-3}Å^{-1}). q^{-1} is then usually large even compared to the sizes of giant micelles. In this limit, it is convenient to develop the $\sin qr/qr$ factor in (2.5) in a power series as $1 - q^2 r^2/3! + q^4 r^4/5! + \cdots$. $P(q)$ then becomes:

$$P(q) = 1 - \frac{q^2 \overline{r^2}}{3!} + \frac{q^4 \overline{r^4}}{5!} \qquad (2.7)$$

with

$$\overline{r^n} = \frac{1}{v} \int_0^\infty 4\pi r^2 \gamma_1(r) r^n dr \qquad (2.8)$$

For centrosymmetric particles, $\overline{r^2}$ is simply related to the radius of gyration $\langle R_G^2 \rangle$:

$$\overline{r^2} = 2 < R_G^2 > \qquad (2.9)$$

So, finally, in the low q, high dilution limit, $I(q)$ can be approximated by:

$$I(q) \; \alpha \; CM \left(1 - \frac{q^2 \langle R_G^2 \rangle}{3}\right) \qquad (2.10)$$

This result is sometimes expressed in the equivalent, asymptotic, form:

$$I(q) \; \alpha \; CM \exp\left(-\frac{q^2 \langle R_G^2 \rangle}{3}\right) \qquad (2.11)$$

which is the well known Guinier approximation [10].

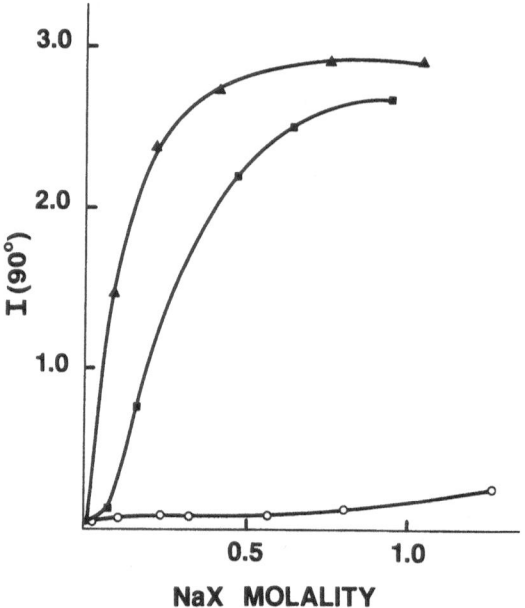

FIGURE 2.3. Scattering at 90° by solutions with a constant CPBr concentration (0.01 M) and variable NaX concentration: O, $X = BrO_3^-$; □, $X = Br$; Δ, $X = ClO_3^-$. From [9].

Two separate characteristic quantitites of the micellar object can be extracted in the above way from the $I(q)$ profile. The first is M, the mass of the micelle, obtained by extrapolating $I(q)$ to $q = 0$. The second is $R_G (\equiv \langle R_G^2 \rangle^{1/2})$, its radius of gyration, derived from the angular dependence of $I(q)$.

Figures 2.3 and 2.4 have been taken from Ref. 9 as illustrative examples of spectacular micellar growth, as revealed by the dependence of the scattered light intensity and the scattered light angular dissymmetry on the molality of different added salts, in a solution of CPBr micelles at constant (low) surfactant concentration. These data indicate an aggregation number growing from 100Å when NaBrO$_3$ is the added salt, up to about 11,000Å when 0.5M NaClO$_3$ is added; the dissymmetry data show a corresponding increase of R_G up to several 10^2Å.

The *time* fluctuations of the scattered light intensity can be analyzed using photon beating spectrometry. The normalized temporal autocorrelation function $g(\tau)$ of the intensity can be experimentally determined [11,12] according to

$$g(q, \tau) = \frac{\langle I(q, t) I(q, t + \tau) \rangle}{\langle |I(q, t)|^2 \rangle} \qquad (2.12)$$

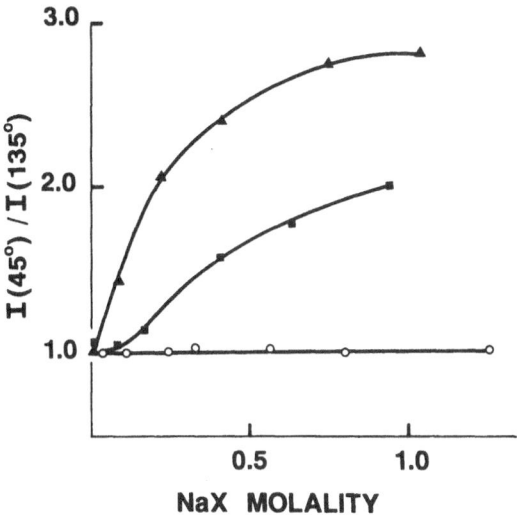

FIGURE 2.4. Angular disymmetry of the light scattered by solutions with a constant CPBr concentration (0.01 M) and variable NaX concentration. Same notations as in Fig. 2.3. From [9].

At sufficiently low q (i.e., $2\pi/q$ larger than the mean interparticle distance) it describes the dynamics of sinusoidally varying concentration fluctuations of wave vector q in the solution. They are expected to decay via a simple diffusion process characterized by a diffusion coefficient D_c:

$$g(q,\tau) \, \alpha \exp(-D_c q^2 \tau) \tag{2.13}$$

In sufficiently dilute dispersion with no strong interparticle interaction, D_C can be identified with the free particle diffusion coefficient D_0:

$$D_c = D_0 = \frac{k_B T}{6\pi\eta R_H} \tag{2.14}$$

where the hydrodynamic radius R_H is the radius of a spherical particle having the same diffusion coefficient D_0 as the investigated particle. R_H is thus related to both the size and shape of the micelles. As an example, Fig. 2.5 gives the variations with temperature of R_H of SDS micelles in solutions with fixed surfactant concentration at different NaCl salinities [13]. At low salt and high temperature, the hydrodynamic radius is essentially constant, equal to 21Å which is the geometrical size of globular SDS micelles. Very strong increase of the size at higher salinities and lower temperature are reflected by the variations of R_H up to about 200Å for 0.6M NaCl and 25° C.

These techniques have been extensively used in the literature to characterize the formation of giant micelles in a large variety of systems [13–26].

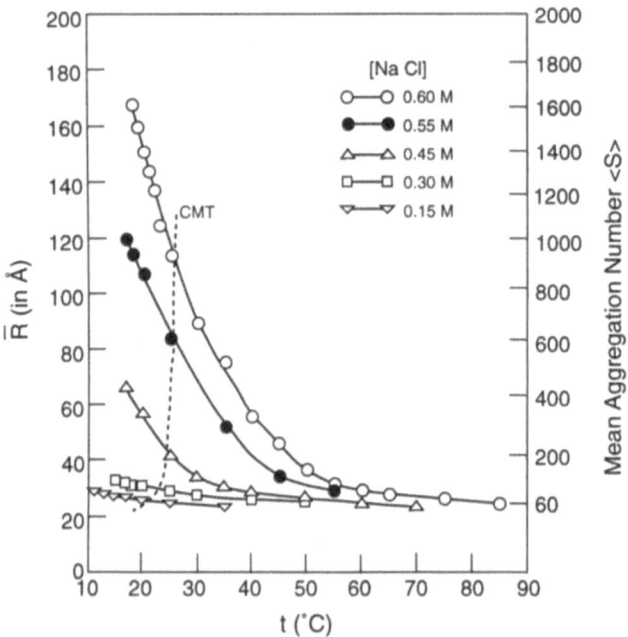

FIGURE 2.5. Temperature dependence of mean hydrodynamic radius for 6.9×10^{-2}M SDS in various NaCl concentrations (0.15M–0.6M). Vertical axis on right provides a scale of mean aggregation numbers. From [13].

From a systematic survey, the general trends can be summed up as follows:

1. Quite generally, usually for single-chain ionic surfactants, the micelles are globular in dilute binary (i.e., pure surfactant in water) solution. Only few exceptions have been reported [19].

2. For some systems micellar elongation can be induced by increasing the surfactant concentration [25]. But in this case, the shape transformation generally occurs at high surfactant concentration ($\phi \gg 0.1$), not far from the conditions of formation of the hexagonal phase. The micellar elongation can often be induced by addition of salt into an initial solution of surfactant in pure water. Classical examples are SDS + NaCl [13–14] and CTAB + NaBr [9]. The shape transformation depends strongly on its chemical nature; see for instance Fig. 2.3. This specificity is discussed in [26].

3. Other additives, in particular short chain alcohols (from butanol to

decanol), are also very efficient in promoting micellar elongation at high dilution [3].

4. When the micelles are large, their size increases with increasing added salt concentration and decreases with increasing temperature, at constant surfactant concentration. At constant temperature and salt concentration, *the micellar size increases monotonically with increasing surfactant concentration* (see, for example, Eq. (1.45)).

The above described experimental methods essentially probe either macroscopic (viscosity) or large spatial scale (light scattering at small wave vector) properties of the solution. Although the probed collective behavior of the aggregates indeed reflects their individual structural transformations, these methods do not provide *direct* structural characterization of the micelles.

In order to check one particular aspect of a structure associated with a characteristic length Λ, it is crucial to collect scattering data on a sufficiently large wave vector range around the corresponding q value $q = \Lambda^{-1}$. As we have just seen, the total average size (R_G) of giant micelles is often large enough to be characterized through data obtained at very low q by light scattering. As we shall see below the formation of giant micelles corresponds to a unidimensional growth of the aggregates. To observe this essential feature, we must investigate the structure at a length scale of the order of the diameter of the cylinder, typically a few nanometers. The appropriate q range thus cannot be reached by light scattering; incident radiation of smaller wavelength λ_0 must be utilized, notably X rays ($\lambda_0 \approx 1$ to 2Å) or neutrons ($\lambda_0 \cong 2$ to 10Å). The pioneering work in this field is that of Reiss-Husson and Luzzati [27] who were the first to demonstrate the unidimensional structure of giant micelles using X-ray scattering.

2.2.2   LOCAL STRUCTURE

The basic principles of X ray and neutron scattering are essentially the same as those of light scattering, so that the set of relations (2.1)–(2.11) still holds subject to only minor modifications [28–30].

The refractive index n is very close to 1 for these radiations and q is simply expressed:

$$q = \frac{4\pi}{\lambda_0} \sin(\theta/2) \qquad (2.15)$$

For X rays, the prefactor in (2.3) takes the form

$$K_0 = I_e(z_m - v_m\rho_0)^2 \qquad (2.16)$$

where I_e is the "effective cross section of the electron" or Thomson factor ($I_e = 7.9 \times 10^{-26} cm^2$), z_m is the number of electrons in the particle, v_m

is the specific volume of the particle, and ρ_o is the electron density of the solvent. For neutrons, we have a similar expression for K_0:

$$K'_0 = (b_m - v_m \rho'_0)^2 \qquad (2.17)$$

where b_m is the scattering length of the particle and ρ'_0 is the scattering length density of the solvent.

For experimental conditions where the micelles maintain a globular shape, their dimensions in all three directions of space are typically 20 to 40Å. The condition of applicability of the Guinier approximation—Eq. (2.11) ($qR_G < 1$)—falls within the q-range accessible to a standard X ray scattering set-up. Indeed, this data treatment has been used many times to derive the size of globular micelles. On the other hand, when the micelles are very large ($R_G \cong 100$Å), the Guinier regime is shifted towards the low q's, and most of the scattering data collected on a standard set-up is strongly affected by the higher order terms in (2.7). However, another asymptotic form for $I(q)$ in this q-range can be derived when at a local scale the particles have the shape of a thin stiff rod. Recalling the definition of $\gamma_1(r)$ in (2.5), one can show that

$$\gamma_1(r) \; \alpha \; r^{-2} \qquad (2.18)$$

for r values in the range $r_c < r < l_p$ where r_c is the radius of the cross section of the rod and l_p is the maximum length below which the rod appears stiff, i.e., the *persistence length*. This r^{-2} dependence of $\gamma_1(r)$ implies, via the Fourier transform of (2.5), a q^{-1} dependence of $P(q)$ and hence $I(q)$ in the corresponding q range: $l_p^{-1} < q < r_c^{-1}$. Conversely this dependence is characteristic of locally rod-like objects. In fact, expression (2.18) is obtained upon neglecting the finite thickness of the rod. Taking the thickness into account introduces a correction factor in the q^{-1} dependence of $I(q)$ which is well approximated by an exponential decay in the range $q < r_c^{-1}$. More precisely (the calculations are detailed in Ref. 29) the scattered intensity varies as

$$I(q) = \pi q^{-1} C M_L \exp\left(-\frac{r_c^2 q^2}{4}\right) \qquad (2.19)$$

where M_L is the mass per unit length of the thin rod.

In the same manner, if the micelles have the shape of large disks, the asymptotic dependence of $\gamma_1(r)$ at large r would be r^{-1} implying a q^{-2} dependence for $I(q)$. But here, again, finite thickness effects introduce an exponential correction factor so that we obtain in this case [29]

$$I(q) = 2\pi^2 q^{-2} C M_A \exp\left(-\frac{q^2 e^2}{12}\right) \qquad (2.20)$$

where M_A is the mass per unit area of the disk and e its diameter.

Comparing Eqs. (2.11), (2.19) and (2.20) shows how scattering data in the appropriate q range can help to discriminate between the three basic types of aggregation: globular micelles, unidimensional growth, and bidimensional growth (see the detailed discussions in Chap. 1). In practice the $I(q)$ profile is plotted as:

$-\ln I(q)$ vs. q^2 (globular)

$-\ln qI(q)$ vs. q^2 (rods)

$-\ln q^2 I(q)$ vs. q^2 (plates)

In principle, a linear decay is obtained in only one of these plots, indicating one dominant type of aggregation. Then the comparison of the geometrical dimensions $(R_G, r_c$ or $e)$, as derived from the slope of the profile, with the specific mass $(M, M_L$ or $M_A)$, as derived from extrapolation of the plot (in absolute units) to zero-q, helps to check further the consistency of the model of aggregation. The main practical limitation of this approach is that it assumes $S(q) = 1$ in the investigated q-range. Hence only very dilute dispersions can be studied according to this procedure.

A wide variety of micellar systems, where the formation of giant micelles was suspected from other macroscopic measurements, have been studied by the Bayreuth group using small angle neutron scattering [31–35]. The conclusion of unidimensional growth has been derived in a great majority of the systems investigated. The data could be accurately fitted by the exponential approximation appropriate for rods—see (2.19)—with radius and mass per unit length mutually consistent. In only one case [34] (diethylammonium perfluorononanoate in D_2O) could the scattering function *not* be attributed to rod-like aggregates, agreeing rather with the theoretical function (2.20) calculated for platelets (or, possibly, vesicles).

Similar studies of micelles in high ionic strength solutions have also been performed by various authors [36-39]. The globule to rod transformation induced by increasing concentrations of salt in dilute SDS/brine (NaCl) system was demonstrated by Cabane *et al.* [36]. The transformation occurs once the NaCl concentration exceeds about 0.5M, as suggested by an earlier, light scattering study, by the MIT group [14]. The experimental approximation (2.19) was found appropriate for samples with salinities 0.6M and 0.8M, whereas large deviations (indicating globules) were observed for the 0.1M NaCl sample (Fig. 2.6).

The same shape transformation—but induced by addition of hexanol to CPCl brine (0.2M NaCl)—was confirmed via the same approach [37] (see viscosity data on this system in Fig. 2.1). In the absence of hexanol, the scattering data are accurately fitted by the Guinier approximation appropriate for globules. An average dry radius $r_s = 26.5$Å and mass $M = 6 \times 10^{-20} g$ are derived, both consistent with an average aggregation number $n_g = 115$. After addition of hexanol, the Guinier plot for globules becomes inappropriate, whereas good fits were obtained with the exponential approximation for rods with radius 20Å and mass per unit length 1.2×10^{-13} g/cm. These values are consistent with the radius of

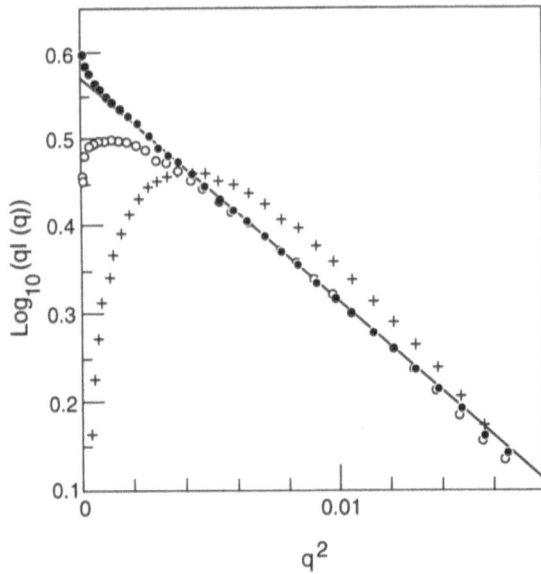

FIGURE 2.6. Scattering curves plotted in the appropriate representation for cylinders, for 2% SDS solutions at 25°C with various ionic strengths. Crosses: NaBr 0.1M; Circles NaBr 0.6M; Dots: NaBr 0.8M. From [36].

the cylinders in the hexagonal phase of this system prevailing at very high surfactant concentrations.

In a recent work, Quirion and Magid [38] showed that the formation of long micelles in aqueous CTAB solutions occured once the NaBr concentration exceeds about 0.04M; the radius of the cylinders is then about 23Å which is again consistent with the X ray measurements by Ekwall et al. [1] in the hexagonal phase of CTAB/H_2O. Similar formation of long micelles in short chain lecithin solutions has been reported by Chen et al. [39].

We note finally that the (improbable) possibility that long micelles could be formed by linear aggregation of globular micelles (retaining their own shape) into a "pearl necklace" has been ruled out by high q neutron scattering data, as shown by Cabane et al. for the system SDS/brine [36].

2.2.3 THERMODYNAMICS OF THE GROWTH

The main purpose of a theory of micellar growth is to reproduce the sensitive variations of the average size with the externally tunable experimental parameters: surfactant concentration, additive (salt or alcohol) concentration, and temperature.

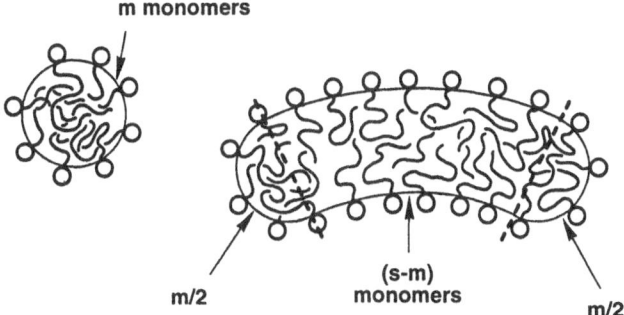

m monomers

m/2

(s-m)
monomers

m/2

FIGURE 2.7. Structure of a spherical micelle with m monomers and of a sphero cylinder with aggregation number s.

The basic version of the theory, for dilute solutions, has been worked out independently by the MIT group [22] and by the Canberra group [40]. This version neglected one essential feature of the large micelles structure: their flexibility. The elasticity of giant micelles is discussed in detail in the next section where emphasis is put on the analogies between solutions of giant micelles and of high polymers. Subsequently the theory was extended in order to account for micellar flexibility and the possibility for closed rings to form spontaneously [41]. We briefly describe it here using the same concepts as in the Chap. 1 discussion.

A large micelle with aggregation number s is assumed to consist of $s - m$ monomers in the cylindrical body and $m(= \text{constant})$ monomers in the two (more-or-less) hemispherical end caps (Fig. 2.7).

The standard free energy of any one given monomer depends only on its local environment (predominance of short range forces). Thus all monomers residing in the cylindrical body bring the same contribution $\tilde{\mu}^o_{cyl}$ to the standard free energy μ^o_s of the s-micelle. In contrast, the monomers in the hemispherical caps experience a different environment (different local curvature of the polar heads surface) and their contribution $(\tilde{\mu}^o_{sph})$ to μ^o_s is thus somewhat different. Then, as discussed in Sec. 1.2.4.1, μ^o_s takes the simple form:

$$\mu^o_{s,rod} = (s - m)\tilde{\mu}^o_{cyl} + m\tilde{\mu}^o_{sph} = s\tilde{\mu}^o_{cyl} + k_B T \delta, \ s > m \qquad (2.21)$$

where the constant term $\delta = m(\tilde{\mu}^o_{sph} - \tilde{\mu}^o_{cyl})/k_B T$ represents the excess standard free energy due to the end caps. So, when two micelles with aggregation numbers s_1 and s_2 combine together to form a larger $(s_1 + s_2)$ micelle the standard free energy of the dispersion is changed by an amount equal to $(-kT\delta)$, and one expects the $kT\delta$ excess term to be the driving force of the growth (see discussion in Chap. 1).

This is, however, another way to gain the excess energy [41]. It is simply to bend homogeneously the long micelle and stick the ends to each other so that a closed ring is obtained. The price paid in this process is the extra elastic energy involved in the homogeneous bending of the cylinder. It can be shown [41] that this elastic term decreases quickly with the micellar size (like s^{-1}) and can be safely neglected provided that some minimum aggregation number $m' > m$ is introduced for the closed rings. Accordingly we write:

$$\mu^0_{s,ring} = s\tilde{\mu}^o_{cyl} , \quad s > m' \tag{2.22}$$

Eqs. (2.21) and (2.22) give the standard contribution of each type of micelle, open rods and closed rings.

Since we deal with the equilibrium size distribution we must also consider possible internal entropic contributions that may vary with s and in particular the contribution related to the various bent configurations of the flexible cylinders. Assuming that the bending fluctuations of two parts $\tilde{AB}$ and $\tilde{BC}$ of one long $\tilde{AC}$ micelle are independent we have:

$$\Omega(\tilde{AC}) = \Omega(\tilde{AB}) \cdot \Omega(\tilde{BC}) \tag{2.23}$$

where the Ω's are the numbers of configurations. Then for an open rod s-micelle we have:

$$\Omega_{rod}(s) = \omega_o^s \quad (\omega_o = \text{constant}) \tag{2.24}$$

For a *closed* ring this number of configurations is drastically diminished by the fact that one end point, initially randomly distributed in a volume of the order of R^3 (where R^2 is the square average end-to-end distance), is now stuck to the other end. This problem has been discussed in detail in the literature, in the context of polymers in solution [42]. $\Omega(s)$ for closed rings takes the asymptotic form:

$$\Omega_{ring}(s) \simeq s^{-5/2} s_c^{-3/2} \omega_o^s. \tag{2.25}$$

where s_c is a constant term related to the actual flexibility of the cylinder (it is, in fact, the number of monomers in one persistence length of the cylinder, as introduced in the next section).

The respective mole fractions X_s(rods) and X_s(rings) of s-micelles of each type are just the probabilities of formation of these objects at a given place. They involve three factors: the probability that n monomers come to close vicinity, a Boltzmann factor related to the energy of formation of the micelle, and the number of distinguishable bent conformations. So:

$$X_s(rods) = (X_1)^s \exp\left(s\frac{\mu_1^o - \tilde{\mu}^o_{cyl}}{k_B T}\right) \exp(-\delta) \cdot \omega_o^s \tag{2.26}$$

and:
$$X_s(rings) = (X_1)^n \exp\left(s\frac{\mu_1^o - \tilde{\mu}_{cyl}^o}{k_BT}\right) s_c^{-3/2} s^{-5/2} \omega_o^s \tag{2.27}$$

where X_1 and μ_1^o are the mole fraction and the standard chemical potential of the free monomers in solution. Eqs. (2.26) and (2.27), together with the relation expressing the conservation of the total amount of surfactant X,

$$X = \sum_{s=m}^{\infty} sX_s(rods) + \sum_{s=m'}^{\infty} sX_s(rings) \tag{2.28}$$

(where we neglect X_1, at practical concentrations $X \gg$ CMC), entirely determine the size distribution of the micelles once X and δ are given.

Introducing
$$Y = X_1\omega_o \exp\left(\frac{\mu_1^o - \tilde{\mu}_{cyl}^o}{k_BT}\right) \tag{2.29}$$

we obtain:
$$X = \exp(-\delta) \sum_m^{\infty} sY^s + s_c^{-3/2} \sum_{m'}^{\infty} s^{-3/2} Y^s \tag{2.30}$$

The convergence of the first infinite summation requires $Y < 1$. Then the second infinite summation is strictly bounded from above by

$$s_c^{-3/2} \sum_{m'}^{\infty} s^{-3/2} \tag{2.31}$$

Evaluating this term with values of s_c and m' compatible with the most flexible micelles (see next section) shows that *the population of closed rings is always small* and can be neglected at all practical concentrations ($X \gg$ CMC). The basic reason for this is the entropically unfavorable reduction of the number of configurations for a closed ring compared to that of an open rod—this effect over-compensates the energetic gain $kT\delta$ associated with the ring formation. Further analysis of the first infinite summation in (2.30) then allows us to derive easily the micellar size distribution (see Sec. 1.2). Two extreme situations must be distinguished, corresponding respectively to small globules and to long rods.

If the excess term δ is small and globular ($X \exp \delta < m^2$) the micelles remain small ($\langle s \rangle \approx m$ where the brackets stand for number average), and the size distribution is narrow. If δ is large ($X \exp \delta > m^2$) micellar growth takes place, i.e., $\langle s \rangle$ increases strongly with X (see Eq. (1.45)):

$$\langle s \rangle \approx \sqrt{X \exp \delta} \tag{2.32}$$

The size distribution is extremely broad with an exponential dependence on s as indicated in (2.26); more precisely we find

$$X_s(rods) = \frac{X}{\langle s \rangle (\langle s' \rangle + 1)} \left[\frac{\langle s \rangle}{\langle s' \rangle + 1}\right]^{s-m} \tag{2.33}$$

where $\langle s' \rangle$ is defined by $\langle s' \rangle = \langle s \rangle - m$ (see also Sec. 1.2.4).

So, consistent with experimental observations, the theory predicts that extremely large micellar growth can take place provided the experimental conditions determine a sufficiently large value for δ. However, as long as the intermicellar interactions are negligible, the micellar size remains finite no matter how large δ. This is indeed relevant to the well known fact that a true phase transition cannot occur in a purely unidimensional system (see Chap. 1). This is in contrast with bidimensional aggregation for which Israelachvili *et al.* have worked out the basic theory [40]. For disklike aggregates the excess term in the standard free energy arises form the contribution of the rim of the disk:

$$\mu_s^o = s\tilde{\mu}_{disk}^o + \delta' k_B T s^{1/2} \qquad (2.34)$$

which leads to

$$X = \sum_{s=m}^{\infty} s Y^s \exp\left(-\delta' s^{1/2}\right) \qquad (2.35)$$

For a sufficiently large value of δ', the series will converge to a value $X \ll 1$ for $Y = 1$. Once this concentration is exceeded the amphiphiles will assemble spontaneously into infinite bilayers. Experimental illustrations of bidimensional growth leading to the formation of infinite bilayers are detailed in Sec. 2.4.

The above theory involves the very restrictive simplifying assumption that bending fluctuations of different parts of one given elongated micelle are statistically independent. Because of excluded volume interactions, this is certainly not exact when the micelle is very large and has the shape of a random coil [42]. Also, the derivation of Eqs. (2.26) and (2.27) implies that different micelles are far apart and do not interact significantly with each other. More realistically one expects that at practical concentrations, when the micelles are very large, there are many entanglement points between different Gaussian coils.

A more rigorous approach [43–46] addressing a similar problem where a unidimensional aggregation process is involved (fluid to viscous transition at high temperature in liquid sulphur [47–49]) has been derived recently. This approach employs lattice gas models, high temperature expansion of the partition function and renormalization group calculations. The results obtained happen to agree roughly with the prediction of the simple theory. At the level of accuracy of the presently available experimental studies, the assumption of the simple theory seems justified.

The first systematic attempt to test the theory has been performed by the MIT group [14–22]. These authors have measured, using quasi-elastic light scattering, the mean value R_H and the variance σ_H of the hydrodynamic radius distribution of SDS micelles at various surfactant concentrations(X), salinities of the brine ($H_2O+NaCl$), and temperature T. Using a simple geometrical description of the micellar shape they derived the mean value $\langle s \rangle$

and the polydispersity of the aggregation number distribution from $\langle R_H \rangle$ and σ_H data. The main prediction of the theory—Eqs. (2.32) and (2.33) above—could then be checked and confirmed, in particular the square root dependence of $\langle s \rangle$ on X at given δ. Later on [21], the same experimental procedure was applied to a different system–cetylpyridinium bromide/brine (NaBr), and similarly good agreement was found. More recently [39], a neutron scattering study of micellar growth in solutions of short chain lecithins also revealed comparably good agreement with theory.

The theory outlined above helps us to understand micellar growth in terms of reversible polymerization [50] (i.e., as a chemical association-dissociation process; see Chap. 1). Note, however, that it has only a limited predictive power. Two parameters—X and δ—appear in expression (2.32): X is indeed accurately known in well controlled experimental conditions, but δ is introduced (cf. (2.21)) phenomenologically. Its actual value depends on the details of the environment of the monomers in the micelle. A microscopic description of the forces between monomers is presently out of reach. Furthermore, although δ does not depend on X at high dilution, its dependence on the other parameters defining the experimental conditions cannot be predicted a priori.

Conversely, however, one can interpret data on micellar size in terms of the growth parameter δ, based, e.g., on (2.32). The purpose of the authors [21,22,26] who have performed this type of analysis was to get new insights into the thermodynamic factors governing the globule to rod transformation. From the "measured" dependence of δ on temperature Missel et al. [22] could derive the "enthalpic" and "entropic" contributions of the hydrophobic interactions and concluded that, at variance with prevailing views, the enthalpic contribution was largely dominant. In the same manner, the dependence of δ on salt concentration is in principle indicative of the importance of electrostatic interactions between monomers. However, other authors [21,26] obtained equally good agreement with experimental data starting from different descriptions of the ionic environment of the micelle, indicating the incompleteness of the theory.

As a last remark, note that the theory is derived for the case of micelles consisting of only one surfactant species. As we saw above, micellar growth can be induced by addition of short chain alcohols. An extension of the theory to mixed micelles has been worked out in Ref. 51, which predicts that $\langle s \rangle$ should then scale as $X^{5/2}$ (for other treatments of mixed aggregates, see [52]). As discussed in Chap. 1, several attempts to generalize the theory to situations where interactions between micelles become important have also been reported [25,53], but no experimental verification of these predictions has yet been reported (see Sec. 1.4 for discussion of interaction effects).

2.3 Polymerlike Behavior

2.3.1 FLEXIBILITY

The possibility that long micelles could be highly flexible was first considered by Stigter on the basis of an analysis of viscosity data obtained for solutions of dodecylammonium chloride in NaCl brine [15]. However, this question remained controversial for quite a while, as reviewed in [54].

The finite resistance to local bending of a thin thread is best represented by its elastic rigidity modulus K which relates the excess bending free energy dE to the local curvature c of the thread:

$$dE = \frac{1}{2}Kc^2 dl \tag{2.36}$$

c is the inverse of the radius of curvature at the point of coordinate l along the thread. Due to thermal motions, the very long micelle will experience a large number of different bent configurations. Simple statistical methods allow us to calculate the mean values of quantities describing its curvature [55]. In particular, introducing $\vec{t}(l_1)$ and $\vec{t}(l_2)$ as the unit vectors along the tangent to the thread at two points separated by a distance $\Delta l = (l_1 - l_2)$, one obtains:

$$\langle \vec{t}(l_1)\vec{t}(l_2) \rangle = \langle \cos\theta(\Delta l) \rangle = \exp\left(-\Delta l \frac{k_B T}{K}\right) \tag{2.37}$$

Here, $\langle \cos\theta(\Delta l) \rangle$ is the orientational correlation function along the thread. Its exponential dependence on Δl allows us to introduce naturally the orientational correlation length $l_p = K/k_B T$ which is more familiar as the *persistence length*:

$$\langle \cos\theta(\Delta l) \rangle = \exp(-\Delta l/l_p) \tag{2.38}$$

From Eq. (2.37) one derives other mean quantities such as the mean square distance $R(\Delta l)$ (measured in a straight line) between two points l_1 and l_2

$$\langle R^2(\Delta l) \rangle = \left| \int_{l_1}^{l_2} \vec{t}(l)dl \right|^2 = 2l_p(\Delta x - 1 + \exp(-\Delta x)) \tag{2.39}$$

where $\Delta x = \Delta l/l_p$. From (2.39), one then computes the radius of gyration $\langle R_G^2 \rangle^{1/2}$ that characterizes the overall size of the thread of contour length L:

$$\langle R_G^2 \rangle = L^2 \left(\frac{1}{3x} - \frac{1}{x^2} + \frac{2}{x^4}(x - 1 + e^{-x}) \right) \tag{2.40}$$

with $x = L/l_p$.

The importance of the persistence length is best illustrated when considering the two limiting cases:

i. Rigid limit: $l_p \gg L$, implying $\langle \cos\theta(L) \rangle = 1$, $\langle R_G^2 \rangle = L^2/12$ and the "thread" behaves as a rigid rod.

ii. Flexible limit: $l_p \ll L$, implying $\langle \cos\theta(L) \rangle = 0$, $\langle R_G^2 \rangle \simeq Ll_p/3$ and the radius of gyration scales as $L^{3/2}$ and the average configuration of the thread is that of a random walk.

We see that the persistence length l_p is that length over which the axis of the cylinder is constant in direction. For length scales shorter than l_p the cylinders are rigid, and for length scales longer than l_p the cylinders wander randomly in space.

Since the flexibility is so directly related to a well defined characteristic length it should in principle be best evidenced through the analysis of scattering data in the appropriate wave vector range $(q \sim l_p^{-1})$. A detailed discussion of the form factor $P(q)$ of large gaussian coils is given in [29]. The general lines can be sketched as follows.

Three different q ranges are distinguished, corresponding to the three different length scales r_c, l_p and $\langle R_G^2 \rangle^{1/2}$:

i. $(2l_p)^{-1} \ll q < r_c^{-1}$: at the corresponding length scales the threads are rigid. The approximation (2.19) holds so that $I(q) \propto q^{-1}$

ii. $R_G^{-1} < q < (2l_p)^{-1}$: at this spatial resolution the scattered radiation mainly probes the correlations between distant parts of the thread inside the particle. At such distances r the function $\gamma_1(r)$ scales as r^{-1} and, correspondingly, $I(q) \propto q^{-2}$ in this range [42].

iii. $q < R_G^{-1}$: the Guinier regime obtains, allowing one in principle to estimate $\langle R_G^2 \rangle$ according to relation (2.11) $(I(q) \propto q^0)$

This general sketch is illustrated in Fig. 2.8, which represents low q neutron scattering data collected for a dilute solution of CPBr micelles in 0.8M NaBr brine [56]. The product of $I(q)$ and the scattering vector q is here plotted against q. At comparatively higher q values we observe a nearly horizontal dependence consistent with the q^{-1} variation of $I(q)$ in the rigid rod domain.

When going to a lower q $(2l_pq < 1)$, a steep upward deviation is observed at some value q_e. This can be interpreted as the onset of the q^{-2} intermediate regime. Significantly, the estimation of l_p by the relation [29]

$$q_e l_p = 1.91 \tag{2.41}$$

yields $l_p \approx 150\text{Å}$, in good agreement with estimates based on light scattering and magnetic birefringence [57] data (see below) for the same system. At still lower q the onset of the Guinier regime related to the finite size of the particles is revealed by the abrupt fall-off to zero.

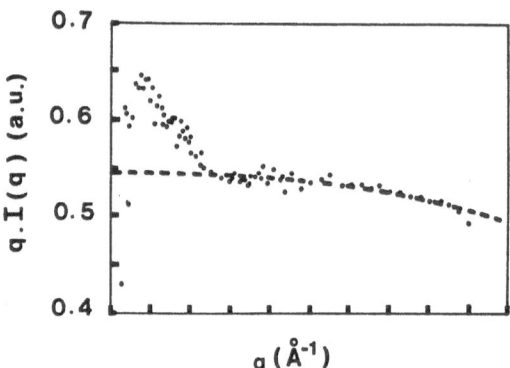

FIGURE 2.8. Very low q scattering profile for a dilute CPBr solution (2% volume fraction) in 0.8M NaBr brine.

For an unambiguous characterization of the structure to be obtained by this method, however, the different characteristic lengths must be sufficiently different in magnitude, so that each $I(q)$ regime can be identified over a large enough q-range. In the present case R_G was measured by light scattering and found to be $R_G = 520$Å, which is not very large compared to the measured persistence length l_p. This explains why in Fig. 2.8 the Guinier regime overlaps the q^{-2} regime which has too small an extension towards low q. This is the main practical limitation of the method and more sensitive experimental methods must be used to explore these flexibility effects.

Any measurable quantity sensitive to orientational correlations within the thread may be appropriate for this purpose. Amongst these, magnetic birefringence has been used [57]. The principle of the method is to submit the micellar dispersion to an intense magnetic field H and then to study its optical anisotropy. Due to its local cylindrical structure, the micelle presents a local anisotropy $\Delta\chi$ of its diamagnetic susceptibility. The effect of the field is therefore to align the micelle either along or perpendicular to its direction (depending on the sign of $\Delta\chi$), thereby offsetting partially the orientational anisotropy due to thermal motions. The measured optical birefringence Δn induced in the dispersion is found to be very different in the flexible and the rigid limits. In the rigid limit, the field reorients the particle as a whole and the optical response is essentially proportional to the length of the micelle,

$$\frac{\Delta n}{\Delta\chi H^2} = AL \qquad (2.42)$$

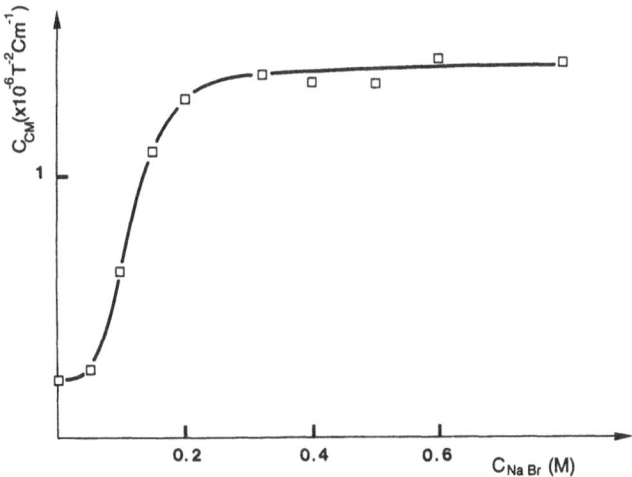

FIGURE 2.9. Magnetic birefringence (C_{CM}) of dilute (0.01M) CPBr solution as function of added NaBr concentration (0-0.8M). From [57].

where A is a system-dependent prefactor. In the flexible limit, parts of one given micelle separated by contour length much larger than l_p are not correlated in orientation and are aligned independently just like unconnected small rods,

$$\lim_{L \to \infty} \left(\frac{\Delta n}{\Delta \chi H^2} \right) = A \frac{l_p}{3} \qquad (2.43)$$

The optical response then becomes independent of the thread length L.

In Ref. 57 the authors report magnetic birefringence measurements of dilute CPBr solutions as a function of the added NaBr concentration (Fig. 2.9). The optical response actually saturates when the NaBr concentration exceeds about 0.2M, while the average total size of micelles increases steadily up to at least [NaBr] = 0.8M, as revealed by the variations of the hydrodynamic radius R_H (quasi elastic light scattering) in Fig. 2.2. Using model calculations the authors have attempted a quantitative estimation of l_p from these data. They obtained $l_p = 200$ Å, in reasonable agreement with the result mentioned above from the low q neutron scattering data.

Using light scattering alone (elastic and quasi-elastic) it is possible to measure both the average radius of gyration R_G, cf (2.10), and hydrodynamic radius R_h, cf (2.14). These quantities are not equally affected by the flexibility of the micelles and model calculations can thus be used to derive l_p from their relative variations over a wide range of micellar size. In this way Mazer et al. [23] obtained a value of 700Å for the persistence length of SDS micelles in concentrated NaCl brine. In Refs. 58 and 59 $l_p \approx 200$Å was again found by this method for the CPBr micelles, in correct agreement with the values obtained by the previous methods.

For unidimensional aggregates, the mass M is proportional to its contour length L. Thus for very large flexible micelles we expect $R_G \sim M^{1/2}$.

Both M and R_G have been measured by Ikeda *et al.* [16] for dimethyl oleiamine oxide micelles in NaCl brine using the elastic light scattering results (2.10) and (2.11). The data obtained agree with $\langle R_G^2 \rangle \simeq L l_p / 3$. Quantitative analysis yields $l_p = 1760 \text{Å}$ in 5×10^{-2} NaCl brine. A similar set of data measured by the same group, for solutions of dimethyl ammonium chloride and bromide in brine, has been treated quantitatively by van de Sande and Persoons [61], and the best fits were obtained for flexible rods of radius 17Å and persistence length 450Å. Note however that these data treatments do not take into account the polydispersity of micellar sizes and the l_p values obtained are probably strongly overestimated.

From this brief survey we conclude that flexibility has been demonstrated for a large set of systems where long micelles are present. The persistence length however varies markedly from one system to another. The flexibility is a local property, and thus it depends on all short range forces between neighboring monomers in the micelles. But the experimental determinations available are not systematic enough to be interpreted in terms of molecular interactions. A molecular theory of the rigidity of cylindrical micelles has not been explicitly formulated so far (see, however, Sec. 1.3.4).

2.3.2   THE SEMI DILUTE REGIME

The above structural description indicates strong analogies between solutions of giant micelles and solutions of macromolecules; the primary structures are in both cases unidimensional and flexible. But in contrast to conventional polymer solutions the size of the giant micelles is not fixed and results from a reversible polymerization process. The polydispersity of the size distribution is very large and the average size is strongly influenced by the experimental conditions. All these features are relevant for the physical properties of the solution in the dilute regime where the micelles are far apart from each other.

At higher concentrations a different regime exists—usually called the semi dilute regime in polymer science—where the overall size of the coils becomes irrelevant for the thermodynamic properties of the dispersion. Hence, as first pointed out by Candau *et al.* [66-69], the main differences between micellar and polymer solution disappear in the semi-dilute regime. The semi dilute regime is defined as the concentration range which is large enough for the threads to overlap but also small enough that the standard free energy of the dispersion involves only one interaction parameter. At these concentrations, the dispersion can be pictured (Fig. 2.10) as a random network of entangled threads, and the relevant parameter is the so called correlation length ξ introduced first by Edwards [62]. ξ can be intuitively

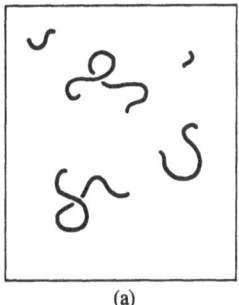
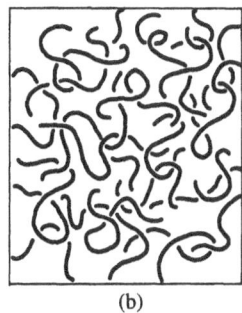

(a)   (b)

FIGURE 2.10. Intuitive representation of a) the dilute regime and b) the semi dilute regime of a dispersion of flexible giant micelles.

identified as the average distance between two entanglement points in the solution. Two distinct parts of one given thread separated by a distance smaller than ξ are related to each other as if the thread were isolated. When the separation distance is larger than ξ, the interaction between the two parts is *screened* by entanglements with neighboring threads.

The remarkable point is that ξ does not depend on the total size of the threads. It is entirely defined by the persistence length and the volume fraction ϕ of threads in the dispersion. The theoretical studies of de Gennes [42] and Des Cloizeaux [63] have shown that ξ obeys the following scaling law:

$$\xi \propto \phi^{\nu'/(1-3\nu')} \tag{2.44}$$

where $\nu' = 0.5$ in the absence of excluded volume effects (θ solvent) and $\nu' = 0.588$ if excluded volume interactions are dominant (good solvent).

In the semidilute regime, the individuality of any given thread in the network is lost at all length scales larger than ξ. In the corresponding q range ($q\xi \gg 1$), the factorization of the scattered intensity $I(q)$ in terms of the intraparticle correlation function $\gamma_1(r)$, see (2.5), and the interparticle correlation function $g(r)$, see (2.16), becomes inappropriate. To derive $I(q)$ for this case, we must turn back to the general expression (2.2) which involves the total correlation function $G(r)$ expressing the distribution of distances r separating subunits, no matter whether they belong to the same thread or not:

$$G(r) = \frac{1}{\phi^2} \left(\langle \phi(0)\phi(r) \rangle - \phi^2 \right) \tag{2.45}$$

Due to the screening effect that destroys the correlations at distances larger than ξ, it is believed that $G(r)$ follows a simple Ornstein-Zernicke [42,64] form at large r,

$$G(r) \simeq \frac{\xi}{r} \exp\left(-\frac{r}{\xi}\right) \tag{2.46}$$

which gives by Fourier transform

$$G(q) \simeq \frac{\xi}{q^2 + \xi^{-2}} \quad (q\xi < 1) \tag{2.47}$$

Thus $I(q)$ should scale as:

$$I(q) \propto \frac{\phi^2 \xi^3}{1 + q^2 \xi^2} \quad (q\xi < 0) \tag{2.48}$$

The *dynamics* of the concentration fluctuations are associated with the elastic modes of the transient network. These modes are overdamped by the viscosity η_0 of the solvent and a simple diffusion process is expected, characterized by a cooperative diffusion coefficient D_c

$$D_c = k_B T / 6\pi \eta_0 \xi_H \tag{2.49}$$

Here ξ_H is a hydrodynamic correlation length which for very long chains varies according to the same power law as ξ; see (2.43). Eqs. (2.47) and (2.48), together with (2.43), imply

$$I(q \to 0) \propto \phi^{-0.31} \tag{2.50}$$

and

$$D_c \propto \phi^{0.77} \tag{2.51}$$

In the semi dilute regime of giant micelle solutions, the general trends for the intensity of the scattered light, for its angular disymmetry and for the inverse decay rate of its time autocorrelation function, should be just opposite to what they are in the dilute regime. At high dilution, all these quantities increase with ϕ, reflecting the increase of the micellar size (molecular mass M, radius of gyration R_G and hydrodynamic radius R_h), whereas at higher volume fraction they should decrease, reflecting the ϕ dependence of ξ (or ξ_H). Accordingly, a maximum is expected at some "cross-over" value ϕ^* of ϕ where the giant micelles begin to overlap.

Actually, evidence of this behavior can be ovserved in the data of several authors [9,16,65] who did not propose, at the time, an interpretation in terms of polymer like properties. Candau et al. [66,69] have undertaken a systematic study of CTAB solutions in 0.1M KBr brine at different volume fraction of surfactant using light scattering. At 30°C the cross over volume fraction ϕ^* occured at $\phi \approx 0.01$. In the dilute regime, R_G as measured from the angular dependence of the light scattered, and R_H as derived from the first cumulant of its time autocorrelation function, were found in the ratio: $R_H/R_G \approx 0.6$ typical of flexible coils. Figure 2.11 shows a logarithmic plot of $I(0)$ and D_c against ϕ in the semi dilute regime ($\phi > 0.01$). Both quantities exhibit a power law dependence with the respective exponents, -0.34 for $I(0)$ and 0.81 for D_c, in fairly good agreement with the theoretical

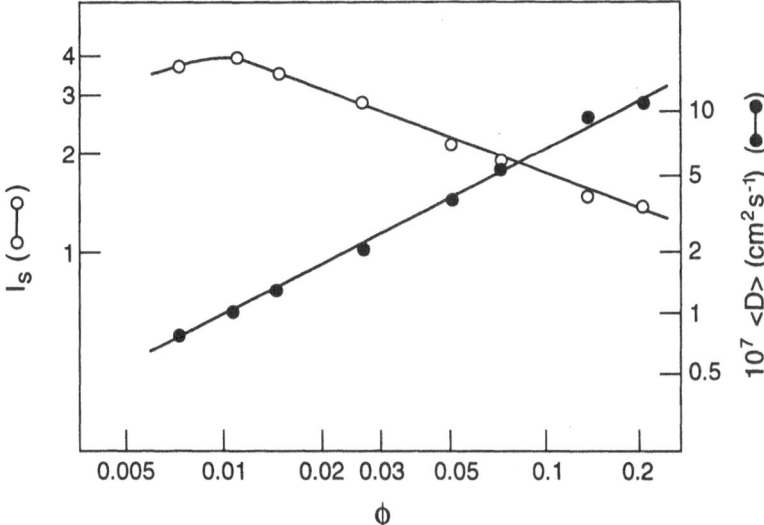

FIGURE 2.11. Variation of the scattered intensity and the average diffusion coefficient for semi dilute aqueous micellar solution of CTAB in presence of 0.1M
KBr. From [66].

values [52,53]. Paradoxically, the agreement is even better than that of true
polymer solutions where the observed exponent for D_c is about 0.67 for
most systems investigated.

This agreement provides a decisive argument in favor of the flexible rod
model of giant micelles. The theory of *rigid* rods dispersions in the semi
dilute regime has been established by Doi and Edwards [70], and its predictions are in marked contrast to those for flexible coils: in particular, D_c
should be independent of ϕ and equal to half its value at infinite dilution.
The data of Candau *et al.* [66-69] thus allows us to discard the rigid rod
model for concentrated micellar solutions.

A promising field for further investigations seems to be the rheological
properties in the semi dilute regime for which a firm theoretical background
has been established [71,72]. For threads of constant size, the viscoelastic
properties of the transient network are described by the "reptation" model
where each individual thread is considered to creep through the holes left
between the other chains. The characteristic time τ_R of this process strongly
depends on the aggregation number s. The theory of de Gennes [71] predicts
the following scaling laws for the low shear viscosity η_s and the longest
viscoelastic relaxation time τ_R:

$$\eta_s \propto s^3 \phi^{3.9} \tag{2.52}$$

and

$$\tau_R \propto s^3 \phi^{1.6} \tag{2.53}$$

The dependence of both quantities on s should help examine whether micellar growth with ϕ continues to take place in the semi dilute regime. In particular, if s still increases like $\phi^{0.5}$, we expect η_s and τ_R to vary with ϕ according to

$$\eta_s \propto \phi^{5.4} \tag{2.54}$$

and

$$\tau_R \propto \phi^{3.1} \tag{2.55}$$

However, in micellar solutions another characteristic time τ_B must also be introduced, associated with the possibility for a micelle to break into pieces and recombine within the time of the experiment. Eqs. (2.54) and (2.55) should not be modified as long as $\tau_B \gg \tau_R$. The opposite limiting case ($\tau_B \ll \tau_R$) has been considered in a recent theory by Cates [73]. Provided that the average micellar size still varies like $\phi^{0.5}$ in the semidilute regime, the equations given by Cates for this limiting case lead to the following scaling laws:

$$\eta_s \propto \phi^{3.6} \tag{2.56}$$

and

$$\tau_R \propto \phi^{1.5} \tag{2.57}$$

Candau *et al.* have investigated the rheological properties of CTAB micelles in H_2O 0.1M KBr and H_2O 0.25M KBr in the semi dilute regime using a magneto rheometer. This set up allows one to measure separately the low shear viscosity η_s and the longest viscoelastic relaxation time τ_R. Their study has shown that at sufficiently low temperatures—i.e., when the micelles are very large—both η_s and τ_R vary according to (2.55) and (2.56), supporting the picture of rheology controlled by the breakage and recombination process ($\tau_B \ll \tau_R$). Furthermore, the stress relaxation function was found to be a single exponential, in agreement with the predictions of Cates's theory: the fast exchange process blurs out the effect of the equilibrium size polydispersity. At the same time, Messager *et al.* [74] have undertaken measurements of the self diffusion coefficient D_{self} of the giant CTAB micelles under similar experimental conditions using the technique of "fluorescence recovery after photo bleaching". Their results also exhibit a single exponential decay, and the scaling of D_{self} with ϕ is found to be very close to the theoretical prediction

$$D_{self} \propto \phi^{-1.6} \tag{2.58}$$

Note however that good agreement between theory and experiment are observed only at high added salt concentrations while substantial deviations are seen at lower salinities. A qualitative explanation of these deviations may be given as follows: at low salinities the quantity of free ions in the solution due to the surfactant itself is comparable to that brought by the added salt. The effective ionic strength of the brine solvent then increases significantly with increasing surfactant concentraion inducing a ϕ dependence of the excess term $kT\alpha$ in the thermodynamics of the growth (see Sec. 2.2.3). The dependence of the average micellar size ($\langle s \rangle$) on ϕ is then expected to be somewhat steeper than $\phi^{0.5}$, thereby introducing the observed upward deviations from the exponents computed theoretically for constant α. Further theoretical and experimental investigations are currently being carried out to get more quantitative insights into this delicate point.

There is yet another type of experimental situation where the analogy between micelles and high polymers is striking: that is when a phase separation occurs in the solution. Experimental evidence of a lower critical solution temperature has been shown [75,76] in high salt (4M) aqueous solutions of cetylpyridinium chlorate. The same system at lower salt concentration (few 10^{-1}M) consists of giant unidimensional micelles [26]. The initial monophasic high salt solutions undergo a separation into two phases of different surfactant concentrations when the temperature is raised above about 50°C. A similar phase separation, but with the opposite temperature dependence (higher critical solution temperature) was found [77] for the ternary system CPBr/hexanol/brine (0.2M NaBr).

The physical mechanism driving the phase separations is unclear at the present time. Indeed, one might guess that at high salt concentration the electrostatic repulsion between charged micelles is sufficiently screened so that van der Waals attractions become dominant: the brine could thus become a bad solvent for the micelles beyond some definite experimental conditions. But another possibility has been considered [3]: at high salinity and/or alcohol content the wormlike micelles could *branch*, thus leading to the formation of a randomly connected network throughout the sample. Suitable variations of salinity and/or temperature would then increase the volume density of connection points between rods, thereby inducing spontaneous shrinking of the network and expelling excess brine. Such randomly connected network actually corresponds to a plausible intermediate morphology between cylindrical micelles and flat bilayers (see Secs. 2.4 and 2.5).

It should be noted that the geometry of the consolution curve is rather insensitive to the details of the physical mechanism driving the phase separation; it depends on the structure of the dispersed objects in the mixture [79]. In particular, a distinctive feature of similar phase separations in high polymer solutions is that the critical concentration is very low, and consequently the coexistence curve has a very asymmetrical shape. This feature is well accounted for by the classical Flory Huggins theory of polymer solutions which predicts that the critical concentration ϕ_c should vary with the polymer size according to

$$\phi_c \propto s^{-1/2} \tag{2.59}$$

where s is the number of chain segments.

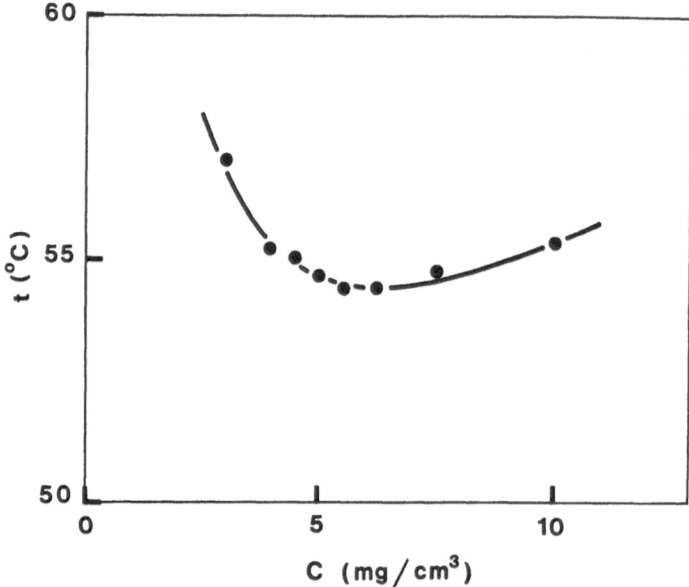

FIGURE 2.12. Coexistence curve for $CPClO_3/H_2O + NaClO_3(4M)$ solutions. From [75].

It is qualitatively clear that the critical concentration in a dispersion of random coils corresponds to the situation where the coils just come into close contact [42]: so $\phi_c \approx \phi^*$.

Both $CPClO_3$/brine (Fig. 2.12) and CPBr/hexanol/brine systems exhibit a very low value of the critical concentration (respectively $\phi_c = 0.006$ and $\phi_c = 0.0075$) compared to what occurs when the phase separation takes place in dispersion of globular micelles [78,79] where ϕ_c is typically of the order of 0.1. Also, light scattering studies of these systems show unusual behavior. ξ_c, the critical correlation length, as derived from the fit of the scattered light asymmetry to the Ornstein Zernike law, appears to be unusually large (up to 4500Å at more than 1°C from the critical temperature). In the critical region ξ_c scales according to

$$\xi_c = \xi_0 \left(\frac{T - T_c}{T_c} \right)^{-\nu} \tag{2.60}$$

with $\nu = 0.53$ for both systems. The scaling prefactors are $\xi_0 = 240$Å for the $CPClO_3$ system and $\xi_0 = 90$Å for the CPBr/hexanol/brine system. These unusually high ξ_0 values also support the analogy with polymer solution where ξ_0 values much larger than in classical molecular solutions are currently reported [80].

2.4   Bidimensional Aggregation

Large domains of stability of lamellar phases are very commonly found in
the concentrated parts of the phase diagrams of binary and ternary systems (see Chaps. 4, 5 and 6). Surfactant molecules thus have the ability to
aggregate into bilayers under appropriate conditions, as described briefly
in Chap. 1. In spite of this common notion, the possibility that bidimensional aggregation could take place in very *dilute* mixtures has not been
investigated systematically until recent years.

By analogy with the case of unidimensional micelles, one *a priori* expects bidimensional growth to induce the formation of large disks or closed
vesicles [81]. However, it was shown by Israelachvili *et al.* [40,81], based on
thermodynamic arguments, that large disks are always unstable with respect to infinite bilayers (see Sec. 1.2). Later on, it was argued further [41]
that infinite bilayers should collapse into finite size vesicles at sufficient
dilution. Yet recent experimental studies on the swollen lamellar phase
(L_α) and the "anomalous isotropic phase"—also called the "sponge," or
"sheety," phase—(L_3) suggest that the concentration where the infinite bilayers collapse into a random distribution of vesicles is extremely low. So, at
practical dilutions ($\phi > 0.01$) bidimensional growth yields instead bilayers
with infinite lateral extension. As shown by Benton and Miller [82] both the
swollen L_α phase and the anomalous L_3 isotropic phase are often encountered in the dilute part of the phase diagrams of quasi ternary systems (ionic
surfactant/alcohol/brine). The phase diagrams of these systems present a
simple common geometry in the brine rich range ($\phi_w > 0.8$ where ϕ_w is the
volume fraction of the brine). Figure 2.13 represents such a typical phase
diagram [83]: the system is cetylpyridinium chloride/n-hexanol/brine (0.2M
NaCl). Three different phases take place successively for increasing values of
the alcohol to surfactant ratio ϕ_A/ϕ_S. The L_1 phase is the classical micellar
phase and corresponds to the lower alcohol contents. The morphologies in
this phase have been described earlier in this chapter, where the globule to
rod transformation was shown to occur upon increasing amounts of alcohol.
At higher ϕ_A/ϕ_S, one finds the domain of stability of the birefringent L_α
phase. It exhibits focal conic textures in polarized light microscopy, characteristic of lamellar structures. At still higher ϕ_A/ϕ_S, one finds a very narrow
domain of stability for an isotropic phase with quite special macroscopic
properties: unusually large turbidity and strong streaming birefringence.
This phase has been denoted L_2^*, L_3 or C in the literature and is commonly
called the "anomalous isotropic phase". Both L_α and L_3 are stable against
very high dilution (at least up to $\phi_w = 0.995$ in the case of the system
of Fig. 2.13). As we discuss below, they both correspond to bidimensional
aggregation.

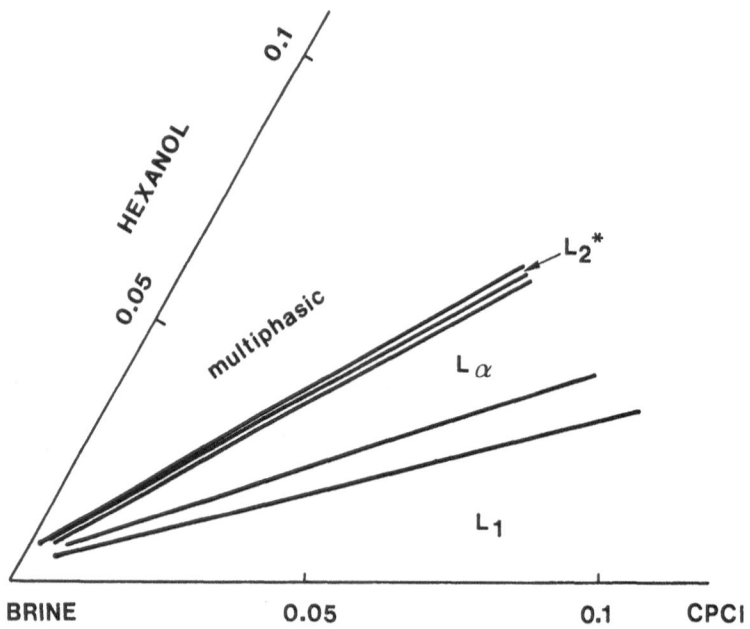

FIGURE 2.13. Brine rich part of the phase diagram of CPCl/n-hexanol/brine (0.2M NaCl).

2.4.1   THE SWOLLEN LAMELLAR PHASE (L_α)

A very striking property of the L_α phase is that it exhibits quasi long range positional order even at high dilution. (The leading role of the "Helfrich steric interaction" in this matter is discussed in Chaps. 5 and 6.) Such persistence of a well defined spatial periodicity provides a very favorable situation for an investigation of the morphology of the individual bilayers. Well aligned monocrystalline samples can be studied in the conventional X ray or neutron scattering geometry [83,86,87]. Scattering patterns collected along the direction q_z normal to the lamellar stacking and along the direction q_{xy} parallel to the bilayers (Fig. 2.14) can thus be interpreted separately. The absence of significant scattered intensity in the q_{xy} direction is a direct indication that the individual bilayers are compact and do not present a noticeable density of structural defects such as pores or fractures. This feature has been further confirmed by electron spin resonance spectroscopy [88]. The intense Bragg maximum observed in the q_z direction is related to the regularity of the lamellar stacking. Its peak position q_B allows an accurate determination of the average distance d between the

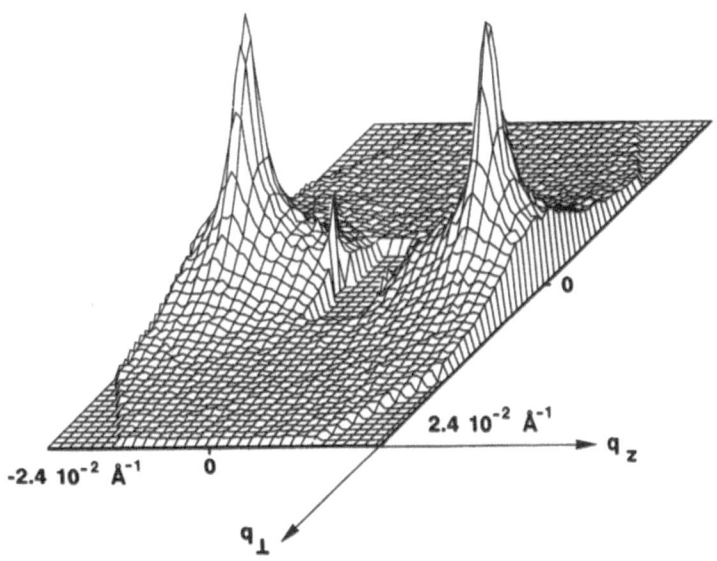

FIGURE 2.14. Neutron scattering pattern of a lamellar sample (system CPCl/hexanol/brine) with average periodicity $d = 360$Å. The $q_\perp$ direction here corresponds to q_{xy} cited in the text.

parallel bilayers,

$$q_B = \frac{2\pi}{d} = \frac{2\pi}{d_0}(1 - \phi_w) \qquad (2.61)$$

which, in turn, is simply related to the dry thickness d_0 of the individual bilayers and to the dilution ϕ_w. The remarkable linear variations (Fig. 2.15) of q_B as a function of ϕ_w for the above mentioned system indicates that d_0 remains fairly constant over an extremely large range of dilution: $d_0(\text{CPCl/hex}) = 26.5$Å.

2.4.2   THE ANOMALOUS ISOTROPIC PHASE (L_3)

The anomalous isotropic phase presents no long range order and the appropriate procedure to determine its structure is the same as discussed for the isotropic dispersions of long micelles. Scattering data must be collected in a very large q-range in order to evaluate the different characteristic lengths of the structure. In Ref. [83] it is shown that the scattering profiles at all dilutions exhibits a broad maximum at low q and a long decreasing tail at

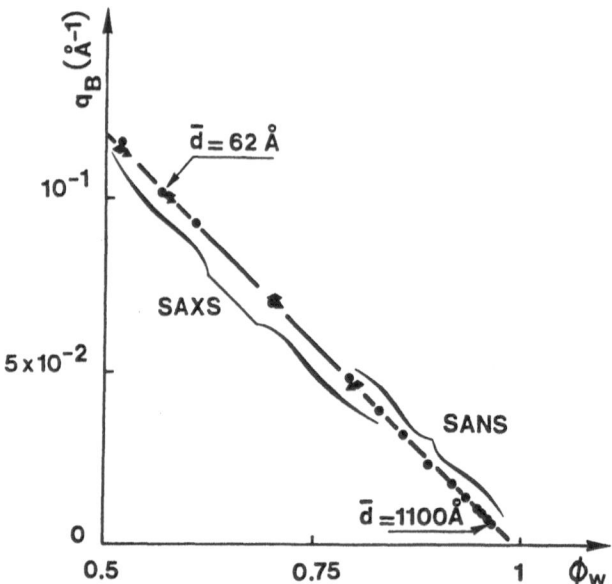

FIGURE 2.15. Bragg peak position q_B versus ϕ_w for the L_α phase as function of ϕ_w for the system CPCl/hexanol/brine.

high q (Fig. 2.16). The quantitative data treatment of the high q decay of the most dilute samples shows that $I(q)$ in this range scales accurately as

$$I(q) \propto q^{-2} \exp\left(-\frac{q^2 d_0^{*2}}{12}\right) \tag{2.62}$$

indicating that *at local scale* the structure consists of bilayers of thickness $d_0^* = 25\text{Å}$ (very similar to d_0 in the L_α phase) with random orientations. The position q_M of the broad maximum at lower q is of the same order as that of the Bragg wave vector q_B in the lamellar phase at the same dilution. q_M can therefore be interpreted as the signature of the average distance d^* between the bilayers in the L_3 structure:

$$q_M = \frac{2\pi}{d^*} \tag{2.63}$$

The variation of q_M with ϕ_w is demonstrated in Fig. 2.17.

Introducing $d_0^* = 25\text{Å}$ (the thickness of the bilayer as measured above), the corresponding dilution relation between d^* and ϕ_w can be written:

$$d^* = d_0^* \alpha (1 - \phi_w)^{-1} \tag{2.64}$$

where α is a numerical constant: $\alpha = 1.4 \pm 0.1$.

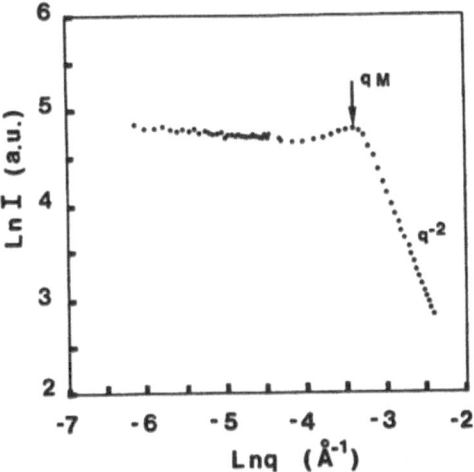

FIGURE 2.16. Typical scattering profile for L_3 samples in the system CPCl/hexanol/brine.

As discussed in [83] the dilution relation between d^* and ϕ_w is very specific to the topology of the structure. Thus, for example, if the aggregates have a finite size (such as disks or vesicles) the above proportionality between d^* and $(1 - \phi_w)^{-1}$ would imply that their mean size increases with ϕ_w, an unreasonable result in view of the discussion of self-assembly thermodynamics in Sec. 2.2.3. Other topologies involving infinite bilayers are discussed in [83], e.g., *locally*-lamellar structure, *foam*-like structure, and *bicontinuous* structure having the same topology as the cubic structure drawn in Fig. 2.18. Each topology corresponds to a particular value of the numerical constant α. Amongst those, only the bicontinuous topology ($\alpha = 1.5$) is quantitatively consistent with (2.63). It consists of an infinite bilayer bent everywhere with a saddle-like curvature so that it separates itself into two self connected, but mutually disconnected, interwoven infinite brine domains.

L_α and L_3 are thus different phases of bidimensional aggregates. Their structures involve the formation of bilayers with extremely large (presumably infinite) spatial extention, thus providing a strong support to the idea of Israelachvili *et al.* [40,41] that finite disks should "explode" into infinite aggregates. The word "micelle", which means "small piece", while appropriate to unidimensional growth, becomes too specific for bidimensional growth. (Note, however, that there are also many surfactant solutions in which the micellar disks remain small and finite all the way up through high concentrations, in particular into nematic and "lamellar"—smectic A!—states: see Chap. 3). The main difference between L_α and L_3 is that

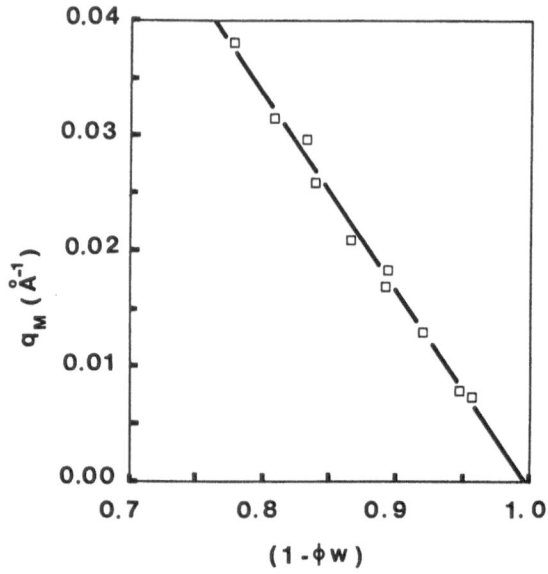

FIGURE 2.17. Position q^M of the maximum of the scattering profiles of the L_3 phase as function of ϕ_w for the system CPCl/hexanol/brine.

the initially flat bilayers in L_α are deformed into saddle like surfaces in L_3. This morphological transformation is interpreted in terms of the curvature elasticity of the bilayer in the next section.

2.5   The Physical Basis of Micellar Polymorphism

Throughout this chapter we encountered situations where the aggregation of surfactant molecules took very different shapes. The different morphologies can be roughly classified into three basic types: globular micelles, long flexible cylinders, large bilayers. In this section, we try to get some physical insights into the basic mechanisms that are believed to be at the origin of this spectacular polymorphism in micellar systms. Since shape transformations are known to occur even in very dilute solutions, we focus on the *intra*-aggregate mechanisms and neglect the eventual complication of interaggregate interactions (see Sec. 1.4 for discussion of interaction effects).

The basic unit of amphiphilic structures is the monolayer. It is usually regarded as a two dimensional incompressible liquid, implying two main properties. First, the molecules within the film have a high diffusion co-

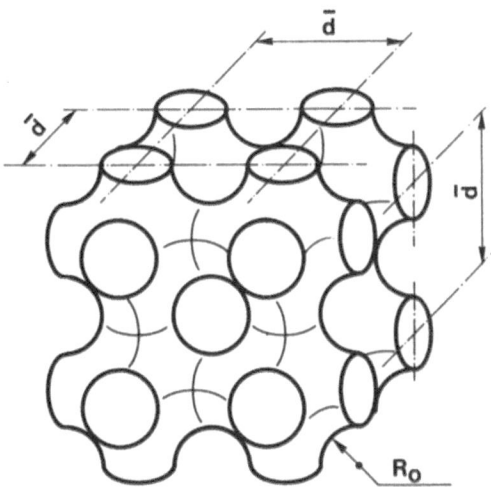

FIGURE 2.18. An example of cubic structure with the bicontinuous topology (here the symetry is $Im3m$).

efficient typical of fluids. Second, to change the film area (and thus the optimal area per molecule) either by compression or stretching, costs a large free energy: i.e., at fixed thermodynamic parameters (temperature, salinity etc.) the area per molecule in the film is essentially constant. The freedom remains to *splay* the surfactant assembly without changing the overall area per molecule: this results in *relative* changes in the area per head and the area per tail.

The most general expression for the elasticity associated with these deformations has been introduced by Helfrich [89]:

$$g = \frac{1}{2}k(c_1 + c_2 - c_0)^2 + \bar{k}c_1 c_2 \qquad (2.65)$$

where g is the elastic free energy per unit area, k and $\bar{k}$ are elastic moduli (of dimension energy), c_1 and c_2 are the two principal local curvatures, and c_0 is the spontaneous curvature. So k represents the energy cost to bend the film with a spherical splay deformation while $\bar{k}$ is associated with the cost of saddle splay deformation (see Sec. 1.3.4). The preferred mean curvature c_0 describes the inherent tendency of the interface to bend toward either the polar medium (water) or the non polar medium. Conventionally, convex curvatures (toward the polar region) are counted as positive. It is elementary to derive the following, equivalent, expression for g

$$g = \frac{1}{2}\left(k + \frac{\bar{k}}{2}\right)\left(c_1 + c_2 - c_0 \frac{2k}{2k + \bar{k}}\right)^2 - \frac{\bar{k}}{4}(c_1 - c_2)^2 , \qquad (2.66)$$

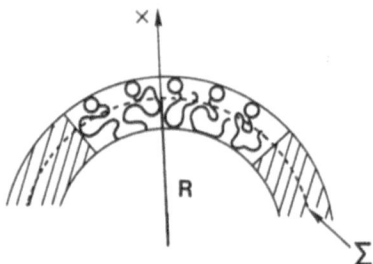

FIGURE 2.19. Schematic representation of a homogeneously bent surfactant monopayer. Σ represents the surface of inextension of the film and x is the coordinate normal to the film.

where the following condition for a stable configuration arises explicitly:

$$-2k \leq \bar{k} \leq 0 \tag{2.67}$$

Then the optimum local configuration corresponds to:

$$c_1 = c_2 = c_0 \frac{k}{2k + \bar{k}} = \frac{1}{2} c_{eq} \tag{2.68}$$

When $\bar{k} \ll -2k$ the monolayer "spheres up" without limit, while if $\bar{k} > 0$ elastic energy is released by increasing saddle like deformations.

In [89], Helfrich has given an intuitively appealing interpretation of the general expression (2.65) in terms of the stress distribution across the monolayer. For a flat configuration, shear stresses are absent due to the isotropy of the monolayer within its plane, and only isotropic tensions or compressions are allowed, represented by the function $s(x)$. Note that the stress profile $s(x)$ depends on the "depth" coordinate x within the layer thickness (Fig. 2.19) but not on the in–plane coordinates y and z.

The condition for stretching equilibrium is

$$\sigma = \int s(x)dx = 0 \tag{2.69}$$

Helfrich also shows that

$$kc_0 = -\int xs(x)dx \tag{2.70}$$

and

$$\bar{k} = \int x^2 s(x)dx \tag{2.71}$$

(see, also, Sec. 1.3.4 where these relations are discussed from a molecular approach).

Expression (2.69) is the mathematical translation of the fact that because of different optimum areas per molecule at the height of the polar head on the one hand, and at the level of the hydrocarbon chain on the other, there is a lateral torque in the planar geometry which is relaxed in a bent configuration with curvature c_0. Expression (2.71) shows that, although often neglected for the sake of simplicity, the Gaussian elastic modulus $\bar{k}$ is not necessarily small in magnitude compared to k. Note however that (2.70) and (2.71) are valid for a plane reference state only, and that the quadratic approximation (2.65) is valid only for small curvature deviations from the optimum configuration. In spite of these practical limitations, the concept of curvature elasticity, and especially of spontaneous curvature, proved very useful at least for a qualitative interpretation of the morphologies in surfactant systems. All the existing theories of amphiphilic structures such as the R–theory of Winsor [90] or the geometric packing models of Israelachvili *et al.* (see Sec. 1.3.2) [40] can be cast into the formalism of Eq. (2.65).

The main role of the added salt in an aqueous dispersion of micelles is to relieve (i.e., screen) electrostatic repulsions between charged sites. The optimum area at the level of the charged head groups then decreases while that of the aliphatic tails remains unaffected. The added cosurfactant [91–93] acts somewhat differently since it incorporates into the micellar structure. Its role is believed [94–95] to act as a "spacer" by interposing itself between the charged surfactant head groups, reducing the surface charge denisty, and thereby relieving part of the electrostatic strain due to head group repulsion. So both additives have the same effect on the spontaneous curvature: increasing their concentration induces a monotonic decrease of c_0.

For systems which do not contain oil, another essential requirement for the micellar shape must be considered: in one direction at least, the aggregate must have a dimension smaller than twice the length l of the surfactant molecule (in order to avoid the appearance of holes in the hydrophobic core (see Sec. 1.3 for detailed discussions)). This geometrical constraint usually prohibits the film from assuming the spherical curvature of minimum elastic energy. As analysed more rigourously in Chap. 4, this conflicting situation (frustration) between the elastic reaction and the geometric constraint is thought to be at the origin of the large variety of structures exhibitied by surfactant systems. Qualitatively, one easily visualizes how minimizing the elastic energy subject to the geometrical constraints will generate the various basic shapes of micelles (globules, cylinders and bilayers), depending on the value of c_0. Globules are expected if c_0 is such that $(c_{eq})^{-1}$ is less than $2l$. They become unstable with respect to the formation of long cylinders when $(c_{eq})^{-1}$ increases up to about l. Flat bilayers will prevail for still larger values of $(c_{eq})^{-1}$ such that a flat configuration becomes more favorable than a cylinder of radius l.

Note that for cylindrical micelles the two principal curvatures are very different. If $\bar{k}$ is large, the cylinders may be less favorable than spheres and

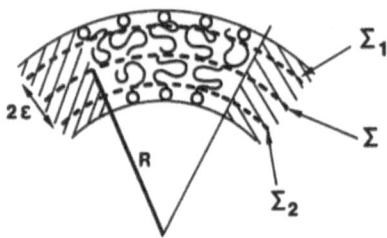

FIGURE 2.20. Schematic representation of a homogeneously bent bilayer. Σ is the mid surface and Σ_1, Σ_2 the surfaces of inextension of each monolayer.

bilayers for all values of c_0 [96]. In this case we expect a direct globule-to-bilayer transformation upon decreaseing c_0 (without formation of giant cylindrical micelles). The experimental evidence of such a situation found for the quasiternary system sodium octylbenzene sulfonate/pentanol/brine (0.5 M NaCl) [97] could be interpreted in this way.

We saw in the preceding section that, in many quasiternary systems, the structural transformation from regular lamellar phase (flat bilayers) to the anomalous isotropic phase (bicontinuous topology) was triggered by increasing the alcohol content. This fact can also be regarded as due to a continuous variation of the spontaneous curvature c_0 of the two monolayers constituting the complete bilayer. Neglecting for simplicity $\bar{k}$ in the elastic energy of the *monolayer*, we write

$$g_{mon} = \frac{1}{2}k(c_1 + c_2 - c_0)^2 \qquad (2.72)$$

In the bilayer, the curvature c_1 and c_2 of each monolayer are those characterizing their surface-of-inextension (Σ_1 and Σ_2 in Fig. 2.20) i.e., at a distance ε from the mid-surface (Σ in Fig. 2.20). Then it is straightforward to show that for the *bilayer* [98]:

$$g_{bil} = k(c_1 + c_2)^2 - 8k\varepsilon c_0 c_1 c_2 \qquad (2.73)$$

and

$$\bar{k} = -8k\varepsilon c_0 \qquad (2.74)$$

Here g_{bil} is the elastic energy per unit area of the mid-surface Σ, and c_1 and c_2, are the principal curvatures of the bilayer at the mid surface Σ. Consistent with its symmetrical structure, the bilayer presents no spontaneous curvature (i.e., $c_{0,bil} = 0$ as opposed to c_0 of the *monolayer* which appears in (2.72)–(2.74) and which is, generally, non-zero). Note, however, that the sign of the bilayer's saddle splay rigidity modulus is determined

FIGURE 2.21. Schematic representation of a surface with a "handle".

by the sign of the spontaneous curvature of each of the two constituting monolayers. As long as c_0 is positive, $\bar{k}$ is negative and according to (2.67) the flat bilayer is stable. If, on the other hand, c_0 becomes negative–i.e. the spontaneous tendency of each monolayer is to bend with convex curvature towards the hydrophobic medium, then $\bar{k}$ is positive and the flat bilayer becomes unstable with respect to spontaneous saddle splay deformations. The integral of $c_1 c_2$ over a given surface is known [89] to be independent of the shape of the surface as long as the deformation does not require cuts (Gauss-Bonnet theorem). It is a topological quantity because its value changes only with the topology (or genus) of the surface, in steps of $\pm 4\pi$, as may be shown by comparing the integral over a sphere (4π) with that over a torus (0). So, starting from a flat bilayer there is an energetic gain of $-4\pi k$ each time a "handle" is created, as schematized in Fig. 2.21. This is reminiscent of the topology of the L_3 phase (see Fig. 2.18) where the formation of self connected brine channels implies a large density of such "handles". In the quadratic approximation the handle formation process is predicted to progress without limit. However, the higher order terms $((c_1+c_2)^4, (c_1 c_2)$, etc.) in the deformation free energy are expected to stop the process at some equilibrium state. Also, the entropy of distribution of handles should influence the final equilibrium. Notwithstanding these still unexplored complications, the concept of spontaneous curvature of the monolayer provides a qualitatively convincing explanation of the complete morphological sequence determined in quasi binary (surfactant-brine) or quasi ternary (surfactant-alcohol-brine) systems: globules→cylinder (L_1)→flat bilayers (L_α) →saddle surface bilayers $(L_2^* = L_3)$.

2.6   Connection with Microemulsions

The distinctive feature of microemulsions, as compared to the micellar systems investigated in the preceding sections, is the presence of significant amounts of oil in the system (see Chaps. 7–9). Due to the amphiphilic character of the surfactant molecule this extra hydrophobic component can be dispersed in the aqueous solvent to form, on a macroscopic scale, an isotropic transparent homogeneous phase. On microscopic scales, however, the hydrophilic (brine) and the hydrophobic (oil) components segregate into domains separated from each other by the surfactant film. In contrast to micellar systems, the constraint of a maximum thickness ($\approx 2l$) for the hydrophobic medium is removed, since the hydrophobic region is now swollen with oil [99]. It is replaced by the much less stringent *global* constraint of conservation of the total amount of oil in the mixture. Otherwise it is reasonable to expect that the relevant properties of the surfactant film are still well represented by a bending elasticity with a spontaneous curvature c_0. Then two opposite situations can be encountered.

If $k, \overline{k} \gg k_B T$, the fluctuations in curvature of the film are very small and the entropy associated with them can be neglected. The actual morphology is derived by minimizing the elastic free energy (integrated over the whole film area A_s) subject to the constraint of constant total volume V_0 for the oil. This situation has been investigated by Safran et al. [100,101]. For a given positive value of c_0 and for decreasing amounts of oil they obtained the same sequence of shape changes as in classical micellar systems. When the oil volume fraction is large enough, the optimum configuration corresponds to a monodisperse distribution of spheres of radius $2(c_{eq})^{-1}$, the excess oil being rejected as a separate phase ("emulsification failure"). When the oil content becomes smaller at constant surfactant concentration, the area to volume ratio decreases, as does the radius of the sphere. At some stage the film becomes too bent in the spherical configuration and formation of infinite cylinders which accomodate a lower curvature for the same area to volume ratio becomes more favorable. At still lower oil contents, the same mechanism will favor the formation of flat lamellar structures.

The microemulsions studied by the Australian group using cationic double chain surfactans [102,103] are believed to provide typical examples of surfactant films with high rigidity. Actually, X ray scattering data seem to support the idea of initial globules opening progressively to form a network of branched cylinders so that the curvature of the film remains constant over the whole range of the composition variations.

Safran et al. (see Chap. 9) have further discussed the case where the cylindrical configuration is optimal and have computed the size and flexibility of the structures formed. The orders of magnitude obtained correspond to very long chain-like objects wandering randomly in space with a persistence length l_p much smaller than their average total length, suggesting again a polymer like behavior at scales larger than l_p. The striking analogy between the shape sequences predicted in high rigidity microemulsions and observed in micellar systems has a simple explanation. Extra degrees of freedom are in principle introduced by the replacement of the local constraint on the (bilayer's or cylinder's) thickness by the global constraint on the hydrophobic volume. On the other hand, due to its high rigidity, the film tends to have everywhere a homogeneous constant curvature. The global ratio between the total area of the film and the total volume of the hydrophobic medium as fixed by the composition of the sample is therefore respected on a local scale. So, finally, the physical situation turns out to be basically the same as in micellar systems.

Very different trends are expected in the low rigidity limit: $k, \overline{k} \leq k_B T$. Large local deviations from the optimum curvature involve a moderate energetic price and the entropy associated with the bending fluctuations of the film becomes the dominant contribution to the overall free energy of the mixture. Correspondingly, the appropriate approach is to maximize this entropy subject to the constraints of area and volume conservation. This procedure forms the basis of the lattice gas models (within the random mixing assumption) which are detailed in Chaps 7 and 8. Here we just note that the main effect of the fluctuations is to smear out the well defined morphologies characteristic of the high rigidity limit.

References

1. P. Ekwall, Adv. Liq. Cry. **1**, 1 (1975).

2. A. Skoulios, Ann. Phys. **3**, 421 (1978).

3. R. Gomati, J. Appell, P. Bassereau, J. Marignan and G. Porte, J. Phys. Chem. **91**, 6203 (1987).

4. B. Lindman and H. Wennerstrom, Top. Curr. Chem. **87**, 1 (1980).

5. P. Mirallas, Thesis, Montpellier, France (1987).

6. E. W. Anacker, in *Cationic Surfactants*, E. Jungermann, ed. (Marcel Dekker, New York, 1970), p. 251.

7. G. B. Benedek, in *Polarization, Matter and Radiation* (Presses Universitaires de France, Paris, 1969).

8. M. Kerker, *The Scattering of Light and Other Electromagnetic Radiations* (Academic Press, New York, 1969).

9. E. W. Anacker and H. M. Ghose, J. Am. Chem. Soc. **90**, 3161 (1968).

10. *Photon Correlation and Light Beating Spectroscopy*, H. Z. Cummins and E. R. Pike, eds. (Plenum Press, New York, 1974).

11. B. J. Berne and R. Pecora, *Dynamic Light Scattering* (Wiley Interscience, New York, 1976).

12. *Scattering Techniques Applied to Supramolecular and Nonequilibrium Systems*, S. Chen, B. Chu and R. Nossal, eds. (Plenum Press, New York, 1980).

13. N. A. Mazer, M. C. Carey and G. B. Benedek, in *Micellization, Solubilization, and Microemulsions* (Proc. Int. Symp., 1977), p. 359.

14. P. J. Missel, N. A. Mazer, G. B. Benedek, C. Y. Young and M. Carey, J. Phys. Chem. **84**, 1044 (1980).

15. D. Stigter, J. Phys. Chem. **70**, 1323 (1966).

16. S. Ikeda, S. Ozeki and M. Tsunoda, J. Colloid Interface Sci. **73**, 27 (1980).

17. P. Debye and E. W. Anacker, J. Phys. Colloid Chem. **55**, 644 (1951).

18. S. Ikeda, S. Hayashi and T. Imae, J. Phys. Chem. **84**, 744 (1980).

19. H. Hoffmann, G. Platz and W. Ulbricht, J. Phys. Chem. **85**, 1418 (1981).

20. H. Hoffmann, G. Platz, H. Rehage and W. Schorr, Ber. Bunsenges. Phys. Chem. **85**, 877 (1981).

21. G. Porte and J. Appell, J. Phys. Chem. **85**, 2511 (1981).

22. C. Y. Young, P. J. Missel, N. A. Mazer, G. B. Benedek and M. C. Carey, J. Phys. Chem. **82**, 1375 (1978).

23. P. J. Missel, N. A. Mazer, G. B. Benedek and M. C. Carey, in *Solution Behavior of Surfactants*, Vol. 1, E. J. Fendler and K. L. Mittal, eds. (Plenum Press, New York, 1982), p. 1264.

24. P. J. Missel, N. A. Mazer, G. B. Benedek and M. C. Carey, J. Phys. Chem. **87**, 1264 (1983).

25. G. Porte, Y. Poggi, J. Appell and G. Maret, J. Phys. Chem. **88**, 5713 (1983).

26. G. Porte and J. Appell, in *Surfactants in Solution*, Vol. 2, K. L. Mittal and B. Lindman, eds. (Plenum Press, New York, 1984), p. 805.

27. F. Reiss-Husson and V. Luzzati, J. Phys. Chem. **68**, 3504 (1964).

28. A. Guinier and G. Fournet, *Small Angle Scattering of X-Rays* (Wiley Interscience, New York, 1955).

29. O. Glatter and O. Kratky, *Small Angle X-Ray Scattering* (Academic Press, London, 1982).

30. B. Cabane, in *Surfactant Solutions: New Methods of Investigation*, Surfactant Science Series, Vol. 22, R. Zana, ed. (Marcel Dekker, New York, 1987).

31. H. Thurn, J. Kalus and H. Hoffmann, J. Chem. Phys. **80**, 3440 (1984).

32. H. Hoffmann, J. Kalus, H. Thurn and K. Ibel, Ber. Bunsenges. Phys. Chem. **87**, 1120 (1983).

33. J. Kalus, H. Hoffmann, K. Reizlein and W. Ulbricht, Ber. Bunsenges. Phys. Chem. **86**, 37 (1982).

34. H. Hoffmann, J. Kalus, K. Reizlein, W. Ulbricht and K. Ibel, Colloid Polym. Sci. **260**, 435 (1982).

35. L. Herbst, H. Hoffmann, J. Kalus, H. Thurn, K. Ibel and R. P. May, Chem. Phys. **103**, 437 (1986).

36. B. Cabane, R. Duplessix and T. Zemb, in *Surfactants in Solutions*, B. Lindman and K. Mittal, eds. (Plenum Press, New York, 1984).

37. G. Porte, J. Marignan, P. Bassereau and R. May, J. de Physique **49**, 511 (1988).

38. F. Quirion and L. Magid, J. Phys. Chem. **90**, 5435 (1986).

39. S. H. Chen, T. L. Lin and C. F. Wu, in *Physics of Amphiphilic Layers*, J. Meunier and D. Langevin, eds. (Springer Verlag, 1987).

40. J. M. Israelachvili, D. J. Mitchell and B. W. Ninham, J. Chem. Soc. Faraday Trans. 2 **72**, 1525 (1976); see also P. J. Mukerjee, J. Phys. Chem. **79**, 565 (1972).

41. G. Porte, J. Phys. Chem. **87**, 3541 (1983).

42. P. G. de Gennes, *Scaling Concepts in Polymer Physics* (Cornell University Press, 1979).

43. J. C. Wheeler, S. J. Kennedy and P. Pfeuty, Phys. Rev. Lett. **45**, 1748 (1980).

44. J. C. Wheeler and P. Pfeuty, Phys. Rev. A **24**, 1050 (1981).

45. J. C. Wheeler and P. Pfeuty, Phys. Rev. Lett. **46**, 1409 (1981); J. C. Wheeler and P. Pfeuty, J. Chem. Phys. **74**, 6415 (1981).

46. P. Pfeuty and J. C. Wheeler, Phys. Rev. A **27**, 2178 (1983).

47. A. W. Tobolsky and A. Eisenberg, J. Am. Chem. Soc. **81**, 780 (1959).

48. A. V. Tobolsky, A. Rembaum and A. Eisenberg, J. Polym. Sci. **45**, 347 (1960).

49. R. L. Scott, J. Phys. Chem. **69**, 261 (1965).

50. F. Oosawa and S. Asakura, *Thermodynamics of Polymerization of Proteins* (Academic Press, London, 1975).

51. W. M. Gelbart, W. E. McMullen, A. Masters and A. Ben-Shaul, Langmuir **1**, 101 (1985).

52. A. Ben-Shaul, D. H. Rorman, G. V. Hartland and W. M. Gelbart, J. Phys. Chem. **90**, 5277 (1986).

53. W. M. Gelbart, A. Ben-Shaul, W. E. McMullen and A. Masters, J. Phys. Chem. **88**, 861 (1984).

54. R. Nagarajan, J. Colloid Interface Sci. **90**, 477 (1982).

55. L. Landau and E. Lifshitz, *Statistical Physics*, 3rd Edition, Part 1 (Pergamon Press, 1980), p. 396.

56. P. Bassereau, J. Marignan, G. Porte and R. May, unpublished results.

57. G. Porte, J. Appell and Y. Poggi, J. Phys. Chem. **84**, 3105 (1980).

58. J. Appell and G. Porte, J. Colloid Interface Sci. **81**, 85 (1981).

59. J. Appell, G. Porte and Y. Poggi, J. Colloid Interface Sci. **87**, 492 (1981).

60. N. A. Mazer, unpublished results.

61. W. Van de Sande and A. Persoons, J. Phys. Chem. **89**, 404 (1985).

62. S. F. Edwards, Proc. Phys. Soc. **88**, 265 (1966).

63. J. des Cloiseaux, J. de Physique **36**, 281 (1975).

64. L. S. Ornstein and F. Zernike, Proc. Akad. Sci. (Amsterdam) **17**, 293 (1914).

65. H. Hoffmann, M. Lobl and H. Rehage, in *Physics of Amphiphiles: Micelles, Vesicles and Microemulsions*, V. Degiorgio and M. Corti, eds. (North Holland, Amsterdam, 1985), p. 237.

66. S. J. Candau, E. Hirsch and R. Zana, J. de Physique **45**, 1263 (1984).

67. S. J. Candau, E. Hirsch and R. Zana, J. Colloid Interface Sci. **105**, 521 (1985).

68. E. Hirsch, S. J. Candau and R. Zana, Proceedings of the 5th International Symposium on Surfactants in Solutions, Bordeaux, July 9–13 (1984).

69. S. J. Candau, E. Hirsch and R. Zana, in *Physics of Complex and Supramolecular Fluids*, S. Safran and N. Clark, eds. (Wiley, New York, 1987), p. 569.

70. M. Doi and S. F. Edwards, J. Chem. Soc. Faraday Trans. 2 **74**, 560 (1978).

71. P. G. de Gennes, J. Chem. Phys. **55**, 572 (1971).

72. P. E. Rouse, J. Chem. Phys. **21**, 1272 (1953).

73. M. E. Cates, Macromolecules **20**, 2289 (1987).

74. R. Messager, A. Ott, W. Urbach, D. Chatenay and D. Langevin, Phys. Rev. Lett. **60**, 1410 (1988).

75. J. Appell and G. Porte, J. de Physique **44**, L689 (1983).

76. G. Porte and J. Appell, in *Physics of Amphiphiles: Micelles, Vesicles and Microemulsions*, V. Degiorgio and M. Corti, eds. (North Holland, Amsterdam, 1985), p. 461.

77. F. Dauverchain, Thesis, Montpellier, France (1988).

78. O. Abillion, Thesis, Paris, France (1984).

79. R. Kjellander, J. Chem. Soc. Faraday Trans. 2 **78**, 2025 (1982).

80. C. Ishimoto and T. Tanaka, Phys. Rev. Lett. **39**, 474 (1977).

81. J. N. Israelachvili, S. Marceljas and R. G. Horn, Q. Rev. Biophys. **13**, 121 (1980).

82. W. J. Benton and C. A. Miller, J. Phys. Chem. **87**, 4981 (1983).

83. P. Bassereau, J. Marignan and G. Porte, J. de Physique **48**, 673 (1987).

84. F. C. Larche, J. Appell, G. Porte, P. Bassereau and J. Marignan, Phys. Rev. Lett. **56**, 1700 (1986).

85. C. R. Safinya, D. Roux, G. S. Smith, S. K. Sinha, P. Dimon, N. A. Clark and A. M. Belocq, Phys. Rev. Lett. **57**, 2518 (1986).

86. G. Porte, R. Gomati, O. El Haitamy, J. Appell, and J. Marignan, J. Phys. Chem. **90**, 5746 (1986); see also ref. 3.

87. G. Porte, P. Bassereau, J. Marignan and R. May, in *Physics of Amphiphilic Layers*, J. Meunier and D. Langevin, eds. (Springer Verlag, 1987), p. 145.

88. J. M. di Meglio and P. Bassereau, unpublished results.

89. W. Helfrich, Z. Naturforsch **28C**, 693 (1973), in *Physics of Defects*, R. Balian, M. Kleman and J. P. Poirier, eds. (North Holland, Amsterdam, 1981), p. 713.

90. P. A. Winsor, in *Liquid Crystals and Plastic Crystals*, Vol. 1, G. W. Gray and P. A. Winsor, eds. (Ellis Horwood Ltd., Chichester, 1974), p. 80.

91. J. Charvolin and Y. Hendrikx, in *Liquid Crystals of One and Two Dimensional Order*, W. Helfrich and G. Hepke, eds. (Springer Verlag, Berlin, 1980).

92. Y. Hendrikx, J. Charvolin, M. Rawiso, L. Liebert and M. C. Holmes, J. Phys. Chem. **87**, 3991 (1983).

93. Y. Hendrikx, J. Charvolin and M. J. Rawiso, J. Colloid Interface Sci. **100**, 597 (1984).

94. D. E. Mather, J. Colloid Interface Sci. **57**, 240 (1976).

95. V. A. Parseghian, Trans Faraday Soc. **62**, 848 (1966).

96. L. A. Turkevich, in *Physics of Amphiphilic Layers*, J. Meunier and D. Langevin, eds. (Springer Verlag, 1987), p. 298.

97. J. Marignan, P. Bassereau and P. Delord, J. Phys. Chem. **90**, 645 (1986).

98. A. G. Petrov and A. Derzhanski, J. de Physique Colloq. **37**, C3-155 (1976).

99. H. Hoffmann and W. Ulbricht, in *Physics of Amphiphilic Layers*, J. Meunier and D. Langevin, eds. (Springer Verlag, 1987), p. 334.

100. S. A. Safran, L. A. Turkevich and P. A. Pincus, J. de Physique **45**, L69 (1984).

101. L. A. Turkevich, S. A. Safran and P. A. Pincus, in Proc. 5th Symp. "Surfactants in Solutions," K. L. Mittal and P. Bothorel, eds. (Plenum Press, New York, 1988).

102. T. Zemb, S. T. Hyde, P. J. Derian, I. S. Barnes and B. W. Ninham, J. Phys. Chem. **91**, 3814 (1987).

103. B. W. Ninham, I. S. Barnes, S. T. hyde, P. J. Derian and T. N. Zemb, Europhys. Lett. **4**, 561 (1987).

3

Micellar Liquid Crystals

Neville Boden[1]

3.1 Introduction

The work by groups such as those of Ekwall *et al.* [1], Luzzati *et al.* [2] and Winsor [3], beginning in the 1930's, laid the basis for our current perception of the phase behavior and structure of lyotropic amphiphilic liquid crystals. In particular, examples of mesophases with one-(lamellar), two-(hexagonal and rectangular), and three-(cubic) dimensional translational order were established. But the translationally invariant nematic phase, which is characterized by long range orientational order of nematogenic particles, was conspicuous by its absence. This has had both conceptual and practical implications for the manner in which our present day knowledge of amphiphilic systems has developed. For example, the difficulty of obtaining macroscopically aligned samples of the translationally ordered mesophases has imposed severe constraints on definitive experimental studies.

However, in 1967, Lawson and Flautt [4] demonstrated that a solution of sodium decylsulphate in water containing small amounts of decanol and sodium sulphate was homogeneously aligned by the magnetic field in NMR experiments. Rosevear [5] then observed that the same solutions exhibited a Schlieren microscopic texture (Fig. 3.1), a characteristic signature of a nematic [6]. These solutions were, therefore, referred to as lyotropic nematic phases. These and other quite similar mixtures were used subsequently by NMR spectroscopists for the determination of the molecular geometry of small polar solutes [7]. But their inherent complexity [8] seemed to act as a deterrent to serious investigations of their structures and properties. Indeed, it was not until 1979 that Charvolin, Levelut and Samulski [9], using X-ray diffraction, confirmed the earlier speculation by Radley and Saupe [10] that these solutions consisted of rod or disc shaped micelles with long range orientational ordering of their symmetry axes. In the same year Boden and coworkers [11] reported the first *binary* system and its phase diagram which showed the nematic phase coexisting with its neighboring phases in a thermodynamically consistent manner. Thus, the existence of

---

[1]Department of Physical Chemistry, University of Leeds, Leeds LS2 9JT,UK

154   Neville Boden

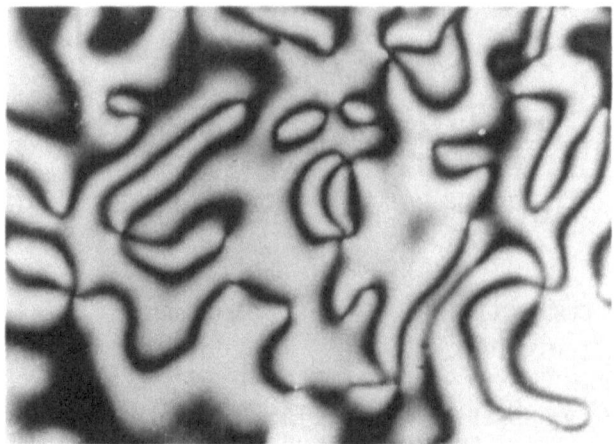

FIGURE 3.1. Thin film (200 μm) of nematic micellar solution viewed between crossed polarizers showing a Schlieren texture, the characteristic opto-micrographical signature of a nematic phase.

what are now referred to as micellar nematic phases was firmly established. The following year Yu and Saupe [12] discovered the hitherto elusive biaxial nematic phase in a potassium laurate/decanol/water mixture [13-17]. This is the fully anisotropic nematic micellar solution: the micelles are essentially biaxial platelets [18-21], statistically oriented along the three directions in space.

The contemporary picture is that nematic phases are to be found intermediate to an isotropic micellar solution at high temperatures/low concentrations and a smectic phase (lamellar for disks or hexagonal for rods) at lower temperature/higher concentrations as illustrated in Fig. 3.2. They appear to be stable within the concentration interval 0.1 to 0.5 volume fraction of amphiphile and below about 350 K. Across this concentration interval the axial ratio is typically in the range 2-4 for rod shaped micelles, and 0.25 to 0.5 for disk shaped micelles and the maximum dimension of the micelle is always of the order of the separation of the centres of mass. Accounting for these empirical facts is central to obtaining an understanding of the origin of micellar nematic phases. Of special importance are the following factors.

First, consider the size of the micelles. The diameter of the rods or the thickness of the disks is typically 2.0 nm which means their maximum dimension is in the range 4.0 to 8.0 nm. That such *small* micelles are stable in concentrated solutions of surfactants is quite surprising, especially as they are obtained by addition of either salts or long chain alcohols to ordinary surfactants. These are the very additives which are used to prepare *giant* micelles in dilute solutions (see previous chapter). Thus, we need to

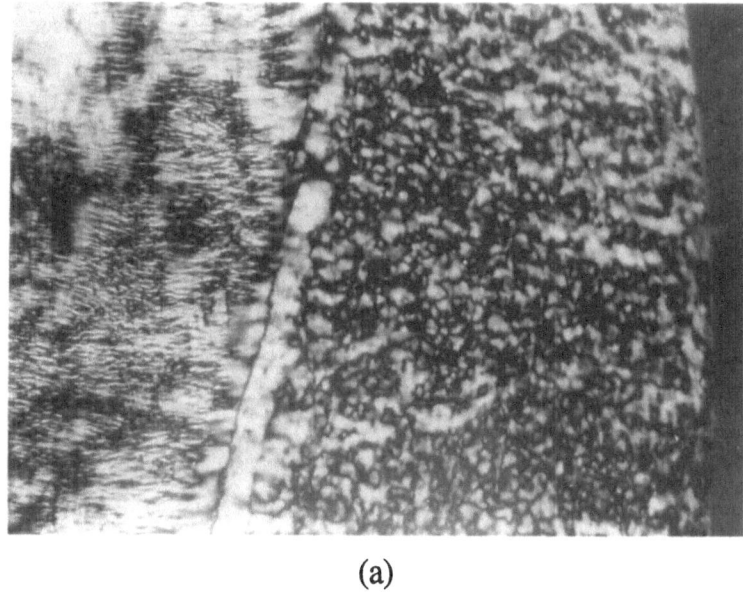

(a)

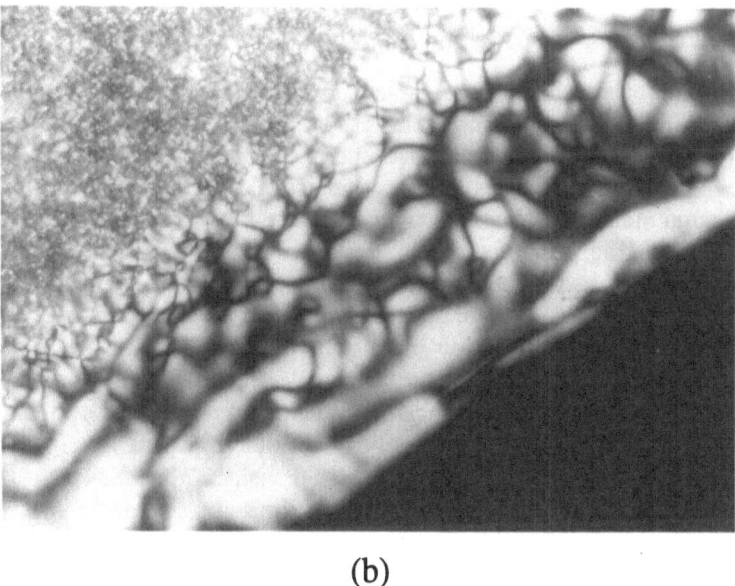

(b)

FIGURE 3.2. Photomicrographs of the liquid crystalline mesophases in the mixtures (a) myristyltrimethlyammonium bromide/water and (b) myristyltrimethylammonium bromide/water/decanol/NH$_4$Br, as observed in 0.1 mm thick parallel sided glass cells at 293 K and magnification 200x. The concentration gradient was established by allowing water to be sucked into the cell by capillary action; from left to right (a) hexagonal, nematic and isotropic phases and (b) lamellar, nematic and isotropic phases.

understand the factors which govern the size and shape of micelles in *concentrated* solutions of surfactants. Secondly, the relationship between the dimensions of the micelles and the separation of their centers of mass can be more explicitly expressed: for rods $\rho L^2 D \approx 1$ and for disks $\rho \sigma^3 \approx 1$, where ρ is the number density of micelles, L the average length of the rods of diameter D, and σ the average diameter of the disks of thickness D. These relationships tell us that the isotropic-to-nematic phase transition can be qualitatively understood in terms of Onsager's theory [22]–see Sec. 1.4.2. In this respect they resemble the colloidal nematic phases obtained in concentrated solutions of rod-like polymers [23]. But they differ from these systems in that the size and shape of the nematogenic particles, that is the micelles, is not fixed. As discussed already in the last section of Chap. 1, the size and shape are expected to vary with concentration and temperature and to depend upon the nature of the forces between the micelles and also upon any coupling to the symmetry and order of the mesophase. This will have a marked effect on the phase behavior. Thus, any theory of micellar nematic phases must include the self-assembly behavior of surfactants in aqueous solution. Conversely, the development of an understanding of micellar nematic phases will tell us much that is new about self-assembly.

The primary aim of this chapter is to consider the extent to which the phase behavior, structure and properties of micellar nematic phases can be understood in terms of the structures and interactions of the individual micelles.

A system of nomenclature is described in the next section.

The third section deals with the preparation of stable nematic phases. This in effect requires an understanding of the factors that govern the size, shape and interactions between the micelles in concentrated solutions, on the one hand, and those that govern order-disorder transitions in such solutions, on the other hand. But such an insight was not available to earlier workers in this field. They tended to follow the lead of Lawson and Flautt and to search for nematic phases in classical surfactant/water mixtures, modified by the addition of small amounts of a long chain alcohol or an inorganic salt [24-26]. But in these systems nematic phases have tended to be stable over only narrow intervals of concentration. However, from a consideration of their phase behavior, an empirical principle can be developed as an aid to the design of novel surfactants which give rise to nematic phases stable over wide ranges of concentration and temperature in binary mixtures of either ionic or non-ionic surfactants in water.

In Sec. 3.4 we explore the extent to which the empirical facts can be understood in terms of the theoretical models outlined in Chap. 1 for the self-assembly behavior of surfactant solutions. These theories have included, for the first time, the important feature that the shape and size of the micelle is a function of concentration, temperature and inter-micellar interactions. It will be seen that this theory accounts for some, but not all, of the observed phenomena.

One of the major challenges to the experimentalist is the characterization of the size and shape of the micelles in aqueous solution. This is facilitated in the nematic phase by the opportunity to produce homogeneously aligned samples. This enables X-ray or neutron scattering patterns to be obtained in directions parallel and perpendicular to the nematic director. These experiments, together with the information they have provided, are described along with the uses of optical microscopy and NMR Spectroscopy in Sec.3.5.

Section 3.6 deals with the physics of the rich variety of phase transitions and critical points which are to be found in micellar soutions. This situation obtains because of the opportunity to vary the concentration over a very wide range, with concomitant changes in both the structure and the forces between the micelles. The presence of biaxial nematic phases is just one example.

The issues raised for future research are discussed in the summary in Sec. 3.7.

3.2 Nomenclature

There are three distinct structural varieties of micelles. These, and the nematic phases they give rise to, are depicted in Fig. 3.3. Rod shaped micelles may be prolate ellipsoids or actual rods with semi-spherical end caps. The corresponding nematic phase (Fig. 3.3a) is designated as "canonic" (from the Greek $\kappa\alpha\nu\omega\upsilon = rod$) and denoted N_C. The disk shaped micelles may be flat disks with hemi-toroidal rims, or oblate ellipsoids. Their nematic phases are designated as discotic (from the Greek $\delta\zeta\sigma\chi o\xi = quoit$) and denoted N_D (see Fig. 3.3b). The third possibility is that micelles may be asymmetric ellipsoids or biaxial platelets which can form biaxial nematic phases denoted as N_B (Fig. 3.3c).

Let $\hat{a}$, $\hat{b}$, and $\hat{c}$ be a set of unit vectors denoting a coordinate frame fixed on the micelle (the micelle frame). In the N_C and N_D phases there will be a preferred direction of orientation of the symmetry axis c. This is referred to as the nematic director $\hat{n}$. The axes a and b will be randomly distributed about $\hat{n}$. Thus, the mesophase will possess uniaxial symmetry and this will be reflected in all of its physical properties. For example, the partially averaged values (denoted by the upper tilde) of the refractive index measured along $\tilde{n}_{\parallel}$ and perpendicular $\tilde{n}_{\perp}$ to the director will be different. This difference $\Delta\tilde{n} = \tilde{n}_{\parallel} - \tilde{n}_{\perp}$ is called the optical birefringence and determines the optical properties of the mesophase. Characteristically, N_D and N_C are, respectively, optically positive ($\Delta\tilde{n} > 0$) and negative ($\Delta\tilde{n} < 0$).

Now there will be a random distribution in the orientation of $\hat{c}$ about $\hat{n}$, that is, the orientational order is not complete. The order of a nematic

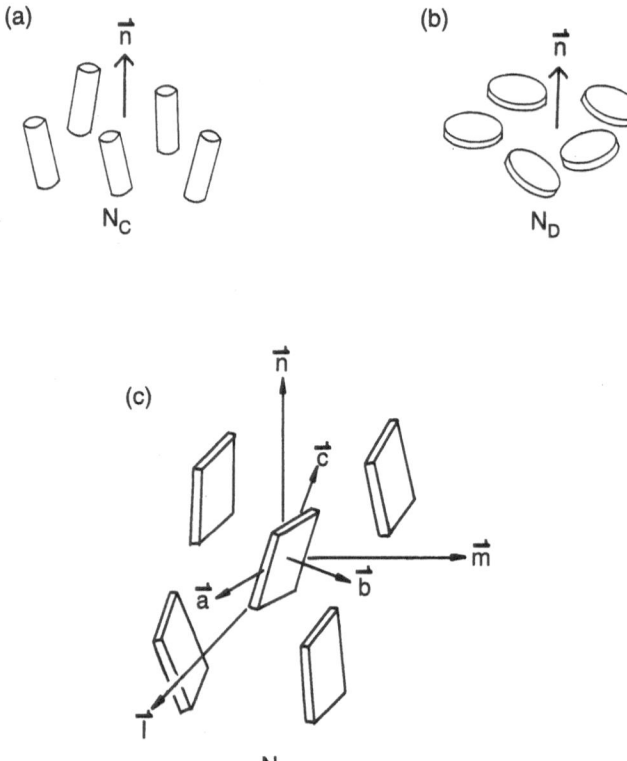

FIGURE 3.3. Schematic diagrams of (a) canonic N_C, (b) discotic N_D, and (c) biaxial N_B micellar nematic phases.

phase is described in terms of the Saupe [27] order parameter:

$$S^{ij}_{\alpha\beta} = \frac{1}{2} < 3I_{i\alpha}I_{j\beta} - \delta_{ij}\delta_{\alpha\beta} > \qquad (3.1)$$

The angular brackets indicate an ensemble average and the $I_{i\alpha}$ are the direction cosines describing the orientation of the micellar frame $i, j = \hat{a}, \hat{b}, \hat{c}$ with respect to the director frame $\alpha, \beta = \hat{l}, \hat{m}, \hat{n}$ fixed in the mesophase. In the case of N_D and N_C phases, there is only one order parameter:

$$S = \frac{1}{2} < 3\cos^2\vartheta - 1 > \qquad (3.2)$$

where ϑ is the angle between the axes $\hat{c}$ and $\hat{n}$. Complete *disorder* and complete *order* correspond to $S = 0$ and $S = 1$, respectively, with typical nematic values being on the order of a few tenths.

In the biaxial nematic phase, there is a preferred orientation of the micellar frame with respect to the director frame (Fig. 3.3c). Five order parameters are required to describe the orientational order. Fully co-operative rotations of the micelles around the directors $\hat{m}$ or $\hat{n}$ will give, respectively, the N_D or N_C phases. Indeed, this situation obtains in the potassium laurate/decanol water system where the sequence of transitions $N_D \rightarrow N_B \rightarrow N_C$ occurs without any major change in micelle structure [18-21]. A description of the orientational order of a biaxial micelle in a uniaxial phase requires two order parameters. Strictly, such phases should be distinguished from the N_D and N_C phases as depicted, respectively, in Figs. 3.3b and 3.3a.

Optical microscopy, NMR spectroscopy and X-ray scattering are all gen-erally used to characterize the structures of micellar nematic phases. A brief description of each of these methodologies is given in Sec. 5. In all of these experiments magnetic fields are employed to produce samples having a uniform orientation of the nematic director. The response to an applied magnetic field $\mathbf{H} = kH_o$ is determined by the anisotropy of the diamagnetic susceptibility: $\tilde{\chi}_a = \tilde{\chi}_{\parallel} - \tilde{\chi}_{\perp}$, where $\tilde{\chi}_{\parallel}$ and $\tilde{\chi}_{\perp}$ are, respectively , the mag-netic susceptibilities measured parallel and perpendicular to the nematic director. The magnetic energy will be a minimum with the directors aligned parallel to $\mathbf{H}$ (Fig. 3.4a) when $\tilde{\chi}_a$ is positive, and perpendicular to $\mathbf{H}$ when $\tilde{\chi}_a$ is negative (Fig.3.4b). In the latter case, a macroscopically aligned sam-ple can be obtained by rotating the sample about the axis perpendicular to the direction of $\mathbf{H}$ (Fig. 3.4c).

The rate of rotation of the director is characterized by the relaxation time τ_D as given by

$$\tau_D = \frac{\mu_0 \lambda_1}{\tilde{\chi}_a H_o^2} \qquad (3.3)$$

where λ_1 is the viscosity coefficient for the rotation of the director with re-spect to a static mesophase. τ_D may have values varying from a few seconds to many hours. This is an important variable as it governs the ease with which macroscopically aligned samples can be prepared. For a given $\mathbf{H}$ the value of τ_D is governed by $\tilde{\chi}_a$. The latter is dominated by the diamagnetic properties of the constitutent molecules, rather than by the shape of the mi-celle [28]. This is evident from the magnetic responses of N_D phases formed by fluorocarbon and hydrocarbon surfactants: the former are diamagneti-cally positive and the latter negative. This corresponds to the signs of the diamagnetic anisotropies of, respectively, fluorocarbon and hydrocarbon chains. The $\tilde{\chi}_a$ for N_C phases are opposite to those of the corresponding N_D phases. Fluorocarbon systems [29] generally have much shorter values of τ_D than hydrocarbon based ones [30]. This is a consequence of the rel-atively small magnetic anisotropy of the hydrocarbon chain. Indeed, the anisotropies of nematic phases of hydrocarbon surfactants can be changed in sign [28] or set equal to zero [31,32] by substituting aromatic based coun-terions such as benzene sulphonate (OSO_3^-) as illustrated by the data in Table 3.1. This cation is located at the micelle surface with the benzene

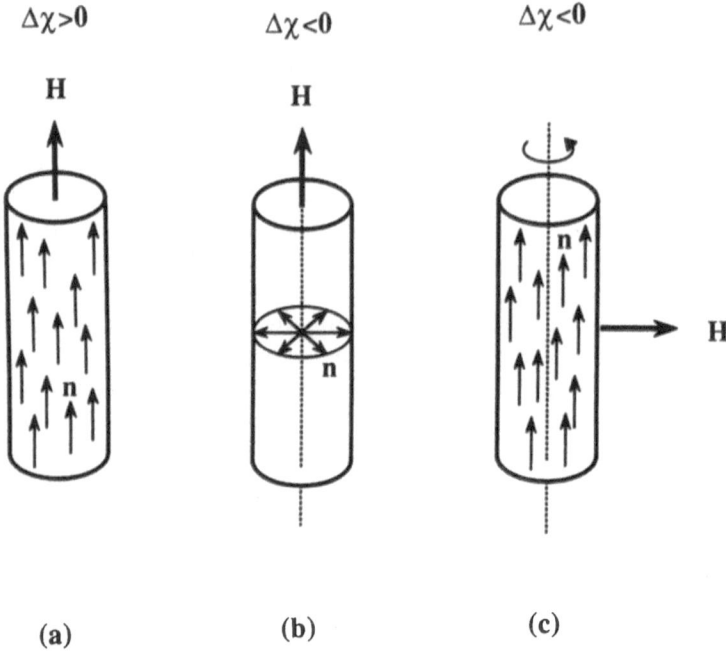

FIGURE 3.4. Orientation of nematic directors n̂ in a magnetic field H for (a) $\chi_a > 0$, (b) $\chi_a < 0$, and (c) $\chi_a < 0$, but with the sample container spinning about an axis perpendicular to the direction of H.

ring directed inwards such that its diamagnetic anisotropy opposes that of the hydrocarbon chain.

The sign of the diamagnetic anisotropy of a nematic phase is denoted by appending the symbol N with a + or a −. This results in four classes of uniaxial nematic: N_C^+, N_C^-, N_D^+, and N_D^-. This classification [28] is now preferred to the earlier one [33] which divides nematic phases into classes Type I or Type II according to whether χ_a is positive or negative.

3.3   Preparation of Stable Nematic Phases

The preparation of stable nematic phases [34,35] can be reduced to the preparation of discrete canonic or discotic micelles (Fig. 3.5) which are stable against explosive growth into, respectively, the infinite cylindrical aggregates of the hexagonal phase or the bilayer aggregates of the lamellar phase. The size of the requisite micelles can be estimated from $\rho L^2 D \approx 1$ for N_C phases or $\rho \sigma^3 \approx 1$ for N_D phases. Since ρ is equal to the volume

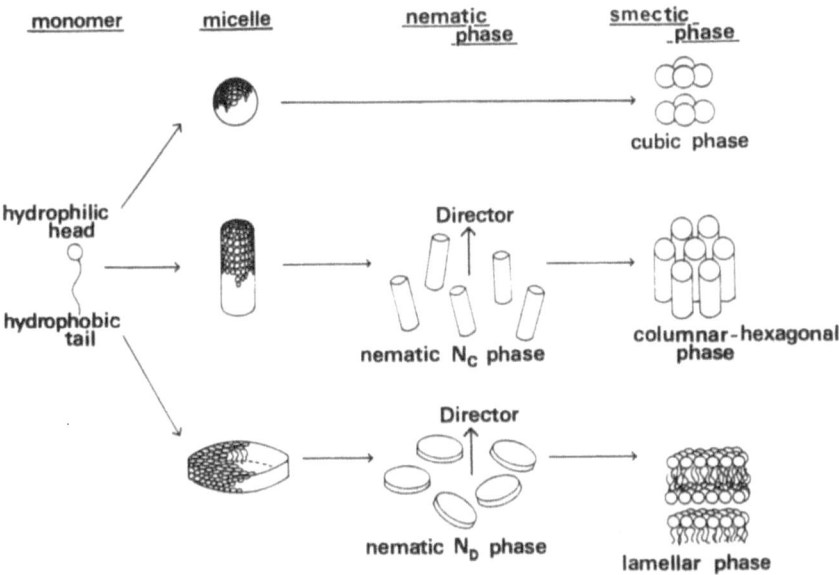

FIGURE 3.5. Schematic representation of possible structures of aggregates and their associated mesophases formed by soap-like surfactants in water. The concentration of surfactant should be read as increasing from left-to-right and top-to-bottom.

fraction of micelles ϕ divided by the average volume of a micelle V_m, we have: for N_C phases, $\phi_N \approx \rho\pi(D/2)^2 L \approx D/L$ and for N_D phases $\phi_N \approx \rho\pi(\sigma/2)^2 D \approx D/\sigma$. Typically, $D \approx 2.5$ nm, so that for ϕ in the range 0.1 to 0.5, the corresponding values of L or σ required to form nematic phases must lie between 25 nm and 5 nm. But, in this concentration range, soaps and soap-like surfactants tend to form translationally ordered mesophases.

It seems quite remarkable that classical soaps (i.e., salts of fatty acids) and soap-like surfactants exhibit an essentially universal phase behavior in water (Fig. 3.6), irrespective of the structure of the polar head group. The dominant feature is that the transition from one mesophase to the next is associated with a dramatic change in the topology of the aggregate. The interpretation of this universal phase behavior can, therefore, be reduced to obtaining an understanding of the factors that govern the change in stability of the various aggregates with concentration. This can be achieved using the elegant continuum model of Charvolin and Sadoc [36] (see Chap. 4). It considers the interplay between the forces normal to the interface, which tend to maintain the surfactant films at constant separation, and the forces which arise from the packing of the polar head groups at the interface and

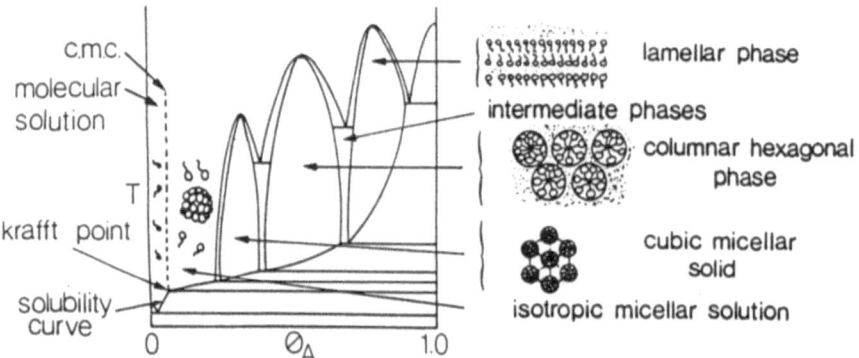

FIGURE 3.6. Schematic phase diagram (volume fraction of surfactant ϕ versus temperature T) for a soap-like surfactant in water.

the chains within the film and which act parallel to the interface and determine the surface curvature. These two forces can be reconciled with the idealized structures of the lamellar, bicontinuous, hexagonal and spherical micellar cubic phases being stable in optimum concentration domains. Away from these domains there is a build up of frustration which may be relaxed by the appearance of local fluctuations or defects in the classical structures.

We can now understand, in terms of this model, why it is generally necessary to add either salt or a long chain alcohol, or both of them, in order to prepare nematic phases from the classical surfactants. It appears that the role of these additives is to shift the lamellar or hexagonal phases away from their optimum concentration domains and thereby introduce local structural fluctuations which lead to finite micelles of the appropriate sizes. This can be seen in the phase behavior of the sodium decylsulphate (SDS)/decanol/water system [37] as given in Fig. 3.7. Here the decanol is seen to stabilize the hexagonal and lamellar structure to lower concentrations of surfactant. Inorganic salts can have a similar role to long chain alcohols. For instance, the addition of NH_4Cl to the decylammonium chloride/water system [38] suppresses the hexagonal phase and enhances the stability of an N_D phase (Fig. 3.8), which in the binary system is only stable for w between 0.42 and 0.49. These are two of the classic systems which have been widely studied, though in both cases the magnetic relaxation times are rather long and sodium decylsulphate undergoes hydrolysis at elevated temperatures [39]. Though it is generally necessary to add a third component, this is not exclusively so: Table 3.1 shows that tetradecyltrimethylammonium bromide forms an N_C phase on dissolution in water alone.

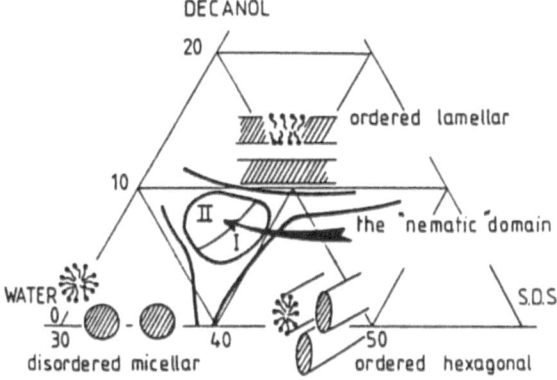

FIGURE 3.7. Partial phase diagram of the sodium decylsulphate (SdS)/decanol/water system from [37].

TABLE 3.1. Structures and magnetic properties of nematic phases of myristyltrimethylammonium bromide and benzene sulphate

					Composition/% w/w		
χ_a	Micellar structure	Desig -nation	MTABr	MTAϕSO$_3$	Water	Decanol	NH$_4$Br
+	canonic	N_D^+	36.0	-	64.0	-	-
-	discotic	N_D^-	26.6	-	63.3	3.8	6.3
-	canonic	N_D^-	-	38.0	62.0	-	-
+	discotic	N_D^+	15.6	14.1	59.2	3.7	7.4

MTA = myristyltrimethylammonium. Temperature was 293 K.

The propensity of soap-like surfactants to pack into aggregates with a variety of structures (so as to relieve the frustration described above) makes them inappropriate for the preparation of stable nematic phases. Consequently, when they *can* be prepared, they tend to be stable over only limited concentration ranges. This behavior stems from the flexibility of the hydrocarbon chain which enables the surfactants to pack into aggregates with varied topologies. So a first requirement is to suppress this degree of structural freedom by replacing the hydrocarbon chain by a less accommodating moiety. This will impose packing constraints on the surfactant molecule and will, in turn, limit the structure of the aggregates. This will not necessarily lead to stable nematic phases *per se*, but, if they are formed, we can expect them to be stable over a much wider concentration range. Importantly, nematic phases will only form provided there exists some mechanism which puts a 'brake' on the growth in size of the aggregates into which the surfactants can assemble.

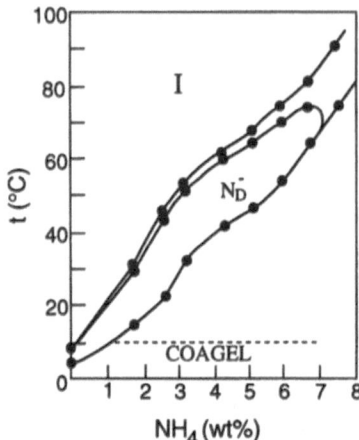

FIGURE 3.8. Phase diagram of the decylammonium chloride (DACl)/NH$_4$Cl/- water system from [38]. The mole ratio of DACl to water is held constant at 0.862.

To optimize the stability of N$_C$ phases it is necessary to choose an amphiphile which can only assemble into canonic aggregates and with the appropriate self-assembly properties to form *finite* sized micelles. This requires a discotic amphiphile consisting of a hydrophobic core with lateral hydrophilic groups. An example of such discotic mesogens is 2, 3, 6, 7,10, 11 - hexa - (1, 4, 7 - trioxaoctyl) - triphenylene, abbreviated to TP6EO2M[40,41]. In water, this amphiphile exhibits a mesomorphism (see Fig. 3.9), which is quite analogous to that normally associated with thermotropic discotic liquid crystal mesogens with increasing volume fractions of water taking on the role of increasing length of aliphatic side chains. But there are two principal differences. First, the nematic phase is rarely observed in thermotropic systems, and when it is it tends to occur usually as a re-entrant phase on cooling a D_{hd} phase. This is because, unlike in the lyotropic N_D phase, the thermotropic molecules are stacked randomly as opposed to one-a-top the other: this constrains the entropy of the side chains, rendering the nematic phase stable only at low temperatures. Secondly, the D_{sd} phase is not observed for thermotropics, suggesting that water is needed to pack the space between the columns in the primitive two-dimensional lattice.

In designing lyotropic discotic mesogens, it is essential to obtain the appropriate hydrophobic/hydrophilic balance. If the ethyleneoxy chains are too short the compounds are not water soluble (for example TP6EO1M), while if they are too long, the compounds become too water soluble, and self-assembly into aggregates is weaker (for example TP6EO3M). On replacing the ether linkages in TP6EO2M with -OCOCH$_2$O-, a N_C phase is

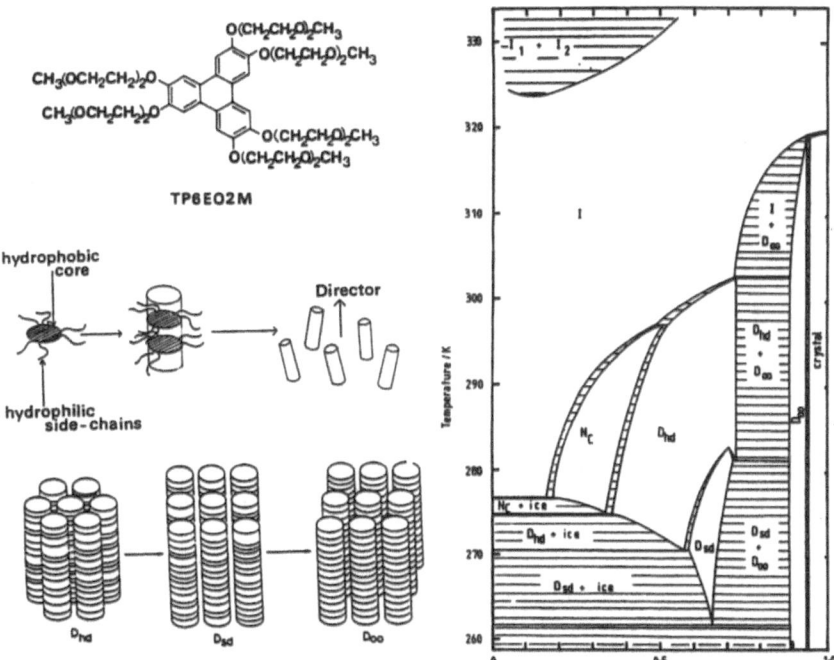

FIGURE 3.9. Phase diagram of the TP6EO2M/^{2}H$_2$O system illustrating the self-assembly behavior of discotic amphiphiles in water: N_c, canonic nematic phase: D_{hd}, hexagonal disordered phase; D_{sd}, possibly the primitive square phase, though a rectangular disordered phase D_{rd} cannot be ruled out; D_{oo}, oblique ordered phase.

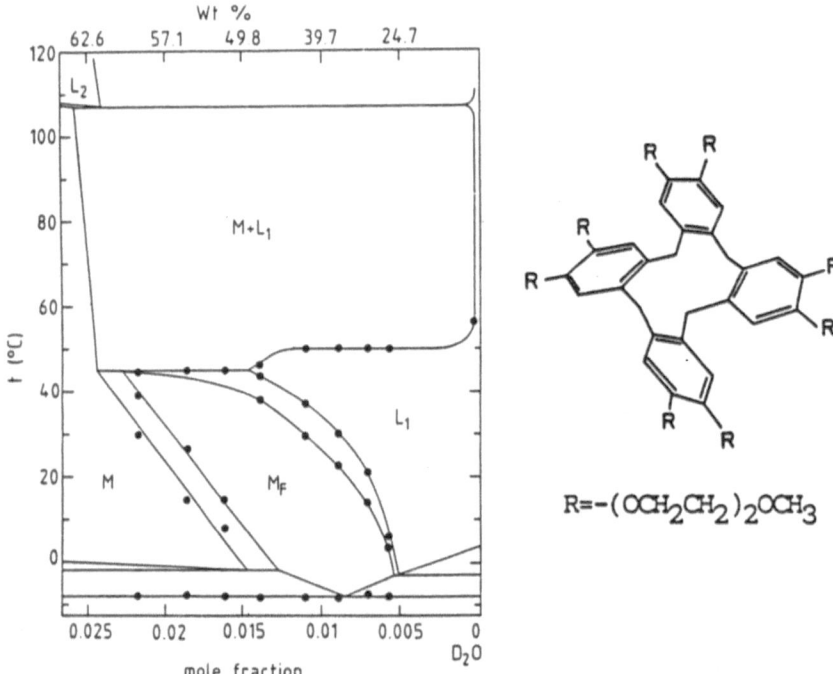

FIGURE 3.10. Phase diagram of 2,3,6,7,10,11,14,15-octa-(1,4,7-trioxactyl) tetra-benzocyclododecatetraene in water, taken from [44]. The M phase extends to the pure surfactant axis. ($M = D_{dh}$; $M_F = N_C$; $L_1 = I$, and L_2 is isotropic inverted micellar phase.)

now only obtained [34] at lower temperatures so as to compensate for the enhanced solubility. The use of $H(OCH_2CH_2)_2NHCOCH_2O-$ side chains gives rise to a D_{hd} phase, but no N_C phase (see page 120 of [43]). The only other discotic amphiphile which, at the present time, is known to give a lyotropic discotic mesophase is tetrabenzocylodecatetraene [44], which has eight $CH_3(OCH_2CH_2)_2O-$ side chains: this gives both an N_D phase and also thermotropic and lyotropic D_{hd} phases (Fig. 3.10).

To optimize the stability of N_D phases, it is necessary to choose an amphiphile which can only pack into two-dimensional planar aggregates. There are two possibilities as depicted in Fig. 3.11. The first is an amphiphile with a rigid, rod-shaped hydrophobic moiety. Fluorocarbon chains are far less flexible than their hydrocarbon counterparts and it is, therefore, quite rea-

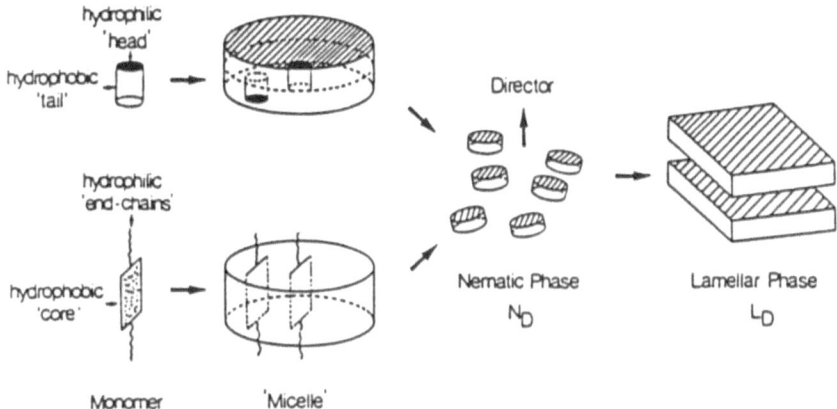

FIGURE 3.11. Schematic of structures of aggregates and mesophases formed by rod-shaped or lath-shaped amphiphiles in water with volume fraction of amphiphile increasing from left-to-right.

sonable to expect fluorocarbon surfactants to self-assemble as rigid rods. Cesium pentadecafluorooctanoate (CsPFO) is an example of such a surfactant. In heavy water (^2H_2O) it forms a N_D^+ phase over a wide range of concentration (0.225 to 0.632 weight fraction w of CsPFO) and temperatures (285.3 to 351.2 K) (Fig. 3.12) [11,45]. This phase lies between an isotropic phase I to higher temperatures and a lamellar phase L_D to lower temperatures. The transition from the L_D-to-N_D^+ phase changes from second order to first order at a tricritical point T_{cp}. The phase diagram in ordinary water (H_2O) is qualitatively the same [46]. The main effect of changing from ^2H_2O to H_2O is to lower the temperature of the L_D-to-N_D^+ to I transitions by roughly 3 K at $\phi \approx 0.1$ and by 1K at $\phi \approx 0.4$; in both cases, Tcp occurs at $\phi \approx 0.25$ with the temperature being 2.7K higher for the ^2H_2O system. Changing the counterion has a much greater effect on the phase behavior as can be seen from the phase diagram for the ammonium salt (APFO) in Fig. 3.13. The N_D^+ phase appears to be much more restricted (w: 0.395 to 0.581; temperature: 292.1 to 338.1 K), however, in T versus ϕ space, the two phase diagrams are quite similar, but with the transition temperatures for the cesium salt some 22 degrees higher. The Rb$^+$ salt also forms N_D^+ phases, while the Na$^+$ and K$^+$ salts are not sufficiently soluble to form liquid crystals. Surprisingly, the Li$^+$ salt behaves more like a classical soap and exhibits both hexagonal and lamellar phases, but no nematic phase [47], and the parent acid exhibits solely a classical lamellar phase [48]. The corresponding salts of heptadecafluorononanoic acid yield qualitatively similar phase diagrams, but the phase transition temperatures are some 25 K higher [48–50]. This tends to make these systems a little less convenient for experimental studies.

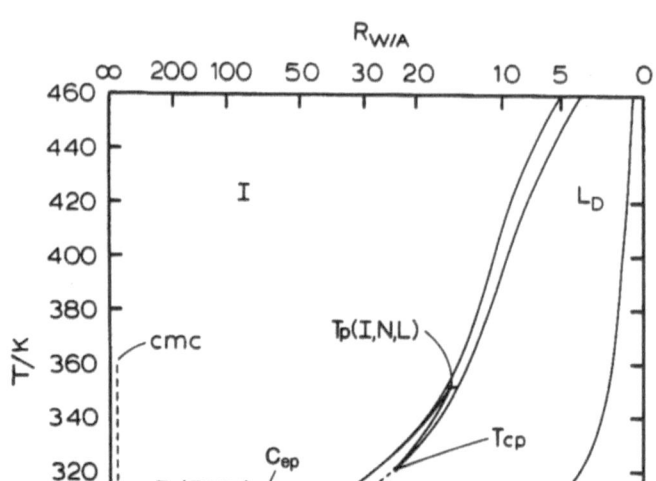

FIGURE 3.12. Phase diagram for the CsPFO/^{2}H$_2$O system taken from [45]. Nomenclature: K, crystal; L_D, discotic lamellar phase—we note that the L_D phase must eventually evolve to a classical bilayer phase at high concentrations w in order to minimize the short-range, inter-layer repulsive forces; N_D^+, nematic phase; I, isotropic micellar solution; HI, heavy ice; Tcp, the lamellar-nematic tricritical point; $Tp(I, N, L)$, the isotropic micellar solution-nematic-lamellar triple point; $Tp*(N, L, K)$, C_{ep}, the nematic-lamellar-critical end point; $Tp(I, N, K)$, the isotropic micellar solution-nematic-crystal triple point; $T_p(HI, I, K)$, the heavy ice-isotropic micellar solution-crystal triple point; Kp, the Kraft point; T_p, the solubility curve for CsPFO.

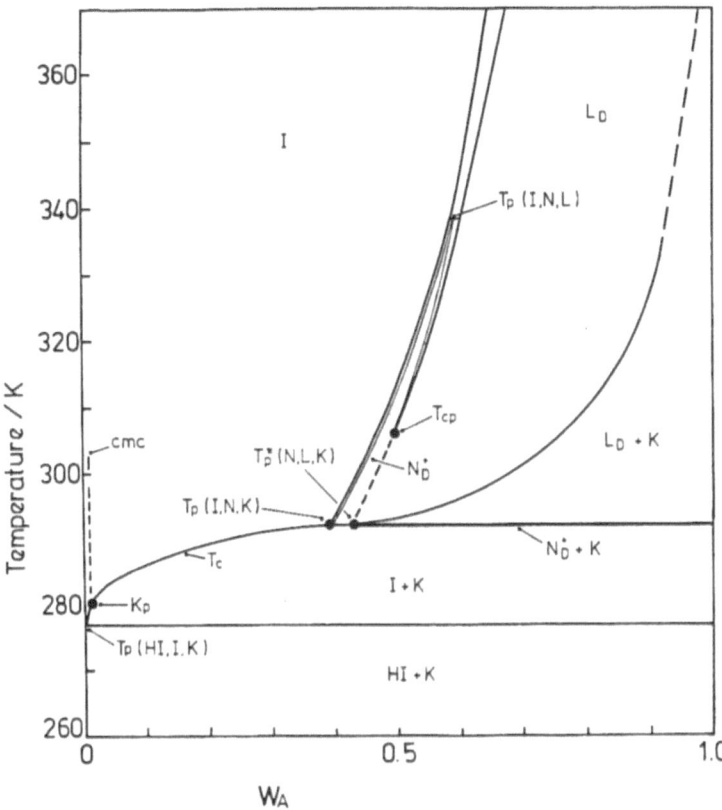

FIGURE 3.13. Phase diagram for the AFPO/^{2}H$_2$O system. Nomenclature is as for Fig. 3.12.

The interaction of counterions with the micelle surface and their role in mediating intermicellar interactions are clearly key factors in stabilizing discrete micelles in concentrated solutions. But before discussing these factors, let us consider the second scenario depicted in Fig. 3.11. That involves using a bipolar surfactant consisting of a fairly rigid lath-shaped core to which are attached hydrophilic end-chains. All attempts to prepare nematic phases from this kind of surfactant have failed: they seem to prefer to form phases analogous to the thermotropic smectic phases (Fig. 3.14). The corresponding unipolar surfactant 3,6,9,12,15,18,21-heptaoxydocosyl-(4'-allyloxy)-4-biphenyloxyacetate does, however, form a N_D phase [52].

In summary, classical soap-like surfactants are not structurally appropriate for the preparation of micellar nematic phases. It is generally necessary to add either an inorganic salt or a long chain alcohol to destabilize the lamellar or hexagonal phase (see [26] for a tabulated summary of nematic systems through 1986). The resulting phases tend to be stable only over

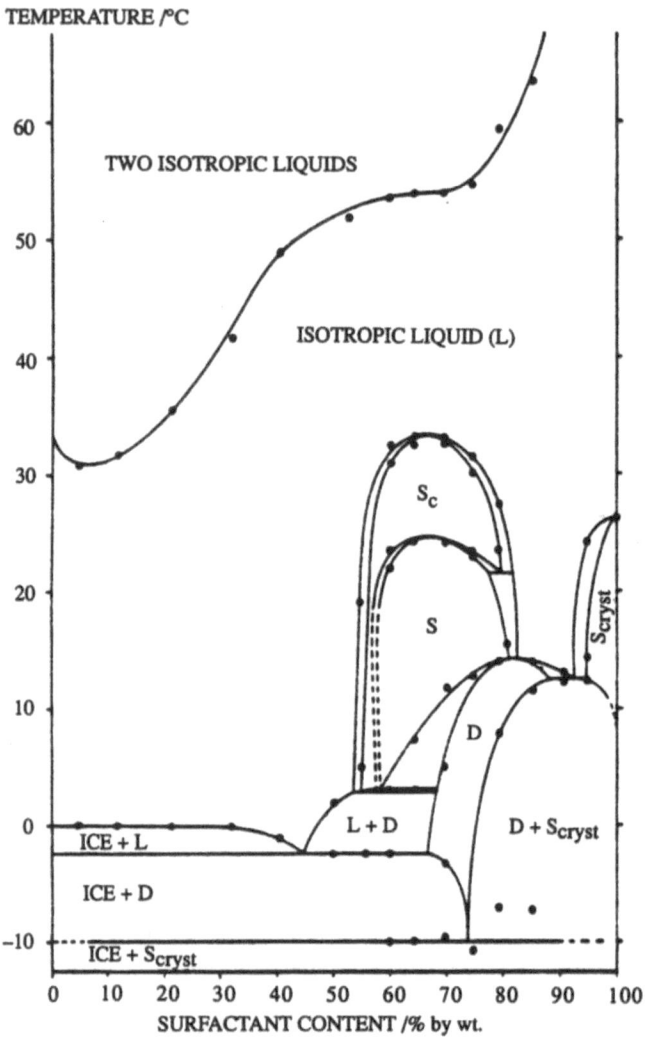

FIGURE 3.14. Phase diagram of a bipolar non-ionic surfactant in water from [51].

narrow concentration ranges. Some progress has been made by using surfactants which are constrained to pack into aggregates having either a cylindrical or planer topology. But it is also essential that there exists some "brake" on the growth of the micelle. This is a complex, yet fascinating problem which is not yet understood. An insight into the factors involved may, however, be obtained by considering the relationship between phase behavior and micelle shape and size.

3.4   Factors Governing Size and Shape of Micelles in Concentrated Solutions

For N_C phases formed by soap-like surfactants L/D is typically 2-4 [9,37,53]. A far more detailed small angle X-ray scattering study of the size of micelles in N_D phases has been practicable with the availability of the CsPFO/ ^2H_2O system. The X-ray diffraction pattern is characteristic of fairly monodispersed discotic micelles in all of the three phases: the axial ratio D/σ is in the range 0.23 to 0.55 with $D = 2.2$ nm [54-56]. The isothermal variation of the micelle size with concentration (Fig. 3.15) is

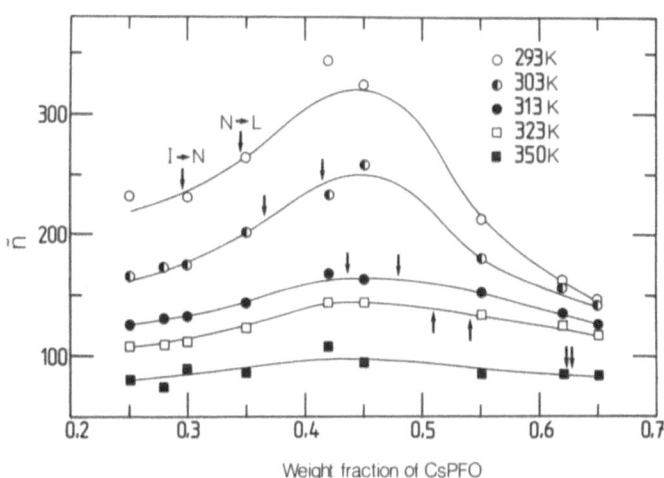

FIGURE 3.15. Average aggregation number $\bar{s}$ as a function of weight fraction w of CsPFO at fixed temperatures. The vertical arrows denote the concentrations at which the isotropic-to-nematic and nematic-to-lamellar transitions occur. Taken from [56].

qualitatively the same at all temperatures. The average aggregate number $\tilde{s}$ initially increases with concentration, reaches a maximum at a weight fraction $w \approx 0.45$ (1.42 mol dm^{-3}) and then decreases. The aggregation number is seen to decrease with increasing temperatures at a rate which is roughly inversely proportional to $1/\sigma$. This also has the effect of weakening the concentration dependence of $\tilde{s}$: in fact, it varies by only 20 per cent at 350 K.

There is only a small (10 to 30 per cent) increase in the value of $\tilde{s}$ at the I-to-N_C^+ transition, as predicted by statistical mechanical models [57-59] discussed at the end of Chap. 1. But, significantly, none is detected at the transition from the N_D^+ to lamellar phase, that is, in the lamellar phase (which is just formed, though not necessarily at higher concentrations), the small discotic micelle is stable with respect to the "classical bilayer". This is an unexpected discovery which is not understood at the present time. The demonstration of similar discotic micellar lamellar phases (denoted L_D) for the tetramethylammonium heptafluorononanoate/water [60] and declyammonium chloride/ammonium chloride/water [61] systems, suggests that such phases may be of wider occurrence. The lamellar layers are essentially two-dimensional fluids of discs and as such they are of fundamental interest. The L_D phase is unique in that it is the only example, to date, of a discotic liquid crystal with one-dimensional translational ordering!

The value of $\tilde{s}$ along the lower boundary T_{NI} of the isotropic-to-nematic transition, summarized in Table 3.2, decreases markedly with increasing concentration. Above 350 K, $\tilde{s} \approx 75$ and, although the micelles are still discotic, the strength of the anisotropic intermicellar interaction is now too weak to seed a nematic phase (Fig. 3.12). We see that along the nematic-

TABLE 3.2. The average aggregation numbers $\tilde{s}$, densities ρ and reduced densities $\rho_{NI}^* = \rho\sigma_{NI}^3$ along the lower boundary T_{NI} of the isotropic to nematic transition of CsPFO/^{2}H$_2$O solutions.

w	ϕ	T_{NI}/K	$\tilde{s}^a$	$\rho_{NI}/10^{24}m^{-3}$	ρ_{NI}^*	$D\sigma$
0.625	0.410	350.0	75	15.19	1.72	0.455
0.55	0.347	330.2	142	8.86	1.79	0.330
0.45	0.262	312.8	170	4.33	1.67	0.302
0.42	0.239	309.0	190	3.54	1.63	0.285
0.35	0.186	299.0	200	2.67	1,32	0,278
0.30	0.153	293.3	220	2.01	1.15	0.265
0.28	0.144	290.8	270	1.51	1.18	0.239
0.25	0.125	287.7	260	1.35	0.98	0.244

The axial ratios D/σ have been calculated assuming the micelle is an oblate ellipsoid with $D = 2.2$ nm.

[a] $\pm$ 7.5 per cent

to-isotropic transition line the value of $\rho_{NI}^* = \rho \sigma_{NI}^3$ is not constant, but, for the assumed ellipsoid micelle, it varies from 0.98 up to 1.72 as the weight fraction of CsPFO is increased from 0.25 to 0.625. These values are consistently smaller than the value of 4.07 predicted by the hard thin disk (i.e., infinite axial ratio) model [62]. This result is hardly surprising in view of the small axial ratio and the undoubted existence of a more complex intermicellar interaction.

To understand the variation of $\tilde{s}$ with concentration it is necessary to extend to discotic micelles the theory of Gelbart, Ben-Shaul and co-workers for the aggregation behavior of amphiphiles into rod-shaped micelles [57–59]. Only the outline is given here since full details can be found in reference 56–see also the discussion in Sec. 1.4.

The average chemical potential of a surfactant molecule in a micelle (assumed to be monodisperse: $X_s = X$) of size s is written (in units of kT) as in the final section of Chap. 1:

$$\tilde{\mu}_s = \tilde{\mu}_s^0 + \frac{1}{s}\ln\frac{X}{s} + \tilde{\mu}_s^\theta + \frac{1}{s}\sigma_s + \frac{1}{s}\chi_s \qquad (3.4)$$

The first term represents the average chemical potential of an amphiphile in a micelle of size s which has a particular orientation and is at a particular position in the solution. It is conveniently expressed as

$$\tilde{\mu}_s^0 = \tilde{\mu}_{\infty,d}^0 + \alpha_d(s)f_d(s) \qquad (3.5)$$

where $\tilde{\mu}_{\infty,d}^0$ is the chemical potential of a molecule in the "flat" portion of the micelle (here, for simplicity, considered to be a cylindrical disk with hemitoroidal rim). The second term in Eq. (3.5) is the chemical potential of a molecule in the rim with respect to that of one in the body. The functions $\alpha_d(s)$ and $f_d(s)$ can be found in reference [56], where it is shown that, by making a number of reasonable approximations, $\alpha_d(s)f_d(s)$ reduces to $\alpha'g(s)$, where $g(s)$ is a function of micelle size, and α' contains the interfacial tension between the fluorocarbon and aqueous phases. Since the latter quantity is quite large ($\approx 60\ mNm^{-1}$), α' is also large, and there is a strong tendency for the system to reduce surface curvature and for the micelle to grow.

The second term in Eq. (3.4) is the entropy of mixing contribution, with X_s denoting the total mole fraction of surfactant incorporated into micelles of size s. This term is negative, and it will decrease as $\tilde{s}$ increases. Consequently, it acts in opposition to the first term and provides a mechanism for keeping the micelles small in dilute solution.

The third term, $\tilde{\mu}_s^\theta$, represents contributions arising from the translational and rotational degrees of freedom of the micelle. It can be written [58]

$$\tilde{\mu}_s^\theta = \frac{1}{s}\ln\left(\Lambda_1^3\rho_m\right) - \frac{3}{2}\frac{1}{s}\ln s - \frac{1}{2}\frac{1}{s}\ln\left\{R_A(s)R_B(s)^2\right\} \qquad (3.6)$$

where $\Lambda_1 = (h^2/2\pi\, m_1 T)^{1/2}$ is the de Broglie wavelength of a surfactant molecule of mass m_1 and length l, and ρ_m is the number density of all molecules in solution. $R_A(s) = l_A(s)/m_1 l^2$ and $R_B(s) = l_B(s)/m_1 l^2$ are the dimensionless moments of inertia of the micelle about its minor and major axes. All three terms in Eq. (3.6) are small since they are proportional to $1/s$. The second and third terms have the same functional form as the entropy of mixing term and reinforce the role of the latter in favoring small micelles in dilute solution.

The fourth term, σ_s/s, represents the contribution arising from the entropy loss due to the establishment of long range orientational order of the micelles in the nematic phase: $\sigma_s/s = (\ln\alpha - 1)/s$, where α is related to the order parameter of the micelle S $[\equiv (\langle 3\cos^2\beta\rangle - 1)/2$, where β is the angle between the symmetry axis of the micelle and the direction of the nematic director] via $S \approx 1 - 3/\alpha$. This term accounts for the small, yet detectable, increase in the size of the micelle at the isotropic-to-nematic transition (for a sample with $w = 0.55$, the values of $\tilde{s}$ are 88 at T_{IN} and 113 at T_{IN} [55]. But its effect is quite small compared with the overall effects of concentration and temperature (Fig. 3.15).

The term χ_s/s represents the contribution arising from the forces between the micelles, and it appears to be responsible for the diminution in the micelle size at high concentrations. But before analyzing this behavior, the growth in the micelle size observed at low concentrations will be considered in the light of Eq. (3.4).

At $low\ concentrations$, where the micelles are sufficiently far apart to neglect the intermicellar interactions, Eq. (3.4) can be written

$$\tilde{\mu}_s = \tilde{\mu}^0_{\infty,d} + \alpha' g(s) + \frac{1}{s}\ln\frac{X_s}{s} + \frac{1}{s}\ln\left(\Lambda_1^3\rho_m\right) - \frac{3}{2}\frac{1}{s}\ln s - \frac{1}{2}\frac{1}{s}\ln\left\{R_A(s)R_B(n)^2\right\} \qquad (3.7)$$

For flat disks with square edges, $R_A(s) = s^2 v_1/4\pi l^3$ and $R_B(s) = s^2 v_1/8\pi l^3$, where v_1 is the volume of a single amphiphilic molecule. On minimizing Eq. (3.7) we obtain

$$s = X_s^p \exp\left(p\alpha' h'(s) + A\right) \qquad (3.8)$$

where $h'(s) = -s^2\, dg(s)/ds$ and A is a numerical constant. The power law relating s and X_s depends critically upon the form and number of terms included in the starting equation. When only the entropy of mixing term (in addition to the first two) is included, $p = 1.0$, while when the translational entropy contribution is included, too, $p = 0.4$, which coincidentally is close to the gradient in Fig. 3.16. But what is of greater significance here is the fact that s is predicted to increase with X_s in the absence of intermicellar interactions, since p is always greater than zero. From Eq. (3.8) we see that this requires $X_s\alpha' dh'(s)/ds < 1$, a condition readily satisfied for the

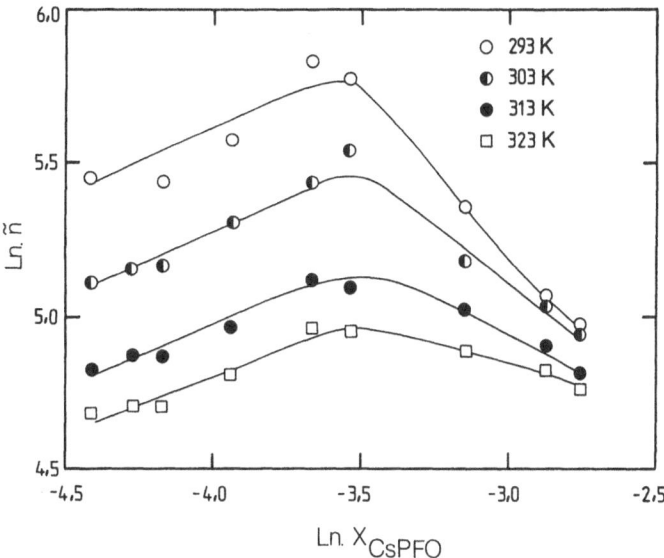

FIGURE 3.16. Log $\tilde{s}$ as a function of log X_{CsPFO}. The continuous lines represent linear fits of the low-concentration data (slopes are 0.44 at 293 K, 0.42 at 303 K, 0.40 at 313 K, and 0.38 at 323 K; on the high concentration side of the maximum they have been drawn simply to aide the eye). Taken from [56].

CsPFO/water system. Indeed, Fig. 3.16 shows that at low concentrations, $\ln \tilde{s}$ is proportional to $\ln X_{CsPFO} \approx \ln X_s$ ($X_{CsPFO} = 4.5 \times 10^{-4}$ at the c.m.c. at 298 K): in fact, to a good approximation, $\tilde{s} \, \alpha \, X_{CsPFO}^{0.4}$ at all temperatures.

We now turn our attention to an explanation for the diminution in the size of the micelles which occurs at *high concentrations*. Here the term involving the intermicellar interactions becomes significant and is expected to dominate the entropy of mixing and the translational and rotational entropy terms. Eq. (3.4) may, therefore, be rewritten as

$$\tilde{\mu}_s \approx \tilde{\mu}_{\infty,d}^0 + \alpha' g(s) + \frac{1}{s}\chi_s \tag{3.9}$$

which, upon minimization, gives

$$\frac{ds}{dX_s} = \frac{d\chi_s}{dX_s}\left(\frac{s^2}{\alpha' h'(s)}\right) \approx \frac{d\chi_s}{dX_s}\frac{s^2}{52} \tag{3.10}$$

Since the factor in brackets is always positive, the sign of ds/dX_s is solely determined by that of $d\chi_s/dX_s$. If the force between the micelles is attractive, then on increasing X_s, χ_s must become more negative so $d\chi_s/dX_s$ and, consequently, ds/dX_s are negative. Conversely, if the force between the micelles is repulsive, χ_s becomes more positive upon increasing X_s, and hence $d\chi_s/dX_s$ and ds/dX_s are positive. Fig. 3.15 shows that at high concentrations ds/dX_{CsPFO} is negative, consistent with a net attractive force between the micelles. It could be argued that the same behavior could obtain from a repulsive force which became less repulsive at higher concentrations. This behavior would be expected if, for example, double layer coulombic repulsions enhance the excluded volume dimension of the micelle [63]: at higher concentrations the Debye length would decrease, resulting in a shorter range repulsive force. But such an interpretation would not appear to be consistent with the behavior observed at low concentrations where there is no evidence for any initial rapid growth in $\bar{s}$. Thus, it seems reasonable to conclude that at high concentrations the growth in size of the micelle, driven by the intramicellar interactions, is suppressed by a net attractive intermicellar force which appears to become stronger with increasing concentration. The intermicellar force will, of course, be anisotropic and possibly with the edge-to-edge force being attractive and the face-to-face force repulsive. This attractive edge-to-edge force is envisaged to be mediated by counterions which give rise to a secondary minimum in the intermicellar potential energy curve at the small edge-to-edge separations.

Intuitively, we might expect the size of the micelle to be simply related to the fraction of bound cesium ions β_{Cs}, the idea being that ions are intercalated between neighboring negatively charged carboxylate groups in the body but not in the rim, lowering the chemical potential of surfactant molecules in the former with respect to those in the latter. The net effect would be to increase the magnitude of the second term in Eq. (3.5) and thereby effect an increase in size of the micelle. The size of the micelle does indeed increase with β_{Cs} as predicted (Fig. 3.17), but only up to $\beta_{Cs} \approx 0.45$, and thereafter it decreases. This is consistent with the notion that up to $w = 0.45$ the micelle size is governed by intramicellar interactions, while at higher concentrations some other factor becomes predominant. The decrease in size of the micelles with increasing temperature (Fig. 3.15) parallels a decrease in the value of β_{Cs} (at $w = 0.55$, β_{Cs} decreases from 0.52 to 0.46 as the temperature is increased from 308 to 328 K). Clearly, the value of the second term in Eq. (3.5) depends not only upon the curvature free energy, but also on the detailed arrangement of the counterions at the surface of the micelle. Of course, the variation of β_{Cs} with concentration and temperature could simply be an indirect effect of changes in micelle size and shape. A more dramatic demonstration of the importance of the role of the counterion is given by the very different phase behavior of CsPFO and LiPFO. This clearly has its origin in the much greater hydration energy of the smaller Li^+ ion. As a consequence, the hydrated Li^+ ion is less likely to be intercalated between neighboring carboxylate groups. This will result in an expansion and concomitant curvature of the interface.

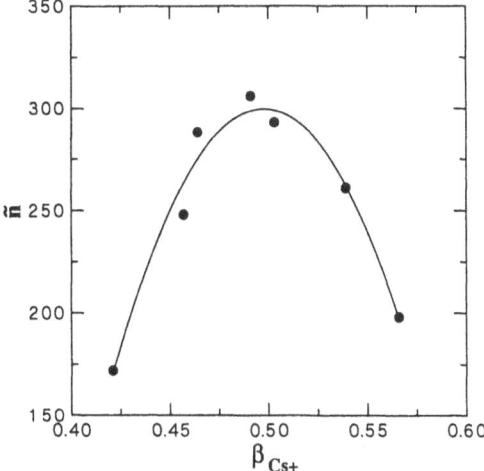

FIGURE 3.17. Average micellar size versus fraction of bound caesium ions β_{Cs} for the CsPFO/^{2}H$_2$O system.

Much less is known about the structure of N_C phases. The TP6EO2M/water system (Fig. 3.9) is a promising one for a detailed investigation. Apart from exhibiting a N_C phase over a fairly extensive concentration interval, and having a very short magnetic relaxation time, it is the archetypal one-dimensional self-assembling system. It may , therefore, be used as a model to represent the behavior of many polyaromatic drug [64] and dye [65] molecules which behave similarly (see Fig. 3.18 for an example).

One dimensional self-assembling systems are inherently polydisperse. The distribution of aggregate sizes and its variation with both concentration and temperature has been characterized recently by NMR spectroscopy [112]. The results are surprising in that the number average axial ratios $\bar{L}/D$ vary from ≈ 18 to ≈ 45 as the concentration increases along the isotropic-to-nematic transition line in Fig. 3.9. This is in marked contrast to Onsager hard-particle theories, which require that $\bar{L}/D$ becomes smaller with increasing concentration. Indeed, Taylor and Herzfeld [113], using a scaled particle treatment of fluid configurational entropy and a cell description of periodic columnar density modulation, have obtained recently a phase diagram for a system of polydisperse self-assembling disks which has many features in common with the experimental phase diagram for TP6EO2M. To reconcile the discrepancies between theory and the experimental size distribution it is necessary to invoke that the aggregates are flexible and become more so as the temperature is raised. The effects of flexibility of micellar rods on the isotropic-to-nematic transition have been discussed by Odijk [114] and, more recently, by Hentschke [115].

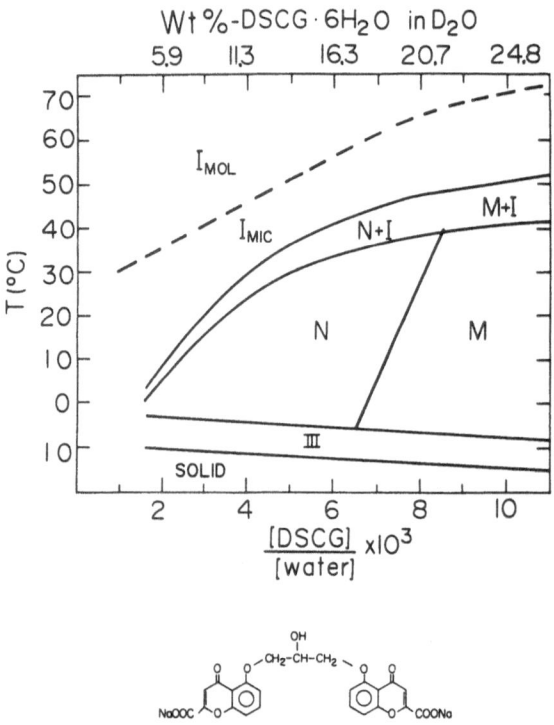

FIGURE 3.18. Phase diagram of the disodium cromoglycate/^{2}H$_2$O system from [64].

3.5   Experimental Characterization of Nematic Phases

The primary aims here are to map the phase diagram, to establish the shapes of the micelles and the sign of the diamagnetic anisotropy, to determine the sizes of micelles and their variation with concentration and temperature, and finally, to measure the nematic order parameter. The latter presents new problems which are not associated with thermotropic liquid crystals, and will be discussed in the following section. Here we focus on the use of optical microscopy, NMR spectroscopy, and X-ray diffraction in establishing the other properties. Of course, there are other techniques which could be employed. But these three are generally widely accessible, relatively easy to apply, and, when used sequentially, provide a routine methodology.

3.5.1 POLARIZING OPTICAL MICROSCOPY

This is a simple and inexpensive technique which can be used to identify the mesophase, micelle structure, sign of the diamagnetic anisotropy, and to map phase diagrams [66].

In the absence of external forces, a thin film (100 to 200 μm) of nematic phase, viewed between crossed polarizers, exhibits a Schlieren texture (Fig. 3.1). This is the characteristic optical signature of a nematic phase. The shape of the micelle and the sign of the diamagnetic anisotropy can be deduced from the response of the texture to shear, surface and magnetic forces. For example, when a N_C is drawn into a rectangular optical capillary, the long axis $\hat{c}$ of the micelle, and consequently the nematic director $\hat{n}$, will tend to align parallel to the direction of flow (the x axis in Fig. 3.19). For perfect alignment (planar film) the film will appear uniformly bright (Fig. 3.19a). In contrast, the corresponding texture for a N_D phase (Fig. 3.19b) has a mottled appearance. This arises because the nematic director now tends to align *perpendicular* to the direction of flow. The texture of the N_C phase is invariant on standing, while that of the N_D phase becomes uniformly black (pseudo-isotropic texture) due to the alignment of the micelles with their $\tilde{a} - \tilde{b}$ planes parallel to the surface (homeotropic film). The sign of the susceptibility anisotropy $\tilde{\chi}_a$ can be determined by observing the response of these two textures to a magnetic field suddenly applied along the z direction. For both N_D^- and N_C^- mesophases, the effect of the magnetic field is simply to reinforce the existing alignment of the director, while for N_D^+ and N_C^+ mesophases, Fréedericksz (i.e., alignment reorientation) transitions with associated dynamic effects [67] are observed (Figs. 3.19c and d).

Microscopy experiments may also be conducted using glass slides and coverslips, though controlling concentration and carrying out experiments involving magnetic fields are more difficult. Moving the coverslip from side-to-side shears the film, giving textures similar to those in Figs. 3.19a and b. "Water penetration" experiments (see Fig. 3.20) are more easily conducted with this experimental arrangement and are invaluable for screening the mesogenic properties of surfactants.

Observations of micrographic textures of samples in sealed capillaries provide an excellent method for mapping phase diagrams. The method does not, however, have sufficient resolution to accurately define the widths of biphasic regions, nor to ascertain the relative amounts of coexisting phases.

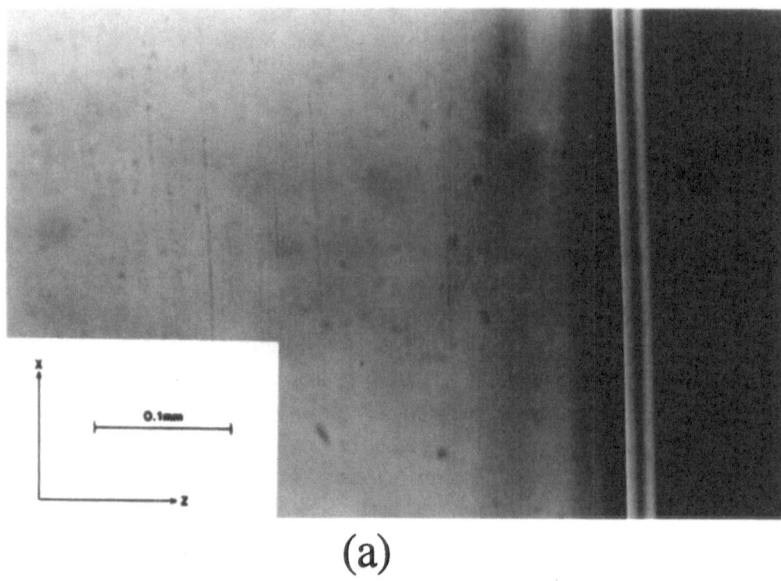

(a)

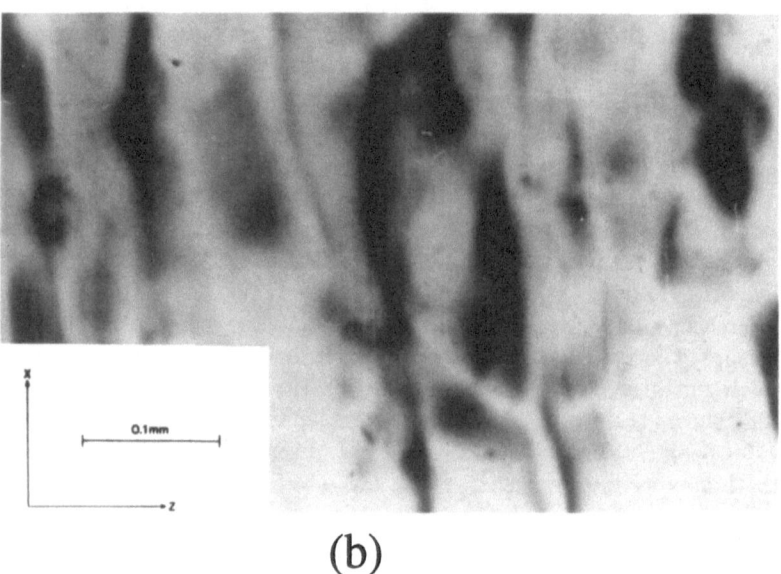

(b)

FIGURE 3.19. (a, b) Photomicrographs of (a) $N_C^\pm$ and (b) $N_D^\pm$ nematics immediately following shear flow and (see below) of the transient textures observed for
(c) N_C^+ and (d) N_D^+ nematics on applying a magnetic field (0.47 T) along the z
direction. The nematic phases used are similar to those described in Table 3.1.
Taken from [66].

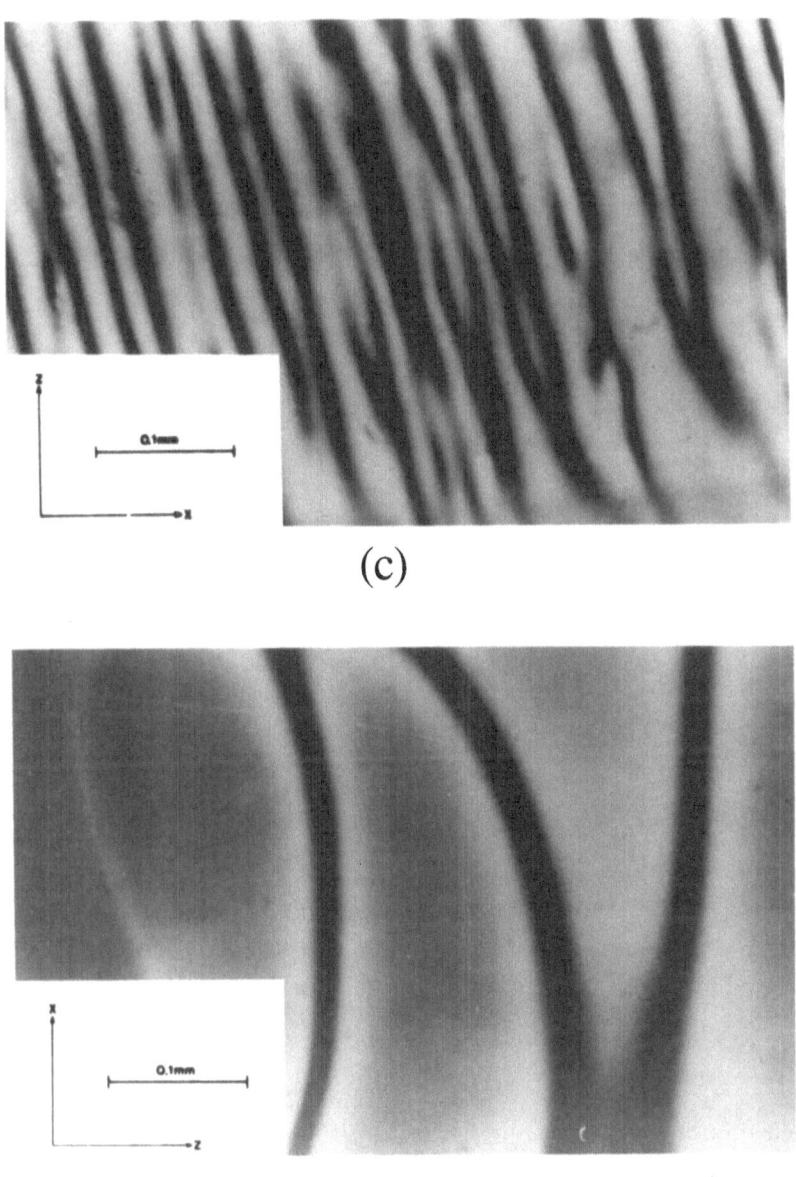

FIGURE 3.19. (c, d)

(a)

(b)

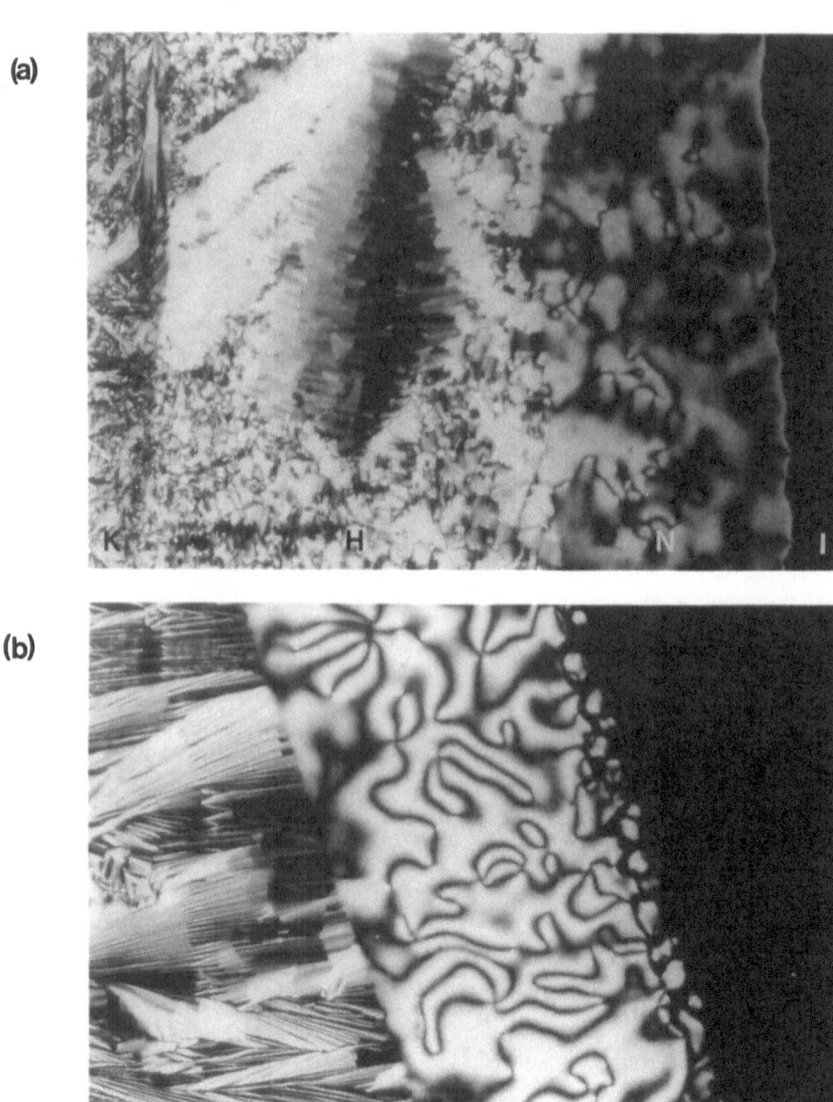

FIGURE 3.20. (a, b) Photomicrographs of textures observed for TO6EO2M/ water mixtures at room temperature and with magnification *times* 100: (a) sequence of phases observed when a concentration gradient is established by allowing water to penetrate into an array of crystals; (b) as in (a), but concentration gradient established by allowing water to evaporate out of the sample cell; (next page) (c) focal conic texture of hexagonal phase (magnification x200). Nomenclature: K, crystal; H, hexagonal phase; N, nematic phase; I, isotropic micellar solution. Taken from [40].

(c)

FIGURE 3.20. (c)

3.5.2   NMR SPECTROSCOPY

NMR measurement of the quadrupole splittings of, for example, ^{2}H in labelled ^{2}H$_2$O(D$_2$O) in the CsPFO/water system, is unparalleled as an experimental method for mapping phase diagrams and, in particular, for characterizing phase transitions. This is because the quadrupole splitting is a characteristic signature of a mesophase and is also a sensitive function of composition and temperature. It can, therefore, provide detailed information about the uniformity of composition and temperature in bulk samples. This is particularly useful information when traversing phase coexisence regions since in most instances it enables the relative amounts of coexisting phases to be monitored. Good temperature control (within 5 mK) is, of course, essential in such applications. It can also be used to monitor the orientational distribution of the mesophase director and to distinguish between uniaxial and biaxial mesophases. The utility of the method will be illustrated by considering its application to, first, the N_D^+ mesophase of the CsPFO/water system [45] and, secondly, the N_C^- mesophase of the TP6EO2M water system [34].

In the case of the CsPFO/water system there are two different species of spin which can be studied: ^{2}H in labelled water (it is preferable to use pure heavy water so that the system is well defined) or ^{133}Cs in the counterions (^{13}C and ^{19}F spins can be studied, but are less useful for the purposes discussed here). In a macroscopically aligned uniaxial mesophase (nematic, lamellar, or hexagonal), the first order spectrum for any spin I($>$ 1/2) will consist of 2I equally spaced lines with separation, referred to as the quadrupole splitting, given by [68]:

$$\Delta\tilde{\nu}(\theta) = \frac{3}{2I(2I-1)}|\tilde{q}_{zz}|_s SP_2(\cos\theta) \tag{3.11}$$

where the upper tilde denotes partially averaged quantities. In Eq. (3.11), θ is the angle between $\hat{n}$ and H. S is an orientational order parameter and represents the ensemble average of the orientational fluctuations of the micellar axes M(a,b,c) with respect to the direction of H. $|\tilde{q}_{zz}|_s$ is the partially averaged component of the nuclear-quadrupole-electric-field gradient interaction tensor measured parallel to $\hat{n}$ in a perfectly ordered mesophase; it is given by (the summation running over all sites n)

$$|\tilde{q}_{zz}|_s = \sum p_n Q_n \{S^n_{qq} - \eta_n \frac{1}{3}(S^n_{oo} - S^n_{pp})\} \tag{3.12}$$

where the S_{ij} are the elements of the Saupe ordering matrix–see Eq. (3.1)– for the principal axes $(\hat{o}, \hat{p}, \hat{q})$ of the nuclear quadrupole interaction tensor at the nth site which has statistical weight p_n. Q_n is the corresponding quadrupole coupling constant, and η_n is the asymmetry parameter. The actual values for Q_n and η_n, in addition to being dependent upon the nucleus being observed, will vary from site to site and will thus be determined by the detailed structure of the micelle.

The spectrum for a ^{2}H spin (I=1) in labelled water will be a symmetrical doublet with separation

$$\Delta\tilde{\nu}(\theta) = \frac{3}{2}|\tilde{q}_{zz}|_s SP_2(\cos\theta) \tag{3.13}$$

with

$$|\tilde{q}_{zz}|_s = < P_2(\cos\alpha) >_s Q_D(x_a/x_w)n_b S_{OD} \tag{3.14}$$

To transform Eq. (3.12) into (3.14) it has been assumed that all of the Q_n are identical (Q_D) and that for water molecules not bound to the surface $|\tilde{q}_{zz}|_s = 0$. n_b is the number of water molecules bound per molecule of amphiphile, x_a and x_w are respectively, the mole fractions of amphiphile and water, and S_{OD} is an "order parameter" which represents the averaging due to the local reorientational motion of bound water molecules. The quantity $< P_2(\cos\alpha) >_s = < 3/2\cos^2\alpha - 1/2 >_s$, where α is the angle between the normal to the surface and the symmetry axis of the micelle and the angular brackets denote an average accounting for the diffusive motion of the molecule over the micellar surface. In contrast, the spectrum for ^{133}Cs (I=7/2) consists of seven equally spaced lines of relative intensities 7:12:15:16:15:12:7 and with separation given by

$$\Delta\tilde{\nu}(\theta) = \frac{1}{14}|\tilde{q}_{zz}|_s SP_2(\cos\theta) \tag{3.15}$$

with

$$|\tilde{q}_{zz}|_s = < P_2(\cos\alpha) >_s Q_{Cs}\beta_{Cs} \tag{3.16}$$

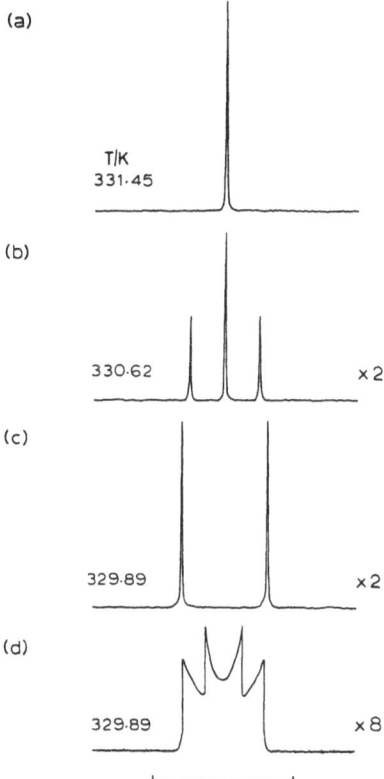

FIGURE 3.21. ^{2}H NMR spectra of ^{2}H$_2$O observed when the sample CsPFO ($w = 0.55$)/^{2}H$_2$O is cooled from the isotropic micellar solution phase (a), across the isotropic-nematic phase coexistence region (b), and into the nematic phase (c). The spectrum in (d) was obtained by spinning the nematic phase at 0.25 Hz about an axis perpendicular to the direction of the magnetic field. Taken from [45].

where β_{Cs} is, as before, the fraction of Cs$^+$ ions bound to the micelle. The field gradient at the cesium nucleus is considered to arise from distortion, of the symmetry of the hydration shell [69]. Furthermore, this distortion, and the surface coverage, are assumed to be independent of the angle α.

The ^{2}H spectra observed on cooling a sample with weight fraction $w = 0.55$ from the I to N_D^+ phase (Fig. 3.12) are shown in Fig. 3.21. The spectrum in the isotropic phase ($S = 0$) is a singlet. In the nematic phase ($S \neq 0$) it is a simple doublet, characteristic of a homeotropic distribution of nematic directors $\hat{n}$ (Fig. 3.4a) throughout the bulk sample (macroscopic alignment with $\theta = 0$). When the tube containing the sample is spun about its axis, which is aligned perpendicular to H, the spectrum

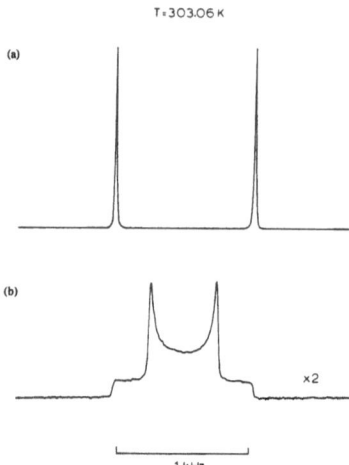

FIGURE 3.22. ^2H NMR spectra of ^2H_2O in the lamellar phase of CsPFO $(w = 0.55)/^2H_2O$ at 303.06 K with (a) a homeotropic distribution of lamellar directors (normals to the planes of the lamellae) and (b) an isotropic distribution of directors. Taken from [45].

obtained is characteristic (Fig. 3.21d) of a two-dimensional distribution of nematic directors in a plane perpendicular to the axis of the rotation (c.f. director distribution in Fig. 3.4b). This confirms that the mesophase has positive diamagnetic anisotropy. When the spinning is stopped, the spectrum relaxes quickly ($\approx 200ms$) to a single doublet. The lamellar phase, in contrast, has an infinite valued rotational viscosity coefficient and the lamellar director (perpendicular to the planes of the lamellae) is locked into the mesophase. This behavior is nicely illustrated by the spectra for the same sample in Fig. 3.22. The spectrum (Fig. 3.22a) obtained when the sample is slowly cooled in the spectrometer magnet from the nematic into the lamellar phase, is a simple doublet whose splitting is proportional to $P_2(\cos\theta)$, consistent with a homeotropic director distribution. If the experiment is now repeated, but the sample is cooled outside of the magnet, a "Pake" spectrum (Fig. 3.22b) is obtained, consistent with an isotropic distribution of director orientations.

It is thus possible to distinguish between isotropic, nematic and lamellar phases by their characteristic deuterium NMR spectra and thereby to locate the phase boundaries. But to locate precisely the phase boundaries it is essential to make use of the variation with temperature of the magnitude of the quadrupole splitting $\Delta\tilde{\nu}(\phi = 0)$, unless stated otherwise. There are two types of $\Delta\tilde{\nu}$ versus temperature curve depending upon composition. For samples with w greater than the concentration of the tricritical point T_{cp}, separate lamellar and nematic doublets are observed at the lamellar-

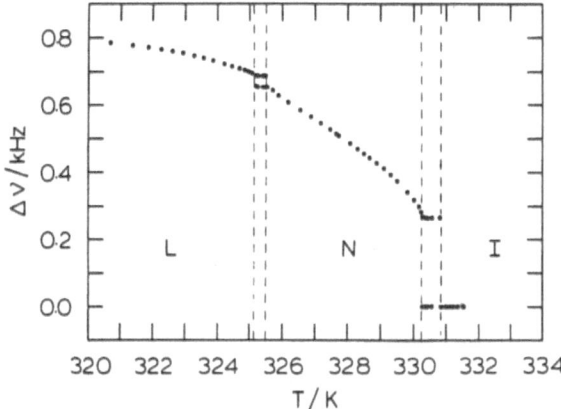

FIGURE 3.23. Partially averaged quadrupole splitting $\Delta\tilde{\nu}$ of ^2H_2O as observed on cooling a sample of $CsPFO/^2H_2O$ ($w = 0.55$) from the isotropic micellar solution phase to the nematic phase and then from the nematic phase into the lamellar phase. Taken from [45].

to-nematic transition, while for samples with lower concentrations only a single doublet is seen.

The behavior at high concentrations is illustrated in Fig. 3.23 by a $w = 0.55$ sample. On cooling from the isotropic micellar solution into the isotropic–nematic biphasic region, a symmetrical doublet from the ^2H_2O molecules in the nematic phase is superimposed on the isotropic singlet arising from the ^2H_2O molecules in the isotropic micellar solution (Fig. 3.21b). As cooling continues, the doublet intensity increases at the expense of that of the singlet, consistent with the variation of the relative amounts of the two phases as described by the lever rule. The quadrupole splitting is seen to change little in the mixed phase region. This is because the value of S is essentially independent of composition along the N_D^+-to-I transition line. At the lower boundary to the transition, the singlet disappears (Fig. 3.21c). Thereafter, in the nematic phase, the doublet splitting increases rapidly at first, then more gradually, reflecting the increase in the nematic order parameter S with decreasing temperature. However, $\Delta\tilde{\nu}$ will increase more rapidly than S, because $|\tilde{q}_{zz}|_s$ also increases as a result of the growth in the diameter of the micelles. (Recall Fig. 3.15.) At the upper boundary of the nematic-to-lamellar transition a second doublet appears, symmetrically disposed with respect to the nematic doublet. The quadrupole splitting in both phases remains fairly constant across the mixed phase region for the same reasons as outlined for the observed behavior at the isotropic-to-nematic transition. The discontinuity in quadrupole splitting is characteristic of a first order lamellar-to-nematic transition. Here it arises mainly from a dis-

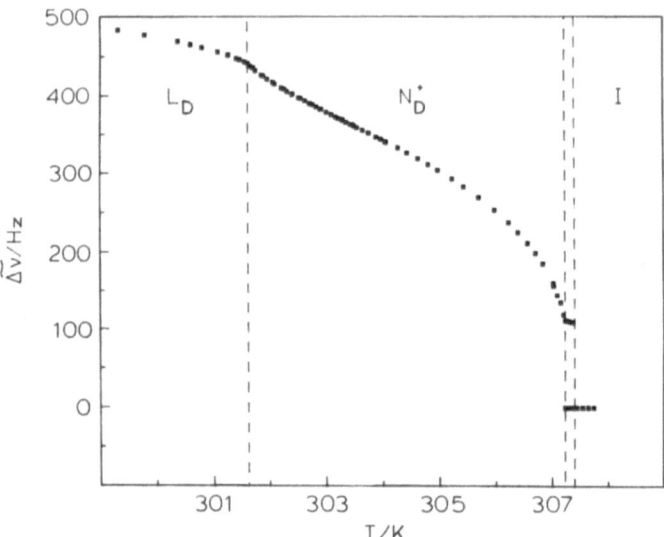

FIGURE 3.24. Partially averaged quadrupole splitting $\Delta\tilde{\nu}$ of ^2H_2O as observed on cooling a sample of CsPFO/^2H_2O ($w = 0.40$) from the isotropic micellar solution phase to the nematic phase and then from the nematic phase into the lamellar phase. Taken from [45].

continuity in S with a smaller contribution from a change in $|\tilde{q}_{zz}|_s$ brought about by the discontinuity in concentration on entering the mixed phase region. There is no change in the structure of the micelles at this transition. At the lower boundary of the transition, the nematic doublet vanishes. The splitting of the remaining lamellar doublet increases relatively slowly with decreasing temperature as compared with the nematic phase, behavior characteristic of the variation with temperature of the micelle size and the very weak temperature variation of the order parameter expected from a smectic–A phase [6].

   The variation of $\Delta\tilde{\nu}$ with temperature for samples with w less than that of T_{cp} is typical of that depicted in Fig. 3.24 for a sample with $w = 0.40$. The behavior at the isotropic-to-nematic transition and in the nematic phase is qualitatively identical to that for the $w = 0.55$ sample. But there is a marked difference at the nematic-to-lamellar transition. Here, there is no discontinuity in $\Delta\tilde{\nu}$, only a discontinuity in its temperature dependence. This means that there is no discontinuity in either $|\tilde{q}_{zz}|_s$ or S at this transition. The former is quite consistent with X-ray measurements which show that at this transition the micelles in the nematic phase simply condense onto the lamellar planes without any detectable change of their structure or size. The variation with temperature of $\Delta\tilde{\nu}$ must reflect solely the variation of S which is characteristic of a second-order nematic-to-smectic–A like transition [67].

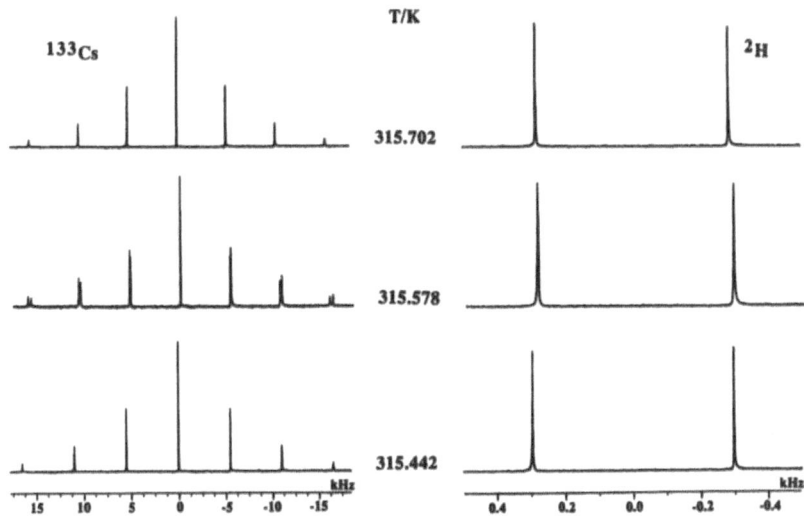

FIGURE 3.25. ^{133}Cs and ^{2}H NMR spectra as observed on cooling the sample CsPFO/^{2}H$_2$O ($w = 0.50$) from the N_D^+ to the L_D phase. The two distinct multiplets for the ^{133}Cs$^+$ spectrum at 315.578 K indicate the presence of a mixed nematic/lamellar coexistence regime. In the ^{2}H spectrum a slight broadening of the doublet lines is the only indication of a mixed phase region. Taken from [46].

In principle, then, the phase boundaries can be located from discontinuities or points of inflection in plots of $\Delta\tilde{\nu}$ versus temperature or from the appearance or disappearance of singlets and doublets. But the precise location of the tricritical point is more difficult as it depends upon being able to resolve the separate signals from the nematic and lamellar phases. The splittings of ^{133}Cs$^+$ are at least an order of magnitude greater than those from ^{2}H, giving ^{133}Cs NMR a greater resolving power for such applications. This is illustrated by the ^{133}Cs$^+$ and ^{2}H spectra (Fig. 3.25) observed on cooling a sample with $w = 0.50$ from the N_D^+ to the L_D phase. The greater $\Delta\tilde{\nu}$ for ^{133}Cs is a consequence of the difference in the values of the two quantities $\beta_{Cs}(\approx 0.5)$ and $n_b S_{OD} x_a / x_w (\approx 0.004)$ which appear in Eqs. (3.16) and (3.14), respectively. As for ^{2}H, the T_{NL} and T_{LN} are obtained from the discontinuities in the temperature dependence of the ^{133}Cs quadrupole splittings at the nematic-to-lamellar transition. At the concentration w of 0.45, the gap $T_{NL} - T_{LN}$ is 20 mK, and is at the limit of the experimental resolution. At lower concentrations, only a single discontinuity in the temperature dependence of the quadrupole splitting is observable. At these concentrations, the transition is taken to be second order. The concentration ($w = 0.43$) of the tricritical point was located by plotting $T_{NL} - T_{LN}$ versus w and extrapolating to zero as shown in

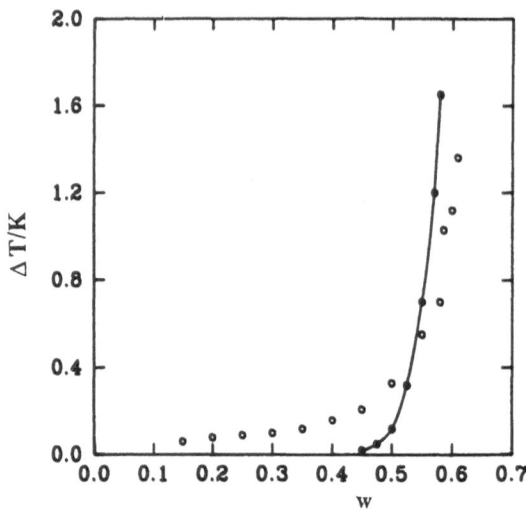

FIGURE 3.26. Plot of the transition gaps $T_{NL} - T_{LN}$ (closed circles) and $T_{IN} - T_{NI}$ (open circles) as a function of the weight fraction w of CsPFO in the system CsPFO/^{2}H$_2$O. $T_{NL} - T_{LN}$ appears to go to zero in the vicinity of $w = 0.43$ which is taken to be the location of the tricritical point Tcp. Taken from [46].

Fig. 3.26. The corresponding temperature (304.80 K) is obtained from the phase diagram in Fig. 3.27. We must now wait to see whether this seemingly second order nematic-to-lamellar transition is indeed so when probed by still more sensitive techniques.

The ^{2}H spectra of ^{2}H$_2$O in the TP6EO2M/^{2}H$_2$O system (Fig. 3.28) illustrates the difference in the behavior of diamagnetically negative nematic phases. The spectrum of the nematic phase is now invariant when the sample is spun about an axis perpendicular to H showing that the director is now uniformly aligned along the axis of rotation (Fig. 3.4c).

The quadruple splittings also contain information about the size and shape of the micelles, their orientational order parameter, and the binding of water molecules and counterions at their surface. The size of micelles in isotropic phases can in principle be estimated from nuclear spin relaxation measurements [70]. But it is preferable, where practical, to use the X-ray diffraction method, as discussed below.

3.5.3   X-RAY DIFFRACTION

X-ray scattering experiments on aligned nematic phases are easily conducted using a simple pinhole camera (Fig. 3.29). The diffraction pattern is characteristic of the type of nematic phase [9]. By way of illustration,

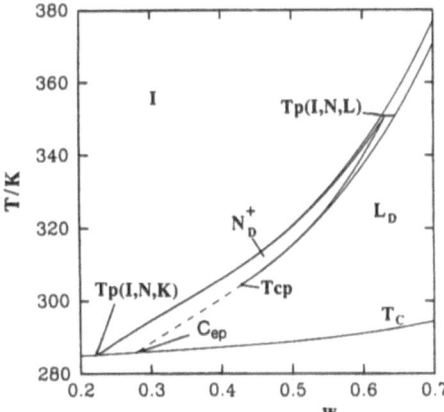

FIGURE 3.27. Partial phase diagram of the $CsPFO/^2H_2O$ system showing the I-to-N_D^+ and the N_D^+-to-L_D transition lines, the solubility curve T_C for the crystalline solid K, the N_D^+-to-L_D tricritical point Tcp, the $I - N_D^+ - L_D$ triple point $Tp(I, N, L)$, the $I - N_D^+ - K$ triple point $Tp(I, N, K)$ and the $N_D^+ - L_D$ critical end point C_{ep}. The temperatures of the phase boundary curves are estimated to have errors of $\pm0.04K$. Taken from [46].

Fig. 3.30 shows a set of diffraction patterns observed in the isotropic, nematic and lamellar phases of the APFO ($w = 0.45$) ^2H_2O system. The scattering pattern of the nematic phase is a continuous oval with a strong enhancement in the intensity along the meridian and a weaker enhancement along the equator. This is consistent with discotic micelles aligned with their symmetry axes parallel to the nematic director (Fig. 3.31). The intense scattering along the meridian arises from the "face–to–face" separation $d_\parallel$ of the micelles parallel to the nematic director. The weak diffuse scattering along the equator arises from the "side–by–side" separation $d_\perp$ of the micelles in the plane perpendicular to the director. This nematic phase must be of the N_D^+ type. The nematic phases of the $CsPFO/^2H_2O$ [55] and the decylammonium chloride/$NH_4Cl/^2H_2O$ [71,72] systems give similar diffraction patterns except that in the latter case the pattern is rotated through 90° (Fig. 3.29b) as it is of the N_D^- type. This interpretation of the X–ray scattering patterns has been confirmed by neutron scattering experiments [53].

We see that as the nematic phase is cooled the meridianal reflection sharpens and becomes more intense, while the equatorial reflection becomes more diffuse. At the nematic-to-lamellar phase transition there is no discernable change in the scattering pattern. The simple interpretation given to this observation is that the transition is associated with the onset of long-range positional ordering of the micelles along the direction of the nematic director without any concomitant change in their structure. That is, the

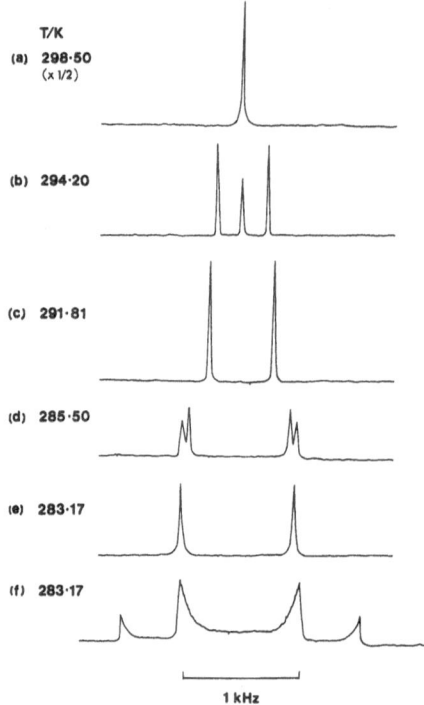

FIGURE 3.28. ^{2}H NMR spectra of ^{2}H$_2$O observed on cooling the sample TP6EO2M/^{2}H$_2$O ($w = 0.406$): (a) isotropic phase; (b) isotropic-nematic two-phase coexistence region; (c) nematic phase; (d) nematic-columnar (hexagonal) two-phase coexistence region; (e) columnar (hexagonal) phase; (f) as in (e) but with the sample tube rotated through 90° about the axis perpendicular to the direction of the applied magnetic field. Taken from [40].

lamellar phase, which is first formed, consists of layers of discotic micelles, rather than "classical bilayers." The decylammonium chloride/ammonium chloride/water system has been shown to exhibit similar X–ray scattering patterns, but they have been interpreted as evidence for lamellae broken by holes or microscopic defects [71]. However, recent freeze fracture electron micrographs [61] suggest the presence of micelles with appropriate dimensions in the lamellar phase. Thus, it appears that in this and in the CsPFO and APFO systems, too, the nematic-to-lamellar transition is a simple order-disorder transition quite analogous to the nematic-to-smectic–A transition observed for thermotropic calamitic mesogens.

With increasing concentration, beyond the N_D^+-to-L_D transition, the structure of the L_D phase must evolve and eventually become a classical bilayer phase so as to minimize the short-range repulsive forces between the approaching lamellae. It is interesting to speculate whether this transformation occurs continuously via an essentially 2-dimensional bicontinuous

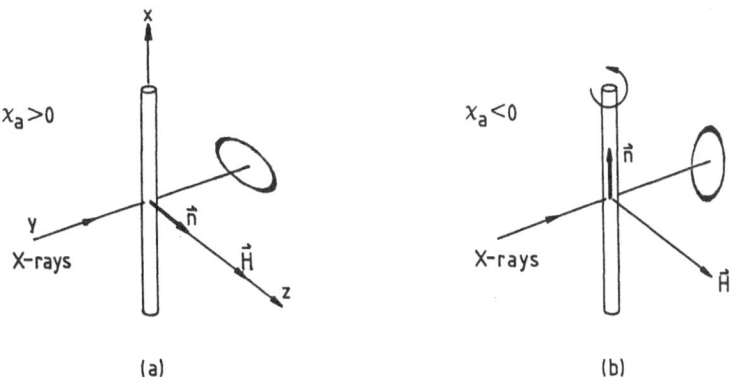

FIGURE 3.29. Geometries used in X-ray scattering experiments on diamagnetically positive (a) and negative (b) nematic phases. The schematic scattering patterns are for N_D phases.

structure, or discontinuously.

The absence of any discernable change in the diffraction pattern at the N_D^+-to-L_D transition implies that the nematic phase, at temperatures just above that of the transition, has a pseudo-lamellar ordering. This can be understood by considering the expression [73]

$$\Delta s = \{(\pi^2\delta^2/2d_\parallel^3)^2 + (1/Nd_\parallel)^2\}^{\frac{1}{2}} \qquad (3.17)$$

for the width Δs of the scattering peak along the meridian. $Nd_\parallel$ is the size of a "domain" in which the micelles have pseudo–lamellar positional ordering with an assumed gaussian distribution, of half-width δ, in the nearest neighbour separation $d_\parallel$. The peak width must be dominated by the first term as there is no discernable change at the transition. This means that there is considerable positional disorder in the packing of the micelles (large δ) coupled with the existence of short range positional correlations (large $Nd_\parallel$). Thus, the N_D^+-to-L_D transition simply involves the onset of long-range positional order along the nematic director without any concomitant change in the short-range structure. The persistence length of these short range positional correlations is estimated to be at least 20 micelle separations at temperatures just above the transition. The width of the equatorial peak does not change at the transition either, indicating liquid-like disorder in the lamellar planes; that is, the lamellae are two-dimensional liquids of discotic particles.

The above behavior contrasts with that of the thermotropic calamitics which show a marked sharpening of the meridianal reflection at the nematic-to-smectic-A transition [74]. In this case, in the nematic phase, δ is small because the molecules are closely packed. The second term in Eq. (3.17) now dominates the peak width. But at the transition N diverges

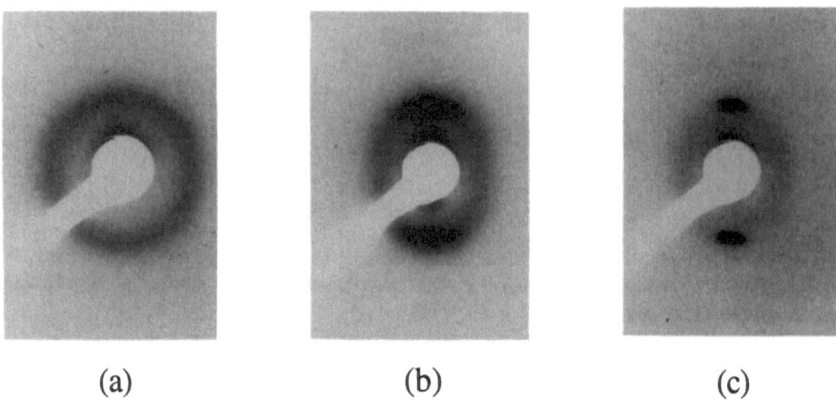

(a)   (b)   (c)

FIGURE 3.30. X-ray scattering patterns of a single domain sample of APFO/^{2}H$_2$O ($w = 0.452$) with the nematic director aligned along the direction of an applied magnetic field: $T_{LN} = 296.70K$, $T_{NI} = 302.70K$, and $T_{IN} = 303.18K$. (a) I phase at 309K, (b) N_D^+ phase at 302K, (c) lamellar phase at 296K.

and the first term takes over as the dominant contribution. Thus, there are significant differences between thermotropic and micellar nematic phases. This is also manifest in the behavior at the nematic-to-isotropic transition. For thermotropic systems, there is no change in the scattering curves in going from the isotropic to the unoriented nematic phase: this demonstrates that the transition is associated with the onset of long-range orientational order without any concomitant change in the short-range orientational order [74]. In contrast, the scattering patterns in the isotropic and unoriented nematic micellar phases are quite different. In the isotropic phase, the scattering pattern is a single broad diffuse ring (Fig. 3.30a). Thus, there is no evidence for short range orientational correlations of the micelles. The micelles would appear to undergo uncorrelated reorientational fluctuations, consistent with their low packing fraction.

In view of the pseudo-lamellar structure of the nematic phase, it seems reasonable to assume the discotic micelles are hexagonally packed on planes and to identify their separation with $d_\parallel$ and the separation of the (11) "planes" with $d_\perp$, cf. Fig. 3.31. The volume V_m of the micelle can then be calculated from $V_m = (2/\sqrt{3})d_\parallel d_\perp^2 \phi$ and hence the average aggregation number from $\tilde{s} = V_s/V_A$, where V_A is the volume of the surfactant molecule. The same procedure is applicable in the L_D phase, while in the isotropic phase it is a reasonable approximation to assume that the X-rays

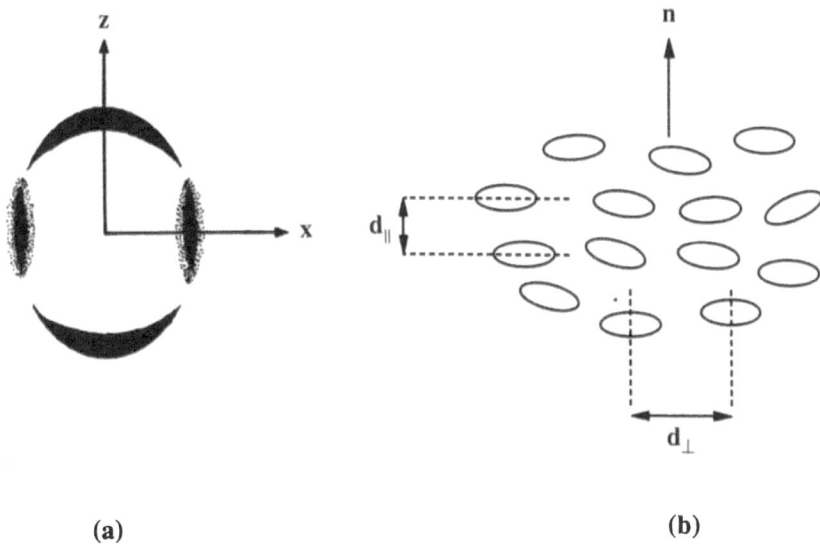

(a)   (b)

FIGURE 3.31. Relationship between the X-ray diffraction pattern (a) and the structure of a discotic nematic phase (b). The scattering angles of the meridional and equatorial reflections are related via Bragg's law ($\lambda = 2d\sin\phi$) to, respectively, the distances $d_\parallel$ and $d_\perp$.

are scattered from the (111) planes of a fcc lattice with separation d_0 such that $V_m = (3\sqrt{3}/4)d_o^3\phi$. The aggregation numbers plotted in Fig. 3.15 were obtained by this procedure.

The scattering patterns of aligned N_C phases might be anticipated to be similar to those of N_D phases except that the distances $d_\parallel$ and $d_\perp$ now correspond, respectively, to the separations along the perpendicular to the symmetry axis of the micelles. However, scattering in the direction of the director is usually very weak or totally absent [40,53]. This is because canonic micelles are far more polydisperse than discotics. The highly curved rim of a discotic micelle or end cap of a canonic micelle are relatively high free energy sites—see Chap. 1. The growth of a discotic micelle requires the generation of more rim and this has a high free energy price which regulates the growth and leads to essentially monodispersed micelles. This is not the case for canonics: the cap is invariant with length and, consequently, growth is less constrained.

3.6   Phase Transitions in Micellar Solutions

The observation that micellar solutions can undergo order–disorder transitions without dramatic concomitant changes in the structure of the aggregates (c.f. classical phase behavior in Fig. 3.6) opens up opportunities for the study of new phase transitions and critical phenomena. The new and important feature is that the size and shape of the micelle, and therefore the forces between the micelles, are functions of concentration and temperature. Thus, the understanding of these phase transitions will require the development of a new statistical mechanics formulation whereby the interacting particles (and hence the forces between them) must be self–consistently determined. The object here is to present a glimpse of the fascinating nature of these transitions by reference to: i) the I-to-N_D^+ and ii) the N_D^+-to-L_D transitions in the CsPFO/water system, iii) the N_D^--to-N_B-to-N_C^+ sequence of transitions in the potassium laurate/decanol/water system, and iv) the nematic-to-cholesteric transition induced by chiral solutes.

3.6.1   I-TO-N_D^+ TRANSITION IN THE CsPFO/WATER SYSTEM

The I-to-N_D^+-to-L_D sequence of transitions exhibited by the CsPFO/water system is quite analogous to the isotropic-to-nematic-to-smectic–A sequence observed for thermotropic calamitic (lath-shaped molecules) liquid crystals [6]. But there is a significant difference in behavior stemming from the low packing fraction of the micelles: 0.11 (and as low as 0.07 in metastable phases [45]) to 0.43 along the N_D^+-to-I transition line (see Fig. 3.27). Thus, we have what amounts to a 'gaseous' liquid crystal and this makes these phase transitions especially interesting to study. In particular, they are very weak. This can be seen from the DSC thermogram in Fig. 3.32. The two peaks, which are not well resolved, are associated with the two transitions. The heat associated with both transitions ($\Delta H = 593 \mathrm{Jmol}^{-1}$ of CsPFO or $\Delta S \approx 2$ JK^{-1} mol^{-1} for $w = 0.45$) is only two per cent of the total heat absorbed, the remainder coming from the large heat capacity of water. While it is not practicable to separate the contributions from the two transitions, the total enthalpy change is found to vary from 240 J mol^{-1} of CsPFO at $w = 0.35$ to 640 J mol^{-1} at $w = 0.75$. Thus, the I-to-N_D^+ transition, though first order, is only weakly so, and appears to weaken as the concentration decreases. The weakening of this transition on dilution is also reflected in the variation of the temperature gap for the transition (Fig. 3.26) [45,46]. This is seen to be proportional to w for concentrations up to 0.35.

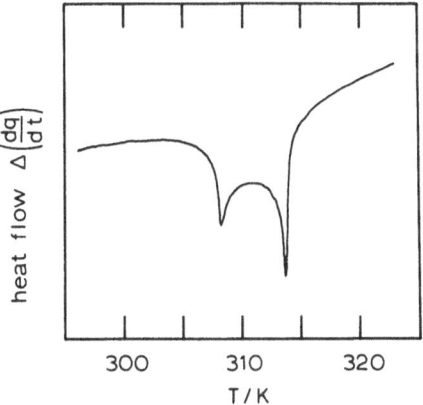

FIGURE 3.32. DSC thermogram observed on heating a sample of $CsPFO/^2H_2O$ ($w = 0.45$) across the L_D-to-N_D^+ and N_D^+-to-I transitions ($\Delta H_{\text{total}} = 593$ J mol^{-1} CsPFO).

A more quantitative measure of the strength of the I-to-N_D^+ transition is given by $T_{IN} - T^*$ where T^* is the extrapolated supercooling limit of the isotropic phase. This quantity is proportional to the free energy barrier of the transition. T^* is obtained from measurements of the orientational order induced by a magnetic field (the order parameter susceptibility χ— not to be confused with the quantities χ_a and χ_s introduced earlier) as the transition is approached from the isotropic phase. Two methods have been used. In one the induced order is detected by optical birefringence $\Delta\tilde{n}$ measurements [75,76], and χ is obtained from the slope of a $\Delta\tilde{n}$ versus H^2 plot at $H = 0$ (the Cotton-Mouton coefficient), cf. Fig. 3.33. The data for a sample with $w = 0.447$ are found to fit [75]

$$\chi = \chi_0(T - T^*)^{-\gamma} \tag{3.18}$$

with $\gamma = 1.01 \pm 0.04$ (Fig. 3.33) consistent with molecular-field theory. A similar result has been obtained [77] for the decylammonium chloride/ammonium chloride/water system for which $T_{IN} - T^*$ is of the order 1 K (as for thermotropic nematics [78]). In contrast, the values of the latter quantity are substantially smaller for the CsPFO/water system and, furthermore, they decrease on dilution (Fig. 3.34). The optical birefringence method ceases to provide reliable values for $T_{IN} - T^*$ for concentrations $w < 0.30$ where they become <25 mK. At these concentrations deuterium NMR spectroscopy of labelled water is the preferred method for monitoring the field induced order [79]. This is because separate NMR signals from the isotropic and nematic phases can be distinguished in the biphasic region (Fig. 3.35) and this enables precise values of both T_{IN} and $T*$ to be

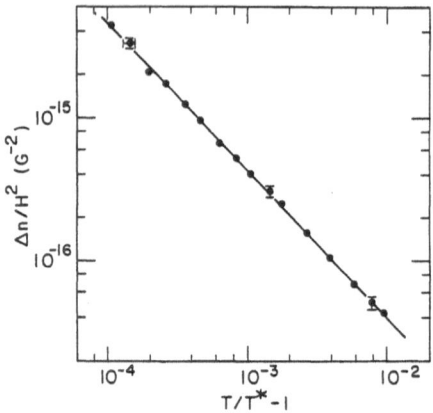

FIGURE 3.33. Susceptibility $\Delta\tilde{n}/H^2$ versus reduced temperature for CsPFO/H$_2$O(w = 0.398). The solid line gives the critical exponent $\gamma = 1.01 \pm 0.04$. Taken from [75].

determined. A molecular-field theory [79] gives

$$\Delta\tilde{\nu} = \frac{3}{2}|\tilde{q}_{zz}|_s \chi_a H^2 / 15k(T - T^*) \tag{3.19}$$

consistent with the plot of $\Delta\tilde{\nu}^{-1}$ versus T (in Fig. 3.36) which is seen to be quite linear: extrapolation of $\Delta\tilde{\nu}^{-1}$ to zero gives T^*. In this way values of $T_{IN} - T^*$ have been obtained down to $w = 0.15$, where it becomes 10 mK (Fig. 3.34).

In view of the classical mean-field behavior on the isotropic side of the transition, one might expect to observe similar behavior on the nematic side. The nematic-order-parameter exponent β

$$S - S^+ = A(T^+ - T)^\beta \tag{3.20}$$

is expected to be 0.5 for classical mean-field behavior, or 0.25 for tricritical behavior [78]. In practice it is found that β is invariant along the N_D^+-to-I transition line and has the avlue 0.34 ± 0.02 for both the CsPFO (Table 3.3) and the APFO (Table 3.4) systems [80]. In contrast, the values of the order parameter at T_{NI} and T^*, the superheating limit of the nematic phase, decrease on dilution.

Before turning to the challenging problem of the measurement of the nematic order parameter in these complex systems, a comment on the signifigance of the universal variation of the order parameter on approaching the nematic-to-isotropic transition is appropriate. A number of theoretical predictions for the value of the critical exponent have been put forward. In particular, the Landau–deGennes model has been extensively explored [78].

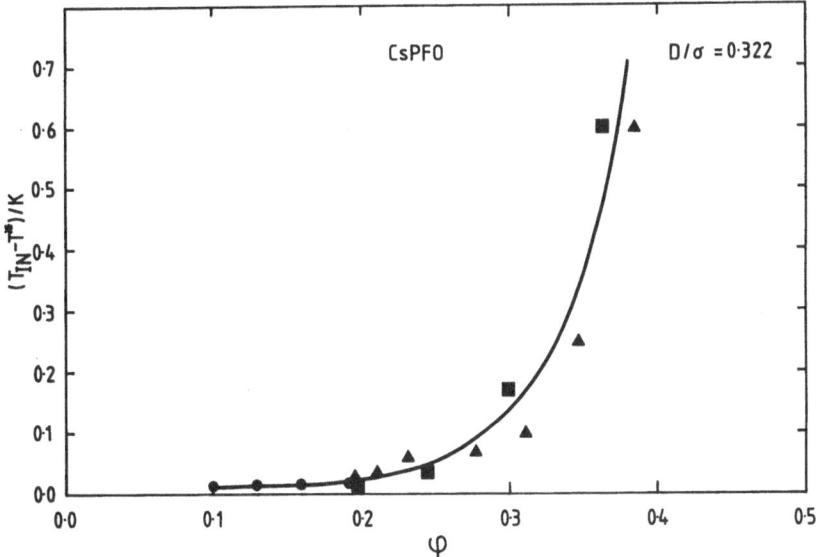

FIGURE 3.34. Plot of $(T_{IN} - T^*)$ versus volume fraction of surfactant ϕ for the CsPFO/water system. The data have been obtained by optical birefringence ($\square$, [75]; $\triangle$, [76]) and NMR ($\bullet$, [79]).

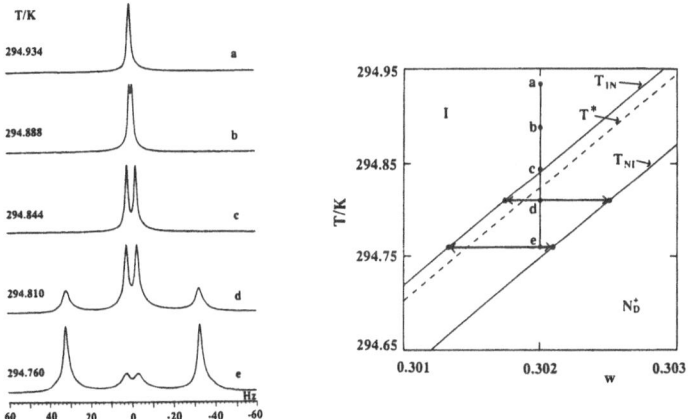

FIGURE 3.35. (a) The sequence of spectra observed on cooling a sample of CsPFO/^{2}H$_2$O with $w = 0.302$ from the isotropic phase into the isotropic/nematic biphaasic region along the isopleth shown in (b). The rapid divergence of the field induced quadrupole splitting in the isotropic phase and its quenching in the biphasic region are clearly seen. (b) Partial phase diagram of the CsPFO/^{2}H$_2$O system showing the upper T_{IN} and lower T_{NI} boundaries to the isotropic micellar solution-to-nematic N_D^+ transition. The dashed line represents the hypothetical supercooling limit T^* of the isotropic phase. The letters along the isopleth refer to the spectra shown in (a). Taken from [79].

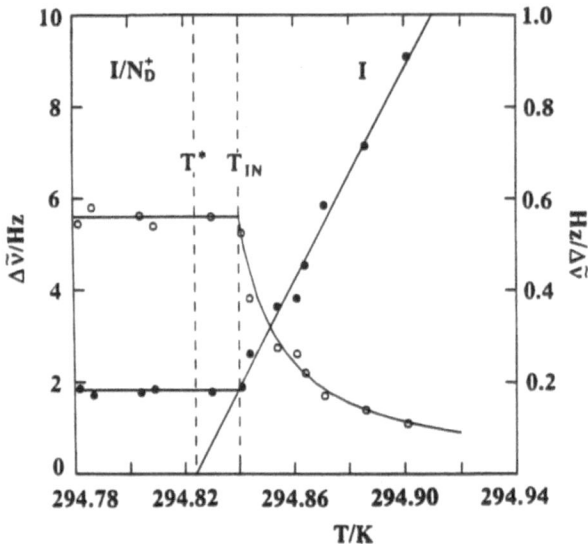

FIGURE 3.36. Plots of the field induced quadrupole splittings $\Delta\tilde{\nu}$ (open circles) and the inverse of the quadrupole splittings (closed circles) versus temperature for ^2H_2O in the isotropic phase I and the isotropic/nematic I/N_D^+ biphasic region of the $CsPFO/^2H_2O$ ($w = 0.302$) system. The discontinuity in the temperature dependence, especially of the $\Delta\tilde{\nu}^{-1}$ plot, clearly identifies T_{IN} (294.840K). Extrapolation of $\Delta\tilde{\nu}^{-1}$ to zero gives T^* (294.824K). The quantity $T_{IN} - T^*$ is estimated to have an error of ± 0.002 K. The solid lines through the experimental measurements were obtained from a least squares fit of the $\Delta\tilde{\nu}^{-1}$ versus T data with Eq. (3.23).

Taking the free energy expansion up to the fourth power of the order parameter gives $\beta = 0.5$, the mean-field value. Considering the sixth-order term leads to the tricritical value of 0.25. Agreement with the latter value has been shown for thermotropic systems [81], but neither prediction conforms with the current values measured for lyotropic liquid crystal systems. Rosenblatt and Litster [82,83] have suggested that a large sixth-order term in the free energy expansion could lead to an exponent intermediate to the mean-field and tricritical predictions. However, given the constancy of the effective exponent along the curve of nematic–isotropic phase transitions, and the fact that the transitions have weakened very considerably from the first to the last entry in Table 3.3, this explanation now seems less likely. Alternative explanations based on the idea that the effective exponents are characteristic of some other fixed point of the renormalization group flow seem more attractive. Thus, in the nematic phase near the transition to the isotropic phase one conceives of a correlation volume of dimension ξ^3(ξ is the bulk correlation length of the phase) where the critical fluctuations producing domains of isotropic phase are characteristic of a certain

TABLE 3.3. Parameters obtained for CsPFO/^{2}H$_2$O

W_A	ϕ_A^*	T_{NI}/K	$(T_{IN}-T_{NI})$K[b]	T^+/K	$(T^+ - T_{NI})$/K	$S(T_{NI})$	s^+	β	a/b(T_{NI})[c]
		±0.04	±0.01	±0.10	±0.11	±0.05	±0.05	±0.02	±0.03
0.558	0.355	331.15	0.60	331.75	0.20	0.35	0.15	0.34	0.37
0.507	0.311	321.99	0.34	321.73	0.08	0.29	0.10	0.35	0.32
0.457	0.270	314.08	0.23	314.96	0.11	0.29	0.08	0.37	0.30
0.411	0.235	307.82	0.16	307.76	0.10	0.23	0.09	0.35	0.28
0.356	0.196	310.12	0.13	301.05	0.06	0.21	0.07	0.34	0.27
0.312	0.167	295.77	0.10	295.72	0.05	0.21	0.07	0.32	0.27
0.252	0.130	289.15	0.08	289.10	0.03	0.21	0.06	0.34	0.26
0.200	0.096	284.38	0.06	284.35	0.03	0.16	0.02	0.34	0.25

a At T_{NI}
b Obtained by NMR measurements[5]
c The axial ratios a/b have been calculated assuming the micelle to be an oblate ellipsoid with minor axis a = 2.2 nm

TABLE 3.4. Parameters obtained for APFO/^{2}H$_2$O

W_A	ϕ_A^*	T_{NI}/K	$(T_{IN}-T_{NI})$K[b]	T^+/K	$(T^+ - T_{NI})$/K	$S(T_{NI})$	s^+	β	a/b(T_{NI})[c]
		±0.04	±0.01	±0.11	±0.11	±0.05	±0.05	±0.01	±0.03
0.490	0.361	310.29	0.62	309.87	0.20	0.24	0.06	0.34	0.38
0.450	0.326	302.22	0.48	301.93	0.19	0.28	0.06	0.34	0.36
0.398	0.281	293.18	0.32	293.03	0.17	0.18	0.06	0.35	0.32

a At T_{NI}
b Obtained by NMR measurements[5]
c The axial ratios a/b have been calculated assuming the micelle to be an oblate ellipsoid with minor axis a = 2.2 nm

universality class. As the transition is approached, the correlation length increases and the order-parameter appears to change as if governed by a critical exponent. Since the transition is weak, the free–energy difference of the two phases is small so that the correlation length will be large at the first–order phase–transition. One would expect the effective exponent to be that characteristic of the appropriate fixed point. One candidate that is compelling on physical grounds is the three-dimensional five-component $(N = 5)$ vector model Landau point [116,117]. While an accurate value of the order-parameter exponent is not yet available, an estimate [82] of 0.39 has been obtained by extrapolation of the resummed $O(\varepsilon^5)$ series that have been calculated for $N = 1$ to $N = 3$. However, the near-universal values of β [0.346(2) for APFO and 0.347(2) for CsPFO] are more consistent with the $O(\varepsilon)$ value for $N = 5$ of 0.342 [116]. Given the near constancy of the measured values, the agreement may not be fortuitous and the order parameter near this weak first-order transition may be dominated by the $O(\varepsilon)$ term of the scaling function. It should be noted, however, that such critical behavior would not seem to be consistent with the magnetic birefringence measurements [75] in the isotropic phase, which yield a mean-field result for the effective susceptibility exponent, an observation which seems to imply that, at least in the isotropic phase, fluctuations do not become very strong. However, since the transition is ultimately first order, one should not necessarily expect the effective exponents to be the same on both sides of the transition.

In principle, measurements of any second rank tensor property can be used to obtain values for the nematic order parameter S [6]. Indeed, optical birefringence measurements have been used for this purpose [82,86]. But such measurements will contain contributions from changes in the size of the micelles and from the reorientational motion of the molecules within the micelle, in addition to the order parameter fluctuations of the micelle itself. Care must therefore be exercised in relating changes in optical birefringence, or any other second-rank-tensor property, directly to changes in nematic order. Moreover, absolute values for the order parameter are not obtainable. A method based on electrical conductivity measurements (Fig. 3.37a) has been discussed [55,87,88] in order to circumvent these problems. It is practicable because the conductivity is solely determined by the diffusivity of the Cs^+/NH_4^+ ions around the micelles. The contribution from the micelles is negligible as their diffusive motion is hindered by strong intermicellar interactions.

For non-conducting, ellipsoidal micelles undergoing reorientational fluctuations with respect to the director of a uniaxial mesophase, the conductivity transforms as a second-rank tensor with a principal axes system coincident with that of the moment-of-inertia tensor. The experiment, therefore, measures the partially-averaged component $\tilde{\kappa}_{zz}$ of the conductivity tensor along the direction of E which is taken to be along the z axis of the laboratory frame $L(x, y, z)$. This is given by

$$\tilde{\kappa}_{zz}(\varphi) = \kappa_i + \frac{2}{3}P_2(\cos\varphi)S(\kappa_{\parallel} - \kappa_{\parallel})_M \qquad (3.21)$$

where φ is the angle between the mesophase director and the direction of E. κ_i is the trace of the conductivity tensor κ as measured in the isotropic phase ($S = 0$), and in the liquid crystalline phases when $\varphi = 54°44'$ (i.e., $P_2(\cos\varphi) = 0$). $(\kappa_{\parallel})_M$ and $(\kappa_{\perp})_M$ are the conductivities measured parallel and perpendicular to the micellar symmetry axis in the frame $M(\hat{a}, \hat{b}, \hat{c})$, and may be interpreted as $\tilde{\kappa}_{zz}(0°)$ and $\tilde{\kappa}_{zz}(90°)$ in a perfectly ordered system ($S = 1$). From Eq. (3.21)

$$\Delta\tilde{\kappa}/\kappa_i = S(\kappa_{\parallel} - \kappa_{\perp})_M/(\kappa_{\parallel} + 2\kappa_{\perp})_M/3\kappa_i \qquad (3.22)$$

where $\Delta\tilde{\kappa} = \tilde{\kappa}_{zz}(0°) - \tilde{\kappa}_{zz}(90°)$. This equation has been used to obtain values of S for the CsPFO/water system (Fig. 3.37c) using values for $(\kappa_{\parallel})_M$ and $(\kappa_{\perp})_M$ calculated by the Fricke equations [89] and micelle dimensions (Fig. 3.37b) obtained from X-ray measurements [55,56].

The nematic order parameter may, in principle, be obtained from the angular intensity distribution (Ψ) of the meridianal reflection (Fig. 3.30), where Ψ is the angle between the radius vector and the z axis. The values

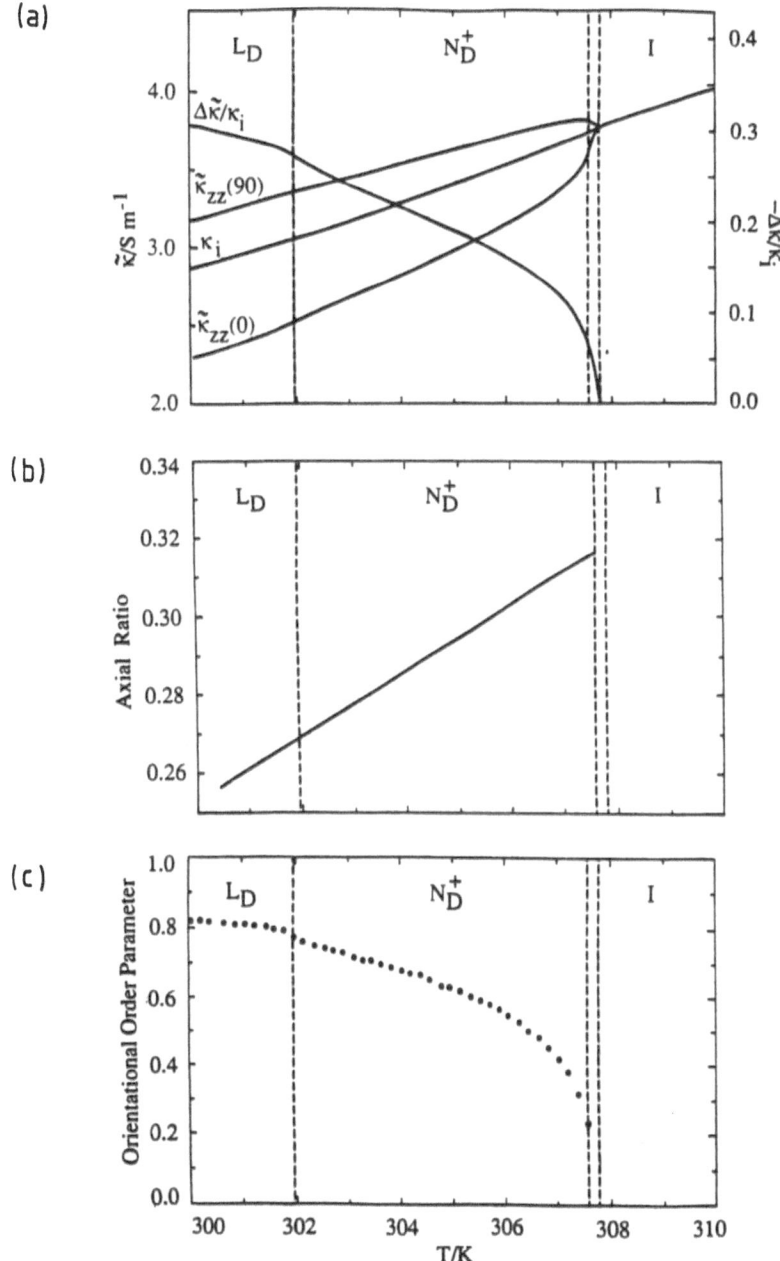

FIGURE 3.37. The electrical conductivity $\tilde{\kappa}_{zz}(\phi)$ as a function of temperature as measured at 30 kHz for CsPFO/^{2}H$_2$O with w = 0.0.411 ($T_{LN} = 301.98K$, $T_{NI} = 307.66$ K, and $T_{IN} = 307.82$ K). $\tilde{\kappa}_{zz}(0°)$ and $\tilde{\kappa}_{zz}(90°)$ are the conductivities measured parallel and perpendicular to the nematic director, while κ_i corresponds to $\tilde{\kappa}_{zz}(54°44')$ in the nematic and lamellar phases. The relative conductivity anisotropy is $\Delta\tilde{\kappa}/\kappa_i = \{\tilde{\kappa}(0°) - \tilde{\kappa}(90°)\}/\kappa_i$.

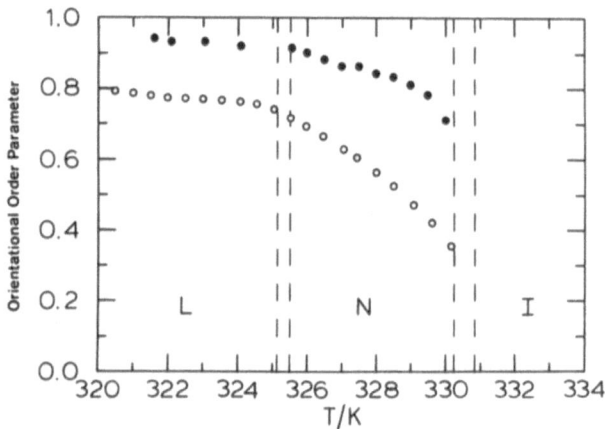

FIGURE 3.38. Orientational order parameters of the micelles in CsPFO $(w = 0.55)/^2H_2O$ as obtained from the intensity distribution of the meridional X-ray reflection (filled circles) and as calculated from electrical conductivity measurements (O).

are typical of those similarly obtained for other micellar nematics [53, 71, 72, 90, 91], but they are significantly larger than those obtained from conductivity measurements (Fig. 3.38). They are also significantly larger than the molecular order parameters for thermotropic nematic phases: $S = 0.3$ to 0.4 at T_{NI} [92]. Thus, the order parameters obtained from the X-ray measurements seem anomalously high. Yet the X-ray method has given reasonable values for thermotropic nematic phases [74], although heavily and justifiably criticized [93]. The discrepancy between the values of the order parameters as obtained by the two techniques must have its origin in the different ways in which they sense the reorientational fluctuations of the micelles. The order parameters obtained from the electrical conductivity measurements represent the ensemble average of the orientational fluctuations of the micellar axes $M(\hat{a}, \hat{b}, \hat{c})$ with respect to the laboratory frame $L(\hat{x}, \hat{y}, \hat{z})$ over a time scale determined by the measurement period (typically 2 x $10^{-3}s$). Now, in contrast, the angular intensity distribution $I(\Psi)$ in the meridional reflection of the X-ray diffraction pattern represents the instantaneous ensemble-averaged orientational distribution of the intermicellar vectors. This will not be affected by any independent reorientational fluctuations of the individual micelles. This is very likely to be a significant reorientational mode in micellar nematic phases in view of the low packing fraction: that is, the micelles are wobbling. The intensity distribution in the X-ray reflection derives from the longer range collective fluctuations.

Taylor and Herzfeld [113] recently have used a hard particle model to calculate the phase diagram for a system of rod-like surfactants (CsPFO) which self-assemble into polydisperse disk-shaped micelles. Their phase diagram is quantitatively similar to those observed for the CsPFO/water and APFO/water systems. Moreover, there is also good quantitative agreement between the theory and experiment along the isotropic-to-nematic line of transitions [112]. This establishes that in these systems the micellar interactions are predominantly hard particle in nature and that the micelles are essentially rigid.

The concentration dependence of $T_{IN} - T^*$, depicted in Fig. 3.34, can be accounted for in the context of the generalized van der Waals theory of nematics [95]. The anisotropic interaction parameter ε is considered to be the sum of an attractive (dispersion) and a shorter range repulsive (steric) component:

$$\varepsilon = J_0\phi + \lambda(\phi)kT \tag{3.23}$$

and this leads to [96]:

$$\frac{k(T_{IN} - T^*)}{J_0} = \frac{\nu\phi}{[1 - \frac{1}{2}\lambda(\phi)][1 - (\frac{1}{5} + \nu)\lambda(\phi)]} \tag{3.24}$$

where $\nu = 1/28$. Values for the steric interaction $\lambda(\phi)$ have been calculated from the known micelle dimensions using a scaled particle formulation for hard right circular cylinders. This enables values for J_0 to be derived from the ε obtained from the Maier-Saupe [97] relationship $kT_{NI}/\varepsilon = 0.2202$; they are roughly two orders of magnitude greater than the corresponding steric interactions, though the latter become more important at higher concentrations. J_0 is expected to be of the form ε_{mm}/V_m where ε_{mm} is the strength of the anisotropic dispersive interaction between two micelles of volume V_m. This is confirmed by the proportionality between ε_{mm} and $\bar{s}^2$ (Fig. 3.39) which is consistent with the force law between two micelles with areas proportional to $\bar{s}$. The continuous line drawn through the experimental points in Fig. 3.34 has been calculated using these results. The rapid rise in $(T_{IN} - T^*)$, commencing at $w \approx 0.45$, has its origin mainly in the strong concentration dependence of $\lambda(\phi)$, rather than in its explicit form. Nevertheless, the results confirm that two interactions, a short and a longer range one, are operative.

A somewhat related observation is the pressure dependence of T_{NI} which is found [98] to be 10 to 100 times smaller than is usually observed for thermotropics. It follows from the Clausius–Clapeyron equation $((\partial p/\partial T)_{x_I} = \Delta H/T\Delta V$, where x_I denotes constancy of the composition of the isotropic phase) that the volume change ΔV must be exceedingly small and positive (recent studies [99] have suggested that Δ V is negative!). This is not

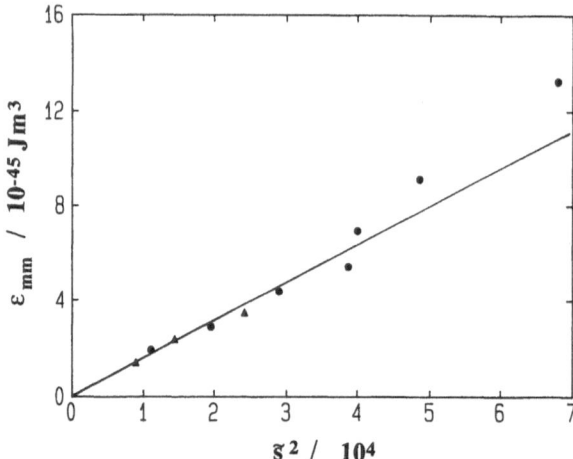

FIGURE 3.39. Plot of ϵ_{mm} versus $\tilde{s}^2$ (O,CsPFO; $\triangle$, APFO). The straight line corresponds to the fit: $\epsilon_{mm} = (1.77 \pm 0.5) \times 10^{-49} \tilde{s}^2$ J m^3.

too surprising in view of the high water concentration. But what is more interesting is that the pressure dependence appears (only four concentrations have been studied) to be concentration dependent in a manner not dissimilar to that of $(T_{IN} - T^*)$.

We can now understand the origin of the narrow range of the nematic phase (typically 6K) and the rapid variation of the nematic order parameter with temperature as compared with thermotropics [92]. $T_{IN} \approx (0.2202/k)\epsilon_{mm}/V_m \sim \tilde{s}$, so that the growth in the micelle as the temperature decreases pushes the nematic phase to lower reduced temperatures until the order parameter reaches the value (≈ 0.75) at which a transition to a smectic–A phase is induced [100].

To sum up, the interest in the nematic-to-isotropic transition of the CsPFO/water system stems from the fact that it is a very weak first–order phase transition for which effective exponents can be measured. This is not practicable with thermotropic nematics. It is also novel in that, in contrast to thermotropics, the volume fraction of interacting particles can be varied over a wide range. There is no a priori reason why micellar nematic phases cannot exist at much lower concentrations provided systems can be invented to give micelles of appropriate sizes and shapes.

3.6.2 N_D^+-TO-L_D TRANSITION IN THE CsPFO/WATER SYSTEM

In contrast to the I-to-N_D^+ transition, studies of this transition are only just beginning. The interest in it stems from the existence of the symmetrical tricritical point T_{cp}, analogous to the $^3He/^4He$ superfluid transition. The

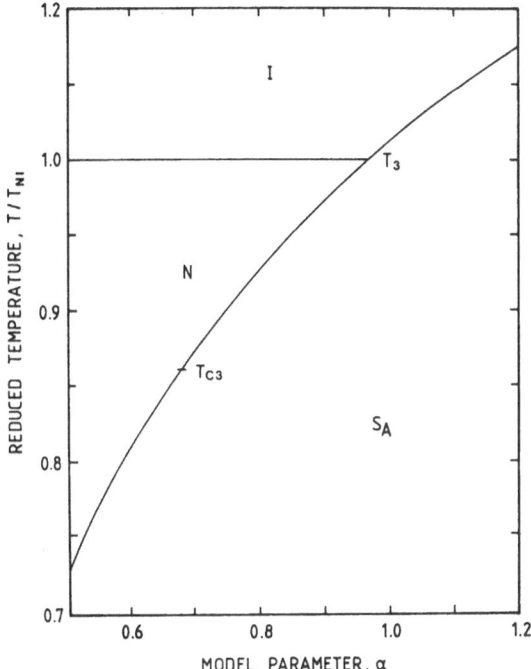

FIGURE 3.40. Phase behavior predicted by the McMillan model [100]: transition temperatures correspond to T/T_{NI} where $T_{NI} = 0.2202\varepsilon/k$.

origin of this tricritical point has been qualitatively explained [34] by a modification of the McMillian model [100] for the nematic-to-smectic–A transition in thermotropics. In the latter case, the interaction driving the transition is envisaged to arise from mutual attraction of the aromatic cores of the calamitic mesogen molecules; this will increase as the ratio of the length to that of the aromatic core, i.e., with the length of the alkyl end chains. In the McMillan theory the strength of this interaction is modelled by the parameter α, and we see from Fig. 3.40 that the transition changes from second to first order at $\alpha = 0.70$, while for $\alpha > 0.98$ the nematic phase is unstable. To establish a correlation between the phase behavior of the CsPFO/water system (Fig. 3.27) and the McMillan model (Fig. 3.40) it is necessary for ϕ and α to be directly functionally dependent.

Intuitively, it seems natural to compare the role of the layers of water separating the lamellae to that of the alkyl end chains in thermotropic smectic–A phases. But this cannot be correct because it would require the $L_D-\text{to}-N_D^+$ transition to change from second to first order as ϕ decreases, exactly the opposite to the observed behavior. Clearly, the strength of the interaction driving the $N_D^+ - \text{to} - L_D$ transition increases as ϕ increases:

its origin must, therefore, be the repulsive force between the micelles in adjacent lamellae rather than in the mutual attraction of neighboring micelles in the same layer (purported to be the origin of the decrease in size of the micelles at high concentrations). This model is consistent with the interpretations given to the $T_{IN} - T^*$ vs ϕ curves. On inclusion of both anisotropic attractive and repulsive terms in the interaction potential, the McMillan model is able to predict the salient features of the $N_D^+ - \text{to} - L_D$ transition [54].

Thus, the tricritical point and the departure from linearity of the $T_{IN} - T^*$ vs. ϕ plot both seem to have their origin in an anisotropic repulsive force whose presence becomes important at higher concentrations of micelles, and which may account for their occurence at the same concentration. The maximum in the $\tilde{s}$ versus w curves (Fig. 3.15) also occurs at the same concentration, but this is not consistent with a strengthening repulsive force. The preferred interpretation, to date, is in terms of a lateral attractive force between neighboring micelles. This would be enhanced as the micelles were pushed into a pseudo–lamellar structure by the anisotropic repulsive force between the faces of neighboring micelles. This would account for the maximum in $\tilde{s}$ at $w \approx 0.45$. The existence of such a repulsive force would also explain the occurence of a lamellar phase as opposed to a discotic columnar hexagonal phase at low temperatures—on the other hand, the intrinsic distribution of the size and diameters of the micelles would counteract any tendency towards a columnar organization: clearly, understanding the nature of the forces between the micelles is a prerequisite to understanding the phase behavior. This is one of the major topics of micellar science which has still to be tackled.

The nature of the nematic–lamellar tricritical point is of fundamental interest. Future studies of the variation of effective exponents on approaching T_{cp} will be of special interest, particularly in comparison with the nematic-to-smectic–A transition in thermotropics. There is an anomalous increase of the nematic order parameter (Fig. 3.37c) on approaching the $N_D^+ - \text{to} - L_D$ transition and this could arise from coupling between the nematic and smectic order parameters. The effects are small, yet measurable. In contrast, the rotational viscosity (Fig. 3.41) is seen to diverge. Expressing it as

$$\gamma_1(T) = \gamma_1^a + \delta\gamma_1 \tag{3.25}$$

with the critical part as

$$\delta\gamma_1 = \gamma_1^b(T)(\frac{T}{T_{NL}} - 1)^x \tag{3.26}$$

has enabled values for the exponent x to be determined [101]. This has given values of $x = -0.68 \pm 0.02$ for all concentrations. This is to be compared with the classical mean-field value of $1/2$ [102], and the value of $1/3$ observed [103] for second-order thermotropic transitions and predicted for the

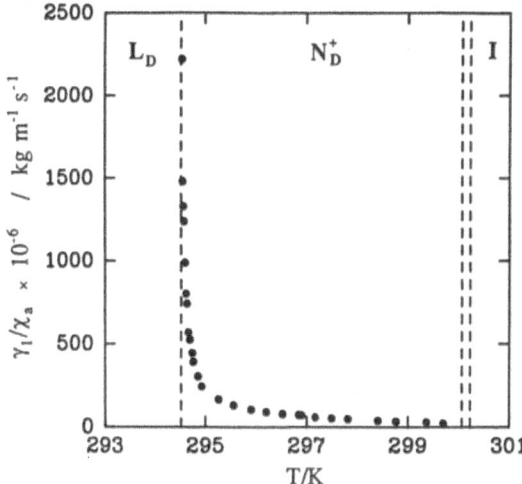

FIGURE 3.41. γ_1/χ_a versus temperature T for CsPFO/^{2}H$_2$O ($w = 0.349$): T_{NL} = 294.52(1) K, T_{NI} = 300.06(1) K, and T_{IN} = 300.19(1) K. Taken from [101].

^4He-like superfluid transition [104] using dynamical scaling laws [105,106]. The difference between the micellar and thermotropic systems has yet to be explained.

3.6.3 N_D^--TO-N_B-TO-N_C^+ TRANSITIONS IN THE POTASSIUM LAURATE/DECANOL WATER SYSTEM

The phase diagram [12] in Fig. 3.42 shows that the N_B phase obtains on either heating or cooling of the N_D^- phase. It occurs as an intermediate phase between the N_D^- and N_C^+ phases and it is stable over a very narrow concentration interval. What is especially intersting is that this sequence of transitions does not involve any significant changes in the structure of the micelle: in all four phases the micelle is a biaxial platelet with estimated dimensions of 8.5 nm, 5.5nm and 2.6 nm [18-21]. The phase transitions are associated only with changes in the rotational motion of the micelle (see Sec. 3.2) effected by changes in concentration or temperature. In other words, we are once again concerned with order-disorder transitions as in the case of the CsPFO/water system.

The phase diagram is very similar to that predicted by molecular field theory [17]. The interesting feature is the intersection of two second order uniaxial-to-biaxial phase transition lines. This is a unique point called a Landau point or an "isolated critical point" [107] where the first order nematic-to-isotropic transition becomes a second order one. The critical behavior at this point is of particular interest as all five order parameters are

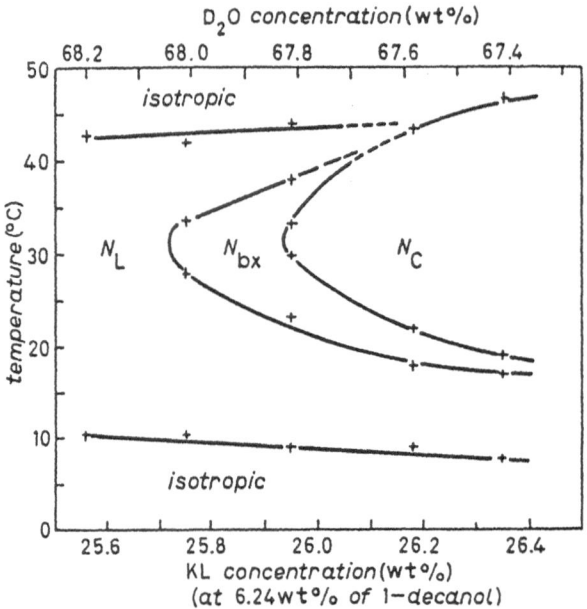

FIGURE 3.42. Partial phase diagram of the potassium laurate/decanol/water system. Taken from [12].

involed [16,108]. Measurements of the nematic order parameter exponent β (as carried out for the CsPFO/water system) along the nematic-to-isotropic transition line, and especially approaching the Landau point, would be especially interesting.

3.6.4   NEMATIC-TO-CHOLESTERIC TRANSITIONS INDUCED BY CHIRAL SOLUTES

The addition of small amounts of chiral compounds such as brucine sulphate or d–tartaric acid have been shown to transform both uniaxial [109] and biaxial [110] nematic phases into cholesteric phases in which the nematic director is twisted about itself as the thread of a screw. The molecular mechanism by which twist is induced is not yet understood. It is quite likely that the chiral solute imparts chirality to the micelle surface and this in turn affects the intermicellar interactions. What is particularly interesting is that these small perturbations are propagated over the fairly large water filled spaces between the micelles.

3.7   Conclusions

The main conclusion which we wish to stress is that small discrete micelles can be stable at high concentrations where they undergo sequences of order-disorder transitions which are quite analogous to those exhibited by thermotropic liquid crystals. Though the L_D phase is the only translationally ordered mesophase to be recognized to date, there is no *a priori* reason why canonic micelles should not form columnar phases—the distribution in their lengths will tend to oppose the formation of smectic phases.

The occurence of these "micellar liquid crystals" poses new questions regarding the factors which govern the size and shape of micelles and the nature of the forces between them in concentrated solutions. Studies of phase behavior in relation to the micelle structure are beginning to shed light on these factors, but they are still in their infancy. It is now emerging that the order-disorder transitions involved can be understood in terms of hard particle theories [113] provided that aggregate flexibility is taken into account [114-115]. The latter seems to be important for rods, but not apparently so for disks. Thus, the precise form of the phase behavior will be very dependent on the factors that govern both the self-assembly and flexibility of the aggregates. With such knowledge it should be possible to design surfactant/solvent systems to optimise the stability of micellar mesophases; this is a challenge embracing synthetic chemistry, the physical chemistry of surfactant solutions, and the physics of liquid crystals. Whether or not such liquid crystals will find technological applications is anyone's guess. But, at first, their study will surely advance our understanding of the fundamentals of self-assembly and self-organization in surfactant solutions..

The phase transitions involved have been seen to be second order or unusually weak first-order ones and to be accompanied by large amplitude fluctuations. Characterization of the associated critical phenomena and comparison with the behavior of thermotropic liquid crystals should provide a good test for theories of universality. In particular, it will be important to establish whether or not critical phenomena are coupled in any way to the self-assembly behavior in surfactant solutions.

Acknowledgements

I would like to thank Dr. K. W. Jolley for carefully checking through the manuscript, correcting the many mistakes and also for his contributions to much of the work described on the CsPFO and APFO systems. The significant contributions of Dr. M. C. Holmes to this work are also acknowledged as are those of Drs. S. A. Corne, G. Ouriques, A. Parbhu, D. Parker, D. J. Reynolds, and M. H. Smith.

References

1. P. Ekwall, Adv. Liq. Cryst. **1**, 1 (1975).

2. V. Luzzati, in *Biological Membranes*, D. Chapman, ed. (Academic Press, London and New York, 1986), Ch. 3, p. 71.

3. P. A. Winsor, *Solution Properties of Amphiphilic Coumpounds* (Butterworth Press, London 1954); Chem. Rev. **68**, 1 (1968).

4. K. D. Lawson and T. J. Flautt, J. Amer. Chem. Soc. **89**, 5489 (1967); P. J. Black, K. D. Lawson and T. J. Flautt, Mol. Cryst. Liq. Cryst. **7**, 201 (1969).

5. F. Rosevear, J. Soc. Cosmet. Chem. **19**, 581 (1968).

6. P. G. de Gennes, *The Physics of Liquid Crystals* (Clarendon Press, Oxford, 1974).

7. C. L. Khetrapal, A. C. Kunwar, A. S. Tracey and P. Diehl, in *NMR-Basic Principles and Progress* P. Diehl, E. Fluck and R. Kosfeld, eds. (Springer-Verlag, Heidelberg, 1975) Vol. 9, pp 1-85.

8. G. J. T. Tiddy, Phys. Rep. **57**, 7 (1980).

9. J. Charvolin, A. M. Levelut, and E. T. Samulski, J. de Physique Lett. (Paris) **40**, L-587 (1979).

10. K. Radley and A. Saupe, Mol. Cryst. Liq. Cryst. **44**, 227 (1978).

11. N. Boden, P. H. Jackson, K. McMullen, and M. C. Holmes, Chem. Phys. Lett. **65**, 476 (1979).

12. L. J. Yu and A. Saupe, Phys. Lett. **45**, 1000 (1980).

13. M. J. Freiser, Phys. Rev. Lett. **24**, 1041 (1970).

14. T. C. Lubensky, Phys. Rev. A **2**, 2497, (1970).

15. C. S. Shih and R. Alben, J. Chem. Phys. **57**, 3057 (1972).

16. R. Alben, Phys. Rev. Lett. **30**, 778 (1973).

17. J. P. Straley, Phys. Rev. A **10**, 1881 (1974).

18. A. M. Figueiredo Neto, Y. Galerne, A. M. Levelut and L. Liébert, J. de Physique Lett. **46**, 499 (1985).

19. Y. Galerne, A. M. Figueiredo Neto, and M. Liébert, J. Chem. Phys. **87**, 1851 (1987).

20. Y. Hendrikx, J. Charvolin, and M. Rawiso, Phys. Rev. B **33**, 3534 (1986).

21. Y. Galerne, Mol. Cryst. Liq. Cryst. **165**, 131 (1988).

22. L. Onsager, Ann. N.Y. Acad. Sci. **51**, 627 (1949).

23. P. J. Flory, Proc. Roy. Soc. **A234**, 60, 73 (1956); P. J. Flory, in *Polymer Liquid Crystals* , A. Ciferri, W. R. Kirgbaum, and R. B. Meyer, eds. (Academic Press, New York, 1982) pp 103-112.

24. B. J. Forrest and L. W. Reeves, Chem. Rev. **81**, 1 (1981).

25. A. Saupe, Nuovo Cim. D. **3**, 16, (1984).

26. M Boidart, A. Hochapfel and M. Laurent, Mol. Cryst. Liq. Cryst. **154**, 61 (1988).

27. A. Saupe, Z. Naturforsch. **19a**, 161 (1964).

28. N. Boden, K. Radley and M. C. Holmes, Mol. Phys **42**, 493 (1981).

29. N. Boden, K. McMullen, and M. C. Holmes, in *Magnetic Resonance in Colloid and Interface Science*, J. P. Fraissard and H. A. Resing, eds. (D. Reidel Publishing Co.) pp 667-673.

30. F. Y. Fujiwara and L. W. Reeves, Can. J. Chem. **56**, 2178 (1978).

31. B. J. Forrest, L. W. Reeves and C. J. Robinson, J. Phys.Chem. **85**, 3244 (1981); M. E. Marcondes Helene and L. W. Reeves, Chem. Phys. Lett. **89**, 519 (1982).

32. A. R. Custodio and F. Y. Fujiwara, Mol. Cryst. Liq. Cryst. **139**, 321 (1986).

33. K. Radley, L. W. Reeves and A. S. Tracey, J. Phys, Chem. **80**, 174 (1976).

34. N. Boden, R. J. Bushby, L. Ferris, C. Hardy and F. Sixl, Liq. Cryst. **1**, 109 (1986).

35. N. Boden, R. J. Bushby, K. W. Jolley, M. C. Holmes and F. Sixl, Mol. Cryst. Liq. Cryst. **152**, 37 (1987).

36. J. Charvolin and J. F. Sadoc, J. Phys. Chem. **92**, 37 (1987).

37. Y. Hendrikx and J. Charvolin, J. de Physique **42**, 1427 (1981).

38. M. R. Rizzatti and J. D. Gault, J. Colloid Interface Sci. **110**, 258 (1986).

39. A. Hochapfel, M. Boidart and M. Laurent, Mol. Cryst. Liq. Cryst. **75**, 201 (1981).

40. N. Boden, R. J. Bushby and C. Hardy, J. Physique Lett. **46**, L-325 (1985).

41. N. Boden, R. J. Bushby, C. Hardy and F. Sixl, Chem. Phys. Lett. **123**, 359 (1986),

42. S. Chandrasekhar, Phil. Trans. R. Soc. A. **309**, 93 (1983).

43. H. Ringsdorf, B. Schlarb, and J. Venzmer, Angew. Chem. Int. Ed. Eng. **27**, 113 (1988).

44. H. Zimmerman, R. Poupko, Z. Luz, and J. Billard, Liq. Cryst. **6**. 151 (1989).

45. N. Boden, S. H. Corne and K. W. Jolley, J. Phys. Chem. **91**, 4092 (1987).

46. N. Boden, K. W. Jolley and M. H. Smith, Liq. Cryst. **6**, 481 (1989).

47. E. Everiss, G. J. T. Tiddy and B. A. Wheeler, J. Chem. Soc., Faraday Trans I. **72**, 1747 (1976).

48. K. Fontell and B. Lindman, J. Phys. Chem. **87**, 3289 (1983).

49. H. Hoffmann, Ber. Bun. Phys. Chem. **88**, 1078 (1984).

50. K. Reizlein and H. Hoffmann, Progr. Coll. and Polym. Sci. **69**, 83 (1984).

51. M. A. Schafheutle and H. Finkelmann, Liq. Cryst. **3**, 1369 (1988).

52. B. Lühmann and H. Finkelmann, Coll. and Polym. Sci. **264**, 189 (1986).

53. Y. Hendrikx, J. Charvolin, M. Rawiso, L. Liebert and M. C. Holmes, J. Phys. Chem. **87**, 3991 (1983)

54. N. Boden and M. C. Holmes, Chem. Phys. Lett. **109**, 76 (1984).

55. N. Boden, S. A. Corne, M. C. Holmes, P. H. Jackson, D. A. Parker and K. W. Jolley, J. de Physique **47**, 2135 (1986).

56. M. C. Holmes, D. J. Reynolds and N. Boden, J. Phys. Chem. **91**, 5257 (1987).

57. W. E. McMullen, W. M. Gelbart, and A. Ben-Shaul, J. Chem. Phys. **82**, 5616 (1985).

58. W. M. Gelbart, W. E. McMullen and A. Ben-Shaul, J. de Physique, **46**, 1137 (1985).

59. W. M. Gelbart, W. E. McMullen, A. Masters and A. Ben-Shaul, Langmuir **1**, 101 (1985).

60. L. Herbst, H. Hoffmann, J. Kalus, K. Reizline, U. Schmeizer and K. Ibel, Ber. Bun. Phys. Chem. **89**, 1050 (1989).

61. M. J. Sammon, J. A. N. Zasadzinski and M. R. Kuzma, Phys. Rev. Lett. **57**, 2834 (1986).

62. D. Frenkel and R. Eppenga, Phys. Rev. Lett. **49**, 1089 (1982); Mol. Phys. **52**, 1303 (1984).

63. C. Rosenblatt and N. Zolty, J. de Physique Lett. **46**, L-1191 (1985).

64. D. Goldfarb, M. M. Labes, Z. Luz and R. Poupko, Mol. Cryst. Liq. Cryst. **87**, 259 (1982).

65. T. K. Attwood and J. E. Lydon, Mol. Cryst. Liq. Cryst. **108** 349 (1984); T. K. Attwood and J. E. Lydon, Mol. Cryst. Liq. Cryst. Lett. **4**, 9 (1986); T. K. Attwood, J. E. Lydon and F. Jones, Liq. Cryst. **1**, 499 (1986).

66. M. C. Holmes, N. Boden and K. Radley, Mol. Cryst. Liq. Cryst. **100**, 93 (1983).

67. S. Chandrasekhar, *Liquid Crystals* (Cambridge University Press, 1977) pp. 150-155.

68. N. Boden and S. A. Jones, NATO ASI Sci. C. Maths Phys. Sci. **141**, 473 (1985).

69. H. Wennerström, G. Lindblom and B. Lindman, Chem. Scripta. **6**, 97 (1974).

70. H. Wennerström, G. Lindblom and B. Lindman, Chem. Scripta. **6**, 97 (1974); U. Henriksson, L. Ödberg, J. C. Eriksson, and L. Westman, J. Phys. Chem. **81**, 76 (1977).

71. M. C. Holmes and J. Charvolin, J. Phys. Chem. **88**, 810 (1984).

72. J. Charvolin, M. C. Holmes and D. J. Reynolds, Liq. Cryst. **3**, 1147 (1988).

73. A. Guinier, *X-ray Diffraction* (Freeman, 1963), p. 72.

74. A. J. Leadbetter, in *The Molecular Physics of Liquid Crystals*, G. R. Luckhurst and G. W. Gray, eds. (Academic Press, 1979), p. 283.

75. C. Rosenblatt, S. Kumar and J. D. Litster, Phys. Rev. A. **29**, 1010 (1984).

76. C. Rosenblatt, Phys. Rev. A. **32**, 1924 (1985).

77. S. Kumar, L. J. Yu and J. D. Litster, Phys. Rev. Lett. **50**, 1672 (1983).

78. See, for example, E. F. Gramsbergen, L. Longa, and W. H. de Jeu, Phys. Rep. **135**, 195 (1986).

79. K. W. Jolley, M. H. Smith and N. Boden, Chem. Phys. Lett. **162** 152 (1989).

80. N. Boden, J. Clements, K. A. Dawson, K. W. Jolley and D. Parker, Phys. Rev. Lett. **66**, 2883 (1991).

81. J. Thoen and G. Menu, Mol. Cryst. Liq. Cryst. **97**, 163 (1983).

82. C. Rosenblatt, Phys. Rev. A. **32**, 1115 (1985).

83. C. Rosenblatt and J. D. Litster, Phys. Rev. A. **26**, 1809 (1982).

84. S. K. Ma, *Modern Theory of Critical Phenomena* (Benjamin, 1976).

85. R. G. Priest and T. C. Lubensky, Phys. Rev. B. **13**, 4158 (1976).

86. B. D. Larson and J. D. litster, Mol. Cryst. Liq. Cryst. **113**, 13 (1984).

87. N. Boden, S. A. Corne and K. W. Jolley, Chem. Phys. Lett. **105**, 99 (1984).

88. N. Boden, D. Parker and K. W. Jolley, Mol. Cryst. Liq. Cryst. **152**, 121 (1987).

89. H. Fricke, Phys. Rev. **24**, 575 (1924); J. Phys. Chem. **57**, 934 (1953).

90. Y. Galerne and J. P. Marcerou, Phys. Rev. Lett. **51**, 2109 (1983).

91. Y. Galerne, A. M. Figueiredo Neto and L. Liebert, Phys. Rev. A. **31**, 4047 (1985).

92. G. R. Luckhurst, in *The Molecular Physics of Liquid Crystals*, G. R. Luckhurst and G. W. Gray, eds. (Academic Press, 1979), p. 85.

93. A. de Vries, J. Chem. Phys. **56**, 4489 (1972).

94. M. Warner, Mol. Phys. **52**, 677 (1984).

95. W. M. Gelbart and B. Barboy, Acc. Chem. Res. **13**, 290 (1980); W. M. Gelbart and A. Ben-Shaul, J. Chem. Phys. **77**, 914 (1982); W. M. Gelbart, J. Phys. Chem. **86**, 4298 (1982).

96. M. R. Kuzma, W. M. Gelbart and Z.-Y. Chen, Phys. Rev. A. **34**, 2531 (1986).

97. W. Maier and A. Saupe, Z. Naturforsch **A14**, 882 (1959).

98. M. R. Fisch, S. Kumar and J. D. Litster, Phys. Rev. Lett. **57**, 2830 (1986).

99. P. J. Photinos and A. Saupe, J. Chem. Phys. **90**, 5011 (1989).

100. W. L. McMillan, Phys. Rev. A **4**, 1238 (1971); **6**, 936 (1972).

101. D. Parker, Ph.D. Thesis, University of Leeds, (1988).

102. W. L. McMillan, Phys. Rev A **9**, 1720 (1974).

103. A. F. Martins, A. C. Diogo and N. P. Vaz, Ann. Phys. **3**, 361 (1978).

104. P. G. de Gennes, Sol. State Comm. **10**, 753 (1972).

105. F. Brochard, J. de Physique **34**, 901 (1973).

106. F. Jähnig and F. Brochard, J. de Physique **35**, 301 (1974).

107. A. Saupe, P. Boonbrahm and L. J. Yu, J. Chim. Phys. **80**, 3 (1983).

108. L. D. Landau and E. M. Lifshitz, *Statistical Physics*, 3rd Edn. Part 1 (Pergamon Press, Oxford, 1985) chap. 14, p.497.

109. T. C. Lubensky and R. G. Priest, Phys. Rev B **13**, 4139 (1976).

110. K. Radley and A. Saupe, Mol. Phys. **35**, 1405 (1978).

111. A. M. Figueiredo Neto, Y. Galerne and L. Liébert, J. Phys. Chem. **89**, 3939 (1985).

112. N. Boden, P. J. B. Edwards and K. W. Jolley, in "Structure and Dynamics of Supramolecular Aggregates and Strongly Interacting Colloids", S. H. Chen, J. S. Huang, and P. Tartagli, eds. (Klure and Dordrect, 1992).

113. M. P. Taylor and J. Herzfeld, Phys. Rev. A, **43**, 1892 (1991).

114. T. Odijk, J. de Physique **48**, 125 (1987).

115. R. Hentschke, Liq. Cryst. **10**, 691 (1991).

116. R. G. Priest and T. C. Lubensky, Phys. Rev. **B13**, 4159 (1976).

117. C. A. Vause, Phys. Rev. **A30**, 2645 (1984).

4

Geometrical Foundation of Mesomorphic Polymorphism

Jean Charvolin[1]
Jean-François Sadoc[2]

4.1 Introduction

The aggregation of amphiphilic molecules in water, and in particular the growth and first steps of micellar organization in the dilute region of their phase diagrams, were discussed in earlier chapters of this book, primarily the first two. In these diluted regions the problem of the formation of structures by amphiphilic molecules in aqueous solution could be treated within a rather simple geometrical framework which can be summarized as follows. First, the "micellar" topology of the organization—an infinite number of finite aggregates separated by a connected film of water—does not change during aggregation and growth. The aggregation and growth processes were therefore described as continuous deformations of aggregates of simple shapes. Second, the aggregates are sufficiently diluted for the solution of micelles to be treated as ideal, except for steric interactions in first order, so that the only interaction present between interfaces takes place *within* the aggregates and constrains their size along one direction to stay constant, with a value of about twice the length of a molecule. The deformations of the aggregates under concentration changes, or the evolution of their interfacial curvature, were then considered to occur subject to this one spatial constraint only.

In this chapter we consider the problems of aggregation and organization in the liquid crystalline and condensed micellar regions of the phase diagrams, *i.e.*, in the so-called *mesomorphic* regions, as their structures show intermediate degrees of order. In these concentrated regions it is no longer possible to use the same simple framework reviewed above. First, structural studies show that the topologies of the organizations change dramatically when moving from one phase to another. For instance, a lamellar phase

---

[1]Laboratoire de Physique des Solides, Université de Paris-Sud, 91405 Orsay, France
[2]Same as footnote 1.

is a set of an infinite number of infinite amphiphilic and aqueous cells, whereas a cubic phase is a set of two infinite cells of one medium separated by an non-intersecting film of the other. The evolutions of the aggregates can therefore not be treated in a continous manner, without processes of rupture and fusion. Second, the aggregates are now so close that we can no longer consider only the steric interactions between them; in particular, the minimum of the interaggregate interaction potential in aqueous solution imposes the additional constraint of contant distance between them and their interfaces. Thus the evolution of the aggregates under concentration changes, or that of their interfacial curvature, is subjected to the more drastic geometrical framework of *two* spatial constraints instead of one. In this chapter, we specifically discuss the role of such spatial constraints in determining polymorphism in the concentrated regions of phase diagrams.

In Sec. 4.2 we recall briefly the broad lines of the classical structural description, in order to emphasize the aspects of the polymorphism which are independent of the details of the chemical structures of the molecules and related to their amphiphilicity only. This makes possible an analysis of these systems as periodic organizations of fluid films. The impossibility for these films to reconcile spatial constraints when their interfaces are not flat leads us to consider them as frustrated systems. In Sec. 4.3, we expose in general terms how such frustration can be resolved. We apply a method which consists in transferring the frustrated structure into a curved space where the frustration can be relaxed; then this curved space is mapped onto the Euclidean space by the introduction of defects of rotation respecting the symmetries of the relaxed structure. This process generates a structure optimizing the frustration in the Euclidean space and which differs from the initial one by the presence of defects. In Sec. 4.4 we apply this method to the case of a periodic system of parallel films and search for its possible geometrical configurations. The topologies of these solutions, their order of appearance when the parameters are varied, and their symmetries are in quite satisfactory agreement with those of the structures presented in the classical description, but for discrete values of the parameters only. The necessity for the system to resolve its frustration for large distributions of values of the parameters, corresponding to the large extensions of the domains of the phases in the phase diagrams, imposes the existence of structural fluctuations within the above configurations. This is commented on in Sec. 4.5, in the light of recent experimental observations which lead to a revision of the classical structural description.

General presentations of the data concerning liquid crystals formed by amphiphilic molecules, as well as compilations of the literaure existing on the subject, can be found in basic review articles [1, 2, 3, 4, 5]. This chapter is limited to a coherent presentation of a set of recent articles [6, 7, 8, 9], devoted to the investigation of the role of spatial constraints to determine the polymorphism in the mesomorphic domain, and to a general discussion of their conclusions.

4.2   Mesophases as Structures of Films

An examination of the literature concerning phase diagrams [2] and structures [1, 3, 5] suggests the idea that general behaviors, common to all amphiphilic molecules, exist in spite of the variety of the details of their chemical structures. Similar structures appear along similar sequences when the dominant parameters of the phase diagrams are varied. This is illustrated in Fig. 4.1, where the phase diagrams of three systems, anionic, cationic and non-ionic, are presented. We are interested in the high temperature phases only. In the two first cases, those of sodium laurate [10] and dodecyltrimethylammonium chloride [11], the parameter determining the polymorphism is the water content. It acts on the degree of ionization of the polar heads and, therefore, on their coulombic interactions within an interface, making the mean area per polar head increase with the water content, and between interfaces. In the third case, that of hexa-ethyleneglycol mono n-dodecyl ether [12], the water content and the temperature intervene. They act on the behavior of the short hydrophilic chains of ethyleneglycol whose solvation, and therefore mean lateral area, is known to increase also when the temperature decreases. Similar sequences of phases are found in the three diagrams when the dominant parameters vary continuously. They go from a lamellar (L_α) phase to cubic (Q'_α) and/or intermediate

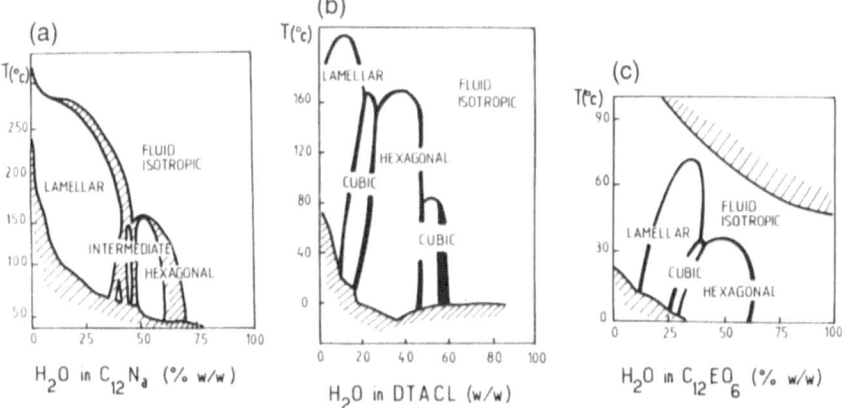

FIGURE 4.1. Schematic representations of the phase diagrams: of the $C_{12}Na/H_2O$ system (a), from reference [10]; of the $DTACl/H_2O$ system (b), from reference [11]; and of the $C_{12}EO_6/H_2O$ system (c), from reference [12]. Broad lines and hatched regions represent polyphasic domains.

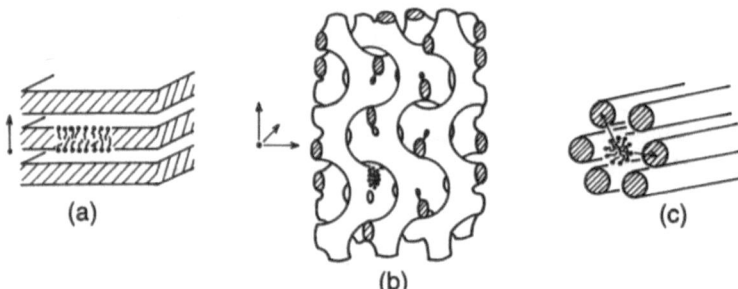

FIGURE 4.2. Drawings of lamellar (a), cubic with space group $Ia3d$ (b), and hexagonal (c) structures.

(I_α) phases, then to a hexagonal (H_α) phase sometimes followed by another cubic (Q''_α), and end with the isotropic micellar solution. Schematic representations of the structures of the L_α, H_α phases and of one particular Q'_α phase are presented in Fig.4. 2. We have not represented the $I\alpha$ and Q''_α phases: the I_α phases are cylindrical phases, as are the H_α ones, but their cylinders have non-circular sections organized on 2-D lattices of rectangular, quadratic or monoclinic symmetry [1-4, 13, 14]; the structures of the Q''_α phases are not clearly established yet [5,8], but proposals describing them as organized packings of finite micelles will be made later in this chapter.

If we just consider the topologies of the structures, the sequences can be described in similar terms. The lamellar phase, which is made up of an infinite number of infinite cells of two media, gives way to cubic phases, which are made up of two infinite cells of one medium separated by a continuous film of the other medium, or to cylindrical phases consisting of an infinite number of infinite cells of one medium separated by one connected film of the other; the sequence ends with micellar phases, which are made up of an infinite number of finite cells separated by one connected film. Here we can insist again on some facts showing that even important chemical variations on the general theme of amphiphilicity leave the structures and their sequence unchanged. For instance: moving from mono-alkyl amphiphiles to di-alkyl amphiphiles, or amphiphiles with branched chains, just exchanges the relative distribution of the amphiphilic and aqueous media [1]; changing the solvent from water to formamide in the cetyl trimethylammonium bromide system does not perturb the phase diagram greatly [15]; and finally, similar structures and sequences are observed with amphiphilic copolymers of molecular weights much larger than those of ordinary amphiphilic molecules [16, 17].

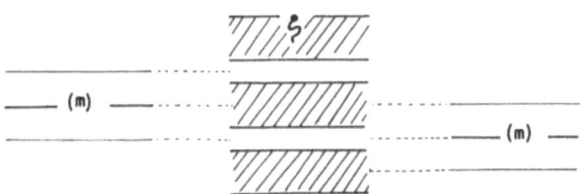

FIGURE 4.3. Definition of the film in a lamellar structure. It can be either a film of amphiphiles limited by two half layers of water, with its middle surface in the bilayer (left), or a film of water limited by two monolayers of amphiphiles, with its middle surface in the water layer (right).

The generality of this topological description calls for an argument which features the aspect common to all amphiphilic molecules only, i.e., their ability to build interfaces organized in symmetric films. This idea is strengthened by the fact that X-ray and NMR experiments have shown the two aqueous and paraffinic media to be hightly disordered, so that the structures can be described, in a distinctly paradoxical way, as ordered entanglements of two disordered liquids separated by one interface [18]. Thus, the structural element is certainly not the molecule, as in ordinary molecular crystals, but rather the interface or, better for symmetry reasons, the film built by two facing interfaces. As shown in Fig. 4.3, this film can be defined in two ways, either by one film of water limited by two monolayers of amphiphile, or by one bilayer of amphiphiles inbetween two half layers of water. These two definitions are complementary: the first is useful to provide an account of direct phases formed by mono-alkyl molecules, where aggregates of amphiphilic molecules are separated by a film of water; the second would be useful to explain the inverted situation known for di-alkyl molecules. In this chapter we limit ourselves to the first situation only (the second one being treatable in a totally symmetric manner).

We consider the amphiphilic film to be made of two identical layers, acted on by several forces which fix the inter- and intra- layer distances: Van der Waals forces, electrostatic interactions between polar groups at the interfaces, forces created by water polarization and charge distributions in aqueous layers, hydrophobic interactions which prevent the presence of water within the layers, and fluctuation-induced forces [19, 20]. These forces are not totally understood at the moment, nor is their interplay [21]. However it is reasonable to imagine that the role of the components normal to the interfaces is to maintain constant distances between them, if they are supposed homogeneous, and that the role of the components parallel to the interfaces is to determine the interfacial curvatures. Owing to the symmetry of the film with respect to its middle surface it is clear that the fact that two facing interfaces may have symmetric curvatures is not always

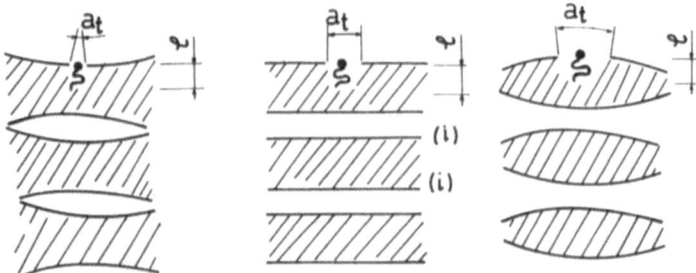

FIGURE 4.4. Schematic representation of a periodic system of films with flat interfaces (center): constant interfacial distances and zero curvature are compatible in R_3. The same with symmetrically curved interfaces (left and right): constant interfacial distances and non zero curvatures are no longer compatible in R_3 and the system becomes frustrated.

compatible with constant distances between the interfaces, if a lamellar-type stacking is kept, as shown in Fig.4.4. The left and right drawings of this figure present situations of conflict between forces normal to the interfaces, which want to maintain them at constant distances, and forces parallel to them, which want to induce a curvature. The forces are therefore obliged to compromise and new structures should arise. Such a situation is a typical case of *frustration*, a very fruitful concept recently introduced in condensed matter physics [22, 23], and we have adapted to the problem of fluid films one general method developed to resolve frustration.

4.3   Frustration, Curved Spaces and Disclinations

4.3.1   PRINCIPLE OF THE METHOD

A frustration is directly related to the structure of the space containing the system under study. As shown in Fig. 4.4, where a system embedded in the usual flat 3-D Euclidean space R_3 is represented, the situation with flat interfaces is the only situation compatible with constant interfacial distances in this space. However, if the system is embedded in an adequately curved space, the situations with curved interfaces may become compatible with constant interfacial distances. This is formally similar to problems of bi-dimensional tilings with regular polygons and tri-dimensional packing of regular polyhedra [24]. We shall show later that ideal lamellar-like structures without frustration, reconciling the two antagonistic constraints, can indeed be built in curved spaces. Such ideal structures have of course no

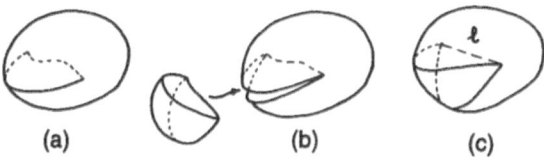

FIGURE 4.5. Creation of a defect in a piece of matter, following a Volterra process. A cut is made (a); the lips of the cut are separated and extra matter is introduced in a manner respecting the symmetry of the matter (b); and the matter relaxes around the line of defect ℓ (c).

physical reality but are nevertheless useful, as starting points, to generate possible configurations in the flat Euclidean space. For this it is necessary to map the curved space onto the flat space. A classical example of such a process with 2-D spaces is that of the mapping of a sphere onto a plane. A sphere can not be deformed into a plane without being torn, except if matter is added during the process. If a structure is present on the sphere, the introduction of matter must respect the symmetry of this structure. This is generally known as the Volterra process shown in Fig. 4.5, and corresponds to the creation of discontinuities, or defects, in the structure. Such defects can be defects of translation (or dislocations), and of rotation (or disclinations) [22, 25]. As the direct displacements on a sphere are rotations, the defects needed to map it onto a flat space are disclinations only [26]. In a similar way the mapping of the 3-D curved space onto the 3-D flat space is obtained by introducing a network of disclination lines respecting the symmetries of the structure without frustration in the curved space, the density of the disclination network being determined by the curvature of the curved space [27]. The possible configurations obtained that way can therefore be seen as structures of disclinations.

It is important at this stage to define precisely the procedure of our approach. We analyze the conflict between antagonistic physical forces in R_3 as a frustration. This frustration is expressed in terms of curvatures and distances; it is relaxed by transferring the structure into a curved space, a Volterra process being used to map this space onto R_3 and to create the disclination structures. The terms and the operations used in this method are purely geometrical, so that the structures of disclinations obtained correspond directly to the possible geometrical configurations imposed by the spatial constraints. They can not be immediately considered as the real structures, since energetic terms - - stretching and curvature elasticities of the film, and entropic terms - - distributions of disclinations and fluctuations, are not yet included. These terms, and their interplay in the free energy, should determine the configurations of the real structures. Nevertheless, we shall see later at the end of the chapter that a striking agreement exists between the results of our geometrical approach and the observed structures.

The application of the method presented above requires abandoning the usual 3-D Euclidean space for a 3-D curved space. Facing the difficulty of pictorial representations in such a space we think it useful to illustrate the method by developing first its application to the case of a 2-D periodic system of parallel strips.

4.3.2   A Simple 2-D Example

We consider the normal section of a periodical stacking of bilayers in the Euclidean plane R_2. It is a periodic system of parallel strips [6, 9]. The frustration appears when the interfacial lines are symmetrically curved, or when their lengths are smaller by equal amounts than that of the middle line of the strip. The frustration is homogeneously relaxed when the flat space R_2 is curved into a spherical space S_2 of homogeneous curvature with radius R if, as illustrated in Fig. 4.6, the middle line of the strip is placed on the equator and the interfaces on parallels to the equator. In this situation the symmetry of the strip is preserved, as the equator is a stationary line separating the sphere into two identical hemispheres. The frustration is relaxed, since the interfacial lengths $2\pi R sin\theta$ are smaller than the middle length $2\pi R$; the interfacial distances are constant, since the parallel and the equator are parallel lines; and, when moving along great circles normal to the equator which are geodesics of S_2, the aqueous and paraffinic media are periodically crossed, corresponding to moving along normals to the interfaces of the infinite system in R_2. The two poles of the sphere could be considered as 2π disclination points for the normal to the strips; they will be compensated by the negative disclinations introduced later in mapping S_2 onto R_2. Finally the symmetry axes of the relaxed structure in S_2 are

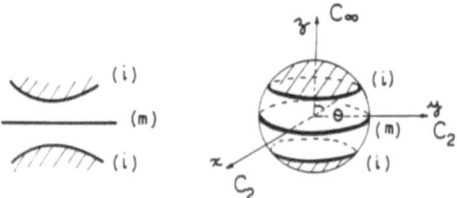

FIGURE 4.6. The 2-D periodic system of parallel stripes: suppression of the frustration in R_2 by mapping it onto S_2.

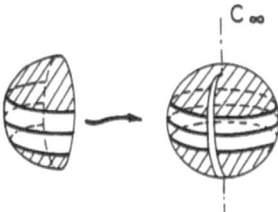

FIGURE 4.7. The 2-D periodic system of parallel stripes: back to R_2 by introducing disclinations around the C_∞ axis following a Volterra process.

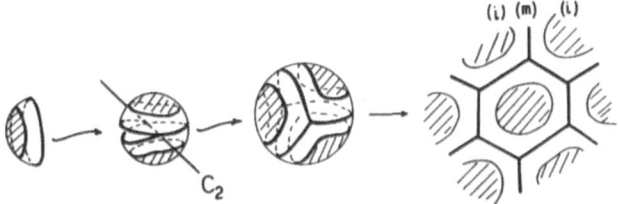

FIGURE 4.8. The 2-D periodic system of parallel stripes: back to R_2 by introducing $-\pi$ disclinations around C_2 axes following a Volterra process.

one C_∞ axis normal to the equatorial plaane and a family of C_2 axes in this plane.

The mapping of S_2 onto R_2 is obtained by introducing the appropriate disclinations around these axes. Disclinations around the C_∞ axis can be of any angle: their effect is to increase the area of the sphere and decrease its curvature locally without changing the topology of the strip, as shown in Fig. 4.7. The result in R_2 is a strip with flat interfaces, but it can not be obtained without important distortions of the distances within the strip and between interfaces, and the process therefore involves stretching energies. Disclinations around C_2 axes are $-\pi$ disclinations because of the C_2 symmetry [22, 25]: they decrease the curvature of the sphere locally, preserving all distances but changing the topology of the structure, as shown in Fig. 4.8. The result in R_2 is a structure defined by a connected strip delimiting closed cells with curved interfaces, and this process involves curvature energies only. The two processes are therefore totally different according to the type of energy they involve; the second is chosen preferably over the first, because curvature energies are classically smaller than stretching ones in liquid crystals.

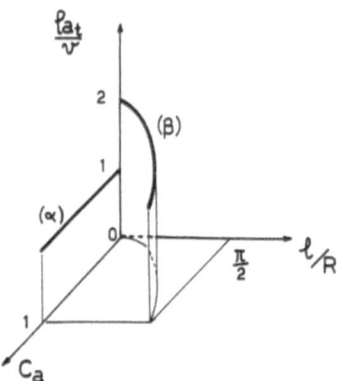

FIGURE 4.9. The 2-D periodic system of parallel stripe: relations between la_t/v and c_a in spherical spaces with curvature radii R. Curve (α) corresponds to the "lamellar" configuration in flat space, and curve (β) to the configuration in curved spaces.

The degrees of frustration, or the possible associations of distances between strips and length ratio between interfacial and middle lines, which can be relaxed that way can be evaluated as follows. If, in Fig. 4.6, the hatched spherical cap of area $2\pi R^2 (1 - cos\theta)$ is filled by n amphiphiles of length l, area v, and mean lateral length per polar head a_t at the interface, and if the rest of the hemisphere is filled by water, the respective surface concentrations of amphiphiles and water being c_a and $c_e = 1 - c_a$, then $na_t = 2\pi Rsin\theta$ and $nv = 2\pi R^2 (1 - cos\theta)$, and the two parameters defining the degree of frustration can be written as $la_t/v = \theta sin\theta/(1 - cos\theta)$ and $c_a = 1 - cos\theta$, with $l = R\theta$, since half the area of S_2 is $2\pi R^2$. In these formulae the constant thickness of the amphiphilic layer is l, that of the water layer is contained in c_e and the interfacial curvature is expressed as the ratio of the mean lateral lengths per polar head a_t and per chain v/l [28]. The relations between la_t/v and c_a in curved spaces S_2, together with that in R_2 where a structure with flat interfaces is geometrically possible whatever c_a, is represented in Fig. 4.9. It is evident from this figure that only certain degrees of frustration, with a definite relation between the mean length per polar head and the water content, can be relaxed in spherical spaces whose radii of curvature increase with the degree of frustration. This implies that solutions of the frustration can be found for a discrete number of values of the parameters only, which appear sequentially when the parameters vary. For instance, working at constant c_a and increasing la_t/v, two solutions only are possible for well-defined values of la_t/v: one with flat interfaces and a lamellar topology, and another with curved interfaces and a closed cell topology, of the type shown in Fig. 4.8. This sequence clearly results from the constraints imposed by the geometries of the different spaces.

After having considered the topologies of the solutions, the first question about them concerns their possible organizations, or the way the disclinations might be ordered. In the new structure the organization of the strips separating the closed cells is connected, and the topological structure of its 2-D flat pattern is imposed by classical laws of topology stability. Each cell must have, on average, 6 edges meeting 3 by 3 at each vertex [29]. That means, using Schlafli's notation [30], that any *ordered* organization must be a {6,3} tiling of the plane, i.e., a hexagonal lattice or any distortion of it, as represented on the right of Fig. 4.8.

4.3.3 CONCLUDING REMARK

The development of the frustration approach in the simple 2-D case of a periodic system of parallel strips shows that the spatial constraints may eventually control three main aspects of the polymorphism : the topologies of the structures, the sequence of these topologies when the parameters of the phase diagrams vary, and the organization of the structures. We now develop the same approach in the case of the 3-D system of parallel films and examine the same points specifically.

4.4   The Periodic System of Parallel Films

We now deal with a periodic stacking of films in the flat Euclidean space R_3. It becomes frustrated when the interfacial area becomes smaller than the area of the middle surface. The search for homogeneous solutions to the frustration in 3-D space is formally equivalent to that in 2-D space. First the frustration is homogeneously relaxed by transferring the stacking into 3-D spaces with homogeneous positive curvatures. Two such spaces have been used, the hypersphere S_3 and the hypercylinder $S_2 * R_1$ which can be defined by their equations in the four dimensional Euclidean space R_4 as $x_1^2 + x_2^2 + x_3^2 + x_4^2 = R^2$ for the first and $x_1^2 + x_2^2 + x_3^2 = R^2$ for the second if it is parallel to the fourth direction; these equations are indeed obtained from those of the sphere and cylinder in the three dimensional Euclidean space R_3 by adding one dimension more (see the appendix). The hypersphere has an isotropic curvature while the hypercylinder does not; in $S_2 * R_1$ there is no curvature along direction R_1 and all the curvature is concentrated on the sphere S_2 orthogonal to this direction [31]. Then the middle surface of the film is transferred to stationary surfaces separating the curved spaces into two equivalent subspaces. There are two in S_3 - - one particular torus with genus one called the spherical torus T_2, and the great sphere S_2 - - and one in $S_2 * R_1$ - - the cylinder $S_1 * R_1$ built on a great circle of S_2. Stereographic projections onto R_3 of the curved spaces and their surfaces with the film are shown in the next few pages in Figs. 4.10, 4.12, and 4.13. In these situations the symmetry of the film is preserved, since the surfaces supporting its middle surface are stationary surfaces separating the spaces into two equal subspaces; the frustration is relaxed, since the interfacial areas are smaller than the middle area and the distances between the interfaces are constant, since the surfaces of a family are parallel surfaces; and the periodicity of the stacking is maintained as it corresponds to cyclic displacements along geodesics normal to the surfaces. (Analytical bases for demonstrating these properties can be found in the appendix and in references [6, 9].) It is interesting to notice here that the polar circles of S_3 and the polar axes of $S_2 * R_1$ constitute 2π disclination lines which are compensated by the negative disclinations needed to return to R_3. Finally, with a view to introducing the disclinations, we examine the symmetry axes of the relaxed structures. They are C_2 and C_∞ axes. The first, which correspond to disclinations involving curvature energies, are geodesic great circles normal to T_2 or in the surfaces of T_2 and S_2 in S_3 and, in $S_2 * R_1$, great circles normal to $S_1 * R_1$ and lines along it [6]. Among them, the axes along the surface of T_2 and those normal to $S_1 * R_1$ will not be considered since they lead to topologies equivalent to those provided by the axes along $S_1 * R_1$ and normal to T_2 respectively [31]. The C_∞ axes, which correspond to disclinations involving stretching energies, will also not be considered here. We now examine only the topologies, sequence, and symmetries of the geometrical configurations in the three possible cases of the spherical torus T_2 in S_3, the cylinder $S_1 * R_1$ in $S_2 * R_1$ and the great sphere S_2 in S_3.

4.4.1 TOPOLOGIES

4.4.1.1 The spherical torus T_2 in S_3

The film on the spherical torus is represented in Fig. 4.10 The $-\pi$ disclinations are introduced around the C_2 axes normal to the surface. A schematic representation, valid only as far as topology is concerned [7], of the introduction of one disclination into a spherical torus in stereographic projection, is given in Fig. 4.11. The torus of genus 1 is cut in half, the lips are separated and half a torus is inserted. The result is a torus of genus 2 which, like the original torus, separates the space in which it is embedded into two identical subspaces. When the set of disclinations needed to map S_3 onto R_3 is introduced, a complex surface of high genus is generated which keeps the important property of separating R_3 into two identical subspaces. The topology of such a configuration is said to be "bicontinuous", in the sense that it can be described as being made up of two labyrinths separated by one film without self-intersection. This is the topology of the cubic phases Q'_α found in the immediate vicinity of lamellar phases.

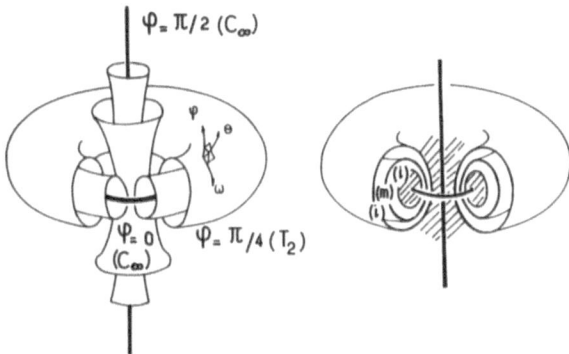

FIGURE 4.10. Stereographic projections onto R_3 of the families of tori in S_3 together with the associated organizations of interfaces (i) and middle surfaces (m) for periodic systems of films. The amphiphilic regions are hatched.

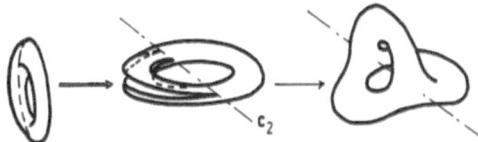

FIGURE 4.11. Introduction of a $-\pi$ disclination around a C_2 axis of a torus, following a Volterra process. The validity of this figure in the T_2 case concerns the topology of the process only; see Fig. 7 of reference [7].

4.4.1.2   The cylinder $S_1 * R_1$ in $S_2 * R_1$

The film on the cylinder is shown in Fig. 4.12. This cylindrical space admits a sphere S_2 as an orthogonal section to the direction defined by R_1. The C_2 axes of the cylinder are therefore the C_2 axes of the equator of the sphere, and the results in R_2 from the discussion of the 2-D example, presented in Fig. 4.8, can be used directly in this case, recalling the extension of the configuration along the direction normal to R_2. The topology created in this way is therefore that of an infinite number of infinitely long cells separated by a connected film. This is the topology of the so-called cylindrical phases.

4.4.1.3   The great sphere S_2 in S_3

The film on the great sphere is represented in Fig. 4.13. The C_2 axes are great circles of S_2 and the introduction of a disclination along one of these axes brings in a third finite subspace, as shown in Fig. 4.14. The introduc-

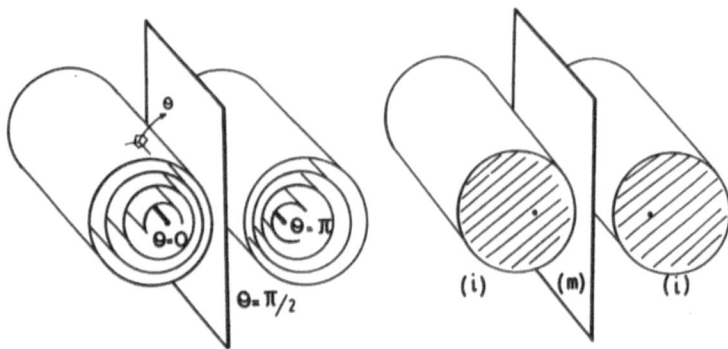

FIGURE 4.12. Stereographic projections onto R_3 of the families of cylinders in $S_2 * R_1$, together with the associated organizations of interfaces (i) and middle surfaces (m) for periodic systems of films. The amphiphilic regions are hatched.

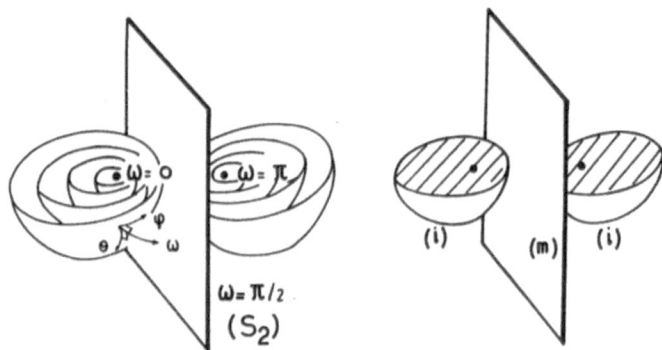

FIGURE 4.13. Stereographic projections onto R_3 of the families of spheres in S_3, together with the associated organizations of interfaces (i) and middle surfaces (m) for periodic systems of films. The amphiphilic regions are hatched.

tion of a set of disclinations just multiplies the number of identical finite subspaces. The topology created in this way is therefore that of an infinite number of finite cells separated by a self-intersecting film. This is the topology of the micellar phases and, most likely, that of the cubic phases Q''_α found in the immediate vicinity of micellar phases.

4.4.2   SEQUENCE [9]

We now establish the relations existing between the two terms of the frustration in these different spaces, similar to those discussed in the 2-D case, and see how they can match with the evolutions of the parameters of the

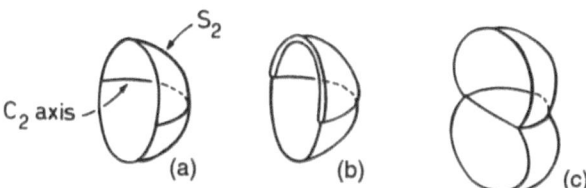

FIGURE 4.14. Introduction of a $-\pi$ disclination around a C_2 axis of a great sphere S_2 in S_3, following a Volterra process. This is a stereographic projection of S_3 onto R_3. For the sake of clarity, only half a sphere S_2 is shown and the film supported by it is not represented. S_2 separates S_3 into two identical sub-spaces (a), S_2 is partly split into two sheets limited by the C_2 axis (b), and a third sub-space, identical to the first two is introduced between the sheets (c).

phase diagrams. We recall that the disclination processes do not change the volume-to-area ratio.

4.4.2.1 The spherical torus T_2 in S_3

In this case the middle surface of the film is supported by the spherical torus and the two interfaces are supported by parallel tori at equal distances of one molecular length l from it. As represented in Fig. 4.10, this distance can also be expressed as the angular variation ϕ along a geodesic of S_3: $l = R\phi$, where R is the radius of the space. The volume limited by each interface contains n molecules of volume v, each polar head occupying a mean area a_t on this interface. It follows that (see the appendix) $na_t = 4\pi^2 R^2 sin\phi cos\phi$ and $nv = 2\pi^2 R^3 sin^2\phi$, and the two parameters defining the degree of frustration can be expressed as $la_t/v = 2\phi/tg\phi$ and $c_a = 2sin^2\phi$, with $l = R\phi$, since half the volume of T_2 is $\pi^2 R^3$.

4.4.2.2 The cylinder $S_1 * R_1$ in $S_2 * R_1$

In this case the middle surface of the film is supported by the cylinder and the two interfaces are supported by parallel cylinders at equal distances of one molecular length l from it, as represented in Fig. 4.12. Since the normal sections are at the equator and two parallels of the sphere S_2, the filling of this cylindrical space by the film is determined by the filling of the sphere S_2 normal to its generatrix. For this purpose we can use directly the formulae established in the case of the 2-D example. They are $la_t/v = \theta sin\theta/(1 - cos\theta)$ and $c_a = 1 - cos\theta$, with $l = R\theta$, or $la_t/v = \theta/tg(\theta/2)$ and $c_a = 2sin^2(\theta/2)$. These are similar to the formulae established for the spherical torus T_2 but, as $\theta = 2\phi$, the radius of curvature R is twice smaller in this case than in that of T_2.

4.4.2.3   The great sphere S_2 in S_3

In this case the middle surface of the film is supported by the great sphere and the two interfaces are supported by parallel spheres at equal distances of one molecular length l from it. As represented in Fig. 4.13, this distance can also be expressed in terms of the angular variation ω along a geodesic of S_3: $l = R\omega$, where R is the radius of the space. The volume limited by each interface contains n molecules of volume v, each polar head occupying a mean area a_t on this interface. It follows that (see appendix) $na_t = 4\pi R^2 sin^2\omega$ and $nv = 2\pi R^3[\omega - (sin2\omega)/2]$, and the two parameters defining the degree of frustration can be expressed as $la_t/v = (2\omega sin^2\omega)/[\omega - (sin2\omega)/2]$ and $c_a = (2\omega - sin2\omega)/\pi$, with $l = R\omega$, since half the volume of S_3 is $\pi^2 R^3$.

4.4.2.4   Confrontation with physico-chemical data

The possible variations of la_t/v and c_a in the different spaces, characterized by their curvature radius, are summarized in Fig. 4.15a. It is clear from this figure that it is not possible to reconcile arbitrary interfacial curvature and arbitrary distance to build an ideal structure. There exist well defined relations between la_t/v and c_a in each of the possible spaces. For instance, for one value of the concentration c_a there are only three compatible values of the mean area per polar head a_t, and the ideal structures exist only as points on the diagram of Fig. 4.15a. These geometrical relations are now compared with the physico-chemical relations involving the parameters a_t and c_a. Once again we do not know the nature of the terms controlling these relations in the various amphiphilic systems, but their exact knowledge is not necessary at this stage. We know from experimental data that, in the case of the direct structures considered here, the mean area per polar head a_t increases with the water content $1 - c_a$ [1]. We can see in Fig. 4.15b, where a reasonable physico-chemical relation extrapolated from experimental data for the $C_{12}Na/H_2O$ system [32] has been drawn, that a sequence of ideal structures is imposed by geometry. They are, starting from the periodic stacking of lamellae in R_3 and increasing the water content: first the spherical torus T_2 in S_3 or the cylindar $S_1 * R_1$ in $S_2 * R_1$; and then the great sphere in S_3. The corresponding sequence of topologies obtained in the Euclidean space R_3 after the introduction of the disclinations is therefore, starting again from the periodic stacking of lamellae in R_3 and increasing the water content, that of the "bicontinuous" topology of the cubic phases or of the "cylindrical" topology of the cylindrical phases, followed by the "micellar" topology of the micellar phases. This order is in agreement with what is observed in phase diagrams, and the different topologies occur in the concentration ranges of the corresponding phases.

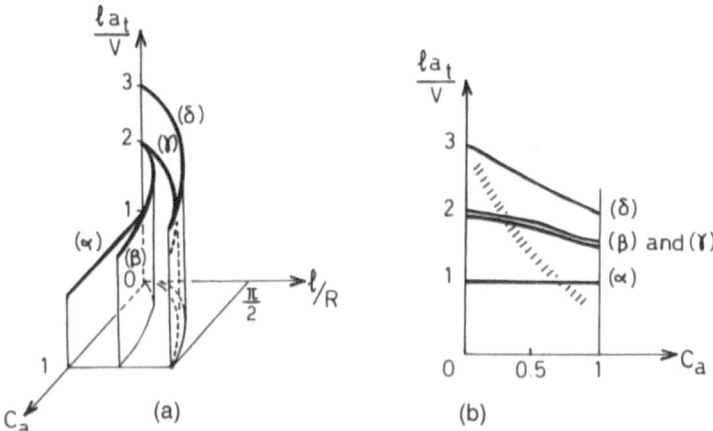

FIGURE 4.15. The 3-D periodic system of parallel layers: relations between la_t/v and physico-chemical relation deduced from reference [32] for the $C_{12}Na/H_2O$ system at 86° C (b). Curve (α) corresponds to the lamellar configuration in flat space, and curves (β, γ, δ) correspond to the toroidal, cylindrical and spherical configurations in curved spaces. In the case of the $C_{12}EO_6/H_2O$ system the thermotropic sequence with decreasing temperature would be obtained at constant c_a for increasing la_t/v.

4.4.3   ORDERED ORGANIZATIONS

We now look for the possible ordered organizations of the disclinations, always within the framework of the spatial constraints imposed by geometry.

4.4.3.1   The spherical torus T_2 in S_3 [7]

We make use of an important property of the spherical torus which is that it can be built by identification of the opposite sides of a square sheet (as this is done in a curved space, the sheet suffers no distortion). From this it can be shown that the only regular periodic tiling possible of the torus is a {4,4} tiling of squares [7,33]. The homogeneous introduction of the disclination network must respect this tiling and its symmetry, so that the elementary aspect of the process is the disclination of a square in a {4,4} tiling. This is represented in Fig. 4.16. The $-\pi$ disclination transforms the square into a hexagon without changing the coordination number of the vertices.

The effect of the introduction of the total set of disclinations is therefore to transform the (4,4) tiling into a (6,4) tiling, which is obviously not

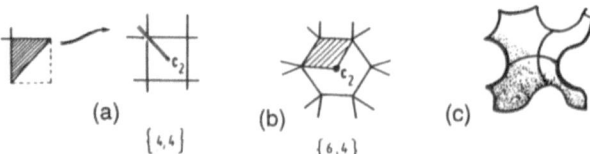

FIGURE 4.16. A $-\pi$ disclination in a square (a) transforms it into a hexagon (b); if the original squares are in a plane with zero Gaussian curvature, the plane is transformed into a hyperbolic surface with negative Gaussian curvature (c).

Euclidean, as hexagons can only be organized on {6,3} in R_2. The surface generated this way has the same local properties as a hyperbolic surface of constant negative Gaussian curvature, or hyperbolic plane [34]. The hyperbolic plane can not be represented in R_3 and we make use of Poincare's model to determine how this surface can be embedded in R_3 [35,36]. This model is built in the Euclidean plane: the hyperbolic plane is represented within a circle—its points at infinity are on the limiting circle, the angles between lines are preserved, and the metric is chosen in order to represent that of the surface, i.e., the distance between two points representing a constant segment on the surface decreases when the two points approach the limiting circle [36]. The {6,4} tiling of the hyperbolic plane we need is represented in Fig. 4.17a, together with its orthoscheme triangles which are the smallest cells, or asymmetric units, from which the whole surface can be constructed by reflections of its sides. It is clear from this representation that the hexagons and the triangles are not Euclidean. The hexagons have angles of $\pi/2$, the angles of the squares before the introduction of the disclinations, so that the orthoscheme triangles have angles of $\pi/2$, $\pi/4$, $\pi/6$. An examination of the possible structures of these trianges readily shows that the possible surfaces in R_3 fall into three categories: the surfaces for which the two sides of the right angle of the orthoscheme triangle are straight lines; those for which the hypotenuse only is a straight line; and those for which all sides are curved. We treat here only the first case; details concerning the other cases can be found in [7]. If the two sides of the right angle of the orthoscheme triangel are straight they build a network of intersecting straight lines whose element is a non-planar quadrangle $(\alpha, \beta, \gamma, \delta)$ with three angles of $\pi/2$ and one angle of $\pi/3$. Such elements can be assembled into a more symmetrical quadrangle $(\delta, \epsilon, \phi, \kappa)$ with four angles of $\pi/3$ and four equal sides having the saddle shape shown in Fig. 4.17b. These quadrangles are regularly organized in the hyperbolic plane and it was shown by

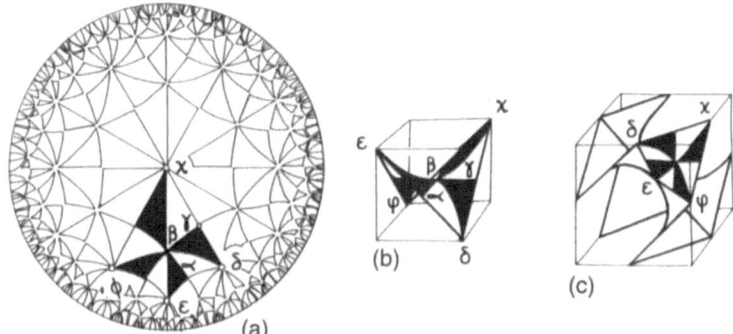

FIGURE 4.17. Poincare's representation of a hyperbolic plane with $\{6,4\}$ tiling and orthoscheme triangles with angles $\pi/2$, $\pi/4$, $\pi/6$ (a), the organization of quadrangles $(\alpha, \beta, \gamma, \delta)$ in the saddle $(\delta, \epsilon, \phi, \kappa)$ in R_3 (b), and the monkey saddle cell with a set of six quadrangles $(\delta, \epsilon, \phi, \kappa)$ (c).

Schoenflies [37] and Schwarz [38] that, when embedded in R_3, six of them build the network having the shape of a monkey saddle shown in Fig. 4.17c, which is the translation cell of a cubic lattice. Schwarz also showed that this periodic network of straight lines is the support of an infinite periodic minimal surface, or IPMS, separating R_3 into two identical subspaces. The translation cell of the surface, together with the two congruent labyrinths separated by it, are represented in Fig. 4.18. Notice that the translation cell of the surface shown in Fig. 4.18 is not exactly that of the network represented in Fig. 4.17c. This is imposed by the periodicity of the normal to the surface [39]. This organization has the Pn3m symmetry of one of the "bicontinuous cubic phase [5,40]. The two other cases of orthoscheme triangles, with straight hypotenuse or all sides curved, lead to organizations with Im3m and Ia3d symmetries [6,39]. "Bicontinuous" cubic structures having these symmetries have also been observed experimentally [5]. These three possible organizations for a film without self-intersection—separating space into two congruent (Pn3m and Im3m) or oppositely congruent (Ia3d) labyrinths—are topologically equivalent. They have the same genus 3, and a study of translation sub-groups in the hyperbolic plane shows that organizations with higher genus might also be possible [39].

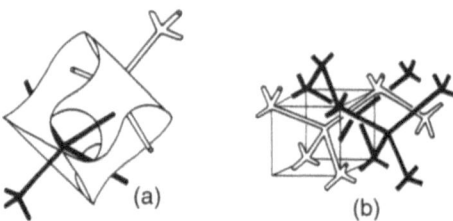

FIGURE 4.18. The translation cell for the F surface of Schwartz (a), and the labyrinths of the $Pn3m$ cubic structure separated by this surface (b).

4.4.3.2   The cylinder $S_1 * R_1$ in $S_2 * R_1$

As the filling of this cylindrical space by the film is determined by the filling of the sphere S_2 normal to its generatrix, we can use directly the results obtained in the case of the 2-D example. The normal sections of the cylinders should be organized on a $\{6,3\}$ tiling of the plane, as suggested in Fig. 4.8. This is often observed experimentally in hexagonal [1], cmm rectangular [13] and monoclinic [14] structures, but not always (as for instance in the case of quadratic structures [3]).

4.4.3.3   The great sphere S_2 in S_3 [8]

We recall that the topology obtained in this case is that of an infinite number of finite cells separated by a self-intersecting film. It is important to notice that the topological stability of the cells implies that the walls limiting them meet three-by-three along common edges which meet four-by-four at each vertex [29, 41]. Because of the identity of the cells, the search for ordered organizations is that of space-filling assemblies of regular polyhedra. Such assemblies are called polytopes.

The above law of topological stability involves the search for polytopes having three faces per edge and four edges per vertex, or having three faces belonging to the same polyhedron around one vertex and three polyhedra around one edge. This means, using Schläfli's $\{$ p,q,r$\}$ notation [30], the polytopes of the $\{$ p,3,3$\}$ family. There are four of them which , unfortunately, exist in curved spaces only [33]:

-in spherical spaces with constant positive Gaussian curvatures $\{$ 3 3,3$\}$, $\{$ 4,3 3$\}$ and $\{$ 5, 3, 3$\}$

-in a hyperbolic space with constant negative curvature $\{$ 6,3,3$\}$.

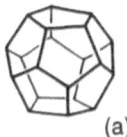 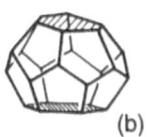 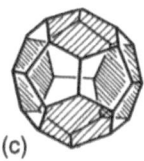

(a)   (b)   (c)

FIGURE 4.19. Transformation of a dodecahedron (a) into a tetrakaidechedron (b) by one $2\pi/5$ disclination around an axis normal to two pentagonal faces, and into a hexakaidecahedron by four half disclinations (c). The hexagons are hatched.

As there is no $\{p,3,3\}$ polytope in a flat Euclidean space, our problem admits no solution with identical regular polyhedral cells [42]. We are therefore driven to search for non-regular but, nevertheless, periodic solutions. For this it is useful to consider the fact that the regular $\{p,3,3\}$ polytopes are found in spaces of decreasing curvatures and therefore arise one after the other during the progressive introduction of disclinations needed to map S_3 onto R_3. They arise at well defined steps of the process, when the curvature of the space is such that it can be tiled by a $\{p,3,3\}$ polytope. Inbetween two steps, when the curvature of the space is not compatible with a regular polytope, the filling of the space cannot be regular. This is the situation we shall be confronted with since our Euclidean space R_3 is not compatible with either $\{5,3,3\}$, which exists in the spherical space of lowest positive curvature for a $\{p,3,3\}$ polytope, or the following polytope $\{6,3,3\}$, which exists in a hyperbolic space of negative curvature. Indeed it can be shown, writing the curvature as a function of p and making it equal to zero, that, in R_3, p = 5.1 and is not an integer [43]. If non-regular solutions exist in R_3 they should be assemblies of polyhedral cells with an average number of edges per face close to 5.1.

The polytope of interest for us is obviously $\{5,3,3\}$ which exists in the space of lowest positive curvature. We must start from it and find the disclination process realizing the final mapping of S_3 onto R_3 which leads to ordered cellular structures. This question was previously addressed in the case of clathrate structures of water and silicon-sodium alloys and, in a related manner, in the case of the dual polytope $\{3,3,5\}$ to analyze the Frank and Kasper's structures of Laves phases and A15 alloys [39]. It was shown that disclinations normal to pentagonal faces of a dodecahedron transform it into the tetrakaidecahedron and hexakaidecahedron shown in Fig. 4.19. These non-regular polyhedra are particularly interesting here as it is known that, when slightly distorted, they can be packed so that they build periodic space-filling assemblies of cells [45]. These assemblies are indeed the orga-

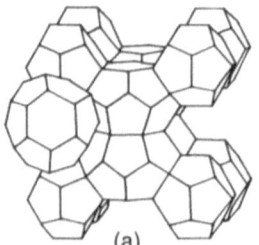 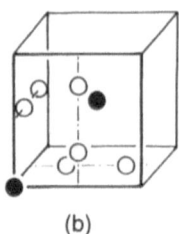

(a)   (b)

FIGURE 4.20. Aggregation of slightly distorted 12-hedra and 14-hedra in type 1 structure with $Pm3n$ space group (a), and the positions of the 12-hedra (•) and 14-hedra (0) in the cell (b), from reference [45].

nizations of cages trapping host molecules in clathrates of water molecules, the vertices being occupied by oxygen atoms and the edges by hydrogen bonds. Coming back to our problem, these organizations should be those of the film and micelles permitted by the properties of our Euclidean space, the film being supported by the faces of the polyhedra and each cage delimited by it containing a micelle. Several structures can be built via the above principle and, among them, we can distinguish two large families according to whether their dihedral and edge angles stay close to 120° and 109°28' or show departures from these values. If we limit ourselves to the first family, whose angles are the closest to those of fluid films balancing their tensions, we are left with two relatively simple structures. One—type I—structure has space group Pm3n and its crystallographic unit cell contains 2 dodecahedra and 6 tetrakaidecahedra, the local arrangement of its polyhedra being shown in Fig. 4.20. The second type II structure has space group Fd3m and its crystallographic unit cell contains 16 dodecahedra and 8 hexakaidecahedra. If larger departures from the classical angles of films are accepted, other structures, involving also pentakaidecahedra, are possible [45]. Cubic structures having these symmetries have been observed in between hexagonal and micellar phases [5]. The topology of the structural models first proposed there to analyze these data was not of the "micellar" type discussed here; but NMR data as well as recent closer analysis of scattering data [51] now support a topology of this type.[46]

4.5   Comments

The results of our approach match well with the broad lines of the classical description of the rich mesomorphic polymorphism cited in our introduction. It provides the topologies of the structures, their sequence in the phase diagrams and, finally, their symmetries, particularly those of the cubic phases. This agreement points up the dominant role played by the conflict between forces normal to the interfaces and forces parallel to them, which must find compromises in a Euclidean space. However this approach cannot account at the moment for certain behaviors and this deserves some explicit comments.

Our first comment concerns the fact that, from Fig. 4.15, it is not possible to discriminate between the occurrences of cubic or cylindrical phases in the region of intermediate degree of frustration, since the two curves la_t/v versus c_a corresponding to them are degenerate. We suspect that this arises from the fact that the two types of structures are obtained from the spherical torus, and that we consider only the properties of the spherical torus to establish the relation between la_t/v and c_a. Differences might be introduced if the different processes used to map the spherical torus onto R_3 were considered. As we have not clearly analyzed their effects at the moment, we ignore these aspects in the rest of this discussion. This result of the model is associated with the fact, emphasized in the introduction, that different systems may exhibit different behaviors in this region of the phase diagrams, often accompanied by important metastabilities. One might say that the systems "hesitate" between the bicontinuous and cylindrical solutions. Indeed, cubic phases and cylindrical phases with ribbon-like aggregates might both be seen as possible solutions for interfacial curvatures intermediate between those of flat bilayers of the lamellar phase and of the cylinders with circular section of the hexagonal phase, the distribution of curvature being more homogeneous in the first case than in the second one. The actual choice between these possibilities is most likely driven by energetic aspects which are not included in our approach. For instance, the bicontinuous and cylindrical solutions are not equivalent in the sense that they exist in spaces with different curvatures and this certainly implies that the real structures deduced from them have different energies of curvature. Thus the cylindrical phases, which come from the space with highest curvature, would involve higher free energy than the cubic phases. This may be the reason why hexagonal phases are always oberved after cubic phases in phase diagrams where both phases are present. (An alternative explanation, involving rigid-object packing efficiencies, was discussed in Sec. 1.4.3 of Chap. 1.) One should also not forget that the details of the *chemical* structures of the molecules might intervene. For instance, the degree of dissociation of ionic molecules may play an important role in this respect, as it introduces molecules in differently ionized states. Cylindrical phases of ribbon-like aggregates, with an important modulation of interfacial curvature, would be more favorable when the range of amphiphilic species is large, in order to accommodate this distribution, while cubic phases, with more homogeneous interfacial curvature, would be preferred when the distribution is limited, as there is no need to accommodate differently ionized molecules [13,14].

Our second comment concerns the values of the parameter la_t/v obtained for the cylindrical and micellar topologies. In the first case, it is smaller than 2, which is the value expected for molecules packed in cylinders with circular section. This suggests that the section of the cylinders is not circular in this range of concentration, as has indeed been observed, not only in rectangular [13], oblique or monoclinic phases [14] with ribbon-like aggregates, but also in hexagonal phases [14, 47]. In the second case, i.e., of "micellar" topology, this parameter is smaller than 3, which latter is the value expected for molecules packed in spherical micelles. This corresponds to the well-known fact that finite micelles are not always spherical but may have rod-like or disk-like shapes under certain condition. This is well illustrated in the concentrated micellar range, corresponding to that of the "micellar" topology discussed above, by the special structures of aggregates in "micellar" cubic phases [8] or by the cylindrical structures observed in concentrated isotropic (see Chaps. 1 and 2) and nematic phases [48].

Our last comment concerns the result that the lamellar, cubic, cylindrical, and micellar phases exist as only a few points in our la_t/v versus c_a diagram, whereas they have extended domains of existence in real phase diagrams. This arises from the fact that our approach gives rise to solutions in which the frustration is optimized in the most homogeneous way. Such solutions are obviously possible for discrete values of the parameters only. However, when moving away from these values, the frustration must be optimized in other ways. This might be obtained by local fluctuations within these homogeneous structures. Such structural fluctuations are currently under study in our group [49]. They appear when approaching the limits of the domains of existence of the phases in their phase diagrams, and they can be described as fragmentations of the bilayers of lamellar phases, ruptures of labyrinths in cubic phases, and modulations along cylinders in hexagonal phases.

4.6   Conclusion

We have analyzed the liquid crystalline structures formed by amphiphilic molecules as periodic stackings of symmetric films delimited by two facing interfaces. The structures are then controlled by two systems of forces: forces normal to the interfaces, which maintain the films at constant distances, and forces parallel to the interfaces, which determine the interfacial curvatures. When the interfacial curvature is zero the periodic stacking exists in the flat Euclidean space R_3 as the classical lamellar structure.

When the interfacial curvature is different from zero it is possible to reconcile curvatures and distances by building the periodic stackings in curved
spaces, where they can be described as "ideal structures". There are three
such structures possible, which are respectively built by a symmetric film
supported by the equatorial cylinder of the cylindrical space $S_2 * R_1$, the
spherical torus, and the great sphere of the spherical space S_3. Networks
of disclinations are then introduced into these curved spaces in order to
map them onto the flat Euclidean space R_3. The organizations of the films
obtained thereby present the topologies and symmetries of the cylindrical
and cubic liquid crystalline phases and micellar phases. This result allows
the real structures to be seen as "structures of disclinations" optimizing
the conflict between curvature and distance in the Euclidean space R_3.
We also show that, owing to the properties of the curved spaces, relations
exist between curvatures and distances which imply that, when the physico-
chemical conditions driving the real system make the curvatures and distances vary continuously, reconciliations in ideal structures can occur for a
discrete number of points only. Their sequence—first a film on the spherical
torus in S_2 or on the equatorial cylinder in $S_2 * R_1$, and then on the great
sphere in S_3—imposes a sequence of "structures of disclinations" in R_3—
first bicontinuous and/or cylindrical, and then micellar—in good agreement
with that generally observed in the phase diagrams. This agreement might
appear to be limited, insofar as the lamellar structure and the structures
of disclinations predicted by our approach exist for discrete values of the
parameters only, while each real structure exists for an extended range of
values in the phase diagrams. But this disagreement is not really significant. Indeed, the structures predicted by our approach are those which
correspond to the ideal structures reconciling interfacial distances and curvatures *homogeneously* over the whole structure, each in its own space. The
fact that this is only possible at a limited number of points of the phase
diagrams suggests that, around these central points, *other* processes must
intervene to solve the frustration. In the immediate vicinity of these points
this can be obtained, *without* perturbing the disclination structures, via
elastic energies; away from these points, this can be obtained by locally perturbing the structures of disclinations. This introduces the idea of "imperfect structures" where local fluctuations, corresponding to disclination processes different from those already considered, take place to help resolve the
frustration. This point of view is consistent with recent experimental studies which show that the structures of the phases, as classically described,
are valid only in limited parts of their domains in the phase diagrams and
that important structural fluctuations appear within each of the phases
when approaching their limits of existence. Thus it is natural to propose
that the lamellar structure and the structures of disclinations introduced
by our approach, and which exist in the phase diagrams only locally, most
likely correspond to the minima resulting from the compromise between
the energies of interaction between interfaces and those of curvature only.

The approach discussed in this chapter was inspired by a problem in the field of physical-chemistry, which is generally more dominated by thermodynamics than by geometry. In our view it is a necessary step due to the constraints imposed by the structure of our space, and is similar to that followed by early crystallographers when they studied the packing of point objects, such as atoms or molecules. In our case the objects are bi-dimensional films and the description of their organization in space requires a specific crystallography [9, 50]. Also, our comments on "disclination structures" and "imperfect disclination structures" introduces the notion of defects in perfect crystals, in analogy with the corresponding notion in the familiar case of ordered atoms or molecules.

Our last remark is to situate the forthcoming chapters with respect to ours. One basic ingredient of our approach is the fact that the forces parallel to the interfaces at different levels of the film vary differently when the parameters of the phase diagrams vary. If, by some physico-chemical means, they can be kept equal and constant, no stress will appear in the film, resulting in a constant interfacial curvature. The swollen lamellar systems considered in Chaps. 5 and 6 are such systems. Their water content, or eventually oil content, can be varied over a very large range without affecting the flat topology of the interfaces. They provide remarkable situations to study the interactions and undulation modes of the amphiphilic films.

4.7   Appendix

The developments presented in this article require the knowledge of some aspects of the geometry of the spherical torus T_2 and the great sphere S_2 in S_3, particularly their axes and centers and the area of the surfaces parallel to them and the volumes they enclose. They can be easily obtained considering that S_3, the 3-D space with positive Gaussian curvature, is also the hypersphere of equation

$$x_1^2 + x_2^2 + x_3^2 + x_4^2 = R^2$$

when embedded in the 4-D Euclidean space R_4. In spite of the simplicity of their equation, these coordinates are not convenient for representing in an easily tractable way the surfaces of S_3 used in this work. For this purpose we must use two other systems of coordinates having the symmetries of the surfaces. They are the toroidal and spherical systems in which the coordinates of one point are expressed as functions of the radius of curvature R and of three adequately defined angles θ, ϕ and ω. This is indeed quite comparable to moving from Cartesian coordinates to cylindrical or spherical ones in the 3-D Euclidean space R_3.

4.7.1 S_3 IN TOROIDAL COORDINATES

The Cartesian coordinates are written as

$$x_1 = Rcos\theta sin\phi, x_2 = Rsin\theta sin\phi, x_3 = Rcos\omega cos\phi, x_4 = Rsin\omega cos\phi.$$

In this system the surfaces at constant ϕ are tori organized around their C_∞ symmetry axes which are the two interlaced great circles,

$$x_1^2 + x_2^2 = R^2 \text{ and } x_3^2 + x_4^2 = R^2,$$

obtained for $\phi = 0$ and $\pi/2$. Among them the spherical torus T_2, obtained for $\phi = \pi/4$, is at equal distances from the two axes and separates S_3 into two equivalent sub-spaces. Any point on a torus is therefore determined by the variables θ and ω and a constant ϕ. Varying the two first by $d\theta$ and $d\omega$ leads to the orthogonal displacements $Rsin\phi d\theta$ and $Rcos\phi d\omega$ on the torus, and changing the third by $d\phi$ leads to a displacement $Rd\phi$ normal to the torus. The infinitesimal length dl on the spherical torus is given by

$$dl^2 = R^2(d\theta^2 + d\omega^2)/2;$$

this is the metric of a Euclidean plane and, therefore, the spherical torus has zero Gaussian curvature. Indeed it can be built by identification of the opposite sides of a square sheet of sides $\sqrt{2}\pi R$ [28,6]. Also the element of area on any torus ϕ is $ds = R^2 sin\phi cos\phi d\theta d\omega$ and the element of volume in S_3 is $dv = R^3 sin\phi cos\phi d\theta d\omega d\phi$, so that the area of the torus is $S = 4\pi^2 R^2 sin\phi cos\phi$ and its volume $V = 2\pi^2 R^3 sin^2\phi$. More particularly, the spherical torus T_2 with $\phi = \pi/4$ has an area $2\pi^2 R^2$ and a volume $\pi^2 R^3$, which is exactly half that of S_3.

4.7.2 S_3 IN SPHERICAL COORDINATES

The Cartesian coordinates are written as

$$x_1 = Rsin\theta cos\phi sin\omega, x_2 = Rsin\theta sin\phi sin\omega, x_3 = Rcos\theta sin\omega, x_4 = Rcos\omega$$

In this system the surfaces at constant ω are spheres S_2 with radii $Rsin\omega$ centered on two points $x_4^2 = R^2$ obtained for $\omega = 0$ and π. Among them the great sphere S_2, obtained for $\omega = \pi/2$, is at equal distances from the two points and separates S_3 into two equivalent sub-spaces. Any point on a sphere is therefore determined by the variables θ and ϕ and a constant ω. Varying the two first by $d\theta$ and $d\phi$ leads to the orthogonal displacements $Rsin\omega d\theta$ and $Rsin\theta sin\omega d\phi$ on the sphere, and changing the third by $d\omega$ leads to a displacement $Rd\omega$ normal to the sphere. The element of area on the sphere ω is therefore $ds = R^2 sin\theta sin^2\omega d\theta d\phi$, and the element of volume in S_3 is $dv = R^3 sin\theta sin^2\omega d\theta d\phi d\omega$, so that the area of the sphere is $S = 4\pi R^2 sin^2\omega$ and its volume $V = \pi R^3(\omega - (sin2\omega)/2)$. More particularly, the great sphere S_2 with $\omega \doteq \pi/2$ has an area $4\pi R^2$ and a volume $\pi^2 R^3$, which is exactly half that of S_3.

References

1. V. Luzzati, Biological Membranes **1**, 71 (1968).

2. P. Ekwall, Adv. Liq. Cryst. **1**, 1 (1975).

3. A. Skoulios, Ann. Phys. **3**, 421 (1978).

4. G. J. T. Tiddy, Phys. Rep. **57**, 1 (1980)

5. P. Mariani, V. Luzzati and H. Delacroix, J. Mol. Biol. **204**, 165 (1988).

6. J. F. Sadoc and J. Charvolin, J. de Physique **47**, 683 (1986).

7. J. Charvolin and J. F. Sadoc, J. de Physique **48**, 1559 (1987).

8. J. Charvolin and J. F. Sadoc, J. de Physique **49**, 521 (1988).

9. J. Charvolin and J. F. Sadoc, J. Phys. Chem. **92**, 5787 (1988) and Physica **A176**, 138 (1991).

10. C. Madelmont and R. Perron, Colloid Polymer Sci. **254**, 581 (1976).

11. R. R. Balmbra, J. S. Clunie, and J. F. Goodman, Nature **222**, 1159 (1969).

12. D. J. Mitchell, G. J. T. Tiddy, L. Waring, T. Bostock and M. P. MacDonald, J. Chem. Soc. Faraday Trans. II **79**, 975 (1983).

13. S. Alperine, Y. Hendrikx and J. Charvolin, J. de Physique Lett. **46**, L-27 (1985).

14. P. Kekicheff and B. Cabane, J. de Physique **48**, 1571 (1987).

15. X. Auvray and C. Petipas, private communication, and X. Auvray, C. Petipas, R. Anthore, I. Rico and A. Lattes, J. Phys. Chem. **93**, 7658 (1989)..

16. A. Skoulios, Adv. Liq. Cryst. **1**, 169 (1975).

17. E. L. Thomas, D. B. Alward, D. J. Kunning, D. C. Martin, D. J. Handlin and L.J. Fetters, Macromolecules **19**, 2197 (1986); H. Hasegawa, H. Tanaka, K. Yamasaki and T. Hashimoto, Macromolecules **20**, 1651 (1987).

18. J. Charvolin and A. Tardieu, Solid State Physics **supp. 14**, 209 (1978), L. Liebert, F. Seitz and D. Turnbull, eds.

19. J. N. Israelachvili, *Intermolecular and Surface Forces* (Academic Press, London, 1985).

20. W. Helfrich, Z. Naturforsch **33a**, 305 (1978).

21. Articles concerning that subject are presented in *Physics of Amphiphilic Layers*, Springer Proceedings in Physics **21** (1987), J. Meunier, D. Langevin and N. Boccara, eds.

22. J. Friedel, Proceedings of the Sixth General Conference of the European Physical Society, page 25, Prague (1984).

23. This term, used in the context of this chapter, contains the already expressed notion of "cumulative strain" acting within a monolayer of the film [1,3] plus the requirement of a periodical stacking.

24. J. F. Sadoc and R. Mosseri, Phil. Mag. **B45**, 257 (1982), and J. F. Sadoc and R. Mosseri, Pour la Science **87**, 10 (1985).

25. W. F. Harris, Sci. Am. **237**, 130 (1977).

26. In other terms, a space with positive Gaussian curvature has an angular deficiency with respect to a flat space. For instance, the integral of the Gaussian curvature over a sphere is 4π, and the filling of this deficiency to transform the sphere into a plane can be obtained by defects of rotation only.

27. J. F. Sadoc and R. Mosseri, J. de Physique **45**, 1025 (1984).

28. The parameter la_t/v was defined as the fundamental surfactant parameter in J. Israelachvili, S. Marcelja and R. Horn, Quat. Rev. Biophys. **13**, 121 (1976).

29. D. Weaire and N. Rivier, Contemp. Phys. **25**, 59 (1984).

30. The Schlafli's notation {p,q} for a tiling of polygons means that the polygons have p edges that meet q by q at vertices; the notation {p,q,r} for a packing of polyhedra means that the polyhdera have faces with p edges, each polyhedron has q faces around one vertex and there are r polyhedra around one edge.

31. It is interesting to note here that an infinite disclination around a C_∞ axis of S_3 transforms it into a space which can be modelled onto $S_2 * R_1$ and which is locally equivalent to it. Such a disclination would also transform the spherical torus into a cylinder.

32. B. Gallot and A. Skoulios, Kolloid Z. u. Z. Polymere **208**, 37 (1966); F. Reiss-Husson and V. Luzzati, J. Colloid Interface Sci. **21**, 534 (1966).

33. H. S. M. Coxeter, *Regular Complex Polytopes* (Cambridge University Press, Cambridge, 1974).

34. This can be understood also by considering that T_2 in S_3 is a surface of zero Gaussian curvature in a space of positive Gaussian curvature so that, when the curvature of the space is decreased to zero, that of the surface becomes negative.

35. D. Hilbert and S. Cohn-Vossen, *Geometry and the Imagination* (Chelsea, New York, 1983).

36. H. S. M. Coxeter, *Introduction to Geometry* (John Wiley, New York, 1961).

37. A. Schoenflies, Compte Rendu **112**, 478 (1882).

38. H. A. Schwarz, *Gesammelte Mathematische Abdhandlungen*, band 1 (Springer Verlag, Berlin, 1890).

39. J. F. Sadoc and J. Charvolin, Acta Cryst.A **45**, 10 (1989).

40. A. Tardieu, Thesis Universite Paris-Sud (1972); W.Longley and J. Mac Intosh, Nature **303**, 612 (1983).

41. N. Rivier, Phil. Mag. **B52**, 795 (1985).

42. Soap bubble froths are macroscopic examples of the impossibility of filling R_3 with {p,3,3}; the solution is random in this case–see W. D'Arcy Thompson, "On Growth and Form", abridged edition by J. T. Bonner, page 88 (Cambridge University Press, 1975), and [41]. Indeed the only regular tesselation of R_3 is the polytope of cubes {4,3,4}.

43. H. S. M. Coxeter, Illinois J. Math. **2**, 746 (1958).

44. J. F. Sadoc, in *Physics of Disordered Materials*, D. Adler, H. Fritzsche and S. R. Ovshinsky, eds. (Plenum Press, 1985), and [24].

45. R. Williams, *The Geometrical Foundation of Natural Structures* (Dover, 1979).

46. T. Bull and B. Lindman, Mol. Cryst. Liq. Cryst. **28**, 155 (1974); P. O. Eriksson, G. Lindblom and G. Arvidsson, J. Phys. Chem. **89**, 1050 (1985) and **91**, 846 (1987).

47. Y. Hendrikx and J. Charvolin, Liq. Cryst. **3**, 265 (1988).

48. Y. Hendrikx, J. Charvolin, M. Rawiso, L. Liebert and M. C. Holmes, J. Phys. Chem. **87**, 3991 (1983), and N. Boden, Chap. 3 of this book.

49. M. C. Holmes and J. Charvolin, J. Phys. Chem. **88**, 810 (1984); P. Kekicheff, B. Cabane and M. Rawiso, J. de Physique Lett. **45**, L-813 (1984); Y. Hendrikx, J. Charvolin, P. Kekicheff and M. Roth, Liq. Cryst. **2**, 677 (1987); and Y. Rancon and J. Charvolin, J. Phys. Chem. **92**, 6339 (1988).

50. A. L. Mackay and J. Klinowski, Comp. and Maths. with Appls. **12B**, 803 (1986).

51. J. M. Seddon, W. A. Bartle and J. Mingins, J. Phys.: Condens. Matter **2**, SA285 (1990); R. Vargas, P. Mariani, A. Gulik and V. Luzzati, J. Mol. Biol., in press; V. Luzzati, R. Vargas, A. Gulik, P. Mariani, J. Seddon and E. Rivas, Biochemistry **31**, 279 (1992).

5

Lamellar Phases: Effect of Fluctuations (Theory)

Didier Sornette[1]
Nicole Ostrowsky[2]

5.1 Introduction

An essential characteristic and common feature of certain amphiphilic systems is the existence of the special *two-dimensional membrane-like structure* of their local order. As seen in previous chapters, these amphiphilic systems can be considered with good approximation as true bidimensional surfaces, at scales larger than their thickness ($\approx$ 10–20Å). Consequently, many of their properties can be related to the geometrical, mechanical and statistical characteristics of simple fluctuating surfaces, independently of the details of their internal structure. Examples are provided by vesicles and lamellar phases, in which the two-dimensional membrane-like structure characterizes the whole system, and by fluid middle-phase isotropic microemulsions [1], whose properties can be understood in terms of the local sheet-like order. Among the extraordinary wealth of phases encountered in lyotropic (ternary, quaternary···) systems, the lamellar phases occupy a surprisingly large place in the phase diagrams [1]. Their importance also relies on their observed coexistence with less ordered phases (microemulsions) [2,3] or more ordered structures (e.g., hexagonal or cubic phases) [4] which place them in a unique intermediary position (and structure) (see Sec. 1.4.2). In the following, we will call under the generic name "membrane" all systems presenting a two-dimensional membrane-like structure in their local order.

The existence of this special *membrane-like* local (and sometimes global) order extends the nature of the local order found in complex fluids. This generalization and its consequences comprise the main motivation for

[1]Laboratoire de Physique de la Matière Condensée, CNRS URA 190, Université de Nice-Sophia Antipolis, Parc Valrose, B.P.71, 06108 NICE Cedex 2, France
[2]Same as No. 1

studying the statistical mechanics of two-dimensional surfaces. Amphiphilic membranes also constitute interesting realizations of random surfaces [5,6] of interest both as a generalization of random walks (one-dimensional lines) [7] and in recent developments of the string theory of elementary particles [8,9]. It is indeed striking that a condensed matter theory of membranes may serve as a toy model for string theories. In particular, interesting connections between the renormalization properties of the 2D-membranes and of the string models of elementary particles have been realized. In this last context, the cosmological constant plays the role of a surface tension and must be taken into account [8].

In this chapter, we focus our attention on the theoretical description of membranes and particularly on their interactions. Such a study might aim to describe the intricate *micro*scopic degrees of freedom of the surfactants in the membranes (see Sec. 1.3.3), but then would not give information on the *macro*scopic stability or phase diagram. Since our goal is to discuss the global stability and phase diagrams of these structures, one needs a treatment which incorporates all degrees of freedom and the contributions of both the energy and entropy to the full free energy of the system. This is the point of view of the statistical mechanics developed here.

5.2   Model of Isolated Membranes

Before studying lamellar systems comprised of several membranes in interaction, it is useful to focus our attention on the physical description of a single membrane. This is a prerequisite for understanding the more complicated case of multilayers since they can be considered as isolated membranes which are put in interaction. Also, since we are interested in the effect of thermal fluctuations, it is useful to visualize the influence of fluctuations at the level of a single membrane.

5.2.1   THE DEFORMATION ENERGY OF
TWO-DIMENSIONAL MEMBRANES

Even at the level of a single membrane, it is necessary to introduce important simplifying assumptions. We will consider essentially idealized membranes and will neglect complications brought about by the internal degrees of freedom (e.g., the influence of heterogeneities in their composition, and the role of small intercalated molecules such as cosurfactants), except on the global continuous properties of the membrane such as the surface tension or the bending rigidity (see Sec. 1.3.4). This is a natural model to start with before facing the full complexity of real structures. Furthermore, it will appear that many features are already captured in the corresponding simplistic models. This statistical mechanical idealization of real sheet-like structures illustrates the general physical approach which is characterized by a quest for "universality," transcending the specificity and complexity of "real life."

In this respect, the most fruitful approach has consisted in describing the membranes as *two-dimensional continuous* objects without internal structure and constructing a statistical mechanics of such idealized "surfaces." This approach is valid for properties which do not involve internal degrees of freedom. However, as we will see, it is often possible to keep such a "mesoscopic" continuous level of description and incorporate also certain features of the internal degrees of freedom such as the liquid or solid order of the lipids forming the membranes.

The elasticity of membranes may be viewed as a special case of the theory of thin elastic shells. A membrane is deformable in several ways and may present the following deformation modes which can be classified according to the nature of the corresponding conjugates stresses:

i. Stretching corresponds to a change in area per lipid molecule due to the application of a tension or tangential stress. At time scales less than the time τ_e necessary for the lipid molecules to exchange with a reservoir (if there is any as, for example, the surrounding fluid), the number of molecules within the membrane is fixed, and stretching is equivalent to an elastic deformation whose energy dependence on changes in the local lipid lateral density 'n' in the membrane is quadratic:

$$F_c = \int dx\, dy \frac{1}{2} k_s \left[(n - n_o)/n_o \right]^2 \qquad (5.1)$$

The integral is carried over the whole surface $S = L^2$ of the membrane. n_o denotes the equilibrium liquid density. Writing $n - n_o = (n - \langle n \rangle) + (\langle n \rangle - n_o)$, where $\langle n \rangle = \int dx\, dy\, n / \int dx\, dy$, expression (5.1) transforms into

$$F_c = \frac{1}{2} L^2 k_s \left[(\langle n \rangle - n_o)/n_o \right]^2 + \frac{1}{2} k_s \int dx\, dy \left[(n - \langle n \rangle)/n_o \right]^2 \quad (5.2a)$$

The first term in the right-hand-side (r.h.s.) of (5.2a) describes the *global* stretching (or compression) of the membrane away from its equilibrium. As we will see below, it couples with the undulations and introduces a non-linear coupling between undulation modes. The second term in expression (5.2a) describes the density fluctuations of non zero wave vector inside the membrane and thus completely decouples from other deformation modes such as the undulations which we are interested in. The modulus k_s may be obtained in the case of bilayers by measuring the swelling of spherical vesicles as a function of the internal excess pressure [10,11,12]. In most cases, the stretching modulus is large, and the corresponding deformations are small.

At time scales larger than τ_e, the lipid molecules have had time to exchange with the aqueous environment. The elastic stretching forces have therefore decreased to zero and are replaced by a surface tension γ. This gives the following expression for the energy of the membrane:

$$F_d = \gamma \int dx\, dy \qquad (5.2b)$$

In (5.2b), the integral defines the whole surface of the membrane. It takes into account the fact that, at long times, an increase of surface is realized only by bringing more molecules from the surrounding fluid to the membrane. The free energy of a molecule in the bulk and within the membrane being different, the difference is accounted for in the surface tension.

ii. A shear deformation can be due to the existence of a shear stress; in the case of fluid membranes on which we will essentially focus our attention, the molecules can freely and rapidly adjust to shape changes of the membrane. Therefore, fluidity amounts to a vanishing shear modulus. However, lipid bilayers often exhibit two or more phases as a function of temperature and (or) composition, and one of them (the L_β phase) is often thought to be associated with in-plane crystalline order. In this case, the shear modulus does not vanish anymore and must be accounted for.

iii. A tilt of the lipid molecules inside the membrane may be due to the existence of a torque density coming, for instance, from an oblique magnetic field acting on magnetically anisotropic lipid molecules [10]. It can also appear in certain phases of bilayers. A precise expression of the tilt deformation energy would require taking into account the various parts of the lipid molecule and not only their *average* tilt. However, tilt should be minute and its elastic energy negligible in most practical cases [10].

iv. A peristaltic deformation of the membrane corresponds to changes of its thickness due to presence of, for example, a compressive pressure on both sides of the membrane. Thermal excitations will in general induce spontaneous fluctuations of a membrane thickness due to the compressibility of the membrane (i.e., of the aliphatic chains). However, thickness fluctuations with short characteristic lengths are small as a result of the large amount of aliphatic/water contact these would entail. Quantitative analysis based on an extension of the treatment for soap films predicts that the root mean square amplitude for fluctuations of wavelength longer than ≈ 10nm is negligible compared to the average membrane thickness for most membrane structures [13].

v. A curvature deformation also involves stresses having the form of torques per unit length. The importance of this term for lipid bilayers, red blood cells and lamellar phases has been emphasized by Helfrich and others. Let us consider a membrane which is distorted from a planar initial position. For any point P on the surface, a local system of cartesian coordinates (Px_1, Px_2, Px_3) exists with origin P such that the equation of the surface in the neighborhood of P reads

$$x_3(x_1, x_2) = \frac{1}{2}\left[\frac{x_1^2}{R_1} + \frac{x_2^2}{R_2}\right] + \dots \tag{5.3}$$

The directions x_1 and x_2 are the principal directions of the surface at P, and R_1 and R_2 are the two principal radii of curvature at P. The average curvature H is defined by [14]

$$H = \frac{1}{R_1} + \frac{1}{R_2} \tag{5.4}$$

and the "intrinsic" Gaussian curvature K reads

$$K = \frac{1}{R_1 R_2} . \tag{5.5}$$

K is called intrinsic because its surface integral is independent of the shape of the membrane as long as its topology is kept fixed (fixed number of vesicles and handles). For a membrane limited by a line boundary, the surface integral of K can be transformed into a line integral performed on the boundary.

In terms of the Monge representation $u(x, y)$ giving the vertical displacement of the surface from a plane reference state (see Fig. 5.1), the "extrinsic" mean curvature H reads

$$H = \text{div}\left\{\frac{\vec{\nabla}u}{\left\{1 + (\nabla u)^2\right\}^{1/2}}\right\} \tag{5.6}$$

which yields

$$H \approx \nabla^2 u \tag{5.7}$$

for small deviations from planarity ($|\vec{\nabla}u| \ll 1$). Here, $\vec{\nabla}u$ denotes the gradient of $u(x, y)$. H is called extrinsic since, in contrast to K, the surface integral of H does depend on the specific membrane shape.

Another equivalent parametrization of the surface deformation is obtained by specifying the orientation of a unit vector $n(x, y)$ which is normal to the membrane surface at every point. The unit vector n is called the *director* in liquid crystal physics. The leading expression of the curvature deformation energy is a quadratic function of the derivative of the components n_x and n_y. Being a two-dimensional fluid, the membrane is rotationally symmetric, and its description should be invariant with respect

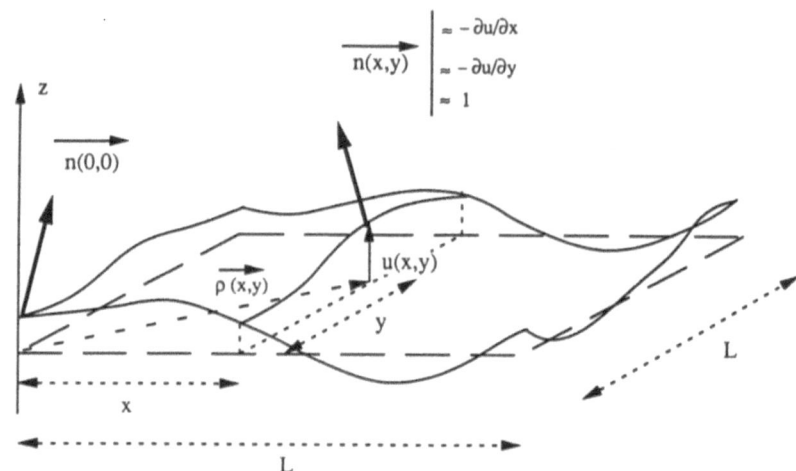

FIGURE 5.1. Schematic representation of an undulating membrane of size L, defining the displacement $u(x, y)$ from the planar average position, and the local normal unit vector $n(x, y)$.

to global rotations of the coordinate system (x, y), so that only those linear and quadratic forms can enter which are independent of the orientation of the $0x$ and $0y$ axes. In this case, a membrane deformation is described in terms of a bending elastic modulus k and a Gaussian modulus $\bar{k}$. Neglecting tilt and peristaltic deformation modes and taking into account membrane tension in the long time limit, the deformation from planarity of the 2D-membrane model is thus controlled by a free energy 'f' per unit surface which can be expressed in terms of the mean H and Gaussian K surface curvature [15]:

$$f = \gamma + \frac{1}{2}k(H - c_o)^2 + \bar{k}K \tag{5.8}$$

c_o is the spontaneous curvature of the membrane, if it exists (for a closed vesicle of average radius R, $c_o = 2/R$). The surface tension γ is often very small in membranes and can even vanish [16], and the membrane shape is uniquely controlled by k and $\bar{k}$. Eq. (5.8) is the basis of the continuum description of idealized two-dimensional membranes. The next term in the r.h.s. of (5.8) is a quadratic function of the extrinsic mean curvature: its surface integral is a function of the local curvature over all the surface of the membrane. In contrast, the last term in (5.8) is a function of the Gaussian curvature. For a given surface without boundaries, its surface integral depends only on the topology of the membrane and not on details of its shape. Indeed, for a closed surface, the Gauss-Bonnet theorem gives the relationship between the surface integral over K and a topological invariant called the Euler characteristic g:

$$\int_S dx\, dy\, K = 4\pi(1-g) \tag{5.9}$$

where g is the number of "handles" of the closed surface ($g = 0$ for a sphere, $g = 1$ for a torus). For an open surface, the surface integral of K is both a function of g and of an integral along its boundary. In sum, for a fixed topology and/or fixed boundary conditions, the second term of (5.8) yields a constant independent of the specific shape of the membrane. At fixed topology, the membrane shape is thus completely determined by the extrinsic mean curvature H.

Expression (5.8) (without the final, intrinsic, term) has thus been used to predict the equilibrium shape of closed vesicles [11]. Many measurements of the bending elastic modulus k have been made [17,18,19,20] (see also Sec. 6.5). The typical biconcave shape of red-blood cells can be obtained by minimizing the extrinsic curvature (bending) energy (5.8) with appropriate constraints (constant enclosed volume and imposed spontaneous curvature c_o). Due to the smallness of the bending energy associated with long wavelength deformations (see below), large transverse thermal fluctuations (undulations) take place. The spectrum of such fluctuations (known in red-blood cells as the "flicker phenomenon" [21]) can be obtained from a simple hydrodynamic theory which includes the energy (5.8) as the dominant contribution. This feature also explains the stability of swollen lamellar lyotropic phases [22] (see Sec. 2.4 and 6.3.1), as we will see, and is involved in the stability of microemulsions [1] (see Sec. 9.4).

Once the deformation energy is known, the statistical properties of the equilibrium structure of the membrane system are determined from the statistical sum ($\beta = 1/k_B T$)

$$Z = \int d\{u(x,y)\} e^{-\beta \int dS\, f} \tag{5.10}$$

performed over all allowed configurations $u(x,y)$ of the membranes weighted by the Boltzmann factor $\exp\{-(k_B T)^{-1} \int f\, dS\}$ where k_B is the Boltzmann constant, T is the temperature and the integral is performed over the area of the membrane. Then, the total free energy of the system is given by

$$F = -k_B T \log Z \,. \tag{5.11}$$

For most cases of interest, explicit calculation of (5.10) is not possible, and one has to resort to "mean field" or harmonic approximations, or to more sophisticated renormalization group treatments better tailored for taking into account the role of thermal fluctuations.

5.2.2   THE HARMONIC APPROXIMATION FOR THE COMPUTATION OF FLUCTUATION AMPLITUDES

The important new fact to recognize is that, due to the low dimensionality of the systems (2D-surfaces in 3D-bulk), the spontaneous thermal fluctuations are very important and therefore change drastically the physics of the membrane compared to the picture established in the absence of fluctuations. Accordingly, the role of thermal fluctuations must be fully addressed to be able to understand the physics of systems made of interfaces, as we discuss below. The undulation mode in lamellar phases was first observed by quasielastic light scattering [23,24] (see also Sec. 6.5).

To develop a feeling for the role of thermal membrane fluctuations, let us consider the leading harmonic approximation of the integral of (5.8), giving the total energy of the membrane:

$$\int f \, dS = F_0 + \int dx \, dy \, \left\{ \frac{1}{2} \gamma (\nabla u)^2 + \frac{1}{2} k (\nabla^2 u)^2 \right\} . \tag{5.12}$$

u is the displacement along z (normal to the average membrane plane) of the point at ρ in the membrane where $\rho(x, y)$ denotes the two-dimensional in-plane vector of coordinates (x, y). In order to derive (5.12), we have used

$$\begin{aligned} dS &= \left(1 + (\nabla u)^2 \right)^{1/2} dx \, dy \\ &\approx \left(1 + \frac{1}{2} (\nabla u)^2 \right) dx \, dy \end{aligned} \tag{5.13}$$

and included the constant term in F_0. Note also that we treat the case of a membrane with zero spontaneous curvature ($c_o = 0$). The Gaussian curvature term in (5.8) is also incorporated in the constant F_0 since we address only the problem of membrane fluctuations at fixed topology and boundary conditions. The terms $\nabla^2 u$ denotes the Laplacian of $u(x, y)$, equal to div(∇u).

Let us introduce the Fourier mode expansion of the deformation

$$u(x, y) = \sum_{q_\parallel} u_{q_\parallel} e^{i(q_x x + q_y y)} \tag{5.14}$$

where $q_\parallel = (q_x, q_y)$ is the wave vector parallel to the membrane average plane. We can use the equipartition theorem valid for harmonic free energies and express the average energy of one mode $q_\parallel$ at a given temperature T for a membrane of surface L^2 as

$$\frac{1}{2} L^2 \left(\gamma q_\parallel^2 + k q_\parallel^4 \right) \left\langle u_{q_\parallel}^2 \right\rangle = \frac{1}{2} k_B T \tag{5.15}$$

where $\langle \rangle$ denotes a statistical average over membrane configurations. From (5.15), it is easy to compute the m.s. (mean square) displacement due to thermal undulations, as $\langle u^2 \rangle = \sum_{q_\parallel} \langle u_{q_\parallel}^2 \rangle$, and with the classic correspondence relating a discrete sum to the corresponding continuous integral $\sum_{q_\parallel} \to (L/2\pi)^{D-1} \int d^{D-1} q_\parallel$ in a space of D dimensions, one obtains, for $D = 3$,

$$\langle u^2 \rangle^{1/2} = \sqrt{\frac{k_B T}{4\pi\gamma} \log \frac{L^2 + \pi^2 L_\gamma^2}{l^2 + \pi^2 L_\gamma^2}} \tag{5.16}$$

where l is a microscopic length related to the surfactant molecule size ($l \approx 20$Å) and $L_\gamma = (k/\gamma)^{1/2}$ is a characteristic cross-over length such that modes with wavelength $2\pi/q_\parallel < L_\gamma$ feel essentially the bending energy whereas modes with wavelength $2\pi/q_\parallel > L_\gamma$ are hindered rather by the surface tension. For a membrane with vanishing surface tension $\gamma = 0$, or if the size L of the membrane is smaller than L_γ, one can neglect γ in (5.16), and the r.m.s. displacement reduces to

$$\langle u^2 \rangle^{1/2} = \sqrt{\frac{k_B T}{4\pi^3 k}} L \tag{5.17}$$

In the regime where γ dominates, (5.16) leads to

$$\langle u^2 \rangle^{1/2} = \sqrt{\frac{k_B T}{4\pi\gamma} \log \frac{L}{l}} \tag{5.18}$$

Eq. (5.17), as contrasted with (5.18), points up the crucial feature of a membrane without surface tension, which exhibits thermal undulations having a magnitude of the order of the membrane size. On the other hand, (5.18) is the usual result of the r.m.s. displacement for, say, a liquid-gas interface showing only a logarithmic and thus very slow divergence of $\langle u^2 \rangle$ as L increases.

5.2.3 THE PERSISTENCE LENGTH

We can characterize more precisely the thermal undulations following de Gennes and Taupin [25] by computing the thermal average of the local orientation of the surface defined by a unit vector n normal to it (see Fig. 5.1) which reads in leading order

$$n_x = \frac{-\partial u}{\partial x}$$

$$n_y = \frac{-\partial u}{\partial y}$$

$$n_z \approx 1 \tag{5.19}$$

In the harmonic approximation valid for small fluctuations $\delta n = (n_x, n_y)$, we can write $|\delta n_{q_\parallel}|^2 = q_\parallel^2 u_{q_\parallel}^2$ which yields with (5.15) in the case $\gamma = 0$:

$$|\delta n_{q_\parallel}|^2 = k_B T / k \, q_\parallel^2 L^2 \tag{5.20}$$

The angular correlation between two points on the surface, whose projected distance on the xy plane is $\rho(x, y)$, is given by

$$
\begin{aligned}
\theta^2(\rho) &= \left\langle |\delta n(\mathbf{O}) - \delta n(\rho)|^2 \right\rangle = \sum_{q_\parallel} 2\left\{1 - \cos(q_\parallel \cdot \rho)\right\} \left\langle |\delta n_{q_\parallel}|^2 \right\rangle \\
&= \frac{k_B T}{\pi k} \int_0^{1/l} \left\{1 - J_0(q_\parallel \rho)\right\} \frac{dq_\parallel}{q_\parallel}
\end{aligned}
\tag{5.21}
$$

Note that $1/l$ acts as a high $q_\parallel$ cut-off and $J_0(q_\parallel\rho)$ is a Bessel function which is of the order of unity for $q_\parallel\rho \ll 1$ and which vanishes for $q_\parallel\rho \gg 1$. Eq. (5.21) yields

$$
\theta^2(\rho) \sim \frac{k_B T}{\pi k} \log \frac{\rho}{l}.
\tag{5.22}
$$

For small θ, it is useful to present this result in the form

$$
\langle \cos\theta \rangle \approx \left\langle 1 - \frac{\theta^2}{2} \right\rangle \approx \exp\left\{\frac{-\langle\theta^2\rangle}{2}\right\} \sim \left(\frac{l}{\rho}\right)^n
\tag{5.23}
$$

where the exponent $n \approx k_B T/2\pi k$ varies continuously with T, a frequent feature of two-dimensional fluctuations [26]. The power law (5.23) shows that there is no true long-range orientational order since $\langle\cos\theta\rangle$ eventually vanishes at large distances; it is characteristic of a *quasi* long-range orientational order [27]. From this formula, De Gennes and Taupin define a *persistence length* ξ_K, such that at distance ρ smaller than ξ_K the angle θ is small on the average, while at distance $\rho > \xi_K$ it is large. Choosing, for instance, $\langle\cos\theta\rangle = 1/e$ as the cross over value, leads to

$$
\xi_K \approx l \exp\left\{\frac{2\pi k}{k_B T}\right\}
\tag{5.24}
$$

Note that this heuristic derivation of the orientation correlation decay (5.23) and of the persistence length is confirmed by a renormalization group treatment [28] (see [29] for a detailed account of the calculation of [28]). The persistence length ξ_K is extremely sensitive to the value of the rigidity constant k. For example, for bilayers in water, $k \approx 10^{-12}$ erg ($\approx 25 k_B T$), yielding $2\pi k/k_B T \approx 10^2$ and $\xi_K \approx 10^{30}$ l! In some lamellar phases found in microemulsions, $k \approx k_B T/2$ which yields $\xi_K \approx 10\ l$. Note that these estimates depend crucially on the numerical factors in the exponential of (5.24) which have not been computed precisely. In sum, the membrane is stiff and planar at scales shorter than ξ_K, whereas it is soft and crumpled at larger scales. ξ_K therefore sets the scale beyond which the membrane structure no longer exists (see Fig. 5.2).

5.2.4   Renormalization of Bending Rigidity k by Fluctuations

From the previous computation, it is important to distinguish two regimes:

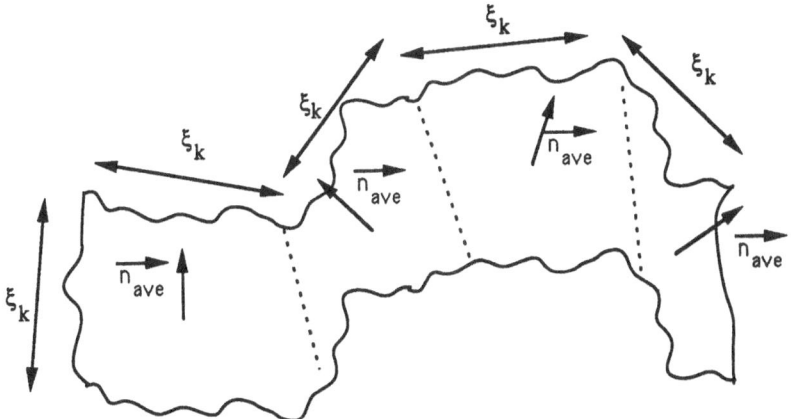

FIGURE 5.2. Schematic representation of an undulating membrane at large scale ($\gg \xi_K$). The persistence length ξ_K is defined as the typical size of domains having, on the average, a given orientation of their normal.

i. At scales larger than ξ_K, normals to the membrane are completely uncorrelated. This means that the membrane is crumpled and can be seen as equivalent to a random surface constituted of plaquettes of typical size ξ_K which are free to reorient with respect to one another. As a consequence, the energy of a macroscopic bending deformation should be very small and essentially dominated by entropic effects similar to those involved in the elasticity of polymers in good solvents beyond the polymer persistence length [7]. In this regime, valid at scales ρ such that $\xi_K \ll \rho$, the effective deformation free energy is no longer given by (5.12). By coarse-graining (or averaging) at scales larger than ξ_K, one obtains a deformation energy of the form (5.8) but with an effective bending rigidity $k^{\text{eff}} \approx k_B T$ essentially governed by random surface conformation effects or, in other words, by entropic effects.

ii. At scales smaller than ξ_K, normals to the membrane are still correlated. This means that it is possible to consider an average plane for the surface defined as the plane perpendicular to the average normal direction. The fluctuations of the membrane away from this reference plane can be described as in Fig. 5.1 within the Monge representation $u(x, y)$ giving the vertical displacement of the surface from the planar reference state. In this representation, the general expression (5.8) for the deformation energy simplifies into (5.12), in the harmonic approximation. This formulation holds for scales ρ such that $l \ll \rho < \xi_K$.

It should be stressed that, even in this relatively "quiet" regime, non-trivial effects occur due to membrane fluctuations. For instance, in Sec. 5.3 we analyze the effect of thermal fluctuations on interactions between membranes and show that long-range entropic repulsive forces appear as a result of steric hindrance of undulations. Furthermore, even for a single membrane, undulation can have a surprising effect: it has recently been discovered [30,28] that non-linear coupling between undulation modes, stemming from surface geometry, leads to a decrease of the effective rigidity of the membrane at large scales ρ (but still smaller than ξ_K) according to

$$k(\rho) = k^0 \left\{ 1 - \frac{\alpha k_B T}{4\pi k^0} \log \left(\frac{\rho}{l} \right) \right\} \text{ for } \rho \ll \xi_K \tag{5.25}$$

The numerical factor is given by $\alpha = 1$ [30,31,32] or $\alpha = 3$ [28] depending upon the treatments. Recent precise computations seem to confirm the value $\alpha = 3$ [33,34]. Note that the logarithmic dependence of $k(\rho)$ as a function of ρ reflects the fact that two-dimensional membranes in a three dimensional space are at their lower critical dimension. A simple heuristic derivation of (5.25) is presented in the first apendix (Sec. 5.5.1).

As a consequence of this additional "weakening" of the bending curvature rigidity at large scales (but still smaller than ξ_K), given by (5.25), the correlation of the normals to the membrane appears to decrease exponentially rapidly [28,29], instead of only algebraically [25]. The characteristic length of the exponential decay of the normal orientation correlation is given by

$$\xi_K = l \exp \left\{ \frac{4\pi k^0}{\alpha k_B T} \right\} \tag{5.26}$$

which can naturally be identified with the persistence length (5.24).

These results hold in absence of long-range forces between distant parts of a membrane. Since two-dimensional membranes in a three dimensional space are at their lower critical dimension, long-range forces may trigger the existence of a crumpling transition [28]. This crumpling transition would separate a "low temperature" rigid phase ($\xi_K \to +\infty$) from a "high temperature" crumpled phase corresponding to the situation found in absence of the long-range force. Indeed, it has been found that a crystalline or a hexatic order of the internal degrees of freedom (lipid molecules) of the membrane creates this kind of long-range force between distant parts of the membrane via a nonlinear coupling between in-plane elastic strain and curvature, as discussed in the elastic continuum theory of thin shells [35,36]. Finite size features of this crumpling transition have been observed in Monte Carlo simulations of crystalline tethered surfaces [37]. Furthermore, it has recently been proposed [38,39] that long-range forces can also emerge from the steric interaction between membranes (which we will describe below in Sec. 5.3) as occurs in swollen lyotropic lamellar phases, a fact which would result in the possible observation of the hardening of membranes by continuous swelling of lamellar phases.

5.3   Membranes in Interaction

5.3.1   STATEMENT OF THE PROBLEM AND ANALOGY WITH WETTING AND INCOMMENSURATE-COMMENSURATE TRANSITIONS

Up to now, we have only considered isolated membranes. However, it is often the interactions between pieces of membranes which determine the equilibrium properties of amphiphilic systems. In the sequel, we shall consider fluid membranes which are parallel on the average, such as the walls of two vesicles brought in contact or the lamellae in lyotropic liquid systems (see Chap. 6). This last system can be considered as a pile of parallel membranes separated by water layers (see Fig. 5.3). We will denote by $\langle z \rangle$ the average distance between these membranes.

It is interesting to note that the problem of the wetting of a substrate by a fluid in coexistence with a gas can also be cast into a similar framework analogous to our problem of two membranes in interaction. In the wetting case, one "membrane" is the substrate-liquid interface (i.e., rigid wall), and the other is the liquid-gas interface. The liquid-gas phase equilibrium in presence of the substrate is then determined from the interaction between these two interfaces [40,41]. As we will see in Sec. 5.5, the analogy between wetting transitions and membrane interactions goes further, in particular with respect to the effect of thermal interface (or membrane) undulations on phase equilibrium. The wetting transition, which corresponds to the growth of an infinite film of fluid above the substrate, can be viewed as the "unbinding" of the liquid-gas interfaces, in a way quite similar to the unbinding of membranes described below.

An analogy with two-dimensional incommensurate-commensurate transitions can also be drawn [42,43]. This subject refers to the description

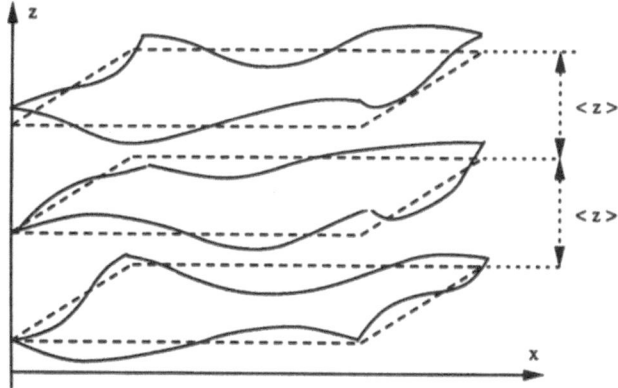

FIGURE 5.3. Stack of membranes, separated by an average distance $\langle z \rangle$.

and phase equilibrium of monolayers of adsorbed atoms or molecules on a substrate, which is not smooth but which exhibits a periodic potential structure at the atomic scale. A rich phase diagram (with commensurate and incommensurate phases) is obtained as a function of

i. the concentration of adsorbed atoms

ii. the "misfit" between the atomic mesh of the substrate surface and the spontaneous equilibrium mesh of the adsorbed atoms on a smooth substrate.

The commensurate phase corresponds to a monolayer made of atoms all sitting at the bottom of the potential-wells created by the substrate. In the incommensurate phase, the adsorbed atoms form a two-dimensional monolayer whose structure is not directly related to the periodic structure of the substrate surface. According to a simple picture of the incommensurate phase, excess atoms or holes are unevenly distributed and concentrated near finite formations called "solitons" or "domain walls," which are actually domain lines. The physics of an incommensurate crystal is thus coincident with the physics of a domain wall system, which is the two dimensional analog of the lyotropic multi-layer smectic systems. The domain wall picture can also be developed to describe three-dimensional incommensurate crystals [44,45]. It turns out that, at finite temperature, entropic steric interactions between "domain walls," of the type that we will describe in the following sections for membranes, control the incommensurate to commensurate phase transition. This transition can be viewed as an "unbinding" of "domain walls" quite similar to the unbinding of membranes described below.

5.3.2 Review of Microscopic Interactions between Membranes (in Absence of Fluctuations)

The forces between membranes have been classified into several contributions [46]. Let us first review briefly these different "direct" (i.e., in absence of thermal undulations) interactions before analyzing the influence of thermal undulations.

a. The van der Waals potential is ubiquitous and comes from induced dipole-dipole electromagnetic interactions. It is attractive for most membrane systems (note that it can become repulsive in some special cases [47,48]. It goes as z^{-2} for z less than the lamellar thickness δ and gradually goes to a z^{-4} dependence for larger spacings in the unretarded case [49]:

$$V_W = -\frac{W}{12\pi} \cdot \left\{ \frac{1}{z^2} - \frac{2}{(z+\delta)^2} + \frac{1}{(z+2\delta)^2} \right\} \qquad (5.27)$$

W is the Hamaker constant $(W \approx 10^{-22} - 10^{-21} J)$ and $\delta \approx 40\text{Å}$ denotes the lamellar thickness. It has thus a rather long range and is often dominant at distances larger than δ.

In the case of the wetting problem, the atom-atom attractive van der Waals force gives an effective repulsive potential on the liquid-gas interface (this means that van der Waals favors the growth of the liquid layer) which decays as z^{-2} for distances less than the London wavelength. When present, it dominates all other contributions and controls the growth of the wetting layer, as the control parameter (distance in chemical potential p from the coexistence curve) is varied [40,41]. (We denote this chemical potential p since it plays the same role as the pressure.)

b. In the membrane problem, the electrostatic forces between charged polar heads can be very short-ranged when screened by added salt [50]. For charged membranes, the electrostatic interaction potential reads

$$V_E \approx A_E \exp\left\{\frac{-z}{\lambda_D}\right\} \tag{5.28}$$

which is valid for z greater than the Debye length $\lambda_D \sim 1/\sqrt{\chi}$, where χ is the ionic concentration in the aqueous solution. A_E can be estimated from the Debye-Huckel approximation or from more elaborate theories [50].

c. At very short distances and in the water part of the system, the "hydration" or "structural" repulsions are extremely short-range but dominate in "the last thirty angströms" [51]. They stem from the cost in free energy associated with the distortion of the water concentration and orientation near the membranes [52]. Generically, they take the form

$$V_H \approx A_H \exp\left\{\frac{-z}{\lambda_H}\right\} \tag{5.29}$$

For membranes, $A_H \approx 0.2 J/m^2$ and $\lambda_H \approx 3\text{Å}$ (see, for example, [53] and references therein).

The same type of forces appears in incommensurate-commensurate and wetting transitions. They also stem from the exponentially decaying distortion of an order parameter in the vicinity of a wall. A_H is a typical coupling energy, and λ_H is of the order of the "soliton" or interface thickness.

d. In the oil part of a membrane system, additional short-range attractive or repulsive forces result from local overlaps between aliphatic chains of surfactant molecules belonging to adjacent layers. The sign of the force depends on whether the oil is a good or a bad solvent for the aliphatic chains. Repulsion (attraction) results in the former (latter) case [2].

To summarize, two rigid membranes feel a long range attractive van der Waals force which tends to bring them together while, at short distances, repulsion occurs due to hydration forces. However, if electrostatic forces are not screened, they will dominate at large distance and the membranes will feel a repulsive interaction at all distances.

Now, fluid membranes exhibit thermally excited undulations. As we shall discuss below, this additional feature introduces a renormalization of the direct microscopic interactions in a non-trivial way. However, in presence of purely *repulsive* direct interactions, membrane undulations amount to replacing the direct microscopic repulsive interaction by a "universal" repulsive long range interaction, called the "entropic undulation" or "steric" interaction [54]. However, at very short distances, one looses the universal regime, and the membranes become sensitive to the direct repulsive microscopic interactions (see appendix 5.5.3).

In presence of an *attractive* van der Waals force, both the repulsive and attractive components of the total direct potential are renormalized by the membrane undulations. As a consequence, the notion of an effective "entropic undulation" interaction does not make sense any more, and the full interplay between microscopic forces and thermal fluctuations must be addressed, bringing into play two different equilibrium regimes: a bound regime where membranes are separated by a finite distance, and an unbound regime where the membranes tend to separate infinitely far away from each other.

In the following sections, we address the case of purely *repulsive* direct interactions and present several different treatments of the long range entropic undulation interactions in the hope that different viewpoints can help clarify and simplify the understanding of this interaction. In Sec. 5.3.7 we turn to the other, more complicated, case where an attractive microscopic interaction is also present.

5.3.3   THE STERIC INTERACTION AS THE IDEAL GAS LIMIT FOR LAMELLAR PHASES

Understanding the complex interplay between all these microscopic interactions and the thermal fluctuations is difficult and has only recently been partially unraveled. For the sake of simplification, let us take the *ideal gas* limit of the lamellar phase problem. Consider therefore a stack of membranes interacting only locally by hard core repulsion and bound together by an external compressive pressure p. The hard core potential is chosen to schematize the short-range repulsive hydration forces. The pressure p, which pushes membranes close together, can be created experimentally by osmosis. In fact, this technique has been used as a trick to measure di-

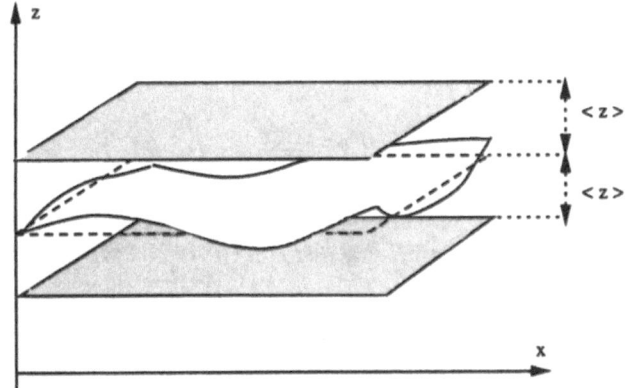

FIGURE 5.4. Single membrane trapped between two rigid walls.

rectly the physical forces between membranes [55]. At zero temperature, the membrane would collapse to a compact state with zero average separation $\langle z \rangle = 0$. At finite temperature, however, membrane undulations appear and introduce a new long range entropic force which swells the lamellar phase, as first discussed by Helfrich [54].

In the following, we simplify the problem even further and assume that a given membrane feels the influence of neighboring ones as if it were constrained between two rigid walls separated by a distance equal to $2\langle z \rangle$ (see Fig. 5.4). This will allow us to derive in a simple manner the expression for the entropic undulation interaction $f_s(\langle z \rangle)$. The mechanical equilibrium of the membrane between the two walls is then obtained by noting that the walls constraining the membrane are held by a pressure p such that $p = \partial f_s / \partial z |_{z = \langle z \rangle}$. This determines $\langle z \rangle$ consistently as a function of the imposed pressure p.

Let us now derive the expression of the entropic undulation interaction, in this model of a single membrane constrained by rigid walls.

5.3.3.1 Ideal gas analogy

The simplest argument making use of the ideal gas analogy is the one discussed by Helfrich and Servuss [56]. They imagine the undulating membrane as a set of independent bodies, each one corresponding to a piece of membrane of area L_z^2 where L_z is such that the r.m.s. displacement of membranes of size L_z is on the order of $\langle z \rangle$, i.e.,

$$\langle u^2 \rangle_{L_z} = \mu \langle z \rangle^2 \tag{5.30}$$

where μ is a number of the order of unity. The subscript L_z of $\langle u^2 \rangle$ in (5.30) refers to the fact that the r.m.s. displacement due to thermal undulations is computed for a piece of membrane of surface area L_z^2 (see 5.16–17)). Treating each body as a particle in an ideal gas, the kinetic gas theory tells us that the average force exerted by such a particle on the boundaries separated by an interval $2\langle z \rangle$ is

$$F = \frac{k_B T}{2\langle z \rangle} \tag{5.31}$$

which is independent of the mass of the "particles" or other details. Eq. (5.31) is nothing else than the ideal gas law $pV = k_B T$, with $p = F/L_z^2$ and $V = 2\langle z \rangle L_z^2$.

Using (5.17) in (5.30) and substituting in (5.31) yields $p = (k_B T)^2/8\pi^3 \mu k \langle z \rangle^3$. Integrating over $\langle z \rangle$ yields the expression of the steric interaction energy per unit membrane surface:

$$f_s = \beta \frac{(k_B T)^2}{k \langle z \rangle^2} \tag{5.32}$$

with $\beta = 1/16\pi^3 \mu$. The most surprising feature is the long range (i.e., $1/\langle z \rangle^2$) nature of this steric interaction for membranes without surface tension, which is the result of the undulations induced by thermal fluctuations of *fluid* membranes.

5.3.3.2   Steric constraint in Fourier space

An alternative argument put forward initially by Nozières in the context of "incommensurate-commensurate" phase transitions [57] introduces a cutoff in the mode wavevectors. Consider a single membrane between two rigid walls separated by the distance $2\langle z \rangle$. The idea is that the steric constraint hinders essentially all fluctuations of wavevector $q_\parallel$ smaller than a cut-off $q_\parallel(z)$, which would yield a contribution to $\langle u^2 \rangle$ larger than $\mu \langle z^2 \rangle$. The cut off is thus given by

$$\sum_{q_\parallel(z)}^{\pi/l_0} \left\langle u_{q_\parallel}^2 \right\rangle = \mu \langle z \rangle^2 \tag{5.33}$$

From (5.15) with $\gamma = 0$, we thus obtain

$$q_\parallel(z)^{-1} \approx \left(\frac{k}{k_B T} \right)^{1/2} \langle z \rangle \tag{5.33'}$$

Note that $q_\parallel(z)^{-1}$ is nothing else than the length L_z introduced in Sec. 5.3.3.1. The idea that $\langle u_q^2 \rangle \approx 0$ for $q \leq q(z)$ means that the fluctuating membrane starts to interact with the walls when some portion of it comes in the vicinity of the wall. This occurs with significant importance when the root mean square amplitude of the thermal undulation becomes of the order of the membrane/wall separation. When such a position of the membrane is brought in contact with one of the walls, it is said to undergo a "collision." The steric interaction is thus

$$f_s = \sum_{q_\parallel=0}^{q_\parallel(z)} \frac{1}{2} k_B T \approx L^2 \frac{1}{2} k_B T \left[q_\parallel(z)\right]^2 \tag{5.34}$$

i.e, the sum of the entropies of the suppressed modes $0 \le q_\parallel \le q_\parallel(z)$. The sum is performed in the $2D - q_\parallel$ space within the disk of radius $q_\parallel(z)$. This recovers expression (5.32) with a slightly different coefficient β.

5.3.3.3 Steric constraint in real space

A third, analogous, approach has been introduced by Fisher and Fisher [44]. The cutoff is introduced in a slightly different fashion. Again, the idea is that the membrane begins to feel the constraining walls when its r.m.s. displacement becomes of the order of $\langle z \rangle$. Then, they argue that the reduction of entropy due to the steric constraint is proportional to the number of "collisions" or contact points between the membrane and the walls. The cutoff is introduced by the assumption that pieces of membrane of area L_z^2, where L_z is such that (5.30) is verified, present essentially one collision point. Assuming an entropy loss $(1/2)k_B T$ per collision yields the steric interaction with the form

$$f_s \sim \frac{1}{2} \frac{k_B T}{L_z^2} \tag{5.35}$$

which recovers expression (5.32), since $L_z \approx (k/k_B T)^{1/2} \langle z \rangle$ according to (5.33').

5.3.3.4 Harmonic theory of the steric interaction

This approach has also been proposed by Helfrich [54] (see also [58], which we follow here). Consider as before (see Fig. 5.4) a single fluid membrane between parallel rigid plates separated by the distance $2\langle z \rangle$ and positioned at $z = \pm \langle z \rangle$. The theory proceeds in two steps:

i. Role of the steric constraint in the membrane deformation.

The thermal undulations of the membrane are restricted by the walls which require that the vertical membrane displacement $u(\rho)$ satisfies, for all ρ,

$$-\langle z \rangle \le u(\rho) \le +\langle z \rangle \tag{5.36}$$

This condition is very difficult to handle mathematically since it implies a complex correlation between the $(L/l)^2$ membrane undulation modes. A simpler constraint is

$$\langle u^2 \rangle = \mu \langle z \rangle^2 \tag{5.37}$$

The purpose of (5.37) is to restrict the local displacement $u(\rho)$ largely (but not strictly) to the interval of width $2\langle z\rangle$, i.e., to the space available between the plates. The factor μ in front of $\langle z\rangle^2$ must be smaller than unity but not too much. The replacement of the inequality (5.36) by the condition (5.37) illustrates the basic simplification of all treatments of the steric interaction, but it is, however, believed to embody the main ingredients of the problem.

ii. Role of the steric constraint in the stresses exerted on the membrane.

One must also describe the influence of the steric constraint on the conjugate of the displacements, i.e., the stresses. In the spirit of a mean field or harmonic approach, one may replace the many-body problem by a one-body problem and envision the stresses exerted by the plates on the membrane as resulting from the action of a harmonic well of spring force m^2. For a given value of m^2, it is easy to compute the r.m.s. displacement of the membrane, and using condition (5.37), to determine the effective spring force consistently. From the value of the spring force m^2, one then deduces the entropic undulation interaction.

The problem of a single membrane between two rigid plates is therefore modeled by the following hamiltonian, which replaces (5.12):

$$F = \int d^2\rho \left\{ \frac{1}{2}\gamma (\nabla u)^2 + \frac{1}{2}k (\nabla^2 u)^2 + \frac{1}{2}m^2 u^2 \right\} \qquad (5.38)$$

together with the constraint (5.37). Now the treatment is straightforward. (5.15) is replaced by

$$\frac{1}{2}L^2 \left(\gamma q_\parallel^2 + k q_\parallel^4 + m^2\right) \langle u_{q_\parallel}^2 \rangle = \frac{1}{2}k_B T \qquad (5.39)$$

For $\gamma = 0$, this yields (after integration over $q_\parallel$)

$$\langle u^2 \rangle = \frac{k_B T}{8k^{1/2}m} \qquad (5.40)$$

The constraint (5.37) allows one to relate self-consistently the spring force m^2 to $\langle z\rangle$:

$$m = \frac{k_B T}{8\mu k^{1/2}} \langle z\rangle^{-2} \qquad (5.41)$$

The entropic undulation free energy reads

$$f_s = -k_B T \log \frac{Z_m}{Z_0} - \frac{1}{2}m^2 \langle z\rangle^2 \qquad (5.42)$$

where

$$\frac{Z_m}{Z_0} = \exp\left\{ -\frac{1}{2}\sum_{q_\parallel} \log \left\{ \frac{\gamma q_\parallel^2 + k q_\parallel^4 + m^2}{\gamma q_\parallel^2 + k q_\parallel^4} \right\} \right\} \qquad (5.43)$$

is the ratio of the partition functions of the membrane with (i.e., $m \neq 0$) and without (i.e, $m = 0$) the steric constraint. The subtraction of the average spring potential energy $(1/2)m^2\langle z\rangle^2$ from the total free energy leads to the steric part f_s of the free energy. This is similar to the case discussed above where an external pressure p equilibrates the repulsive steric interaction, leading to a total free energy $f_t = f_s + p\langle z\rangle$, such that $\partial f_t/\partial z|_{z=\langle z\rangle} = 0$ gives $p = \partial f_s/\partial z|_{z=\langle z\rangle}$ which is the condition of equilibrium discussed in the introduction of Sec. 5.3.3. Thus, in the same spirit, in order to extract the entropic undulation interaction, one must subtract $(1/2)m^2\langle z\rangle^2$ from f_t.

Equation (5.42) thus gives for $\gamma = 0$, after some simplifications,

$$f_s = \frac{(k_BT)^2}{128\mu k}\langle z\rangle^{-2} \tag{5.44}$$

which is the same as (5.32) with $\beta = 1/128\mu$.

Note that this presentation is really equivalent to the treatment of Helfrich [56] in which an "invariable apparent force constant α attributed to the steric effect" is introduced. Indeed, α in [56] is defined by $\alpha = L^2[k_BT/64\mu^2k]\langle z\rangle^{-4}$, which is exactly the value obtained when identification is made between (5.43) of this chapter and Eq. (40) of [56] via the relation $\alpha = L^2m^2/k_BT$.

The arguments presented in this section provide an appealing intuitive understanding of the basic origin of the steric interaction. An increasing number of experiments, discussed in this book, show that, for neutral membranes, the entropic undulation interaction (Chap. 6 and [54,59]) may become the dominant contribution to the effective inter-membrane interactions. In view of its importance, it is useful to develop more precise quantitative theories which we present below. Furthermore, this section (5.3.3) has treated the entropic undulation interaction within the model of a *single* membrane constrained by rigid walls. It is instructive to analyze the alternative model where *all* the degrees of freedom of the lamellae are kept in the theory. In practice, this is possible within the smectic crystal formalism which is now described.

5.3.4   LAMELLAR PHASES: THE SMECTIC LIQUID CRYSTAL ANALOGY FOR THE STERIC INTERACTION

The initial derivation of the entropic undulation interaction presented by Helfrich [54] used the framework of smectic liquid crystals. This is a natural formulation for describing lyotropic multilamellar systems. In the following we will review Helfrich's derivation and an extension [5] which allows one to treat the case of finite size membranes for which a cutoff is introduced by the finite size effect. As a prerequisite, we first recall the basic formalism of the theory of smectic liquid crystals [60], as well as some other relevant results.

5.3.4.1 The smectic liquid crystal analogy

A lamellar phase is a smectic liquid crystal defined, in the harmonic approximation, by the following deformation energy [60] (see also Sec. 6.3.2)

$$\int d^3x f = \int d^3x \left\{ \frac{1}{2} K_1 \left(\nabla^2 u \right)^2 + \frac{1}{2} B \left(\frac{\partial u}{\partial z} \right)^2 \right\} \tag{5.45}$$

where the elastic constant K_1 is related to the bending curvature modulus of a single membrane by

$$K_1 = \frac{k}{\langle z \rangle} \tag{5.46}$$

The operator ∇ is the transverse gradient (i.e., in the plane of the membranes) and the z-axis is perpendicular to the membranes. The nth membrane is placed at an average height above the origin equal to $n\langle z \rangle$ where $\langle z \rangle$ is the average membrane spacing (see Fig. 5.3). From this definition, the term $u(x, y, z)$ is determined by

$$u(x, y, z) = z(x, y) - n\langle z \rangle \tag{5.47}$$

and denotes the displacement along z of the point (x, y) of the nth membrane. Note that the integral in (5.45) extends over the whole volume of the sample.

The coefficient B is the elastic compression modulus of the smectic liquid crystal formed by the stack of lamellae. B takes into account all the interactions between adjacent membranes which bind them together. The minimum of the total potential gives the mean interlamellar distance $\langle z \rangle$, and the curvature of the minimum of the potential introduces the macroscopic compression elastic modulus B in the direction z normal to the membranes.

If one considers a single membrane, $B = 0$. We then recover expression (5.12) with $\gamma = 0$ by integration of (5.45) in the variable z over a length $\langle z \rangle$. This is not surprising since, in (5.45), there is no term analogous to a surface tension. This comes from the condition of rotational invariance [60,61]: in contrast to a single interface for which a rotation modifies its area and therefore its energy, a stack of membranes has a constant energy, whatever its orientation—its energy depends only on the distance of separation between membranes.

It is important to remark that, if expression (5.45) looks very similar to (5.12), the introduction of the compression modulus changes profoundly the influence of the fluctuations. For example, (5.45) yields

$$\frac{1}{2} L^2 \left(B q_z^2 + K_1 q_\parallel^4 \right) \langle u_q^2 \rangle = \frac{1}{2} k_B T \tag{5.48}$$

which is similar to (5.15) for a single membrane. Summing over the modes yields a m.s. undulation amplitude for each layer equal to

$$\langle u^2 \rangle = \frac{L^2 \langle z \rangle}{(2\pi)^3} \int_{\pi/L}^{+\infty} \pi dq_{\parallel}^2 \int_0^{\pi/\langle z \rangle} \langle u_q^2 \rangle \, dq_z \tag{5.49}$$

Let us introduce the dimensionless variable $t = \pi/\{\lambda \langle z \rangle q_{\parallel}^2\}$. λ is the so-called penetration length given by $\lambda^2 = K_1/B$ which describes the "softness" of the lamellar phase and acts as a cutoff wavelength beyond which longitudinal deformation may actually propagate in the multilayer system. With this notation, (5.49) transforms into

$$\langle u^2 \rangle = \frac{k_B T}{8\pi^2 k} \lambda \langle z \rangle \int_0^{L^2/\pi\lambda\langle z \rangle} \frac{dt}{t} \arctan t \tag{5.50}$$

This integral cannot be expressed as a finite combination of elementary functions, but its leading behavior is, to a good approximation, for large L

$$\langle u^2 \rangle^{1/2} = \sqrt{\frac{k_B T}{8\pi k} \lambda \langle z \rangle \log \frac{L}{\sqrt{\pi\lambda\langle z \rangle}}} \tag{5.51}$$

which must be compared with (5.17). Equation (5.51) shows a very weak logarithmic divergence of the mean square undulation displacement $\langle u^2 \rangle$ of a single membrane within a stack of lamellae, which is due to the constraining effect of neighboring membranes. Eq. (5.51) is the well-known result [62,63] which shows that a smectic liquid crystal has no true long range translational order (in agreement with Landau-Peierls [64] arguments for one dimensional and two dimensional systems) but rather a *quasi*-long range order characterized by an algebraic position correlation function (i.e., decreasing to zero as a powerlaw; see Sec. 6.4.3). On the other hand, the interactions between membranes which introduce the compression modulus B change the orientational order from quasi long range as given by (5.22) to true long range since the orientational fluctuations $\theta^2(\rho) = \langle |\delta \mathbf{n}(\rho) - \delta \mathbf{n}(\mathbf{O})|^2 \rangle$ do not diverge logarithmically as for an isolated membrane but have a finite value [62,65]:

$$\theta^2(\rho) = \frac{k_B T}{2\pi k^0} \left\{ 1 + \log \left(\frac{\lambda \langle z \rangle}{\pi l^2} \right) \right\} \quad \text{as } \rho \to \infty \tag{5.52}$$

for a stack of membranes of mean separation $\langle z \rangle$.

5.3.4.2   The smectic liquid crystal analogy for the steric interaction

This approach has been proposed by Helfrich [54]. One considers the steric interaction between layers belonging to a multilamellar system with an average separation $\langle z \rangle$ between membranes. The starting point is the smectic deformation free energy (5.45). The effect of the steric hindrance of neighboring layers is accounted for by the elastic compressive modulus B which is to be determined in a self-consistent way. The general principle of the calculation is the following self-consistent procedure:

i. One first computes the variation f_s in free energy when the layers are brought together from infinity to within a distance of separation $\langle z \rangle$. This variation is a function of the curvature elastic modulus k and the compression modulus B which represents phenomenologically the mutual steric hindrance of the layers.

ii. One must then relate the compression modulus B to the second derivative of f_s with respect to the membrane separation:

$$B = \left. \frac{\langle z \rangle^2 \partial^2 \left(\frac{f_s}{L^2 \langle z \rangle} \right)}{\partial z^2} \right|_{z=\langle z \rangle} \tag{5.53}$$

This leads to a differential equation for f_s whose solution yields the steric interaction.

Let us now proceed to follow this program. f_s is the sum of its Fourier components Δf_q: $f_s = \sum_q \Delta f_q$ with $\Delta f_q = -T \Delta S_q$ and ΔS is the difference of entropy between a phase of free membranes and a phase of interacting membranes whose interaction is accounted for by the compressive modulus B:

$$\Delta S_q = \frac{k_B}{2} \log \frac{K_1 q_\parallel^4}{B q_z^2 + K_1 q_\parallel^4} \tag{5.54}$$

f_s therefore reads

$$f_s = -\frac{L^2 \langle z \rangle}{(2\pi)^3} \frac{k_B T}{2} \int_{\pi/L}^{+\infty} \pi dq_\parallel^2 \int_0^{\pi/\langle z \rangle} dq_z \log \frac{q_\parallel^4}{\frac{q_z^2}{\lambda} + q_\parallel^4} \tag{5.55}$$

Introducing dimensionless variables $Y = f_s/k$, $Z = z/L$ and

$$p = \pi^{-1} \left(\frac{\partial^2 Y}{\partial Z^2} \right)^{1/2} \tag{5.56}$$

we obtain with (5.53) and (5.55)

$$Y = \frac{\pi^2}{16} \frac{k_B T}{k} \left[p - \frac{2}{\pi} \{ \log (1 + p^2) - 1 \} - \frac{2}{\pi} \left\{ p \arctan \frac{1}{p} + \frac{1}{p} \arctan p \right\} \right] \tag{5.57}$$

Expression (5.57) with (5.56) is an implicit second order differential equation. Its solution $Y(Z)$ represents the repulsive steric interaction free energy between membranes of size L separated by $\langle z \rangle$. A general discussion of the solution of (5.57) is found in [5]. Two regimes are worth analyzing:

i. At small separation $(z/L \ll 1)$ such that $p \gg 1$, (5.57) reduces to

$$Y = \frac{\pi}{16} \frac{k_B T}{k} \left[\frac{\partial^2 Y}{\partial Z^2} \right]^{1/2} \tag{5.58}$$

The solution of (5.58), which satisfies $Y/L^2 \to 0$ and $L^{-2} \partial Y/\partial z \to 0$ as $z \to +\infty$ (even though $Z = z/L$ remains small enough for (5.58) to be valid), is

$$f_s(\langle z \rangle) = \beta \frac{(k_B T)^2}{k} \langle z \rangle^{-2} \tag{5.59}$$

which recovers expression (5.32) with a numerical coefficient $\beta = 3\pi^2/128 \approx 0.231$. This is the expression valid in particular for infinite membranes $(L \to +\infty)$.

ii. At large membrane separations $(Z \gg 1, i.e., \langle z \rangle \gg L)$, the steric interaction decreases to zero as the spontaneous curvature fluctuations of each membrane are less and less hindered by the presence of other membranes. In this limit, Y and p are small. The steric interaction is therefore obtained by taking the leading terms of (5.57) in powers of p, which yields

$$Y = \frac{1}{24\pi} \frac{k_B T}{k_c} \frac{\partial^2 Y}{\partial Z^2} \tag{5.60}$$

Its solution satisfying the condition $Y \to 0$ as $Z \to +\infty$ is

$$f_s(\langle z \rangle) = Y_0 \exp \left\{ \frac{-\langle z \rangle}{\xi_{st}} \right\} \tag{5.61}$$

with

$$\xi_{st} = \left(\frac{1}{24\pi} \right)^{1/2} \left(\frac{k_B T}{k} \right)^{1/2} L \tag{5.62}$$

ξ_{st} is essentially the r.m.s. displacement of a single membrane of size L (compare with (5.17)).

We can now match the two solutions (5.59) (valid in an inner region $Z \ll 1$) and (5.61) (valid in an outer region $Z \gg 1$) at a distance Z_0 both for f_s and its first derivative. We thus obtain two equations for the two unknowns Y_0 and Z_0 which yield the solutions

$$Y_0 = \frac{3\pi^2}{128} e^2 \left[\frac{k_B T}{k} \right]^2 Z_0^{-2} \tag{5.63a}$$

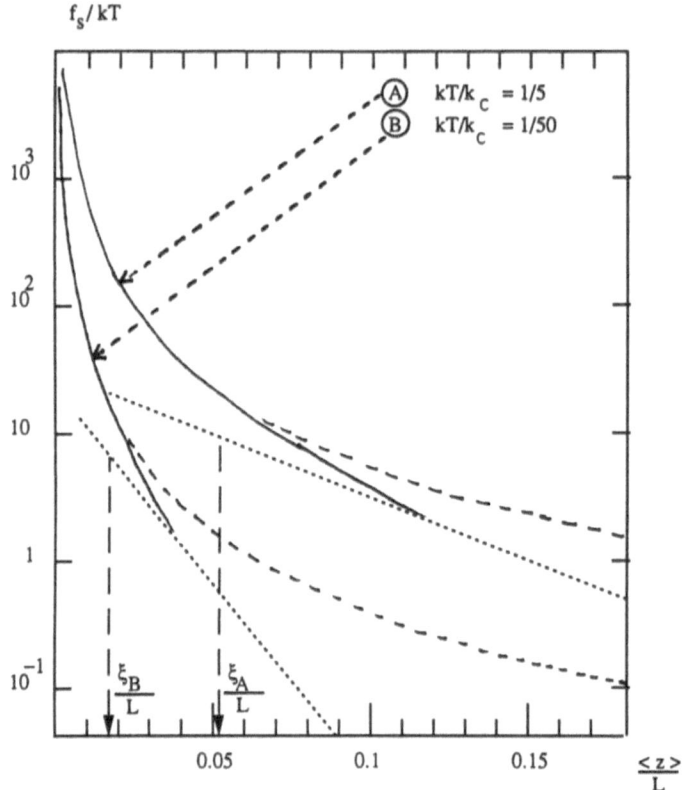

FIGURE 5.5. Steric interaction $f_s(\langle z \rangle)$ for membranes of finite size L.

$$Z_0 = \frac{2\xi_{st}}{L} \tag{5.63b}$$

Eq. (5.63b) gives the range of the steric interaction between membranes of finite size L which is of the order of $2\xi_{st} = 0.23(k_B T/k)^{1/2} L$ and scales as L. The general result is (5.59) for $\langle z \rangle \leq 2\xi_{st}$ and (5.61) for $\langle z \rangle \geq 2\xi_{st}$. Direct numerical solution of (5.57) gives agreement with these analytical results within a few percent in the crossover region $\langle z \rangle \approx 2\xi_{st}$ and cannot be distinguished from the asymptotic expression (5.59) and (5.61) in their region of validity (see Fig. 5.5). Note that the cross over between the two regimes occurs when $\langle z \rangle$ is of the order of the r.m.s. displacement of a free undulating membrane.

The existence of the cutoff length ξ_{st} stemming from the finite size of the membranes can be understood within the general framework of the "finite size scaling" theory of critical phenomena. The scale $L_z \sim q(z)^{-1}$ given by (5.34) can be interpreted as an in-plane correlation length $\xi_\parallel$ as we will see below. Viewing the power law (5.59) as the signature of a critical point reached for $L \to +\infty$ (or $\langle z \rangle / L \to 0$), the $\langle z \rangle \leq \xi_{st}$ regime corresponds to the critical region. For $\langle z \rangle \geq \xi_{st}$, one is outside the critical region, and the algebraic solution (5.59) must be replaced by a finite size scaling ansatz of the form

$$f_s \sim \frac{(k_B T)^2}{k} \langle z \rangle^{-2} g(\langle z \rangle / \xi) \tag{5.64}$$

where $g(x) \to 1$ for $x \to 0$ and $g(x)$ goes to zero exponentially as $x \to +\infty$, which recovers the second regime and the solution (5.61).

5.3.4.3   "Multilamellar" versus "single membrane between plates" approaches

The two approaches, in terms of the "single membrane between two rigid walls" (Sec. 5.3.3) and "the smectic liquid crystal analogy" (Sec. 5.3.4) for the steric interaction, yield essentially the same expression for the entropic steric interaction, apart from a small difference in the numerical factor β. In this section, we point out that a subtle difference between these two geometries exists, which becomes apparent with the existence of anharmonic corrections in the large scale limit.

The difference comes from the requirement that, for a stack of membranes (see Fig. 5.6a), the Hamiltonian should be invariant with respect to global rotations. In the case of a single membrane constrained within two parallel walls, the presence of the walls has broken the rotational symmetry (see Fig. 5.6b).

As a consequence, (5.45) must be modified and $\partial u / \partial z$ must be replaced by [66,61]

$$\frac{\partial u}{\partial z} \to \left[\frac{\partial u}{\partial z} - \frac{1}{2} (\nabla u)^2 \right] \tag{5.65}$$

(5.65) can be justified as follows. A given rotation θ of a perfectly aligned multilamellar structure with spacing $\langle z \rangle$ (see Fig. 5.6c) yields an effective distance between layers, counted in the z direction, given by $\langle z \rangle_{eff} = \langle z \rangle / \cos \theta$. The differential displacement along $z, \partial u / \partial z$, away from the true equilibrium spacing $\langle z \rangle$ is therefore, for small θ,

$$\frac{1}{\cos \theta} - 1 \approx \frac{\theta^2}{2} = \frac{1}{2} (\nabla u)^2 \tag{5.66}$$

where ∇ is the transverse gradient in the plane of the membranes. Consequently, a z-deformation $\partial u / \partial z = (1/2)(\nabla u)^2$ corresponds to a pure rotation and does not involve additional compressive energy. A true z-deformation is therefore the difference $\partial u / \partial z - (1/2)(\nabla u)^2$ which justifies the change (5.65).

The influence of such anharmonic corrections have been discussed in [67,68] and leads to a renormalization of both smectic elastic moduli K_1 and B. The effect of the inter-membrane coupling can be heuristically understood as follows. The anharmonic term (5.65) is such that a tensile

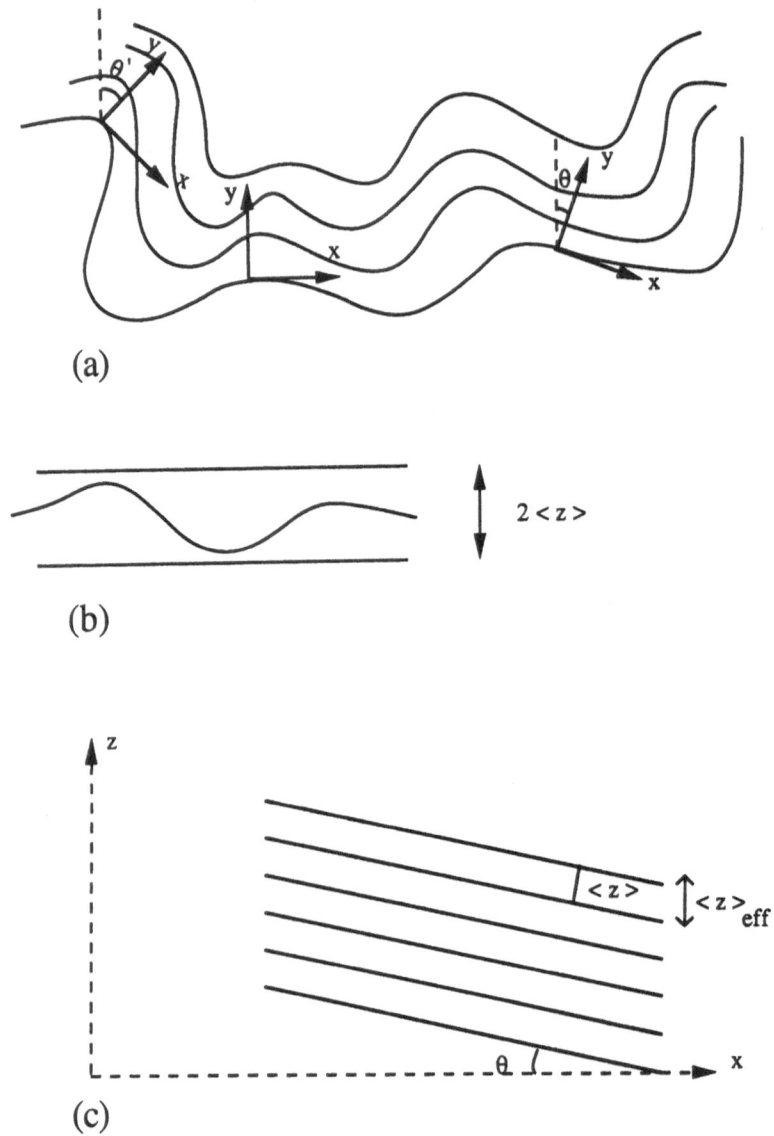

(a)

(b)

(c)

FIGURE 5.6. (a) Side view of a stack of membranes, showing the rotation invariance. (b) Side view of a single membrane between two rigid walls, which is no longer rotationally invariant. (c) Side view of a perfectly aligned multilamellar structure showing the difference between $\langle z \rangle$ and $\langle z \rangle_{eff}$ measured along an arbitrarily defined $0z$ axis.

stress can be relaxed by membrane undulations. This is at the basis of the so-called undulation instability studied in smectics [66,61]. In presence of thermal fluctuations, one therefore expects that B decreases at large scales due to the coupling between tension and undulation. These subtle corrections lead to a logarithmic correction to the $\langle z \rangle^2$ powerlaw dependence of the steric interaction. This correction is of course absent for a single membrane between two rigid walls due to the absence of any coupling term between $q_{\parallel}$ and q_z (as occurs in the multilayer case): in this case, q_z is "stuck" with the constant value $\pi/\langle z \rangle$.

5.3.5 UNIVERSALITY OF THE z^{-2} POWER LAW AND NUMERICAL VALUE OF THE COEFFICIENT β OF THE STERIC INTERACTION BETWEEN MEMBRANES

Scattering techniques (X-rays, neutrons and light, see Sec. 6.4)) applied to swollen lamellar phases allow one to observe the quasi-long-range translational order in lamellar phases which is characterized by an algebraic decay of correlations with an exponent given by [65]:

$$\eta_m = k_B T \frac{q_m^2}{8\pi} \left(\frac{kB}{\langle z \rangle} \right)^{1/2} \tag{5.67}$$

This exponent (5.67) can be obtained by analyzing the powerlaw scattering intensity peaks as a function of scattering wavevector: $S(q_z) \sim (q_z - q_m)^{-(2-n_m)}$ (see Sec. 6.4.3), where q_m is the position of the mth harmonic of the structure factor, i.e., $q_m = m\pi/\langle z \rangle$. Due to the particular form of the compressive elastic modulus B obtained from the steric interaction (but without the logarithmic correction steming from the non linear term (5.66)), η_m turns out to be a pure number, which does not depend on temperature or other physical quantities:

$$\eta_m = \frac{m^2 \pi}{8(2\beta)^{1/2}} \tag{5.68}$$

η_m is a function only of the coefficient β of the steric interaction, defined in (5.32). Several experiments [69,24] have been reported which seem to agree with Helfrich's prediction (5.59) with $\beta = 3\pi^2/128 \approx 0.231$ (see the general discussion in following chapter). With this value, one obtains from (5.68) a value of $4/3$ for the exponent η_m, which appears to be in quantitative agreement with other scattering experiments [70,71]. However, if the data is fitted with a fixed value for β, there is at least one adjustable parameter: the fit usually provides either an estimation of the bending elastic coefficient k [24] or of the finite thickness δ of the membrane [69,70]. It is clear that other choices for β would lead to different quantitative estimations for k or δ. For the sake of comparison with experiments, it is therefore important to have a precise prediction for β.

The treatments presented in Sec. 5.3.3 express β in terms of the unknown parameter μ (of the order of, or less than unity), and we cannot rely on these approaches for a precise determination of β. On the other hand, the *continuous* smectic treatment of Sec. 5.3.4 yields a precise value for β:

$$\beta = \frac{3\pi^2}{128} \approx 0.231 \qquad (5.69)$$

However, as pointed out in [72,73], one could, for instance, have worked with a discrete array of layers in the z direction rather than a continuum as in Sec. 5.3.4. In this case, q_z^2 becomes $[2 - 2\cos q_z\langle z\rangle]/\langle z\rangle^2$, and the final result is similar to (5.59) but with a 30% difference in the value of β [72,73]:

$$\beta = \frac{3}{2\pi^2} \approx 0.152 \qquad (5.70)$$

The belief in a universal value for β has lead several authors to investigate this problem using numerical Monte Carlo simulations [72–77]. The first four articles used an extension of a SOS (solid on solid) model on a square lattice with a *continuous* membrane displacement 'u' in the z-direction. The continuous character of the displacements avoids the problems related to the existence of a roughening transition as in an early version [72]. A finite number N of membranes is placed between two rigid walls. The entropic undulation interaction is obtained by computing the internal energy density of the N membranes as a function of $k_B T/k$, using the Monte Carlo Metropolis algorithm and taking into account the effect of finite membrane size. Janke *et al.* [74] obtain $\beta_1 = 0.079 \pm 0.002$ for a single membrane constrained between two rigid walls. For $N \geq 3$, they do not find any systematic N-dependence for β_N:

$$\beta_N = 0.101 \pm 0.002 \qquad (5.71)$$

where β_N is the value of β for N membranes constrained between two rigid walls. Gompper and Kroll [75] find a small N-dependence: $\beta_1 \approx 0.0798 \pm 0.0003$, $\beta_3 \approx 0.093 \pm 0.004$ and $\beta_5 \approx 0.0966$. Using an analogy with the free-fermion representation describing the steric hindrance between chains [78–80], they propose that the dependence of β_N with N can be used to extrapolate the result

$$\beta_\infty = 0.106 \qquad (5.72)$$

in rather good agreement with the value (5.71).

Lipowsky and Zielinska [76] have studied the behavior of interacting fluid membranes using a vectorized Monte Carlo code. They analyzed a single membrane in the vicinity of a wall and subjected to a pressure p. The lattice was divided into nine sublattices such that each sublattice can be updated independently in a vector loop using the usual Metropolis algorithm. In this way, they analyzed $L \times L$ square lattices (membranes) of mesh 'a' with periodic bounday conditions for $L = 11, 20$ and 41. The entropic undulation interaction was determined for a single membrane by computing the dependence, with respect to an applied pressure p, of the average membrane-wall distance $\langle z \rangle$, of r.m.s. membrane displacement $\langle u^2 \rangle^{1/2}$ and of the average distance between "collisions" L_z (see (5.30)), which they called the in-plane or parallel correlation length $\xi_\parallel$. Minimization of $pz + \beta(k_BT)^2/kz^2$ with respect to z indicates that all these quantities should scale as

$$\langle z \rangle \sim \langle u^2 \rangle^{1/2} \sim \xi_\parallel \sim p^{1/3} \tag{5.73}$$

Their numerical verification of the exponent $1/3$ in (5.73) is the signature of the entropic undulation interaction. From their dimensionless $\underline{p} = a^3 p/(k_BT\ k)^{1/2}$ and $\underline{z} = (k/k_BT)^{1/2}z/a$, their Fig. 1 indicates that

$$\langle z \rangle = (2\beta_1)^{1/3} \left[\frac{(k_BT)^2}{k} \right]^{1/3} p^{-1/3} \tag{5.74}$$

with an estimated value $(2\beta_1)^{1/3} \approx 0.5 \pm 0.05$ yielding $\beta_1 \approx 0.06 \pm 0.02$ in the same range as the two preceding quoted values. From their results, we obtain $\langle \underline{z} \rangle \approx 0.5p^{-1/3}$ and $\langle \underline{u}^2 \rangle^{1/2} \approx 0.23p^{-1/3}$ (the underlining here corresponds to the dimensionless variables used in [76]). This allows an estimation of the parameter μ introduced in Sec. 6.3.3:

$$\mu = \frac{\langle \underline{u}^2 \rangle^{1/2}}{\langle z \rangle^2} \approx 0.2 \tag{5.75}$$

Thus, these results allow one to check the z^{-2} dependence of the steric interactions via one of its consequences (Eq. (5.73)). They also lead to a determination of β and provide an estimate of the constraining parameter μ, used in the mean field approaches to the entropic steric interaction.

In summary, it should be stressed that the form (5.32) of the entropic undulation interaction is independent of the form of the *repulsive* microscopic membrane-membrane (or membrane-wall) potential, as long as it decreases faster than z^{-2} at large distances. In other words, an unpenetrable wall or an exponential repulsion falling off to a constant for $z < 0$ (i.e., upon mutual penetration), for instance, yield the same characteristic z^{-2} powerlaw decay at large distances for the entropic undulation interaction between membranes without surface tension in three dimensions. This result illustrates the concept of *universality* obtained upon *renormalization* of direct microscopic interactions by the thermal fluctuations [81–83]: in the language of the renormalization group (RG), the entropic undulation interaction is the universal "fixed point" potential obtained upon renormalization of short range microscopic repulsive interactions. Note that these ideas hold only in the limit of large membrane-membrane or membrane-wall separation. In [84,58], we have studied the "non-universal" case where the distance between membranes is small and the undulations have not yet renormalized the microscopic potential to their universal behavior given by the entropic steric interaction. In an appendix, Sec. 5.5.3, we discuss briefly this regime where thermal fluctuations renormalize only partially the bare microscopic interaction.

5.3.6   ENTROPIC UNDULATION INTERACTION BETWEEN MEMBRANES OR INTERFACES WITH NON-ZERO SURFACE TENSION γ

We have considered until now steric interactions between membranes with no surface tension. However, the existence of an entropic undulation interactionse has also been discussed in the context of wetting transitions (see [40] and [41] for a review) and incommensurate-commensurate phase transitions [42,43], where surface tension of the interface plays a dominant role.

We showed in Sec. 5.2.2 that, for an isolated membrane with surface tension, the r.m.s. displacement of the membrane is no longer proportional to its size L according to (5.17) but increases only logarithmically with L as seen from (5.18). It is natural to expect that the derivation of the expression of the entropic undulation interaction should be similar to that described in Sec. 5.3.3 and Sec. 5.3.4 for membranes without surface tension, using, however, for $\langle u^2 \rangle^{1/2}$ the expression (5.18) instead of (5.17). The value of the characteristic scale $L_z = 2\pi/q_\parallel(z)$ (see (5.33/)) introduced by the steric constraint should now read

$$L_z \sim l \exp\left\{\frac{4\pi\mu\gamma}{k_B T}\langle z \rangle^2\right\} \tag{5.76}$$

Instead of being proportional to $\langle z \rangle$ for surface tension-less membranes, it is exponentially large in $\langle z \rangle^2$. This reflects the fact that interfaces with surface tension have much smaller thermal fluctuations. Assuming as in Sec. 5.3.3.3 an entropy loss of order $k_B T$ per area L_z^2, this would yield the following form for the steric interaction with $\gamma \neq 0$ (we neglect the influence of k which introduces a minor correction):

$$f_s \sim k_B T \, l^{-2} \exp\left\{-\frac{8\pi\mu\gamma}{k_B T}\langle z \rangle^2\right\} \tag{5.77}$$

Eq. (5.77) predicts a very short range interaction with the form of a Gaussian.

It turns out that the preceding derivation is incorrect, as is the result given by (5.77) [59]. This is due to a subtle effect coming from the fact that the space dimension $D = 3$ is a borderline dimension for interfaces with non-zero surface tension (the analogous border dimension for membranes without surface tension is $D = 5$ [81,83]). This can be seen, for example, from the characteristic logarithmic dependence of $\langle u^2 \rangle$ as a function of the interface size L in (5.18). In this case, the treatments presented in Sec. 5.3 and Sec. 5.4, which are all in the spirit of an harmonic or "mean field" approximation, fail. One must then resort to more sophisticated approaches, such as the renormalization group [85], which take into account non-linear coupling between undulation modes induced by the steric constraint. Usually, one would expect the renormalization group to give non-trivial corrections to the mean field treatment only in strong fluctuation regimes. For an interface with $\gamma \neq 0$, the fluctuations are much weaker than for surface tension-less membranes. However, the mean field theory works well for membranes and fails for interfaces with $\gamma \neq 0$. As explained above, this paradox stems from the subtlety that $D = 3$ is a marginal or cross-over dimension for interfaces with non-zero surface tension.

The correct expression for f_s has been derived within a linear functional renormalization group [59,83], and the result is

$$f_s \sim k_B T \, l^{-2} \exp\left\{ -\left(\frac{8\pi\gamma}{k_B T}\right)^{1/2} \langle z \rangle \right\} \qquad (5.78)$$

This result has been confirmed [86] via a mathematically rigorous derivation based on the determination of upper and lower bounds for the expression of the "mass term" (see Sec. 5.3.3.4 for a definition of the "mass term"). It is interesting to note that the entropic undulation interaction is still relatively short ranged with an exponential decay. However, its range is much longer than would be predicted from the mean field theory leading to (5.77).

We now give two heuristic derivations of expression (5.78).

i. *The case of a non-vanishing surface tension as a finite size effect of the membrane problem*

We now show that (5.78) can be guessed intuitively within the smectic liquid crystal analogy of Sec. 5.3.4 by using the following argument. In Sec. 5.2.2, it was noted that a surface tension $\gamma \neq 0$ introduces a characteristic length $L_\gamma = (k/\gamma)^{1/2}$, such that the effect of the surface tension is felt only at scales larger than L_γ. Viewing L_γ as a finite size cutoff, we may think of the steric interaction between *infinite* interfaces with *non-zero surface tension* as the finite size regime of the steric interaction between *finite* membranes with *zero surface tension* and of size $L = L_\gamma$. Inserting L_γ in (5.62) then yields

$$\xi_{st} = \left(\frac{1}{24\pi}\right)^{1/2} \left(\frac{k_B T}{\gamma}\right)^{1/2} \qquad (5.79)$$

and inserting expression (5.79) in (5.61) recovers (5.78).

This reasoning works because we express the non-zero surface tension interface problem (which poses difficulties in three dimensions within a harmonic theory) in terms of the zero surface tension membrane problem (which can be tackled correctly within the harmonic approximation), the non-zero value of the surface tension appearing as a mere finite size effect.

ii. In Sec. 5.3.3, we discussed various approximate ways to represent and take into account the basic steric constraint (5.36). The main idea is that this constraint introduces a characteristic length L_z which is finite for finite membrane separation. One can trace back the failure of the harmonic field approach, leading to (5.77), to the incorrect estimation of L_z. An improved heuristic treatment, which attempts to cure this mistake, is the following [43].

One starts from a microscopic interface-interface interaction potential of the form

$$V_0(z) = Ae^{-z/\lambda} + pz \qquad (5.80)$$

It incorporates a short range repulsion (for instance, the hydration repulsion discussed in Sec. 5.3) and a pressure term introduced so that the equation $\partial V_0/\partial z = 0$ gives a finite mean interface-interface equilibrium separation $\langle z \rangle$. In the neighborhood of this minimum, $V_0(z)$ can be written

$$V_0(z) \approx \frac{1}{2}m^2z^2, \text{ where } m^2 \approx 2A\lambda^{-2}e^{-z/\lambda} \qquad (5.81)$$

One can now use this expression for the elastic strength m^2 and incorporate it into (5.38) and (5.39), with $k_c = 0$. The determination of the r.m.s. displacement $\langle u^2 \rangle$ of a single interface in the potential well (5.81) involves a summation of $\langle u_{q_\parallel}^2 \rangle$ over the wavevectors $\pi/L \leq q_\parallel \leq \pi/l$. In this summation, we now argue that all modes such that $m^2 < \gamma q^2$ do not feel the wall much and therefore are not constrained, whereas modes such that $m^2 > \gamma q^2$ have their energy dominated by the interaction with the wall. This condition $m^2 > \gamma q^2$ now replaces the constraint (5.30) on the mean square displacement $\langle u^2 \rangle$. The transition between the two regimes defines a wavevector cutoff given by

$$q(z) \approx \left(\frac{m^2}{\gamma}\right)^{1/2} \approx \left\{2A\left(\gamma\lambda^2\right)^{-1}e^{-z/\lambda}\right\}^{1/2} \qquad (5.82)$$

which should replace (5.76) (with $L_z = q(z)^{-1}$). (In the last step of (5.82) we have used the approximation for m^2 given by (5.81).) For $q < q(z)$, the modes feel the wall and are thus sterically constrained. A crude simplification is to set the amplitude of the fluctuations of these modes with $q < q(z)$ equal to zero and let the modes with $q > q(z)$ be free in a similar way as done in Sec. 5.3.3.2. This amounts to defining

L_z as equal to $q(z)^{-1}$ and, from the usual argument $f_s \sim k_B T \, L_z^{-2}$, this leads to the result (5.78) quoted above. The difficulty remains, however, to find the correct value of the range λ of the interaction. The renormalization group [59,83] shows that λ is renormalized from its microscopic value, say λ_H, to a value, independent from λ_H, given by (5.79).

In sum, for the particular case of the borderline dimension $D = 3$ with $\gamma \neq 0$, the steric constraint is not given by $\langle u^2 \rangle \approx \langle z^2 \rangle$ as proposed within the harmonic treatment, but rather by

$$\gamma q^2 \sim \frac{\partial^2 V_R(\langle z \rangle)}{\partial z^2} \tag{5.83}$$

where V_R is a suitable short range potential. This result shows that the steric interaction has a much longer range than that predicted by the harmonic approximation. In particular, L_z, which has been interpreted as an average distance between interface-wall "collisions," is much smaller than the prediction of the harmonic approximation. The dominant contribution to the repulsion actually comes from *relatively rare large excursions of the interface*, since the value of L_z given by (5.79) and (5.82) implies that $\langle u^2 \rangle_{L_z} \ll z^2$. This has important consequences for the nature of the critical wetting transitions [41]: in contrast to the usual critical transitions where only logarithmic corrections are found, the fluctuations change completely the mean field predictions and produce a line of continuous critical transitions with non-universal critical exponents varying continuously with the temperature (in presence of both attractive and repulsive forces) [87]. In the case of incommensurate-commensurate transitions, the amplitude is changed but not the law of divergence of the separation of "solitons" in the vicinity of the transition [42,43].

5.3.7 THERMAL UNDULATIONS OF MEMBRANES: EXTENSIONS

Up to now, we have considered the effect of a *repulsive* microscopic membrane-membrane or membrane-wall interaction on the thermal undulations. The concept of an entropic undulation interaction has been introduced which amounts to a *renormalization* of the bare, direct, repulsive interactions. The case of a membrane constrained by rigid walls, i.e., subjected to a potential which goes from zero to infinity upon penetration, has been considered in Sec. 5.3.3, and the analogous multilamellar case was discussed in Sec. 5.3.4.

In presence of both repulsive *and* attractive forces, this simple picture of a single repulsive entropic undulation interaction may break down. In this case, a naive step would consist in superposing the steric interaction going as z^{-2} (for $\gamma = 0$) at large distance, discussed in Secs. 5.3.3–5.3.4, on the other bare, attractive, interactions described in Sec. 5.3.2. It turns out that this is profoundly in error for membranes with vanishing surface tension and sometimes dangerous even for interfaces with non-zero surface tension. This means that the concept of a repulsive entropic undulation interaction becomes meaningless since all the microscopic interactions (either attractive or repulsive) are renormalized by the thermal fluctuations.

In this case, the effect of the long wavelength undulations is more tricky to evaluate. The generalization of the effect of fluctuations on both attractive and repulsive interactions, however, can be understood qualitatively as follows. In a first approximation, it is reasonable to assume that "collisions" or "contacts" between the membrane and the wall still lead to a *loss* in free energy of order $k_B T$ for the part of the membrane in the repulsive part of the potential: this corresponds to the loss of a single degree of freedom. But, in the presence of an attractive potential, a "collision" also leads to a *gain* proportional to the strength $-W$ of the interaction due to the sampling of the attractive well by the wandering membrane, multiplied by a factor $\approx (k/k_B T)$ which describes the fact that a more rigid membrane entails a larger area of contact and therefore a larger attraction. Thus, the "gain" $-(k/k_B T)W$ increases in absolute value as k increases since a larger portion of the membrane "feels" the attractive potential when in contact. These ideas can be checked by an exact computation carried out for a "similar" problem of a one-dimensional interface in 2D-bulk space near a wall [80]. For this discussion, it is now clear that the competition between attraction and repulsion depends on the ratio $k/k_B T$ and on W.

However, a complete description of these nonlinear couplings between undulations and bare microscopic potentials can only be given with the help of renormalization group techniques. For interfaces with non-zero surface tension, as occurs for wetting transitions, the situation has been clarified several years ago [87,88]. The case of membranes with zero surface tension ($\gamma = 0$) and non-zero curvature rigidity ($k \neq 0$) in 3D, has been discussed more recently. RG computations [81,89,93] show that the full renormalized interaction, i.e., the sum of all renormalized potentials, may be attractive or repulsive at large membrane separation corresponding to a bound or an unbound state of the membranes. These two different states are separated by a phase boundary at which the membranes undergo an unbinding transition [81]. Such a transition, which resemble a wetting transaction [82], can also occur for a membrane interacting with a solid wall. We refer to these references and to several other authors [90,91,76,39].

As an interesting development, the entropic undulation interaction has been recently studied in the case of *elastic* (e.g., polymerized) membranes whose undulations are constrained by the presence of other membranes or by a wall. As discussed in Sec. 5.5.1, these elastic membranes can exhibit several phases. In the low-temperature ordered phase, a Monte Carlo simulation has found that the steric interaction decreases as $z^{-\tau}$ with an exponent $\tau = 3.1 \pm 0.2$ [77]. This result is very different from that predicted and measured for the case of fluid membranes ($\tau = 2$), which has been the subject of this review. It suggests possible experiments on the swelling of polymerized lamellar phases which would demonstrate the existence of an ordered phase below the crumpling transition (see Sec. 5.5.1).

In our remaining discussion, we propose to understand this result ($\tau \neq 2$) within a simple scaling argument which allows the entropic undulation force to be estimated. We first use the fact that the bending rigidity k is renormalized according to [35]:

$$k^R(L) \sim kL^\eta \tag{5.84}$$

at the scale L, for *elastic* membranes. Eq. (5.84) means that an elastic membrane is more rigid at large scales due to the coupling of the r.m.s. displacement of a single membrane; using a wave-vector-dependent bending rigidity $k(q)$ [$\sim q^{-\eta}$ according to (5.84)], one obtains

$$\langle u^2 \rangle^{1/2} \sim \left(\frac{k_B T}{k}\right)^{1/2} L^\zeta \text{ with } \zeta = \frac{2 - \eta}{2} \tag{5.85}$$

For $\eta = 0$, one recovers (5.17). An estimation of the entropic repulsive potential can be obtained in a phenomenological way, in a way similar to those reviewed in Sec. 5.3.3. The idea is to identify its order of magnitude with that of the typical bending energy per unit area in the presence of a rigid wall. Indeed, if the transverse fluctuations of an elastic, ordered membrane are limited to z in the presence of other membranes or a solid wall, then the bending energy per unit area scales as

$$V_{fl} \sim k^R (L_z) \left(\frac{z}{L_z^2}\right)^2 \tag{5.86}$$

In this expression, z/L_z^2 gives an estimation of the curvature at the scale L_z, and L_z is as before the characteristic length scale introduced by the steric constraints, i.e., such that $\langle u^2 \rangle^{1/2}(L_z) \sim z$. Substituting (5.84) in (5.86) and using (5.85) yields [91]

$$V_{fl} \sim k_B T \left(\frac{k_B T}{k}\right)^{\tau/2} z^{-\tau} \tag{5.87}$$

with

$$\tau = \frac{2}{\zeta} \tag{5.88}$$

This expression is in agreement with the finding [77] that $\zeta \approx 2/3$ and $\tau \approx 3$. Note that this same reasoning also works for fluid membranes and recovers the result $\tau = 2$.

5.4   Conclusion

This chapter has shown the importance of thermal fluctuations for the determination of membrane interactions. When only a repulsive microscopic interaction exists between membranes, the undulations produce a renormalization of this direct interaction into a *universal* entropic undulation interaction. We have reviewed in detail different approaches developed to describe this type of interaction, starting from various heuristic arguments, explaining then the generally accepted statistical mechanics derivations and giving some elements of the more generalized approach involving the renormalization group formalism.

A large emphasis was placed on the role of membrane surface tension which governs the undulation amplitudes. For amphiphilic membranes with vanishing surface tension, the undulations are limited by the curvature rigidity k and the wall/interface distance which imposes an infrared cutoff for the deformation in-plane wave vectors.

On the contrary, more rigid interfaces with surface tension are in a "weak fluctuating" regime, and the presence of a nearby wall does not merely result in an infrared cut-off. A full renormalization treatment shows that the membranes are repelled by the wall for distances much larger than those predicted by the harmonic approach and for which only rare excursions of the interface occur close to the wall. The case of fluctuating membranes in two dimensions (in fact, lines), for which exact results can be obtained, is described in Sec. 5.5.4, below.

The realistic case of both repulsive and attractive interactions has been considered for both zero and non-zero surface tension interfaces. In this case, the correct way to handle the coupling between the fluctuations and the bare interactions requires the renormalization group formalism, whose results have been summarized. A linear functional renormalization group is sufficient for the case of membranes with surface tension whereas the correct description of interactions between membranes without surface tension requires a more powerful approach due to strong non-linear couplings. Unbinding transitions between free isolated membranes (or unbound interface) and a lamellar structure (bound interface in the wetting case) are predicted. However, most of these predictions have not yet been explored experimentally and constitute interesting challenges for future researches.

Acknowledgments

During all the course of our interest in this subject, we have benefitted from many stimulating discussions with D. Chatenay, C. Coulon, D. Förster, W. Gelbart, W. Helfrich, J. N. Israelachvili, D. Langevin, S. Leibler, S. Marcelja, J. Meunier, L. Peliti, G. Porte, D. Roux, S. A. Safran and many others.

5.5 Appendix

5.5.1 HEURISTIC DERIVATION OF THE RENORMALIZATION OF k

In this appendix, we present a heuristic derivation of Eq. (5.25) describing the renormalization of the effective bending rigidity of membranes at large scales ρ (but still smaller than the persistence length ξ_K) which makes explicit its physical origin, namely the *non-linear coupling* between thermal undulations. We then briefly mention recent works on *elastic* or tethered membranes.

Let us first recall that (5.25) gives the decrease of the effective bending rigidity $k(\rho)$ of the membrane at large scales ρ according to

$$k(\rho) = k^0 \left\{ 1 - \frac{\alpha k_B T}{4\pi k^0} \log \left(\frac{\rho}{l} \right) \right\} \text{ for } \rho \ll \xi_K \qquad (5.25)$$

We present a heuristic derivation of (5.25) which illustrates the fact that it stems from an interplay of non-linear coupling and fluctuations. We first compute the elastic free energy (5.12) (with $\gamma = 0$) of a small deformation $u(\rho)$ due to the presence of a single long wavelength sinusoidal ripple with a wavevector Q in the x-direction:

$$u(x,y) = u_Q \cos Qx \qquad (5.89)$$

The deformation energy F_Q corresponding to this mode reads to second order

$$F_Q = \frac{1}{4} L^2 k^0 Q^4 u_Q^2 \qquad (5.90)$$

We now consider a membrane profile given by the simultaneous presence of a long wavelength ripple as (5.89) and a short wavelength sinusoidal undulation (of wavevector q): $u_Q \cos(\mathbf{Q} \cdot \rho) + u_q \sin(\mathbf{q} \cdot \rho)$, with $|\mathbf{Q}| \ll |\mathbf{q}|$. The elastic energy of such a profile is, up to order $|u_Q|^2 |u_q|^2$:

$$F = L^2 \left\{ \frac{1}{4} k^0 Q^4 u_Q^2 + \frac{1}{4} k^0 q^4 |u_q|^2 - \frac{1}{2} k^0 |u_Q|^2 |u_q|^2 B(Q,q) \right\} \qquad (5.91)$$

where $|u_q|^2 = |u_Q|^2 + |u_q|^2$ and

$$B(Q,q) = \frac{1}{4} Q^2 \left\{ \frac{5}{2} q_x^2 + \frac{1}{2} q_y^2 \right\} (Q^2 + q^2) \qquad (5.92)$$

The non-linear coupling is represented by the quartic term $|u_Q|^2 |u_q|^2$ obtained from an expansion, beyond harmonic approximation, of Eq. (5.6) for the mean curvature H and of (5.13) for the surface element of the membrane. Eq. (5.91) defines an effective bending rigidity $k(u_Q) = k^0 \{ 1 - 2|u_Q|^2 B(Q,q)/Q^4 \}$ acting on the mode Q.

The strategy is now to calculate the difference ΔF in free energy in presence or in absence of the long wavelength ripple in order to identify the value of the bending rigidity at the scale Q^{-1}. In this goal, we assign an energy $(1/2)k_BT$ to each mode of wavelength q. THe result is $\Delta F = F_Q - T\Delta S$, where F_Q is the elastic energy of the long wavelength ripple given by (5.90), and ΔS is the entropy difference of the modes q in presence and in absence of the mode Q which is given by

$$\Delta S = \frac{1}{2}L^2 k_B = \int \frac{d^2q}{(2\pi)^2} \log\left\{\frac{k^0 q^4}{k^0 q^4 - 2|u_Q|^2 B(Q,q)}\right\} \qquad (5.93)$$

Expanding this expression to first order and integrating out the degrees of freedom of the q-modes in the allowed range $|Q| \leq |q| \leq \pi/l$, we obtain

$$\Delta S = -L^2 \frac{3k_B|u_Q|^2}{16\pi} Q^4 \log(Ql) + (Q^2) \qquad (5.94)$$

The term proportional to Q^2 represents a renormalization of the surface tension and is therefore eliminatd by the introduction of a suitable bare surface tension. The term proportional by Q^4 represents the renormalization of the rigidity which therefore obeys (5.25) with $\rho = \pi/Q$ and $\alpha = 3$. Note that this derivation of (5.25) embodies the entropic effects which can be evaluated in the harmonic approximation plus the crucial effect of non-linear mode coupling. This derivation is justified by more sophisticated approaches [28,29,92] which demonstrate that the whole effect of thermal fluctuations amounts to a renormalization of the surface tension and of the bending rigidity.

It is important to understand that (5.25) cannot be captured within an harmonic treatment, which neglects coupling between thermal undulations. In order to demonstrate this point, we start from the expression of the total free energy of a single membrane as given by (5.11) and from the partition function Z which is given by (5.10). If we use (5.12) for the total free energy of the membrane in (5.10), we obtain Gaussian integrals which can be readily computed and yield, when substituted in (5.11),

$$F = L^2\left\{\gamma + \frac{k_BT}{2l^2}\log\left(\frac{L}{b}\right)\right\} + 2\pi\left\{k^0 + \left(\frac{k_BT}{4\pi}\right)\log\left(\frac{L}{c}\right)\right\} \qquad (5.95)$$

The constants b and c are given by $b = l\{(2\pi)^3 k/(k_BTe^2)\}^{-1/2}$ and $c = l\{(2\pi)^3 k/(k_BTe^2)\}^{1/2}$, where $e = 2.718\cdots$. Eq. (5.95) can be recovered by writing $F = \sum_q\{U_q - TS_q\}$ where $U_q = (1/2)k_BT$ is the energy per mode and $S_q = -P_q \log P_q$ is the entropy per mode with the Gaussian probability distribution $P_q(u_q) = (2\pi\langle u_q^2\rangle)^{-1/2}\exp\{-u_q^2/2\langle u_q^2\rangle\}$ for the deformation u_q of the qth mode. Note that the wavevector q should be denoted $q_\|$ in agreement with the notation adopted in Sec. 5.2.2. In the following, we simplify the notation and keep q instead of $q_\|$.

Eq. (5.95) shows that the effect of the entropy in the harmonic approximation is to renormalize and *increase* both the surface tension and the bending rigidity by logarithmic corrections similar to that in (5.25) but of the opposite sign. Note that anharmonic corrections due to the stretching deformation modes i. of Sec. 5.2.1 have been shown to lead to a renormalization of the surface tension similar to that of (5.95):

$$\gamma = \gamma_0 + \frac{Ek_BT}{4\pi k} \log\left(\frac{L}{l}\right) \qquad (5.96)$$

with a correction term involving the stretching modulus E [16]. These authors have also pointed out a possible origin for the vanishing value of the observed surface tension in certain surfactant systems: if the bare surface tension γ_0 is negative such that γ_0 is exactly equal to $\gamma_0 = -(Ek_BT/4\pi k)\log(L/l) - (k_BT/2l^2)\log(L/b)$, then it is possible to achieve an equilibrium state with a vanishing effective surface tension γ. A negative γ_o means that the unperturbed membrane has a tendency to spontaneously increase its surface: in presence of "surface consuming" undulations, this may result in a zero macroscopic surface tension. In summary, by comparing (5.95) with (5.25), we have shown that the softening of the bending rigidity by thermal fluctuations is not a pure entropic effect: it is due to an interplay between non-linearity and fluctuations.

It is interesting to note that, for a multilamellar system, there is an additional non-linear coupling term, obtained from an expansion beyond the harmonic approximation of (5.6) for the mean curvature H and of (5.13) for the surface element of the membrane. It is again due to rotational invariance but applied to the multilamellar case, as discussed in Sec. 5.3.4.3. It turns out that it leads also to a *positive* logarithmic correction to the bending rigidity k [67,68]. The competition between these two corrections has been studied in [38,39], where it is suggested that a "hardening transition" in lamellar phases may exist, which separates a "soft" phase (i.e., k decreases at large transverse scales $q_{\parallel}^{-1}$) at small membrane separation, from a "rigid" phase (i.e., k increases at large transverse scales $q_{\parallel}^{-1}$) at large membrane separation. This transition could be reached simply by swelling the lamellar system and could explain the extraordinary stability of hyperswollen lamellar phases and their extremely thin scattering Bragg peaks observed by [22].

Up to now, we have considered fluid membranes having no internal order. Other interesting states can occur which exhibit novel behaviors. Indeed, recent numerical [94] and theoretical studies [35,36,95–98] show that the nature of membrane fluctuations is strongly dependent on details of the intrinsic order of its constituent particles (see also [99] for more information). As we have seen, a two-dimensional incompressible fluid membrane fluctuating at constant area at any non-zero temperature is, like a one-dimensional polymer, a crumpled object, having no long range orientational order of normals erected at points on its surface, at scales beyond the persistence length (see Sec. 5.2.3). On the other hand, two-dimensional crystalline membranes, characterized by a fixed internal connectivity of constituent particles, are believed to exhibit more interesting phase behavior [100]. At high temperature (or low bending rigidity) they become, like fluid membranes, crumpled, fractal objects. However, numerical simulations [94] indicate that, at sufficiently high rigidity (low temperature), these membranes exhibit a flat phase, separated, by a presumably second order crumpling transition, from the low rigidity crumpled phase. A theoretical argument supporting these simulations has been proposed by [35], according to which long-range orientation order occurring at high bare rigidities is due to a long range interaction between local Gaussian curvatures mediated by transverse phonons of the crystalline membrane. The effect of this interaction is a strong dependence of the bending rigidity

$$k^R(L) \sim kL^\eta \qquad (5.97)$$

on scale L, at large values of L. As we discuss at the end of Sec. 5.3.7, $k^R(L)$ determines the spectrum of undulation modes and leads to an entropic undulation interaction between membranes which decreases faster than z^{-2}.

5.5.2   STERIC INTERACTION BETWEEN CYLINDERS IN A HEXAGONAL PHASE

As we have seen in Sec. 5.3, the concept of an effective steric interaction applies equally well to 2D-surfaces in 3D or to 1D-lines in 2D. In fact, it is very general and can be useful as soon as an extended object which has a non-zero topological dimension is submitted to some steric constraints acting on its fluctuations.

In this appendix, we present another example of the application of this concept and estimate the mutual steric interaction between fluctuating cylinders which are parallel to each other, on the average. This problem appears in the computation of the phase diagram for direct and inverse hexagonal phases in lyotropic liquid crystals [101]. The idea is that the steric interaction contribution to the total free energy is the analog of the mixing entropy in micellar phases. It therefore involves an important contribution for describing (hexagonal-lamellar) phase transitions.

The derivation follows the procedure described in Sec. 5.3.3. We first consider the case of a single cylinder of radius R and of length L constrained within a larger non-penetrable cylinder of radius d (see Fig. 5.7).

One first computes the displacement 'u' with respect to the equilibrium position of the cylinder. With this goal, one determines the amplitude of the undulation mode $\langle u_{q_\parallel}^2 \rangle$, in a way similar to (5.15), by using the equipartition

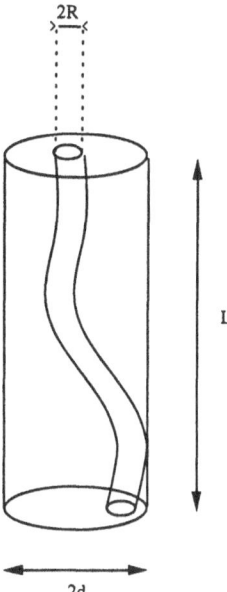

FIGURE 5.7. Undulations of a cylindrical membrane constrained by a rigid cylin-drical wall.

theorem:

$$\frac{1}{2}2\pi RLkq_{\parallel}^4 \langle u_{q_\parallel}^2 \rangle = \frac{1}{2}k_B T \qquad (5.98)$$

$\langle\,\rangle$ again denotes a statistical average over membrane configurations. Note that the membrane surface L^2 is replaced here by the cylinder surface $2\pi RL$. From (5.98), it is easy to compute the m.s. (mean square) displace-ment due to thermal undulations, as $\langle u^2 \rangle = \sum_{q_\parallel} \langle u_{q_\parallel}^2 \rangle$. However, in contrast to the case of a planar membrane, the integration over $q_\parallel$ must be carried out only over wavevectors parallel to the cylinder (we are considering cylin-der undulations at fixed radius). With the classic correspondence relating a discrete sum to the corresponding continuous integral $\sum_{q_\parallel} \to (L/2\pi) \int dq_\parallel$

$$\langle u^2 \rangle \approx \frac{1}{48\pi}\frac{k_B T}{k}\frac{L^3}{R} \qquad (5.99)$$

Note that this problem is different from the one discussed below in Sec. 5.5.4 for one-dimensional lines with surface tension. Here, the cylinders behave as effective one-dimensional objects and their rigidity is controlled by the bending modulus k rather than a surface tension.

The second step is to use the effective steric constraint (5.30) in order to estimate the average number of contacts between the fluctuating cylinder and the wall. Using the fact that the bending fluctuations of the rod are limited by d, the mean distance 'l' between contacts is given by

$$l \sim \left[48\pi \frac{k}{k_B T} R d^2 \right]^{1/3} \tag{5.100}$$

The mean number of contacts per unit length is simply equal to $1/l$. The corresponding loss of entropy per unit length (assuming a loss of $k_B T$ per contact) is

$$\Delta f = k_B T \left[\frac{k_B T}{48\pi k} \frac{1}{R d^2} \right]^{1/3} \tag{5.101}$$

This result (5.101) gives the entropic undulation interaction for a cylinder constrained by a rigid wall which surrounds it. If the cylinder is now placed on a two dimensional hexagonal lattice, its undulations will be less constrained by its neighbors than by a rigid wall which completely surrounds it. If the neighboring cylinders are placed at an average distance d from the central cylinder, they are seen individually from it under an angle of α of the order $2R/d$. One can then argue that the steric hindrance works as if a rigid wall were placed around the central cylinder but only on a portion of its surface, roughly equal to $6\alpha/2\pi \approx 6R/\pi d$. In consequence, the steric free energy can be obtained from the above formula (5.101) by taking into account the geometrical multiplicative factor $\approx 6R/\pi d$ which describes the fact that some large undulations may not lead to a contact with a neighboring cylinder. Note that this factor is important for small R/d. The correct expression for the entropic undulation interaction of an assembly of fluctuating cylinders is thus given by

$$\Delta f \sim k_B T \left[\frac{k_B T}{k} R^2 \right]^{1/3} \frac{1}{d^{5/3}} \tag{5.102}$$

It is used by Roux and Coulon [101] to compute the phase diagram of surfactant molecules, taking into account the competition between lamellar, cylindrical and micellar phases.

5.5.3   SHORT RANGE NON-UNIVERSAL BEHAVIOR

In this review, we have essentially discussed the large distance "universal" regime $z \gg \lambda$, λ being the typical range of the interactions, where the precise value of λ is not relevant. The other case $z \sim \lambda$ is also encountered experimentally, since lamellar phases are often characterized by an inter-lamellar spacing of the order of a few tens of angströms [55]. In this case, the fluctuations also play a very important role, but do not completely erase the form of the microscopic interactions. A specific example has been treated in [84,58,102].

A simple picture of the effect of the thermal membrane undulation can be developed as follows. Consider a bare microscopic exponential repulsive potential

$$V_0(z) = A_0 e^{-z/\lambda} \tag{5.103}$$

The effect of the undulations is to bring some portion of the membrane near the wall. This portion will therefore "feel" a stronger repulsion than the remaining part of the membrane which is farther away. If the potential were linear in z, the net effect of the fluctuations would vanish on averaging. However, due to the non-linear dependence of $V_0(z)$ with respect to z, the fluctuations "renormalize" and increase the effective strength of the repulsive potential. This effect can be mathematically written in terms of a convolution operator which weights a given value of the interaction potential for a spacing z, by the probability that a piece of membrane actually is at the distance z. This proposed ansatz [84,58] has been proven to correspond exactly to the linear approximation of the functional renormalization group discussed in [83]. This regime allows one to give an intuitive meaning to the renormalization group via the convolution ansatz. One finds that the fluctuations renormalize both the strength and range of the potential according to

$$V_R \approx A_H \exp\left\{ \frac{\langle u(z)^2 \rangle}{\lambda_H^2} \right\} e^{-z/\lambda} \tag{5.104}$$

where the r.m.s. displacement $\langle u(z)^2 \rangle^{1/2}$ is determined self-consistently in presence of the renormalized repulsive potential. Typical values are $z \approx 10\text{Å}$, $\lambda \approx 2\text{Å}$ which yield a factor of three in the amplification of the hydration repulsion, when going from the solid state ($\langle u(z)^2 \rangle^{1/2} \approx 0$) to the liquid state ($\langle u(z)^2 \rangle^{1/2} \approx 3\text{Å}$). This provides a qualitative explanation for the measurements of Lis et $al.$ [55].

5.5.4 STERIC INTERACTION IN TWO DIMENSIONS: EXACT RESULTS

In two dimensions, the entropic steric interaction may be studied within many equivalent different approaches (Ising spin models, discrete SOS and related models, Fermion approach, continuous interface model, etc.), but one among them (the random walk analogy) is particularly simple, fully tractable, and we therefore briefly report its main results. In this case, the tractability of the problem is deeply related to the 1D-character of interfaces. A fluctuating interface in a (x, y) plane can now be modeled as a random walker in the (x, t) plane. It should be stressed that the random walk problem describes the wandering of lines with a non-zero tension.

With Fisher [80], we begin by asking for the effective force between a rigid wall and a nearby random walker. With this goal, we compute the full partition function (5.10) for a walk, which starts at a small distance 'a' and reaches the point x. The steric free energy will be derived by taking the logarithm of the partition function according to (5.11). This exact procedure allows one to take into account not only the effect of an average value of the finite size cutoff L_z introduced by the steric constraint but the whole distribution of L_z, corresponding to the set of distances separating "collisions" between the interface and the wall. The partition function is given by the method of images [103]

$$Q_n(x) \sim n^{-1/2} \left[\exp\left\{ -\frac{(x-a)^2}{2na^2} \right\} - \exp\left\{ -\frac{(x+a)^2}{2na^2} \right\} \right]$$

$$\sim n^{-3/2} \exp\left\{ \frac{-x^2}{2na^2} \right\} \tag{5.105}$$

in the large n limit. The most probable distance and the mean distance $\langle z \rangle$ from the wall after n steps are both given by

$$\langle z \rangle \sim a n^{1/2} \tag{5.106}$$

The total partition function for an n-step walk near a wall is

$$Z_n = \int dx \, Q_n(x) \sim n^{-1/2} \tag{5.107}$$

The reduced free energy per step at the nth step is given by

$$f_1 = -\log\left(\frac{Z_{n+1}}{Z_n} \right) = \frac{1}{2} \log\left(1 + n^{-1} \right) \tag{5.108}$$

which can be expressed as

$$f_1(\langle z \rangle) = \frac{1}{2} \frac{a}{\langle z \rangle^2} \quad \text{for } \langle z \rangle \gg a \tag{5.109}$$

by using (5.106). The coefficient in (5.109) is exact! Equation (5.109) is due to the interaction between the walk and the wall and corresponds to an effective repulsive potential $k_B T \, f_1$ per step (i.e., per unit length of the walk in the (x,t) plane). The $k_B T$ dependence of this potential serves as a reminder that the origin of the force lies in the loss of entropy which a walk incurs when the walker approaches but cannot penetrate the wall.

Note that expression (5.109) can be reobtained heuristically with the arguments of Sec. 5.2.2. In two dimensions, with $\gamma \neq 0$, (5.18) is replaced by

$$\langle u^2 \rangle \sim \frac{k_B T}{\gamma} L \tag{5.110}$$

where γ is now a line tension (energy per unit length) of the order of $\gamma \approx k_B T / a$ and L is again the length of the line. This constraint defines the characteristic length $L_z \sim (\gamma/k_B T)\langle z \rangle^2$ and the corresponding steric free energy is

$$f_s \sim k_B T \, L_z^{-1}$$

$$\sim \frac{k_B T}{\gamma} \langle z \rangle^{-2} \tag{5.111}$$

which is exactly (5.109) with a proper definition of $\gamma(\approx k_B T/a)$.

The random walk analogy allows us to test the validity of the concept of an effective steric interaction. Consider a walk which is confined between two rigid walls separated by the distance $2\langle z \rangle$. Supposing that each wall may be regarded as acting independently on the walker, the total effective potential is

$$f_2(z) = f_1(z) + f_1(2\langle z \rangle - z) \qquad (5.112)$$

By symmetry, the most probable position for the walker is at $z = \langle z \rangle$; this location can alternatively be regarded as that which minimizes $f_2(z)$. Using this result that the average position of the walker is at the middle of the strip between the two walls, the total free energy is simply equal to twice the steric energy:

$$f_2(z) \approx 2 \frac{1}{2} \frac{a^2}{\langle z \rangle^2} \qquad (5.113)$$

This expression can be compared with the exact treatment which computes the true partition function for the two-wall problem [80]. The form (5.113) turns out to be quite correct, but one discovers that the coefficient 2 must be replaced by the exact value $\pi^2/4 \approx 2.4674$. The description in terms of two superposable effective steric wall forces is thus qualitatively correct and quantitatively in error by only about 20%. The simplification made in the derivation of (5.113) is to neglect the multiple "images" introduced by the reflection of the second wall on the first one and vice-versa. More challenging tests have been studied which all converge and convince one that the concept of an effective steric entropic interaction gives a good account of the effects of walls or other external potentials on an otherwise random walker (which models a random interface with a non-vanishing surface tension).

References

1. P. Ekwall, Adv. Liq. Cryst. **1**, 1 (1975).

2. D. Roux, Ph.D. Thesis (1984), Bordeaux.

3. *Physics of Amphiphiles: Micelles, Vesicles and Microemulsions*, V. Degiorgio and M. Corti, eds. (North Holland, Amsterdam, 1985).

4. J. Charvolin and A. Tardieu, Solid State Phys. **48**, 209 (1978), F. Seitz and D. Turnbull, eds.

5. D. Sornette and N. Ostrowsky, J. de Physique **45**, 265 (1984).

5. J. Fröhlich, *The Statistical Mechanics of Surfaces*, Lecture Notes in Physics **216**, 151 (1985).

6. *Statistical Mechanics of Membranes and Surfaces*, D. R. Nelson, T. Piran and S. Weinberg, eds., Fifth Jerusalem Winter School (World Scientific, Singapore, 1987).

7. P. G. de Gennes, *Scaling Concepts in Polymer Physics* (Cornell University Press, Ithaca, 1985), 2nd edition.

8. A. M. Polyakov, Nucl. Phys. B **268**, 406 (1986).

9. P. Orland, Phys. Rev. Lett. **59**, 2393 (1987).

10. W. Helfrich, Z. Naturforsch **28c**, 693 (1973).

11. H. J. Deuling and W. Helfrich, Biophys. J. **16**, 861 (1976); H. J. Deuling and W. Helfrich, J. de Physique **37**, 1335 (1976).

12. R. M. Servuss, W. Harbich and W. Helfrich, Biochem. Biophys. Acta **436**, 900 (1976).

13. S. B. Hladky and D. W. R. Gruen, Biophys. **38**, 251 (1982).

14. M. A. Peterson, J. Math. Phys. **26**, 711 (1985).

15. W. Helfrich, in *Physics of Defects* (Les Houches, Session XXXV, 1980), R. Balian et al. eds. (North Holland, Amsterdam, 1981).

16. F. Brochard, P. G. De Gennes and P. Pfeuty, J. de Physique **37**, 1099 (1976).

17. H. Engelhardt, H. P. Duwe and E. Sackmann, J. de Physique Lett. **46**, L-395 (1985).

18. J. F. Faucon, M. D. Mitov, P. Méléard, I. Bivas and P. Bothorel, J. de Physique **50**, 2389 (1989).

19. H. P. Duwe, J. Kaes and E. Sackmann, J. de Physique **51**, 945 (1990).

20. M. Mutz and W. Helfrich, J. de Physique **51**, 991 (1990).

21. F. Brochard and J. F. Lennon, J. de Physique **36**, 1035 (1975).

22. F. C. Larche, J. Appell, G. Porte, P. Bassereau and J. Marignan, Phys. Rev. Lett. **56**, 1700 (1986).

23. J.-M. de Meglio, M. Dvolaitzky, L. Léger and C. Taupin, Phys. Rev. Lett. **54**, 1686 (1976).

24. F. Nallet, D. Roux and J. Prost, Phys. Rev. Lett. **62**, 276 (1989); F. Nallet, D. Roux and J. Prost, J. de Physique **50**, 3147 (1989).

25. P. G. de Gennes and C. Taupin, J. Chem. Phys. **86**, 2294 (1982).

26. *Ordering in Two Dimensions*, S. K. Sinha, ed. (North Holland, 1982).

27. D. R. Nelson, in *Fundamental Problems in Statistical Mechanics V*, E. G. D. Cohen, ed. (North Holland, Amsterdam, 1985).

28. L. Peliti and S. Leibler, Phys. Rev. Lett. **54**, 1690 (1985).

29. D. Sornette, Thèse d'état, Université de Nice (1985).

30. W. Helfrich, J. de Physique **46**, 1263 (1985).

31. W. Helfrich, J. de Physique **48**, 285 (1987).

32. D. Förster, Europhys. Lett. **4**, 65 (1987).

33. F. David and E. Guitter, Europhys. Lett. **5**, 709 (1988).

34. H. Kleinert, J. Stat. Phys. **56**, 227 (1989b); H. Kleinert, Phys. Lett. A **138**, 201 (1989a); H. Kleinert, Phys. Rev. Lett. **58**, 1915 (1987).

35. D. R. Nelson and L. Peliti, J. de Physique **48**, 1085 (1987).

36. F. David, E. Guitter and L. Peliti, J. de Physique **48**, 2059 (1987).

37. Y. Kantor and D. R. Nelson, Phys. Rev. A **35**, 4020 (1987).

38. D. Sornette, J. Phys. Condens. Matter **1**, 1905 (1989); D. Sornette, in *Physics of Amphiphilic Layers*, Les Houches, J. Meunier, D. Langevin and N. Boccara, eds. (Springer-Heidelberg, 1987).

39. L. Golubovic and T. C. Lubensky, Phys. Rev. B **39**, 12110 (1989).

40. P. G. de Gennes, Rev. Mod. Phys. **57**, 827 (1985).

41. S. Dietrich, in *Phase Transitions and Critical Phenomena*, C. Domb and J. L. Lebowitz, eds. (1987).

42. V. L. Pokrovsky and A. L. Tapalov, *Theory of Incommensurate Crystals*, Soviet Scientific Reviews, Physics **1** (Harwood, 1984).

43. J. Villain, in *Structures et Instabilités*, C. Godrèeche, ed. (Les éditions de Physique, 1986).

44. M. E. Fisher and D. S. Fisher, Phys. Rev. B **25**, 3192 (1982).

45. S. G. J. Mochrie, A. R. Kortan, R. J. Birgeneau and P. M. Horn, Z. Phys. B **62**, 79 (1985).

46. J. N. Israelachvili, *Intermolecular and Surface Forces, With Applications to Colloidal and Biological Systems* (Academic Press, 1985).

47. J. Mahanty and B. W. Ninham, *Dispersion Forces* (Academic Press, London, 1976).

48. B. V. Derjaguin, Y. I. Rabinovich and N. V. Churaev, Nature **272**, 313 (1978).

49. S. Nir, Progress. Surf. Sci. **8**, 1 (1976).

50. R. Kjellander and S. Marcelja, Chemica Scripta **25**, 1 (1985).

51. D. F. Evans and B. W. Ninham, J. Phys. Chem. **90**, 226 (1986).

52. N. Ostrowsky and D. Sornette, Colloids and Surfaces **14**, 231 (1985).

53. J. N. Israelachvili and D. Sornette, J. de Physique **46**, 2125 (1985).

54. W. Helfrich, Z. Naturforsch **33a**, 305 (1978).

55. L. J. Lis, M. McAllister, N. Fuller, R. P. Rand, and V. A. Parsegian, Biophys. J. **37**, 657 (1982).

56. W. Helfrich and R. M. Servuss, Il Nuovo Cimento **3D**, 137 (1984).

57. P. Nozières, Course at Collège de France, Paris (1984), unpublished.

58. D. Sornette and N. J. Ostrowsky, J. Chem. Phys. **84**, 4062 (1986).

59. D. Sornette, Europhys. Lett. **2**, 715 (1986).

60. P. G. de Gennes, *Physics of Liquid Crystals* (Oxford Unitersity Press, Oxford, 1974).

61. D. Sornette, J. de Physique **48**, 151 (1987).

62. O. Parodi, in *Colloides et Interfaces*, Les éditions de Physique, A. M. Cazabat and M. Veyssié, eds. (1984).

63. M. Kléman, *Points, Lignes, Parois dans les Fluides Anisotropes et les Solides Cristallins*, **1**, Les éditions de physiques (1977).

64. N. D. Mermin and H. Wagner, Phys. Rev. Lett. **17**, 1133 (1966).

65. A. Caillé, C.R.A.S. (Paris) **274**, 891 (1972).

66. R. Ribotta and G. Durand, J. de Physique **38**, 179 (1977).

67. G. Grinstein and R. A. Pelcovitz, Phys. Rev. Lett. **47**, 856 (1981).

68. G. Grinstein and R. A. Pelcovitz, Phys. Rev. A **26**, 915 (1982).

69. C. R. Safinya, E. B. Sirota, D. Roux, and G. S. Smith, Phys. Rev. Lett. **62**, 1134 (1989).

70. C. R. Safinya, D. Roux, G. S. Smith, S. K. Sinha, P. Dimon, N. A. Clark and A. M. Bellocq, Phys. Rev. Lett. **57**, 2718 (1986).

71. D. Roux and C. R. Safinya, J. de Physique **49**, 307 (1988).

72. W. Janke and H. Kleinert, Phys. Lett. A **117**, 353 (1986).

73. W. Janke and H. Kleinert, Phys. Rev. Lett. **58**, 144 (1987).

74. W. Janke, H. Kleinert and Meinhart, Phys. Lett. B **217**, 525 (1989).

75. G. Gompper and D. M. Kroll, Europhys. Lett. **9**, 59 (1989).

76. R. Lipowsky and B. Zielinska, Phys. Rev. Lett. **62**, 1572 (1989).

77. S. Leibler and A. Maggs, Phys. Rev. Lett. **63**, 406 (1989).

78. P. G. de Gennes, J. Chem. Phys. **48**, 2257 (1968).

79. J. Villain and P. Bak, J. de Physique **42**, 657 (1981).

80. M. E. Fisher, J. Stat. Phys. **34**, 667 (1984).

81. R. Lipowsky and S. Leibler, Phys. Rev. Lett. **56**, 2541 (1986).

82. R. Lipowsky and M. Fisher, Phys. Rev. B **36**, 2126 (1987).

83. D. Sornette, J. Phys. C **20**, 4695 (1987).

84. N. Ostrowsky and D. Sornette, Chemica Scripta **25**, 108 (1985).

85. K. G. Wilson, Sci. Am. **241** (August issue), 158 (1979).

86. J. Bricmont, A. El. Mellouke and J. Fröehlich, J. Stat. Phys. **42**, 743 (1986).

87. E. Brézin, B. I. Halperin and S. Leibler, Phys. Rev. Lett. **50**, 1387 (1983); E. Brézin, B. I. Halperin and S. Leibler, J. de Physique **44**, 775 (1983).

88. D. S. Fisher and D. A. Huse, Phys. Rev. B **32**, 247 (1985).

89. S. Leibler and R. Lipowsky, Phys. Rev. B **35**, 7004 (1987).

90. *Physics of Amphiphilic Layers*, L. Meunier, D. Langevin and N. Boccara, eds. (Springer-Heidelberg, 1987).

91. R. Lipowsky, Europhys. Lett. **7**, 255 (1989).

92. F. David, Europhys. Lett. **6**, 603 (1988).

93. R. F. Kayser, Phys. Rev. A **33**, 1948 (1986).

94. Y. Kantor and D. R. Nelson, Phys. Rev. Lett. **58**, 2774 (1987); Y. Kantor and D. R. Nelson, Phys. Rev. A **36**, 4020 (1987).

95. J. Aronovitz and T. C. Lubensky, Phys. Rev. Lett. **60**, 2634 (1988).

96. M. Paczusky, M. Kardar and D. R. Nelson, Phys. Rev. Lett. **60**, 2638 (1988).

97. Y. Kantor, M. Kardar and D. R. Nelson, Phys. Rev. A **35**, 3056 (1987).

98. M. Kardar and D. R. Nelson, Phys. Rev. Lett. **58**, 1289 (1987).

99. J. Aronovitz, L. Golubovic and T. C. Lubensky, J. de Physique **50**, 609 (1989).

100. J. Toner, Phys. Rev. Lett. **62**, 905 (1989).

101. D. Roux and C. Coulon, J. de Physique **47**, 1257 (1986).

102. E. A. Evans and V. A. Parsegian, Proc. Natl. Acad. Sci. **83**, 7132 (1986).

103. S. Chandrasekar, Rev. Mod. Phys. **15**, 1 (1943).

6

Lyotropic Lamellar L_α Phases

Didier Roux[1]
Cyrus R. Safinya[2]
Frederic Nallet[3]

6.1 Introduction

As previous chapters have described, surfactants in solution lead to a large number of fascinating structures, among which are the lyotropic liquid crystal phases [1]. Liquid crystals exhibit long range order: orientational (nematics, see Chap. 3) and positional (lamellar, hexagonal or cubic—see Secs. 1.4.3–4 and Chap. 4). The lamellar phases consist of stacks of surfactant bilayers separated by solvent, usually water. In this chapter we are concerned only with the lamellar L_α phase. Figure 6.1 shows a schematic representation of such a structure. There is quasi-long-range positional solid-like order along the direction perpendicular to the layers (which we label as the z-direction): the surfactant density is modulated along the z-direction with a well-defined period, the repeat distance d. In the two other, in-plane, directions (x and y) the system is liquid-like and the solvent and surfactant molecules are free to move in this plane. In general, at lower temperatures the L_α phase may undergo a phase transition to the lamellar $L_{\beta'}$ phases where the surfactant molecules exhibit various degrees of in-plane positional and orientational order [2]. The lamellar L_α phase has the symmetry of the smectic A phase encountered in thermotropic liquid crystals [3]; there are nevertheless important reasons for specifically studying such lyotropic smectics. First, their most striking property is that the repeat distance d can be varied continuously, upon addition of an appropriate solvent. In favorable cases, this swelling goes from molecular sizes ($d \approx 20$ Å) to extremely large values (100-1000 Å) [4,5]: this opens the "experimental window," for studying very "soft" smectics with vanishing bulk

---

[1]Centre de recherche Paul-Pascal, CNRS, av. A. Schweitzer, F-33600 Pessac, France

[2]Materials and Physics Departments, University of California, Santa Barbara, California 93106, USA

[3]Same as Footnote 1

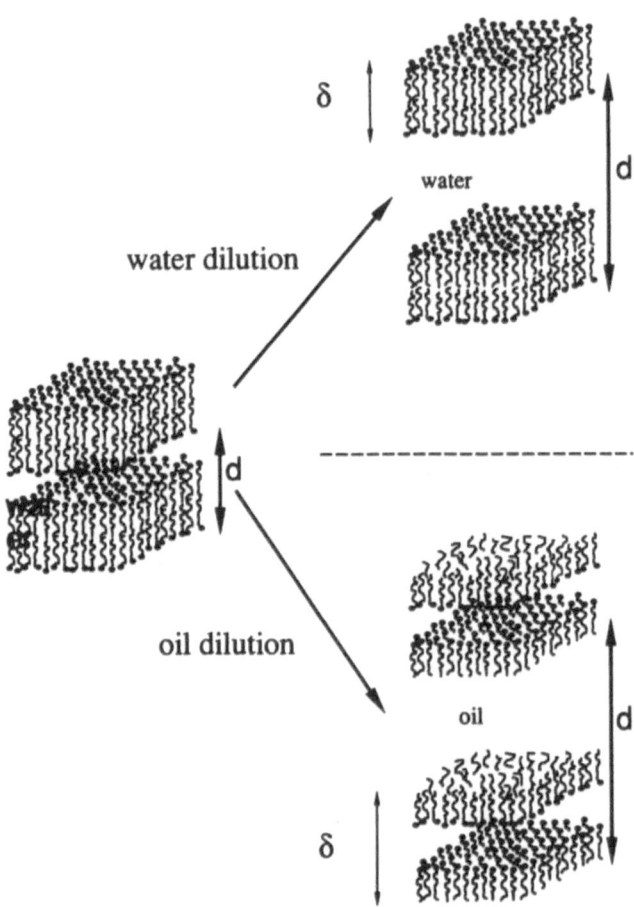

FIGURE 6.1. Schematic drawing of the membrane structure of lyotropic lamellar phases: the lamellar phase can be swollen either with water (hydrophilic solvent) or oil (hydrophobic solvent), leading to direct or inverted bilayers. The membrane thickness is δ; d is the smectic repeating distance.

elasticities as d increases over two orders of magnitude. In a few extreme cases, the smectic period reaches values of the order of optical wavelengths, so Bragg peaks are observed with light scattering [5-7], analogous to the irredescence of colloidal crystals [8]. Second, the L_α phase which consists of multilayer fluid surfaces (although confined between neighboring walls) provide experimental models for studies of the statistical mechanics of fluctuating random surfaces, an area of active current research [9] (see Chaps. 5 and 9). Third, the relevance of lamellar phases to the understanding of cell membranes, which are composed mainly of lipids, a biological surfactant [10], was recognized early on [11].

For a long time, the most investigated surfactant/solvent systems have been "simple" binary surfactant/water systems. The surfactant bilayer is then very often a "rigid" object: its well-defined shape is not sensitive to thermal fluctuations on a length scale comparable to the lamellar spacing. The ability to swell the smectic structure with water was realized early on [1] and was associated with strong electrostatic repulsions. The more recent studies of microemulsions—surfactant/water/oil systems, with specially designed surfactants or, more commonly, with mixtures of surfactants (see Chaps. 8 and 9)—highlighted the relevance of *flexible*, thermally fluctuating, surfactant sheets [12]. In such systems thermal fluctuations have to be large, since they are, in fact, responsible for the very existence of the microemulsion structure. Easily noticed is the presence of a large lamellar domain in the phase diagram of microemulsion-forming systems. This lamellar phase can be swollen either with oil-like solvents [5,13], or with water [14-16], resulting in "direct" or "inverted" bilayers as sketched in Fig. 6.1. The large swelling exhibited by such systems stems from long range repulsive interactions between bilayers, which have a non-electrostatic origin (swelling observed with non-ionic surfactants, apolar solvents or high ionic strength water). The interaction originates in the high flexibility of the surfactant sheets, through a mechanism of entropy reduction proposed by Helfrich [17] and explained in detail in Chap. 5, the *undulation interaction*.

The aim of this chapter is to present experimental results demonstrating quantitatively the crucial role of undulation interactions for flexible membranes and, more generally, to review how dilute lamellar phases, through the physics of the smectic-A state, can be used to study membrane properties.

In the next section we recall general features of the phase diagrams of archetypal surfactant/solvent systems. The third section is devoted to the presentation of the theoretical tools describing, and relating, membranes and lyotropic smectics. The fourth and fifth sections review techniques and results for static X-ray and dynamic light scattering experiments on lamellar phases.

6.2   Phase Diagrams

6.2.1   BINARY SYSTEMS

Figure 6.2 represents the typical phase diagram of a binary system (sodium dodecanoate and water) [1]. Depending on concentration and temperature, three stable phases are found: an isotropic liquid phase (I; micellar phase), a hexagonal phase (H; infinite cylinders organized on a two-dimensional hexagonal lattice) and a lamellar phase (L). The lamellar phase exists only at high concentrations (> 40% of surfactant) corresponding to rather small repeat distances ($d \approx 50$ Å) [1]. In comparison, the two other binary phase diagrams shown in Fig. 6.3 look very different [6,18,19]. In these systems the lamellar phase exists over a wide range of concentrations, including very dilute regions. The corresponding repeat distance can be very large: indeed, it varies continuously from 40 Å to 3000 Å ($C_{12}E_5$ and water, Fig. 6.3b) [6,19] and from 50 Å to 200 Å (AOT and water, Fig. 6.3a) [18]. The existence of a dilute lamellar phase results from interlayer electrostatic repulsions in the AOT system (which is an ionic surfactant with the phase behaviour strongly dependent on the ionic strength of the solvent), and from the undulation interaction in the $C_{12}E_5$ system (which is a non-ionic surfactant,

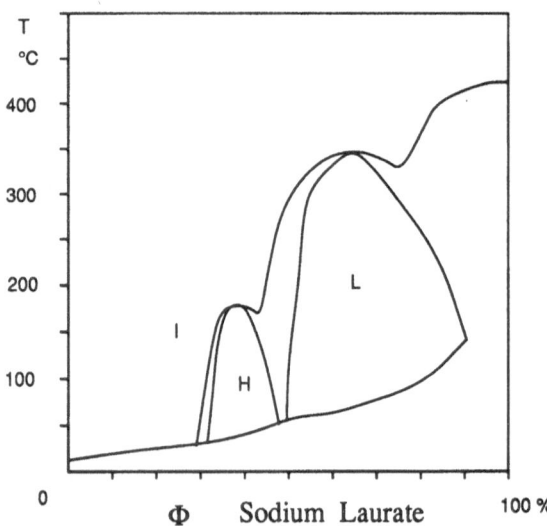

FIGURE 6.2. Temperature-concentration (mass fraction) phase diagram of a typical binary system, sodium dodecanoate-water, from [1]. The three stable phases are: the isotropic liquid phase (micellar, I), the hexagonal phase (H) and the lamellar phase (L). Note that the lamellar phase is stable only at high surfactant concentrations (> 40%).

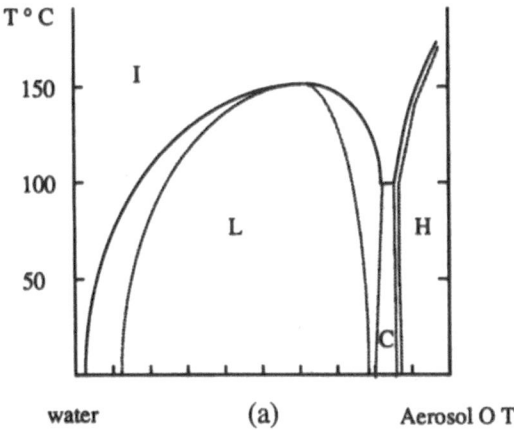

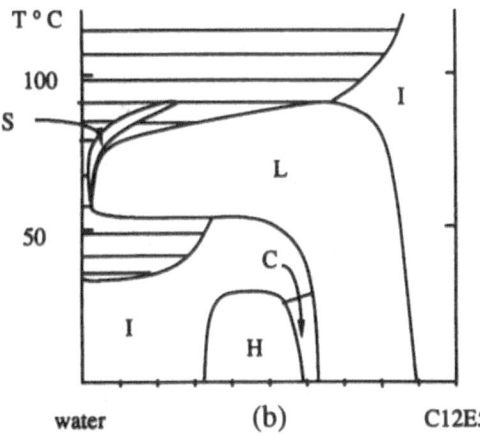

FIGURE 6.3. Phase diagrams of binary systems exhibiting dilute lamellar phases: (a) corresponds to the phase diagram of a charged surfactant (AOT) in water [18]; and (b) to the phase diagram of a nonionic surfactant ($C_{12}E_5$) in water [6, 19] ($C(S)$ stands for a *cubic (sponge)*) structure). High dilutions are made possible by the presence of long range electrostatic (a) or undulation (b) repulsive interactions.

with no qualitative changes of the phase diagram occuring upon adding salt [20]).

6.2.2   MULTICOMPONENT SYSTEMS

For historical reasons, related to a prior interest in microemulsions, commonly studied surfactant-solvent systems are more complex than the ideal binary mixture. They involve usually three or more components and the complete description of the phase diagram may be quite intricate. In what follows, we present results obtained with a quaternary system consisting of water, sodium dodecyl sulphate (SDS), dodecane and pentanol. The phase diagram at constant temperature and pressure has been mapped out [4d,21]; here we only consider the relevant cuts. The phase diagram of the *ternary* water-pentanol-SDS system is shown in Fig. 6.4a exhibiting four phases: isotropic liquid (I), hexagonal (H), rectangular (R) and lamellar (L). The lamellar phase is stable for surfactant concentrations as small as 0.10, which corresponds to a maximum repeat distance of about 100 Å. The SDS molecules being ionized in water (sulphate group), we expect electrostatic interactions to be dominant. Indeed, adding sodium chloride to the system strongly affects the stability of the lamellar phase, though not in the expected way. Fig. 6.4b is the pseudo-ternary phase diagram of water-NaCl (0.5 molar) pentanol-SDS [22]. The lamellar phase is now stable at even lower surfactant concentrations (about 0.01) and the repeat distance reaches 1000 Å. The reason for this striking behaviour is that the mixture of pentanol and SDS leads to very flexible films: there are strong undulation interactions once electrostatic ones are screened out.

The same basic ternary system can also be swollen with a hydrophobic solvent consisting of a mixture of dodecane and pentanol [4d,21]. Fig. 6.5 shows a cut of the quaternary phase diagram at fixed water over surfactant ratio of 1.55 by weight. The maximum swelling of the lamellar phase is at a SDS concentration of 0.05, with a repeat distance about 400 Å.

Another interesting case where dilute lamellar phases may be found is with neutral lipids such as dimyristoyl phosphatidylcholine (DMPC). These compounds in aqueous solution are known to exhibit a lamellar L_α phase at high enough temperatures (above 22°C for DMPC) [2]. This phase can incorporate up to 40% of water and not more. However, with the addition of a suitable cosurfactant such as pentanol the swelling extends to 90% of water (repeat distance above 200 Å [23], cf. phase diagram Fig. 6.6). This is attributed to a significant increase in the flexibility of the membrane, owing to the presence of cosurfactant molecules, with the appearance of undulation repulsions in a system where, without pentanol, the van der Waals attraction was limiting the dilution.

To summarize, one may state that dilute lamellar phases are found either when electrostatic interactions are strong or when membranes are flexible enough for undulation interactions to overcome van der Waals attractions.

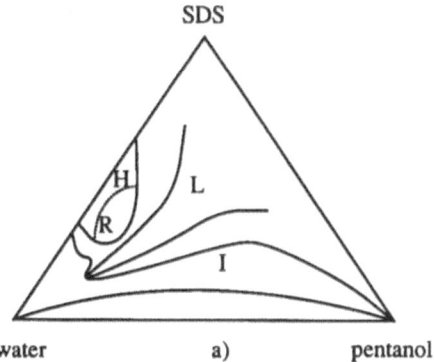

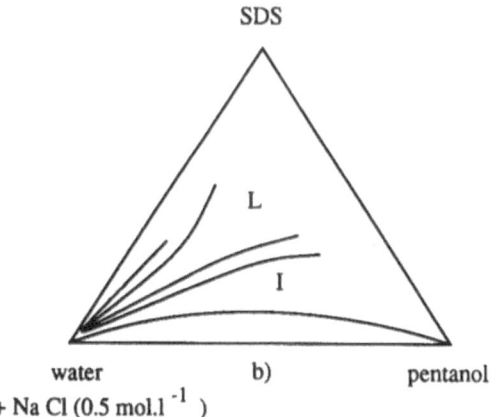

FIGURE 6.4. Ternary and quaternary systems exhibiting a dilute lamellar phase L with a water dilution (see Fig. 6.1). In addition to the previously defined phases one finds a rectangular (R) phase. The water-pentanol-SDS lamellar phase (a) is stabilized by electrostatic interactions when the water-NaCl-pentanol-SDS system (b) is stabilized by undulation forces.

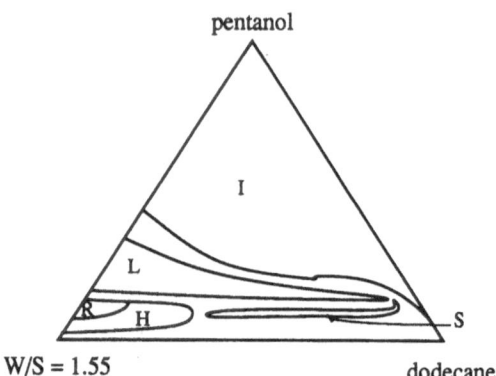

FIGURE 6.5. Cut of the quaternary phase diagram water-pentanol-SDS-dode-cane. The water over surfactant ratio is kept fixed at 1.55 (mass fraction). The lamellar phase L is swollen with a hydrophobic solvent (a mixture of dodecane and pentanol).

The microemulsion phase is then found in the vicinity of dilute lamellar L_α phases.

6.3   Membranes and Smectic Properties

In this section, we first give a discussion of membrane elasticity and membrane-membrane interactions; we then recall the elastic and hydrody-namic theories of the two-component smectic A phase, the generic model of lyotropic lamellar phases; and we allude to the links between macro-scopic, experimentally accessible smectic and microscopic membrane prop-erties.

6.3.1   MEMBRANE ELASTICITY AND INTERACTIONS

6.3.1.1   Membrane flexibility

We consider an isolated surfactant bilayer as a non-breakable, non inter-secting surface with its two sides equivalent. The packing of the surfactant is that of a two dimensional , nearly incompressible liquid, which excludes total area fluctuations. Following Helfrich [24], we therefore characterize the membrane through its curvature energy, with a bending elasticity constant

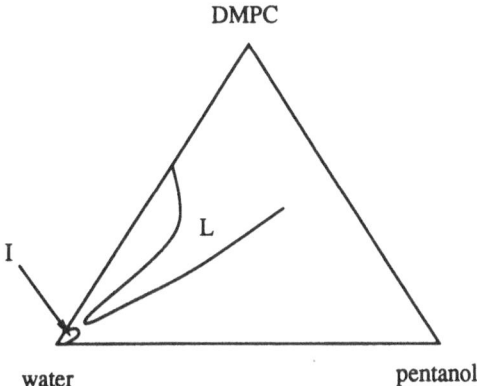

FIGURE 6.6. Ternary phase diagram of an uncharged lipid (DMPC) with pentanol and water. The dilute lamellar phase is stabilized by undulation interactions which are magnified by the pentanol-induced increase of the membrane flexibility.

k (see also Chaps. 1, 2, 5 and 7)

$$E_{curv} = \int dS \frac{1}{2} k H^2 \qquad (6.1)$$

where H is the mean curvature of the surface.

Flexible membranes have a bending energy constant k of the order of the thermal energy $k_B T$ and thus exhibit a high configurational entropic state, while rigid membranes with $k \gg k_B T$ remain at their minimum curvature energy state, the flat plane with $H = 0$ everywhere.

6.3.1.2 Interactions between membranes

We give in the following various estimates of the free energy per unit area for two membranes of thickness δ at a distance r, $V(r)$. We present first the classical contributions to the free energy, classical because they have been extensively studied in the last 20 years and can be considered as well established. We present then the undulation contribution, more recently proposed by Helfrich [17].

When undulations are negligible (we shall come back to this point) the free energy of interaction between flat layers is given by a sum of three terms:

$$V = V_{vdW} + V_{elec} + V_{hyd} \qquad (6.2)$$

The first term, $VvdW$, describes the van der Waals interaction and, neglecting retardation effects, is given by the surface integration of the $1/r^6$

potential. This leads to the following expression [25]

$$V_{vdW} = -\frac{A}{12\pi}\left[\frac{1}{r^2} + \frac{1}{(r+2\delta)^2} - \frac{2}{(r+\delta)^2}\right] \tag{6.3}$$

This interaction is attractive and varies as $1/r^2$ for r smaller than δ and as $1/r^4$ for $r \gg \delta$ (A is the Hamaker constant and is typically of order k_BT).

The second term in Eq. (6.2), V_{elec}, comes from electrostatic repulsions and is relevant only for charged membranes in polar solvents. This interaction is calculated from the one-dimensional Poisson-Boltzmann equation [26]. Two simple limiting cases are worth considering: first, the weak screening regime where there are no other ions in water apart from the counter-ions of the surfactant; and second, the strongly screened regime where salt has been added in such amounts that the Debye length is smaller than the distance between membranes. In the weak screening regime, the exact solution is known [26]; its expansion at large separations reads [16]

$$V_{elec} = \frac{\pi k_B T}{4 L_B r}\left[1 - \frac{a}{L_B r} + \left(\frac{a}{L_B r}\right)^2 + \ldots\right] \tag{6.4}$$

In Eq. (6.4), L_B is the Bjerrum length of the solvent ($L_B \approx 7\,\text{Å}$ at room temperature for water) and a the area per charge. One should notice that the dominant term in Eq. (6.4), varying as $1/r$, is long ranged; that is, there is no *efficient* screening of the interaction by the counter-ions; additionally, the leading term of the interaction does not depend on the area per charge a (this corresponds to the so-called ionic condensation [27]).

In the strongly screened regime where large amounts of salt have been added to the solvent, the interaction exhibits the usual exponential decay [25]

$$V_{elec} = E_0 e^{-r/l_D} \tag{6.5}$$

The screening length l_D depends on the co-ion concentration, through the usual Debye formula.

The last term in Eq. (6.2) has been discovered more recently. It describes the very short range steric repulsion that results when the polar head regions of two layers approach each other [28]. The precise origin of this *hydration force* is unclear. It may be represented empirically by

$$V_{hyd} = F_0 e^{-r/l_h} \tag{6.6}$$

where l_h is a molecular length (typically 2-3 Å); owing to its exponential decay, this interaction is important only for lengths smaller than 10 Å.

As pointed out by Helfrich [17], when the membranes of a multilayer system are flexible, thermal fluctuations have to be included in the free energy. This problem has been discussed in the previous chapter; here, we only recall the results. A membrane in such a multilayer system is confined between its neighbors and therefore loses configurational entropy when its distance away from a confining neighbor is smaller than the persistence length of the membrane ξ_K [12] (ξ_K is the distance over which the normals of the membrane are correlated). This loss in configurational entropy leads to an effective repulsive interaction, somewhat similar to the pressure arising from the confinement in a finite volume of an ideal gas [29], or from a constrained polymer [30]. An estimate of this undulation interaction using the Landau-de Gennes [3] elastic energy for a smectic leads to a contribution to the free energy V_{und} given by [17]

$$V_{und} = \frac{3\pi^2 (k_B T)^2}{128 k} \frac{1}{r^2} \qquad (6.7)$$

This is a long range (power law decay) repulsive interaction, inversely proportional to the bending constant k, which can overcome the van der Waals attraction at large distances r.

6.3.2   ELASTICITY AND HYDRODYNAMICS OF THE TWO-COMPONENT SMECTIC A

The lamellar phases we are interested in are intrinsically multi-component systems. Even if for many practical systems there are more than two components, one may hope that only two effective components, namely the membrane and the solvent, are required for a consistent phenomenological description of the observed experimental properties. This amounts to assuming that components like alcohol or salt have no other effects than modifying the phenomenological constants (bending elasticity k, viscosity $\eta \ldots$) introduced to describe the system. We shall show in particular that in describing inherently dynamical quantities, a further component reduction, with only the membrane left, is too stringent.

6.3.2.1   Elasticity

The elastic free energy density f of a two-component, isothermal and incompressible smectic A, up to second order in layer displacement u and concentration fluctuations δc is (axis z along the optical axis) [3,31]

$$f = \frac{1}{2} B \left(\partial_z u \right)^2 + \frac{1}{2} K \left(\nabla_\perp^2 u \right)^2 + \frac{1}{2\chi} \delta c^2 + C \delta c \partial_z u \qquad (6.8)$$

with B defined as the layer compression modulus (at constant concentration), K as the splay modulus, χ as the osmotic compressibility (at constant layer spacing) and C as the coupling constant between layer displacement and concentration fluctuations (see Fig. 6.7). Of particular significance is the combination $\bar{B} = B - C^2 \chi$, i.e. the layer compressibility modulus at *constant chemical potential*. Indeed, it is related to the interactions between membranes (anticipating 6.3.3: $\bar{B} = d(\partial^2 V / \partial d^2)$).

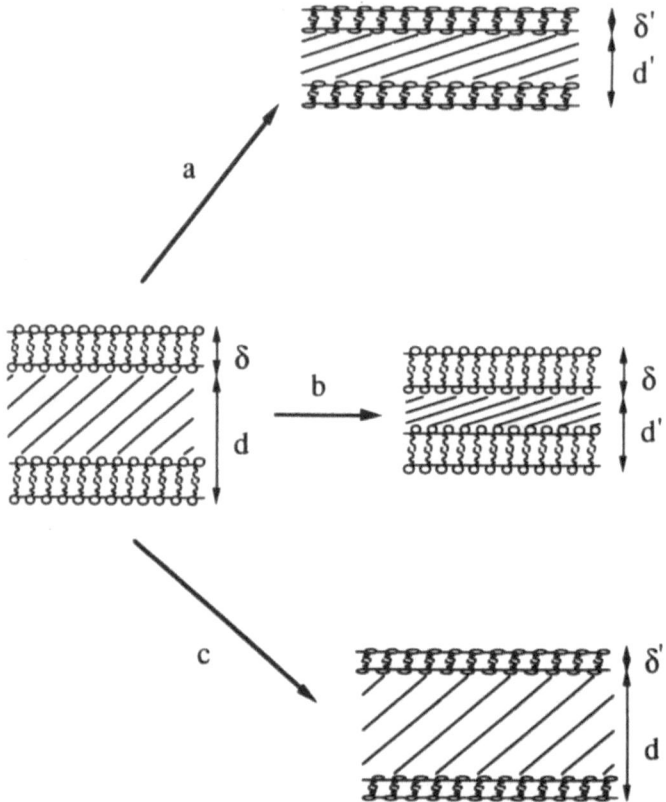

FIGURE 6.7. Schematic drawing of the elastic deformations corresponding to the elastic moduli B (a), $\bar{B}$ (b) and χ (c). B is mainly related to the membrane elasticity (as for χ), while $\bar{B}$ is related to membrane-membrane interactions.

From Eq. (6.8) the long wavelength behavior of the three basic static correlation functions can be obtained [32,33]

$$S_{uu}(\mathbf{q}) = \langle u(\mathbf{q})u(-\mathbf{q})\rangle = \frac{k_B T}{\bar{B}q_z^2 + Kq_\perp^4} \tag{6.9a}$$

$$S_{cc}(\mathbf{q}) = \langle \delta c(\mathbf{q})\delta c(-\mathbf{q})\rangle = \frac{k_B T\chi\left(\bar{B}q_z^2 + Kq_\perp^4\right)}{\bar{B}q_z^2 + Kq_\perp^4} \tag{6.9b}$$

$$S_{uc}(\mathbf{q}) = \langle u(\mathbf{q})\delta c(-\mathbf{q})\rangle = i\frac{k_B TC\chi q_z}{\bar{B}q_z^2 + Kq_\perp^4} \tag{6.9c}$$

These correlation functions may be measured in scattering experiments. We consider first the case of light scattering, where the signal comes from *dielectric tensor* fluctuations. In a uniaxial birefringent medium, mean dielectric constant $\bar{\epsilon}$ and dielectric anisotropy $\Delta\epsilon$ fluctuations have to be considered. For lamellar phases, the first originate in concentration fluctuations, the second in layer displacement ones. The intensity I scattered at a wave vector $\mathbf{q}$, for incoming and outgoing polarizations $\mathbf{i}$ and $\mathbf{f}$, is proportional to [32]

$$\begin{aligned}
I = {} & \left(\frac{\partial\epsilon}{\partial c}\right)^2 (\mathbf{f}.\mathbf{i})^2 S_{cc}(\mathbf{q}) + \Delta\epsilon^2 \left(f_z \mathbf{i}.\mathbf{q}_\perp + i_z \mathbf{f}.\mathbf{q}_\perp\right)^2 S_{uu}(\mathbf{q}) \\
& + 2\frac{\partial\bar{\epsilon}}{\partial c}\Delta\epsilon\mathbf{f}.\mathbf{i}\left(f_z \mathbf{i}.\mathbf{q}_\perp + i_z \mathbf{f}.\mathbf{q}_\perp\right) Im\left(S_{uc}(\mathbf{q})\right)
\end{aligned} \tag{6.10}$$

If we turn now to the case of X-ray, or neutron scattering, one should remember that the signal now comes from the fluctuations in *mass density*. The scattered intensity is then given by the spatial Fourier transform of a correlation function $g(\mathbf{x})$ defined by [33,34]

$$\begin{aligned}
g(\mathbf{x}) = {} & \left|\frac{b_S\bar{\rho}}{m_S}\right|^2 \langle\delta c(0)\delta(\mathbf{x})\rangle \\
& + \frac{1}{2}\left|\frac{b_S\Delta_S}{m_S}\right|\cos(q_0 z)\exp\left(-\frac{q_0^2}{2}\left\langle(u(\mathbf{x}) - u(0))^2\right\rangle\right)
\end{aligned} \tag{6.11}$$

where b_S (m_S) is the surfactant scattering length (molecular mass), $\bar{\rho}$ the mean mass density, and Δ_S the depth of the surfactant density modulation; q_0 is the Bragg peak position, i.e., $q_0 = 2\pi/d$. The Fourier transform of $g(\mathbf{x})$ contains two terms: a small angle part, related to concentration fluctuations and proportional in the small wave vector limit to the function $S_{cc}(\mathbf{q})$ (Eq. (6.9b)); and a Bragg part, related to the smectic ordering, proportional to the Fourier transform of $\cos(q_0 z)\exp(-q_0^2 < (u(\mathbf{x}) - u(0))^2 > /2)$ and in fact controlled in the vicinity of the Bragg peak position by the correlation function $S_{uu}(\mathbf{q})$ [34].

As has been shown by Lei *et al.* [40a], the Bragg part contributes at each reciprocal lattice vector including the origin and in fact (for dilute membranes) is normally larger than the first term for small $q'_z s$. The intensity is scattered *anisotropically*, both at small angles [33,35b] and around the Bragg peak positions. At small angles, this is seen in the following $q_z = 0$ and $q_\perp = 0$ limits of Eq. (6.9b)

$$\langle \delta c(q_z, 0)\, \delta c(-q_z, 0) \rangle = k_B T \frac{\chi B}{\bar{B}}$$
$$\langle \delta c(0, \mathbf{q}_\perp)\, \delta c(0, \mathbf{q}_\perp) \rangle = k_B T \chi \qquad (6.12)$$

This implies that small-angle scattering originating from the first term of Eq. (6.11) is more intense along q_z than it is along $q_\perp$ (recall that, by definition, $\bar{B}$ is always smaller than B). *The Bragg part is similarly anisotropic around the origin.*

In the vicinity of the (first order) Bragg peak position, the signal diverges anisotropically [34]

$$I_B(q_z, \mathbf{q}_\perp = 0) \propto |q_z - q_0|^{-2+\eta}$$
$$I_B(q_z = q_0, \mathbf{q}_\perp) \propto |\mathbf{q}_\perp|^{-4+2\eta} \qquad (6.13)$$

with the exponent η defined by

$$\eta = \frac{q_0^2 k_B T}{8\pi \sqrt{K\bar{B}}} \qquad (6.14)$$

The divergence of the intensity at the Bragg peak position is "weaker" here as compared to the case of ordinary solids (i.e., 3D ordered phases). This stems from the so-called *Landau-Peierls effect* [36,37]: the absence of true long-range order in a 3D body whose density modulation is one or two-dimensional (besides the smectic, the same behavior is expected for the 2D solid [38]). In such systems, because of the strong role of long wavelength phonons, elastic displacement correlation functions diverge as the logarithm of the sample size. For instance one finds, for smectic A phases, by an appropriate Fourier transform of $S_{uu}(\mathbf{q})$ (Eq. (6.9a)), the following asymptotic behaviour [34, 39]

$$\left\langle (u(z, \mathbf{x}_\perp) - u(0, 0))^2 \right\rangle = \frac{k_B T}{4\pi \sqrt{K\bar{B}}} (2\gamma +$$
$$\ln \frac{\pi^2 \mathbf{x}_\perp^2}{l^2} + E_1 \left(\frac{\mathbf{x}_\perp^2}{4\lambda|z|} \right)) \qquad (6.15)$$

with γ the Euler's constant and E_1 the exponential integral (l is a molecular size). In Eq. (6.15), the parameter λ is de Gennes' smectic penetration length [3], defined by:

$$\lambda = \sqrt{K/\bar{B}} \qquad (6.16)$$

Owing to the Landau-Peierls effect, the elastic constants K and $\bar{B}$ can be deduced from a high resolution study of the shape of the scattered signal about the Bragg peak positions. This contrasts with the case of resolution-limited Bragg peaks of 3D solids where the thermodynamic information lies in the very weak thermal diffuse wings about the Bragg peak positions.

Before proceeding to the next section we wish to add a few remarks to our discussion of the two-component smectic A. Our present description is based on the phenomenological free energy density of Eq. (6.8). It is therefore accurate for long wavelength modes only and is not aimed at describing the small angle X-ray or neutron scattering beyond the anisotropic Ornstein-Zernike regime of Ref. [33], or the shapes of the peaks far from the Bragg peak positions. One may in fact calculate, for *all* wave vectors **q**, the static structure factor by means of a microscopic model based on layer displacement variables $u_n(\mathbf{x}_\perp)$ [40a].

6.3.2.2 Hydrodynamics

The hydrodynamics of the two-component smectic A was first worked out by Brochard and de Gennes [31]. In what follows, we recall the mostly equivalent results of Ref. 32. According to the general prescription by Martin, Parodi and Pershan[41], seven hydrodynamic modes are to be expected for a two-component smectic A. They are associated with the relaxation towards equilibrium of small, long wavelength, perturbations in the *conserved* or *continuously broken symmetry* variables describing the macroscopic system: total mass density ρ, momentum density **g**, energy density ϵ, layer displacement field u and composition (here: surfactant mass density) ρc. The seven modes are in principle coupled, but simplifying and innocuous approximations can be made [32]. For instance, in order to describe the low frequency part of the hydrodynamic spectrum, it is reasonable to assume that the system is incompressible and athermal (sound and heat diffusion, which account for three modes, have high frequencies: 2 GHz and 1 MHz, respectively, at a wave vector $q = 10^7 m^{-1}$). Since the momentum density component which is transverse with respect to both the wave vector and the optical axis is always uncoupled, one has finally to deal with three hydrodynamic equations. They couple the transverse momentum density component in the wave vector/optical axis plane, g_t, with the surfactant mass fraction fluctuation δc and with the layer displacement field u [31,32]

$$\partial_t g_t = -\frac{\eta}{\rho}q^2 g_t - i\,C\frac{q_x q_z}{q}\delta c + \left(Bq_z^2 + Kq_x^4\right)\frac{q_x}{q}u \qquad (6.17a)$$

$$\partial_t \delta c = -\frac{\alpha_\perp}{\rho^2\chi}q_x^2\delta c - i\frac{\alpha_\perp C}{\rho^2}q_x^2 q_z u \qquad (6.17b)$$

$$\partial_t u = -\frac{q_x}{\rho q}g_t \qquad (6.17c)$$

These equations are written in Fourier space, with the x-z plane containing both the optical axis (the z-axis) and the wave vector $\mathbf{q}$. The elastic constants appearing here are defined in Eq. (6.8). It has been assumed, for simplicity, that there is only one shear viscosity η and that diffusion of surfactant is permitted only along the membranes, with a diffusion coefficient $\alpha_\perp/\rho^2\chi$; permeation is neglected [42].

These equations can be solved exactly for arbitrary wave vector orientations; however, it is useful to consider first the mode structure in the limiting case $q_z = 0$. In this limit Eq. (6.17) reads

$$\partial_t g_t = -\frac{\eta}{\rho}q_x^2 g_t + K q_x^4 u$$

$$\partial_t \delta c = -\frac{\alpha_\perp}{\rho^2\chi}q_x^2 \delta c \qquad (6.18)$$

$$\partial_t u = -\frac{g_t}{\rho}$$

The concentration fluctuations are therefore decoupled from the layer displacement and transverse momentum ones. Consequently, we recover the one-component smectic A mode structure superimposed on the binary fluid one. The two coupled equations lead to a transverse shear wave and an undulation mode [43]. In the limit $K\rho/\eta^2 \ll 1$, their relaxation frequencies are given by

$$\omega_s = -\frac{i\eta}{\rho}q_x^2 \quad \text{(shear mode)} \qquad (6.19a)$$

$$\omega_u = -\frac{iK}{\eta}q_x^2 \quad \text{(undulation mode)} \qquad (6.19b)$$

The shear mode is a high-frequency mode (about 20 MHz at $q = 10^7 m^{-1}$) that corresponds to $g_t = \eta q_x^2 u$ and $\delta c = 0$: it does not couple to concentration fluctuations and only weakly to layer displacement ones. The undulation mode, with $g_t = (K\rho/\eta^2)\eta q_x^2 u$ and $\delta c = 0$, couples strongly to layer displacement fluctuations (Fig. 6.8a shows a schematic representation of this mode); its frequency is usually much lower (typically 10 kHz).

The last mode ($g_t = 0, u = 0$) does not exist in one-component smectics A. It corresponds to concentration fluctuations, that are schematically displayed as a modulation of the membrane thickness in Fig. 6.8b. It is reminiscent of the well-known peristaltic mode of soap films [44]. The dispersion relation of this *membrane peristaltic* mode is

$$\omega_p = -i\frac{\alpha_\perp}{\rho^2\chi}q_x^2 \quad \text{(membrane peristaltic)} \qquad (6.19c)$$

This high frequency mode (a few MHz at $q = 10^7 m^{-1}$) is difficult to observe [45].

For an *oblique* wave vector (i.e., $q_x q_z \neq 0$), the three variables g_t, δc and u are coupled and, in order to find the three mode frequencies, one

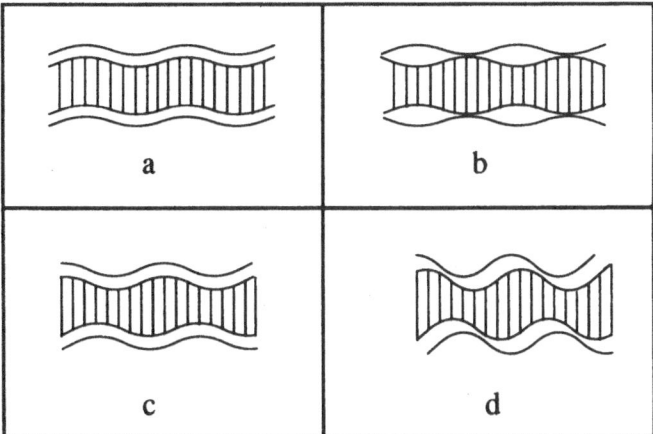

FIGURE 6.8. (a), (b) Schematic drawing of the hydrodynamic modes of a binary smectic A, corresponding to: (a) undulation mode; (b) the membrane peristaltic mode; (c) the baroclinic mode; and (d) the second sound (propagative) mode. Note that it is the second sound that degenerates into the membrane peristaltic mode and the baroclinic mode into the undulation mode when the wave vector q goes from oblique to in-plane orientations.

should solve a cubic equation. It is perhaps more illuminating to focus on the lowest frequency mode. From the previous discussion at $q_z = 0$, it is reasonable to guess that the slowest mode is a layer displacement mode, merging into the undulation mode when the wave vector lies in the layer plane.

The three-to-one reduction runs as follows [46]: with a change in the concentration variable $\delta c \rightarrow \widetilde{\delta c} = \delta c + iC\chi q_z u$, and replacing $\partial_t u$ for g_t in Eq. (6.17a), Eqs. (6.17a) through (6.17c) can be rewritten as:

$$\partial_t g_t - i\frac{\rho^2 C\chi}{\alpha_\perp}\partial_t \widetilde{\delta c} = \left(\frac{\eta q^3}{q_x} + \frac{\rho^2 C^2 \chi^2}{\alpha_\perp}\frac{q_z^2}{qq_x}\right)\partial_t u$$
$$+ \left(\bar{B}q_z^2 + Kq_x^4\right)\frac{q_x}{q}u \tag{6.20}$$

With the assumption that g_t and $\widetilde{\delta c}$ are high frequency modes for all wave vector orientations as compared to the u-mode, Eq. (6.20) may be rewritten with $\partial_t g_t \approx 0$, $\partial_t \widetilde{\delta c} \approx 0$, which leads to a relaxation frequency ω_b for the slow u-mode given by

$$\omega_b = -i\frac{\left(\bar{B}q_z^2 + Kq_x^4\right)q_x^2}{\eta q^4 + \dfrac{\rho^2 C^2 \chi^2}{\alpha_\perp}q_z^2} \tag{6.21}$$

This mode corresponds to a (low frequency) layer *compression* wave at *oblique* **q** and is called the *baroclinic* mode [32]. It should not be confused with the (propagative) second sound mode, a g_t-u mode for *one-component* smectic A phases [43], but usually a $g_t - \bar{\delta c}$ mode for two-component smectics [47]. The modes are schematically sketched in Fig. 6.8c (baroclinic) and Fig. 6.8d (second sound). For $q_z = 0$, the relaxation frequency of the undulation mode, Eq. (6.19b), is recovered from Eq. (6.21): baroclinic and undulation modes are continuously connected, as is intuitively clear from Fig. 6.8. Far from $q_z = 0$, and in the long wavelength limit where the q^4 terms are negligible compared to the q^2 ones, Eq. (6.21) simplifies

$$\omega_b = -i\mu\bar{B}q_x^2 \tag{6.22}$$

with $\mu = \alpha_\perp/\rho^2 C^2\chi^2$.

6.3.3 MICROSCOPIC MODELS FOR THE LYOTROPIC SMECTIC A PHASE

This paragraph is devoted to the description of the lyotropic smectic A phase in terms of rather simple microscopic models that allow the connection between the macroscopic smectic properties, mainly bulk elastic constants, measurable for instance through static or dynamic scattering experiments, with membrane properties: the membrane flexibility k and membrane/membrane interactions V (Sec. 6.3.1).

Let us describe the microscopic structure to be a stack of surfactant bilayers of constant thickness δ [48] and repeat distance d (Fig. 6.1). We first assume the membranes to be perfectly flat, and introduce later the corrections arising from departure from flatness, i.e., undulations. If the area per surfactant head is a, with n_s the number of surfactant molecules in a unit cell, the total membrane area S is $S = n_s a$, the volume of the unit cell V being $V = dS/2$.

The surfactant volume fraction ϕ is then simply

$$\phi = \frac{n_s v_s}{V} = \frac{2v_s}{ad} = \frac{\delta}{d} \tag{6.23}$$

with v_s the surfactant molecular volume (there are two molecules per area a in a bilayer).

It is important to notice that this description is restricted to *flat* membranes, which is valid when electrostatic interactions dominate [49]. When membrane undulations are strong, as they have to be when undulation interactions stabilize the lamellar phase, an excess area, with respect to the flat case, is stored in the surfactant bilayer [29]. The volume fraction/spacing relation now involves the *crumpling ratio* [50], defined as the ratio of the "base" area A_B of the membrane to its total area A (see Fig. 6.9b)

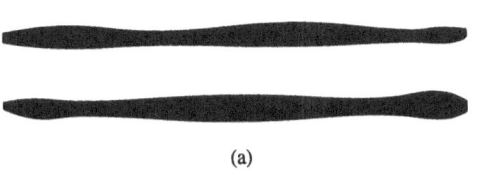

(a)

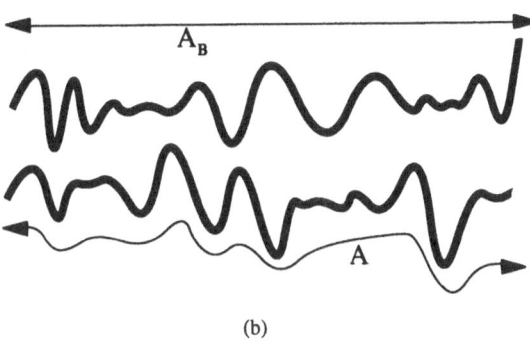

(b)

FIGURE 6.9. Shematic drawing of concentration fluctuations at constant layer spacing: (a) membrane thickness fluctuations (rigid membranes); (b) crumpling ratio fluctuations (flexible membranes).

$$\phi = \frac{\delta}{d}\frac{A}{A_B} \tag{6.24}$$

This is a logarithmic correction to Eq. (6.23) [6,50]

$$\frac{A}{A_B} \approx 1 + \frac{k_B T}{4\pi k}\ln\left(\frac{d}{l}\right) \tag{6.25}$$

where l is a molecular size and k is the membrane bending elastic constant defined in Eq. (6.1).

It is noteworthy that very different mechanisms are involved for surfactant concentration fluctuations, according to whether the membranes are flat or crumpled. At a constant repeat distance d, the only way to change ϕ is by changing δ, when the membranes are flat (Eq. (6.23); see also Fig. 6.9a). This amounts to changing the molecular packing of the surfactant in the bilayer, which obviously requires a lot of energy. On the other hand, it is enough to change the crumpling state of the membrane (Eq. (6.24); Fig. 6.9b), an inexpensive process if k is close to $k_B T$, when membranes are undulated [51]. This has far-reaching consequences on the values of the smectic elastic constants, as we now discuss.

At a given composition (volume fraction ϕ, or equivalently mass fraction c), the equilibrium values for d and δ (flat membranes) or for d and A/AB (crumpled ones) result from two competing effects: repulsive interactions, direct or steric, which favour large repeat distances d; and the selection of an optimum thickness, or crumpling ratio, which prefers a particular value of d for a given composition. A model for the free energy per unit volume of the stacked structure which features this competition is the following sum of two terms

$$f = \frac{V(d-\delta)}{d} + \frac{1}{2}U \cdot (X - X_0)^2 \qquad (6.26)$$

The interaction potential per unit area is $V(r)$ (see Sec. 6.3.1); the second term describes the harmonic approximation, with a characteristic energy per unit volume U, the departure of X (ratio of the membrane thickness to its preferred value, δ/δ_0, or crumpling ratio, A/A_B) from its optimum value X_0 (1 or close to one, up to logarithmic corrections [29,50,51]). For flat membranes the energy U is estimated as the product of a characteristic molecular energy ϵ, presumably of the order of k_BT, by the surfactant number density ϕ/v_s; for crumpled membranes an exact calculation has been performed [51]

$$U = \frac{12\pi^2 k}{\left(1 - \dfrac{4}{\pi^2}\right) d^3} \qquad (6.27)$$

The elastic constants B, χ and C are then obtained by an expansion, up to second order in spacing and composition fluctuations, of the free energy density, Eq. (6.26), around an equilibrium state $\{c, d_{eq}(c)\}$. For flat membranes with weakly screened electrostatic repulsions (Eq. (6.4)) we get [16,32]

$$\chi^{-1} \approx \frac{\bar{\rho}^2 v_S \epsilon d}{m_S^2 \delta}, B \approx \frac{\epsilon \delta}{v_S d}, C \approx \frac{\bar{\rho} \epsilon}{m_S}$$

$$\bar{B} = \frac{\pi k_B T d}{2 L_b (d - \delta)^3} \qquad (6.28)$$

with thus the following relations between elastic constants: $C \approx \chi^{-1} c; B \approx \chi^{-1} c^2$.

For crumpled membranes with Helfrich's undulation interactions (Eq. 6.7) the elastic constants are given by [51]

$$\chi^{-1} \approx \frac{12\pi^2 v_S^2 \bar{\rho}^2 k}{\left(1 - \dfrac{4}{\pi^2}\right) m_S^2 \delta^2 d}, B \approx \frac{12\pi^2 k}{\left(1 - \dfrac{4}{\pi^2}\right) d^3},$$

$$C \approx \frac{12\pi^2 v_S \bar{\rho} k}{\left(1 - \dfrac{4}{\pi^2}\right) m_S \delta d^2}, \bar{B} = \frac{9\pi^2 (k_B T)^2 d}{64 k (d - \delta)^4} \qquad (6.29)$$

The constants are much "softer" than for the previous case (for instance B now scales as $1/d^3$ instead of the $1/d$ law of Eq. (6.28)); nevertheless, the relations $C \approx \chi^{-1} c$; $B \approx \chi^{-1} c^2$ still hold.

For both kind of systems, the smectic splay constant K is related to the membrane bending modulus k by

$$K = k/d \tag{6.30}$$

However, with electrostatic interactions between membranes the possible dependence of the membrane bending modulus on dilution [49] is still an open question.

6.4   Static Scattering Studies of Lyotropic Lamellar Phases

A significant amount of information can be extracted from scattering experiments, in particular from high-resolution synchrotron X-ray scattering studies on dilute lamellar phases. Fig. 6.10 is a schematic representation of the scattered intensity as a function of the wave vector q_z, for an experiment performed on an oriented sample (z along the bilayer normal). The scattering profile exhibits a peak related to the quasi-long-range positional order of the membranes and some small angle scattering. *Structural* information can be obtained from the peak position q_0, i.e., the repeat distance d, with the relation $q_0 = 2\pi/d$. The shape of the small angle and thermal diffuse

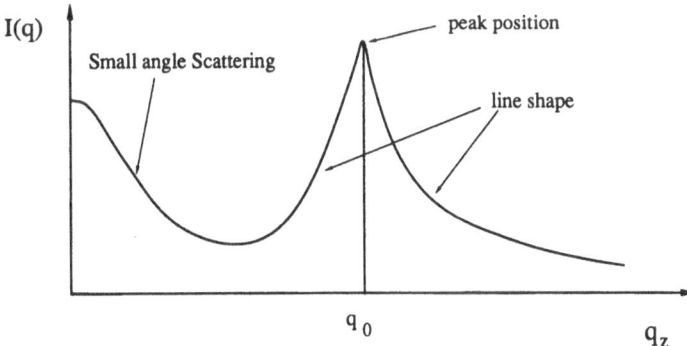

FIGURE 6.10. Schematic drawing of a typical scattering profile, as obtained using neutron or X-ray scattering. The analysis of such curves yields three pieces of information: the repeating distance, from the peak position; concentration fluctuations together with the Bragg component at the origin, from the small angle scattering; long wavelength fluctuations (Landau-Peierls effect), from the line shape.

scattering is related to concentration and layer displacement fluctuations which are controlled by the *thermodynamics* of the phase.

6.4.1 PEAK POSITION ANALYSIS

From the peak position, one first gets the repeat distance d. It is interesting to study it as a function of the dilution since d is related to the composition of the phase (surfactant volume fraction ϕ) through simple geometrical descriptions, Eq. 6.23 and Eq. 6.24. This offers an opportunity to test the physical assumptions on flexibility and interactions that led to these equations. Data on two binary systems, AOT-water [18] and $C_{12}E_5$-water [6,19] (phase diagrams Fig. 6.3) are displayed in Fig. 6.11. One expects flat, electrostatically stabilized membranes in the first case, and crumpled ones with undulation interactions for the latter system. In order to emphasize this difference, Fig. 6.11 is a plot of $d.\phi$ against $\ln(\phi)$: the plot demonstrates clearly the validity of Eq. (6.23) (constant thickness, flat membranes) for AOT and exhibits the logarithmic correction characteristic of crumpled membranes implied by Eq. (6.25) for $C_{12}E_5$. Besides, an estimate of the membrane bending modulus k can be extracted from the data: $k \approx 1.3k_BT$ [6].

6.4.2 SMALL ANGLE SCATTERING

Neutron and X-ray scattering experiments on dilute lamellar phases often exhibit a strong, dilution enhanced, small angle scattering signal [13, 16, 33, 35]. Such a behaviour is illustrated in Fig. 6.12, which displays data obtained on an *oriented* sample [33]. The small angle signal is very anisotropic, located mainly along the q_z direction [33,35]. The concentration fluctuation mechanism, originally proposed by Porte and coworkers [35b], and explicitly taken into account in Eq. (6.11), explains *qualitatively* the basic features of the experimental data: from Eq. 6.12, combined with Eq. (6.28) and Eq. (6.29), a low intensity, dilution-independent, small angle scattering along q_z is predicted for electrostatic systems, to be compared with the strong signal, proportional to $1/\phi$, when undulations are present. For instance, with characteristic energies equal to k_BT, and representative numbers $d = 300$Å, $\delta = 30$Å, one gets: $\left(\chi B/\bar{B}\right)_{elec} \approx 110^{-6}Pa^{-1}$, much smaller indeed than $\left(\chi B/\bar{B}\right)_{und} \approx 510^{-5}Pa^{-1}$, with an anisotropy $\left(B/\bar{B}\right)_{elec}$ about 1, to be compared with $\left(B/\bar{B}\right)_{und}$ about 150.

A satisfactory quantitative description of the small angle signal over the experimentally accessible wave vector range has recently been described by Lei *et al.* [40a]. The small angle scattering results from a combination of finite-size effects and the incoherent thermal fluctuations of the layers (i.e., concentration fluctuations).

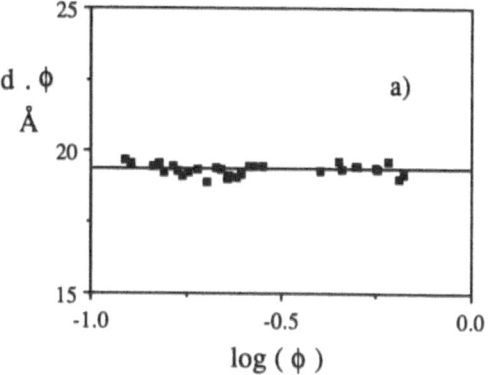

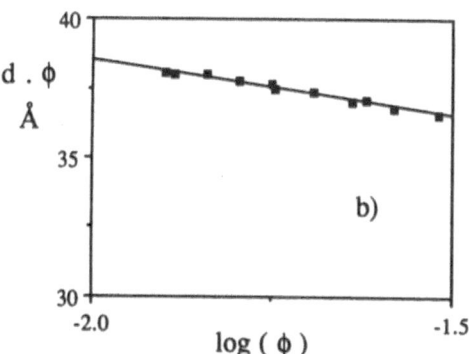

FIGURE 6.11. Dilution behaviour of the repeating distance. AOT system [18]; $C_{12}E_5$ system [6]. The $C_{12}E_5$ data exhibit the logarithmic correction corresponding to the excess area of undulated membranes.

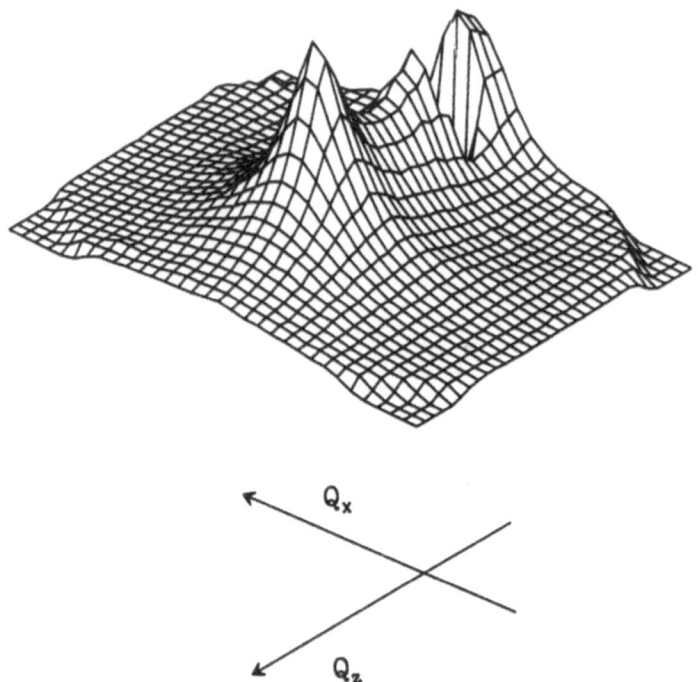

FIGURE 6.12. Three dimensional plot of the (neutron) scattering from an oriented dilute lamellar phase [33].

6.4.3 LINE SHAPE ANALYSIS

As was mentioned in Sec. 6.3.2.1, the Bragg peaks for a smectic A liquid crystal have enormously enhanced wing scattering compared to what is observed for a regular 3D solid. This Landau-Peierls effect [36,37] has been observed on thermotropic (one-component) smectic A phases close to the nematic/smecic A phase transition [39,52], and it allows an accurate determination of the elastic constant related exponent η, Eq. (6.14). The X-ray studies of lyotropic smectics [13–16, 23] have shown that the Landau-Peierls effect is significantly larger than for thermotropic systems, which can be associated with the softening of the bulk elasticities as a function of dilution. In practice the line shape analysis requires the use of a high-resolution X-ray set-up, to enable measurements in the immediate vicinity of the peak ($\Delta q = |q - q_0| \ll q_0$), in a region where the asymptotic power law divergences of Eq. (6.13) are expected to be valid. This is made possible by using as monochromator and analyzer double and triple bounce channel-cut crystals. The experiments that we describe here have been performed on the Exxon Beam line X-10A at the National Synchrotron Light Source at Brookhaven [13, 16, 23]. The configuration yields a very sharp in-plane Gaussian resolution function with a half width at half maximum

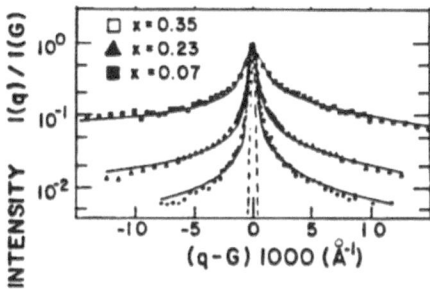

FIGURE 6.13. Typical line shape of a quasi-Bragg peak in a semilog plot along the oil dilution line of [13]. The dashed line corresponds to the experimental resolution function. Note that the tails of the peaks correspond to real thermal diffuse scattering since the intensity falls off much more slowly than the resolution function. (The wave vector at the peak position is G.)

(HWHM) equal to $8 \times 10^{-5} \, \text{Å}^{-1}$ and tails decreasing as $1/q^4$. The out-of-plane resolution is set by narrow slits that yielded a Gaussian resolution function with HWHM$= 10^{-3} \, \text{Å}^{-1}$.

The experiments have been performed on unoriented ("powder") samples, which forbids the use of the asymptotic intensities given by Eq. (6.13). Instead, data have been fitted using the isotropic averaging of the Fourier transform of the correlation function $g(\mathbf{x})$, Eq. (6.11) and Eq. (6.15) (disregarding now the concentration fluctuation part), taking account of the finite size of the ordered domains

$$I(q) \propto \int dz d^2\mathbf{x}_\perp g\left(z, \mathbf{x}_\perp\right) \exp\left(-\frac{\pi R^2}{L^2}\right) \frac{\sin(qR)}{qR} \qquad (6.31)$$

where L^3 is the average finite volume of the oriented domains and $R = \left(z^2 + x_\perp^2\right)^{1/2}$. Recent high resolution synchrotron X-ray scattering data from *oriented* samples are consistent with the powder averaged data in the vicinity of the Bragg peak positions [40b,40c].

The analysis consists of a least squares fit of Eq. (6.31) convoluted with the resolution function (represented as Gaussian functions) to the experimental profiles. In all the following figures, the fits are indicated as solid lines. In Fig. 6.13, the evolution of the line-shape along the dodecane dilution line of Fig. 6.5 is shown and compared with the resolution function (dashed line). The Δq range for the analysis is typically $10^{-2} \, \text{Å}^{-1}$; note that the tails are much higher than the resolution function when the HWHM of the signal is about the same order of magnitude, indicating that the line-shape is controlled by the scattering arising from the long wavelength fluctuations. Fig 6.14a and Fig. 6.14b show log-log data of samples

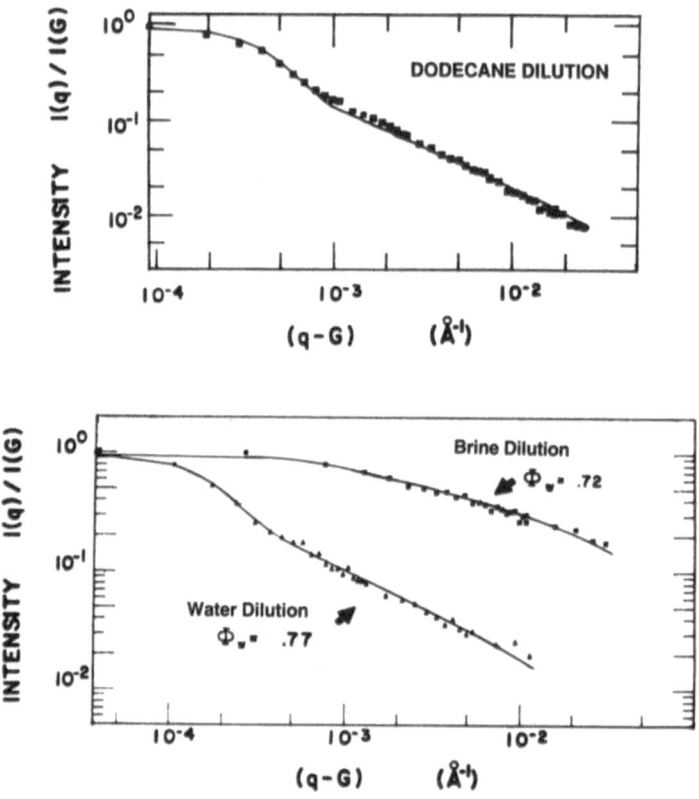

FIGURE 6.14. Typical profile of the first harmonic of a quasi-Bragg peak for mixtures along the oil dilution (top) and brine and water dilution (bottom) lines. The large q region is nearly linear corresponding to the power law behavior expected from the Landau-Peierls effect. The small q region corresponds to a round-off due to a finite size effect. The solid lines are least squares fits using the Caillé expression (Eq. (6.31)).

along the oil, brine and water dilution (on the high-q side of the peak). At small Δq the data are dominated by the finite size of the ordered domains in the powder sample and by the finite resolution, when at large Δq the profile is linear, indicating a power law decay of the scattered intensity, as expected from the isotropic averaging of Caillé's expressions [34a]

$$S(q) \propto |q - q_0|^{-p}, \text{for } |q - q_0| > \frac{2\pi}{L} \qquad (6.32)$$

with an exponent p close to $1 - \eta$ [13, 16, 23].

A more subtle aspect of the data is apparent on a closer inspection of Fig. 6.13: the profiles are slightly asymmetric, with positive $q - q_0$ yielding more intense signals than negative $q - q_0$. This asymmetry comes from the parameter λ (Eq. (6.16)), which together with η, enters the real space correlation function (cf Eq. (6.15). In practice, the powder averaged intensity depends rather weakly on λ only for $\lambda < 2\pi/q_0$; for larger values of λ, it turns out that the asymmetry is negligible [13,16]. Thus the accuracy of the determination of λ, especially for the most dilute samples, where the strong small angle scattering hides the asymmetry, is much less than for η.

An important self-consistency check that the scattering is resulting from the Landau-Peierls effect is to compare the shape of the first and second harmonics. This is possible for the most concentrated samples since two and sometime three peaks may be observed. The generalization of Eq. (6.11) to account for higher order harmonics leads to the replacement of q_0 by $m.q_0$ for an order-m harmonic. This is equivalent to replacing η by $m^2.\eta$. Thus, the exponent is expected to be respectively four (nine) times the value for the first harmonic in the analysis of the second (third) harmonic. Fig. 6.15a and Fig.6.15b show fitted data for the first and the second harmonic of the same sample for mixtures from the oil and water dilution lines. Fitted values of L (finite size) and λ are basically the same for the two sets of data, but η is 4 times larger for the second harmonic than for the first one.

To summarize, the line shape analysis in a high resolution X-ray scattering experiment makes it possible, through the Landau-Peierls effect, to measure η, proportional to $(K\bar{B})^{1/2}$, and at least estimate λ, proportional to $(K/\bar{B})^{1/2}$. We now review more specific experimental results.

6.4.3.1 Electrostatic interactions

We consider here the results on the lamellar phase of ionic surfactants swollen with pure water. Two systems, characterized by different flexibilities k, have been studied: the SDS-water-*pentanol* [16] system and the SDS-water-*hexanol* one [23]. Their phase diagrams are very similar (Fig. 6.4a, pentanol system), the lamellar phase existing at a slightly higher dilution in the latter case (maximum repeat distance $d = 130\,\text{Å}$ instead of $90\,\text{Å}$). Fig. 6.16 shows a series of data for the pentanol system, at various water

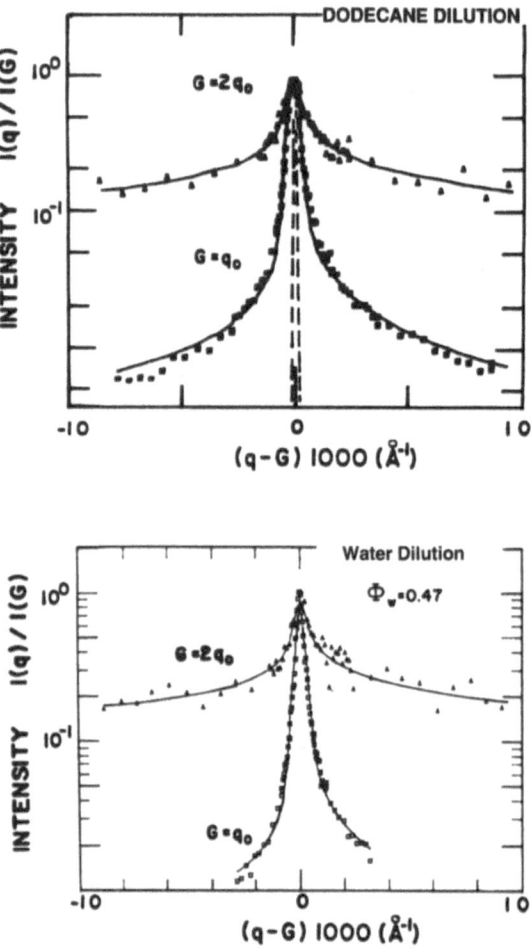

FIGURE 6.15. Comparison between the first and second harmonic peaks of the same sample: along the oil-dilution (top) and along the water dilution (bottom) lines. The independent fits for each peak in the oil-dilution mixture lead to: L_1(finite size) $= 10000$ Å, $\lambda_1 = 8.6$ Å, $\eta_1 = 0.14$, for the first harmonic and $L_2 = 9100$ Å, $\lambda_2 = 8.1$ Å, $\eta_2 = 0.57$ for the second harmonic. The results in the water-dilution mixture give: $L_1 = 12500$ Å, $\lambda_1 = 6 \pm 2$ Å, $\eta_1 = 0.16 \pm 0.01$ and $L_2 = 10400$ Å, $\eta_2 = 4\eta_1$, $\lambda_1 = \lambda_2$. The relation $\eta_2 \approx 4\eta_1$ obtains when L and λ remain approximatively equal as expected.

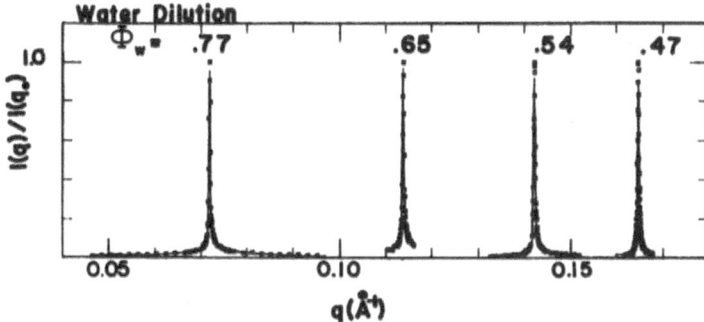

FIGURE 6.16. Evolution of the quasi-Bragg peak as a function of the water dilution for the water-SDS pentanol system. Note that the peaks do not exhibit strong tails even for the most dilute samples, indicating that η remains small.

concentrations. The shape of each peak is analyzed using Eq. (6.31) and values of η for both the pentanol and the hexanol systems as a function of the dilution are plotted in Fig. 6.17. The solid lines correspond to the predictions of the electrostatic model, using Eq. (6.28) and Eq. (6.30) of Sec. 6.3.3:

$$\eta_{elec} = \sqrt{\frac{\pi}{2} \frac{k_B T}{k} \frac{L_b}{d} (1 - \delta/d)^3} \qquad (6.33)$$

with k ranging from 2 to 0.7 $k_B T$ for the pentanol system and from 2 to 1 $k_B T$ for the hexanol one [16,23]. In both cases the values of η remain small all along the dilution line (below 0.35 for pentanol, below 0.15 for hexanol).

6.4.3.2 Undulation interactions

We show here results for three systems, where long range electrostatic interactions are unlikely, that: the *inverted* (Fig. 6.1) bilayer of an ionic surfactant is swollen with a non-polar solvent; the direct bilayer of an ionic surfactant is swollen with a *high ionic strength* water; and *non-ionic* surfactant is swollen with water. The corresponding systems are the oil-rich pentanol-water-SDS-dodecane system (Fig. 6.5); the brine rich pentanol-SDS-water-NaCl system (Fig. 6.4b); and the DMPC-water-pentanol system (Fig. 6.6). Fig. 6.18 shows a series of data for the first system at various oil concentrations [13]. In contrast with electrostatic systems, the tail-to-peak ratio is now much larger, a sign of larger values of η. As already discussed in Sec. 6.4.2, the complication of small angle scattering occurs for the most dilute samples. For unoriented samples this restricts the determination of η to samples with d not much larger than 200Å. In oriented samples η may be determined for spacings as large as 400Å [40a].

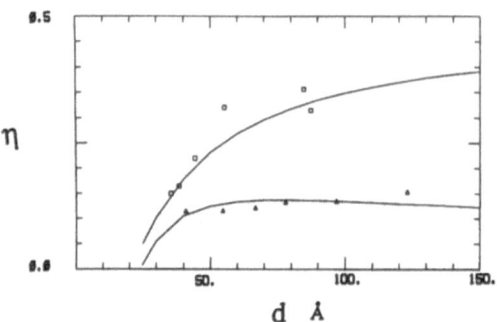

FIGURE 6.17. Evolution of η as a function of the dilution for systems dominated by electrostatic interactions: the squares correspond to the water-pentanol-SDS system while the triangles correspond to the water-hexanol-SDS system. The solid lines are predictions of the Poisson-Boltzmann solution for electrostatic interactions. The larger values of η for the pentanol system are due to a smaller value of $k\ \left(\eta \propto k^{-1/2}\right)$.

In the case of undulation interactions, the prediction of the microscopic model, Eq. (6.29) and Eq.(6.30), is the following particularly simple *universal* expression:

$$\eta_{und} = \frac{4}{3}\left[1 - \frac{\delta}{d}\right]^2 \qquad (6.34)$$

Fig. 6.19 is a plot of all the experimental values of η as a function of the reduced dilution variable $X = (1 - \delta/d)^2$. All the data (three different systems) lie on the same straight line, with the slope 4/3 expected from Eq. (6.34). Such a good quantitative agreement, the first and perhaps most convincing proof [13, 16, 23] of the relevance of Helfrich's mechanism and the universal nature of undulation interactions, might seem surprising: in Helfrich's calculation [17], as in other theoretical investigations [50,51], the interaction free energy V_{und}, Eq. (6.7), is computed up to a somewhat arbitrary multiplicative coefficient. The value 4/3 in Eq. (6.34) is thus not theoretically very firmly grounded, though experimentally less controversial.

The universal nature of η, independent of k (in contrast to electrostatic systems) and dependent only on geometrical parameters such as the bilayer thickness δ and the repeat distance d, demonstrates that high resolution

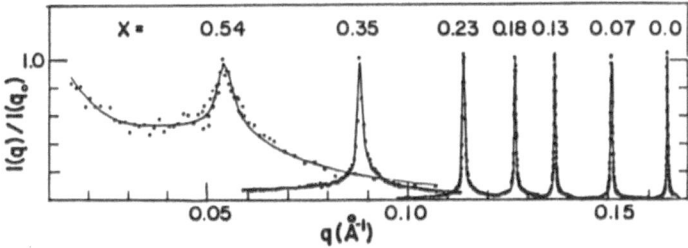

FIGURE 6.18. Evolution of the quasi-Bragg peak along the oil dilution of the water-SDS-pentanol dodecane system. Contrary to systems dominated by electrostatic interactions (see Fig. 6.16), the tail to peak ratio increases strongly with the repeating distance d, indicating a large variation of the coefficient η.

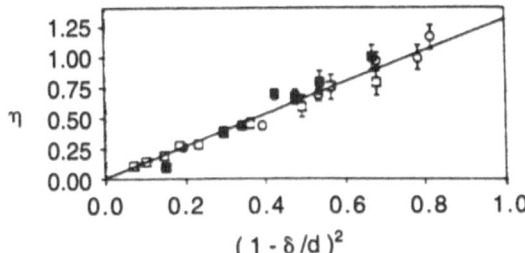

FIGURE 6.19. Exponent η as a function of $(1 - \delta/d)^2$ for three different systems. The solid squares correspond to the DMPC-pentanol-water system while the open circles and squares are SDS-pentanol system respectively diluted with brine (water + NaCl) or oil. The universality of the η behavior appears clearly when data are compared to the prediction for undulation forces (solid straight line).

synchrotron X-ray scattering techniques are ideally suited for the studies of undulation-stabilized systems.

6.4.3.3 Intermediate cases

The systems previously presented belong to clear-cut categories: either electrostatics is the relevant interaction or undulation forces dominate. The question arises regarding the competition between interactions (including van der Waals attraction). For such intermediate cases, while there are some theoretical papers on the competition between van der Waals and other interactions [53,54], no direct calculation is available for the elastic constants. On the other hand, the competition between electrostatic and undulation forces has been investigated [49,55], through the increase of the ionic strength in a system first dominated by electrostatic interactions. Experiments have been done in these intermediate regimes [15] showing behaviors more complex than those described by Eq. (6.33) and Eq. (6.34).

6.4.3.4   Role of the alcohol

A typical lipid-water system leads to a lamellar phase only with high concentrations of surfactant. For instance, the DMPC-water lamellar phase cannot be diluted with more than 40 % of water, with a maximum repeat distance of 45Å. The membrane flexibility k has been estimated for this system, analyzing the thermal fluctuations of giant vesicles [56]. There is still a considerable dispersion in quoted values for k, in the range 10–40 k_BT. Such high values imply that at distances of about 45Å the undulation repulsion is much weaker than the other relevant interactions [57]. As previously mentioned, adding a cosurfactant (pentanol) to the DMPC membrane has a strong effect on the stability range of the lamellar phase, cf. Fig. 6.6: a repeat distance $d > 200$Å can be reached [23]. This is because the undulation forces between pentanol-DMPC membranes dominate the other interactions and this suggests much lower values for k, of the order of k_BT, because of the pentanol. The role of the alcohol in the surfactant bilayer has been studied [23]. The membrane system consists of SDS and alcohol, swollen with water. The alcohol length is varied, from pentanol (five carbons) to dodecanol (twelve carbons). This is an electrostatic system, with η proportional to $k^{-1/2}$, Eq. (6.33). Changing the alcohol length has two effects: firstly it increases the membrane thickness δ; secondly it changes k. Fig. 6.20 shows the result of this series of experiments, k plot-

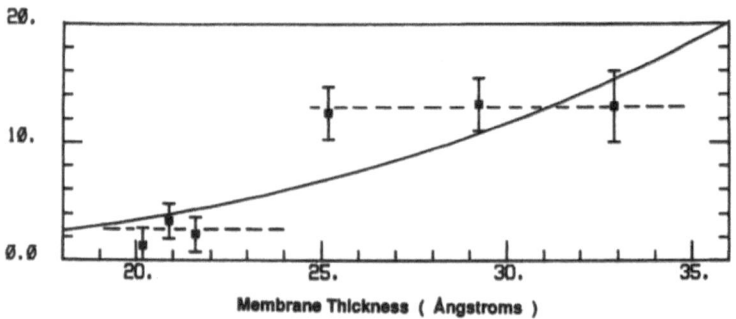

FIGURE 6.20. Evolution of k/k_B as a function of the membrane thickness for seven cosurfactants with varying chain lengths. Note the discontinuous behaviour for an alcohol chain length between 8 and 9 carbons (octanol-nonanol).

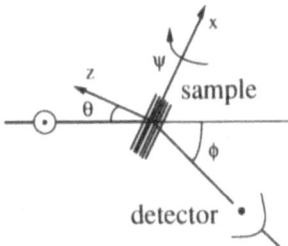

FIGURE 6.21. Schematic drawing of the geometry of the light scattering set-up.

ted against δ. The result is quite surprising since, instead of the smooth behavior expected from microscopic theories [58–61], $k \propto \delta^e$ with e ranging from 2–3, it shows a jump in the value of k for octanol (eight carbons). More work is obviously required in order to understand the mechanism responsible for such an effect.

6.5 Dynamic Light Scattering

The dynamic light scattering technique is also quite useful in the study of dilute lamellar phases. It allows the determination of the characteristic frequency of the lowest frequency hydrodynamic mode, namely the undulation/baroclinic mode discussed in Sec. 6.3.2.2. In such experiments, one has to work with *oriented* samples, because of the strong anisotropy of the dispersion relation of the undulation/baroclinic mode, Eq. (6.21); some care should also be taken in defining the light polarization states and the scattering geometry [32].

Fig. 6.21 is a sketch of the set-up used. The samples are oriented in sealed rectangular glass capillaries (100μm thick, 1 mm wide and 30 mm long). The position of the detector is defined by the scattering angle Φ. Two angles θ and ψ define the direction of the optical axis: the angle θ controls the in-plane rotation (around an axis perpendicular to the scattering plane) and ψ the out-of-plane one (around an axis parallel to the scattering plane). The incident light is always polarized perpendicularly to the scattering plane (polarization vector $\mathbf{i}$). The analyzer is either parallel or perpendicular to the scattering plane (polarization $\mathbf{f}$). With these notations, the orientation

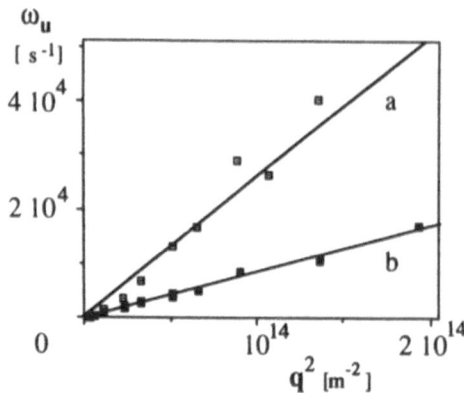

FIGURE 6.22. Undulation mode dispersion relation at different dilutions; $d = 113$ Å(a) and $d = 344$ Å(b).

for $q_z = 0$ corresponds to $\theta = \Phi/2$ and one has to choose $\psi \neq 0$ and polarizer and analyzer at a right angle in order to observe the undulation mode [32].

Some representative experimental relaxation frequencies are plotted in Fig. 6.22 and Fig. 6.23, at $q_z = 0$ as a function of q^2 (undulation mode), and in the limit $q_\perp \ll q_z$ as a function of $q_\perp^2$ (asymptotic baroclinic mode), respectively. Straight lines describe well the dispersion relation for these two cases, as expected from the hydrodynamic predictions, cf. Eq. (6.19b) and Eq. (6.22).

In order to interpret the values of the effective "diffusion coefficients" D_u and D_b, respectively K/η and $\mu\bar{B}$ (and their behaviors under dilution), we have to know something about the friction parameters μ and η and to use the d-dependences of K and $\bar{B}$ of Sec. 6.3.3. Following Brochard and de Gennes [31], we assume that η is the pure solvent viscosity η_S. This is probably realistic for very dilute samples even though the case of concentrated ones is less obvious . As regards μ, we first notice that it appears in both the relaxation frequency of the membrane peristaltic mode (Eq. (6.19c) rewritten as $\omega_p = \mu\bar{B}q_x^2$, with the help of Eqs. (6.28) and (6.29) of the microscopic models in Sec. 6.3.3) and the relaxation frequency of the baroclinic mode (Eq. (6.22)), and then again refer to Brochard and de Gennes [31] to get

$$\mu = \frac{\beta(d-\delta)^2}{12\eta_S} \tag{6.35}$$

with $\beta = (\rho_S d)/(\rho_S(d-\delta) + \rho_M\delta)$ and ρ_S (ρ_M) the mass density of the solvent (the membrane).

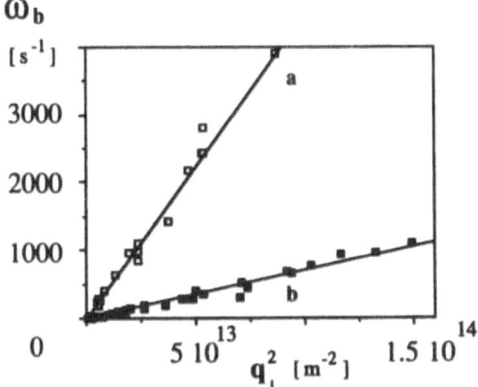

FIGURE 6.23. Baroclinic mode dispersion relation at different dilutions; $d = 61$ Å(a) and $d = 232$ Å(b).

With $K = k/d$ (Eq. (6.30)), we get the dilution dependence for D_u

$$D_u = \frac{k}{\eta_S} \frac{1}{d} \tag{6.36}$$

For undulation interactions, the layer compression modulus $\bar{B}$ is given by Eq. (6.29) and thus we have

$$D_b = \frac{3\pi^2 \beta}{256 \eta_S} \frac{(k_B T)^2}{k} \frac{d}{(d - \delta)^2} \tag{6.37}$$

while for the electrostatic interactions ($\bar{B}$ from Eq. (6.28))

$$D_b = \frac{\pi \beta}{24 \eta_S} \frac{k_B T}{L_B} \frac{d}{d - \delta} \tag{6.38}$$

6.5.1 UNDULATION MODE

The dilution behavior of the undulation mode relaxation frequency has been studied along a dilution line for the SDS-water-pentanol-dodecane system [32]. The results, i.e., $1/D_u$ as a function of d, are shown in Fig. 6.24. The least square fit of the data to Eq. (6.36) is shown as a solid line; a value $k = 0.8k_B T$ is obtained from this experiment (the solvent viscosity is taken as that of dodecane: $\eta_S = 1.35 \, 10^{-3} Pa$). Note that along the dilution range investigated the experimental splay modulus K decreases from about 4×10^{-13} N to 1×10^{-13} N, which is about two orders of magnitude smaller than values commonly encountered in thermotropic smectic A phases ($K \approx 10^{-11}$N). This is partly caused by the intrinsic large flexibility (low k) of the membrane and partly by the dependence on d ("colloidal effect").

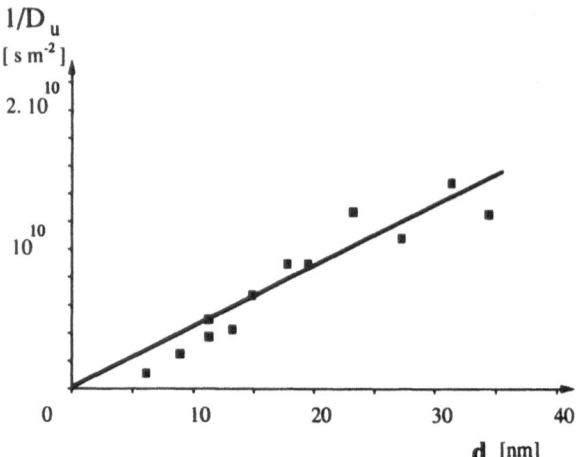

FIGURE 6.24. Inverse slope $1/D_u$ of the undulation mode dispersion relation as a function of the smectic repeating distance d. Filled squares: experimental data; solid line: prediction from Eq. (6.36) with $k = 0.8k_BT$.

6.5.2   BAROCLINIC MODE

The evolution of D_b for the SDS-pentanol membrane diluted with water [62] is shown in Fig. 6.25a. The effective diffusion constant does not vary much over the dilution range investigated. Its value compares within a factor 2 with that of the high dilution limit (with no adjustable parameters) of Eq. (6.38): $D_b = 7.810^{-10} m^2.s^{-1}$.

The evolution of D_b for the SDS-water-pentanol-dodecane system is shown in Fig. 6.25b [32,63]. The solid line is a fit to Eq. (6.37), with only one adjustable parameter, namely k (the membrane thickness δ is fixed to 20 Å; the solvent viscosity is taken as the pure dodecane viscosity). The result is $k = 1.6k_BT$ (to be compared with the value coming from the undulation mode analysis). The layer compression modulus $\bar{B}$ ranges from 5 10^4 Pa to 10^2 Pa: this is about two to five orders of magnitude less than for typical thermotropic smectics. It is mainly because of the large value of the repeat distance $d\left(\bar{B} \propto d^{-3}\right)$.

6.6   Conclusion

We hope that this chapter has demonstrated several important features related to the physics of dilute lamellar phases. First, as illustrated by the strong Landau-Peierls effect and by the exceptionally neat signal obtained in light scattering, these phases are interesting systems for studying smectic properties. Moreover, the wide dilution range over which they are stable

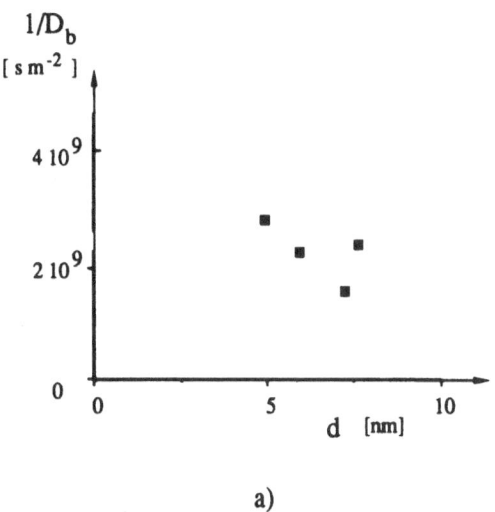

a)

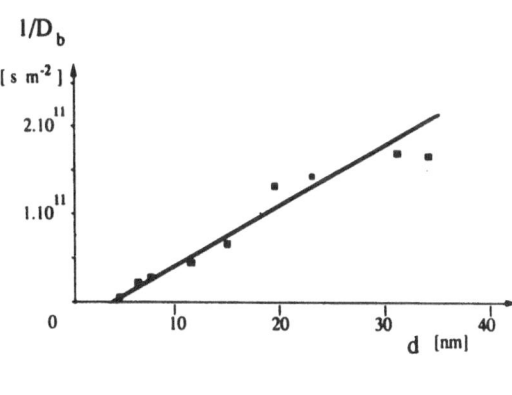

b)

FIGURE 6.25. Inverse slope $1/D_b$ of the baroclinic mode dispersion relation in the range $q_\perp \ll q_z$, as a function of the smectic repeating distance d. Electrostatic interactions (a) and undulation interactions (b).

allows accurate and convincing measurements of interaction forces. The main result is the crucial role of the membrane flexibility (characterized by the bending constant k) and its consequence for interactions between membranes, namely the existence of an undulation interaction.

The experimental techniques described are complementary. The high resolution X-ray technique can be used with unoriented samples, which is obviously of large practical interest. It allows the direct determination of an elastic-constant-related coefficient η. The variation of η illustrates remarkably the universal behavior of the undulation interaction: η is a function only of the ratio between the membrane thickness and the repeat distance and reaches a universal value (4/3) for large repeat distances. The technique allows also in principle the measurement of the smectic penetration length λ, which, combined with η, determines the two elastic constant K and $\bar{B}$. However, the value of λ is based on the analysis of a subtle aspect of the data, the asymmetry of the peak, and is therefore less accurately determined than η. Another limitation in powder samples comes from the overwhelming small angle scattering arising for distances d greater than about 200 Å, which is partly due to defect scattering resulting from the powder nature of the sample and partly to the scattering from concentration fluctuations. Very recent results show that synchrotron small-angle X-ray scattering studies of oriented samples will allow an accurate measurement of λ as well as η for d as large as 400Å [40a,40b].

The dynamic light scattering technique requires oriented samples but can be used for all of them, including the most dilute ones. In fact, this technique is even more appropriate for dilute samples since the static signal is then larger and the characteristic frequencies smaller. Measurements up to a repeat distance of 800 Å have been performed. This technique allows a separate determination of the flexibility (K), through the undulation mode, and of the interactions ($\bar{B}$), through the baroclinic mode. In principle, the other elastic constants (B and χ) could be reached through the study of higher frequency modes, the second sound and the membrane peristaltic modes. However, owing to the large values of the elastic coefficients involved, the experimental study of these modes could be quite difficult. The ability of selecting, using the wave vector and the polarization of light, the concentration from the order parameter fluctuations and the out-of-plane from the in-plane ones, make this technique very promising. Nevertheless, the main drawback of the light scattering experiment is that one has to rely on poorly known microscopic models for the friction coefficients in order to determine K and $\bar{B}$.

Both techniques agree on the main features of the studied systems. When membrane interactions are dominated by electrostatic forces, the Poisson-Boltzman solution V_{elec}, Eq. (6.4), gives a very good representation of these interactions. This confirms a well-established fact using other techniques [26,28]. What is more original is the existence of undulation forces in dilute lyotropic smectics. Indeed, even though this interaction was proposed as a possible membrane-membrane interaction by Helfrich more than ten years ago, its relevance was unclear, mainly owing to a lack of systems and techniques. Both the d and k dependences of this interaction $\left(d^2 k\right)^{-1}$ have been confirmed. The cardinal role of this interaction comes from the large flexibility of those bilayers where cosurfactant molecules are added to regular surfactant (SDS or lipids). Measurements of k indicate that $k \approx k_B T$ and the choice of the cosurfactant is crucial for obtaining thus small value. In particular, the role of the alcohol length has been elucidated.

The relevance of this mechanism for real cell membrane-membrane interactions is not clear. However, even if this system is still very far from a real membrane, the investigation of the effect of pentanol on the interactions between membranes of DMPC is a first step toward the biological relevance of undulation interactions.

Acknowledgments

It is a pleasure to acknowledge our many collaborators over the last six years. These include A.-M. Bellocq, N. A. Clark, P. Dimon, J. Prost, S. K. Sinha, E. B. Sirota and G. S. Smith. In addition, we have had many fruitful discussions with Fyl Pincus, S. A. Safran, D. Andelman, M. Cates, S. Leibler, S. Milner and W. M. Gelbart.

The X-ray part of this work was carried out at the National Synchrotron Light Source at Brookhaven National Laboratory and the Stanford Synchrotron Radiation Laboratory; both are supported by the U. S. Department of Energy.

References

1. P. Ekwall, in *Advances in Liquid Crystals*, G. M. Brown, ed. (Academic Press, New York, 1975).

2. a) G. S. Smith, E. B. Sirota, C. R. Safinya and N. A. Clark, Phys. Rev. Lett. **60**, 813 (1988); b) G. S. Smith, E. B. Sirota, C. R. Safinya, R. J. Plano and N. A. Clark, J. Chem. Phys. **92**, 4519 (1990).

3. P.-G. de Gennes, *The Physics of Liquid Crystals* (Clarendon, Oxford, 1974).

4. a) F. Candau and J. C. Wittmann, Mol. Cryst. Liq. Cryst. **56**, 171 (1980); b) W. J. Benton and C. A. Miller, J. Phys. Chem. **87**, 4981 (1983); c) M. Dvolaitzky, R. Ober, J. Billard, C. Taupin, J. Charvolin and Y. Hendrikx, C. R. Hebd. Acad. Sci. **II45**, 295 (1981); d) A.-M. Bellocq and D. Roux, *Microemulsions: Stucture and Dynamics*, S. Friberg and P. Bothorel, eds. (CRC Press, Boca Raton, 1986); e) J.-M. Di Meglio, M. Dvolaitzky and C. Taupin, J. Phys. Chem. **89**, 871 (1985).

5. F. Larché, J. Appell, G. Porte, P. Bassereau and J . Marignan, Phys. Rev. Lett. **56**, 1700 (1985).

6. R. Strey, R. Schomacker, D. Roux, F. Nallet and U. Olsson, J. Chem. Soc. Faraday Trans. **86**, 2253 (1990).

7. a) N. Satoh and K. Tsujii, J. Phys. Chem. **91**, 6629 (1987); b) C. Thunig, H. Hoffmann and G. Platz, Progr. Colloid Polym. Sci. **79**, 297 (1989).

8. P. Pieranski, Contemp. Phys. **24**, 25 (1983).

9. a) D. R. Nelson, in *Statistical Mechanics of Surfaces and Membranes*, D. R. Nelson, T. Piran and S. Weinberg, eds. (World Scientific, 1989); b) F. David, ibid.; c) S. Leibler, ibid.; d) R. Lipowsky, in *Random Fluctuations and Pattern Growth: Experiments and Models*, H. E. Stanley and N. Ostrowsky, eds. (Kluwer Academic Publishers, Dordrecht, Boston and London, 1988); e) L. Peliti, ibid.; f) D. Roux, ibid.; h) F. Pincus, in *Phase Transitions in Soft Condensed Matter*, T. Riste and D. Sherrington, eds. (Plenum Press, New York and London, 1989); i) C. R. Safinya, ibid.; j) M. E. Cates, ibid..

10. S. J. Singer and G. L. Nicolson, Science **175**, 720 (1972).

11. V. Luzzati, in *Biological Membranes*, D. Chapman, ed. (Academic Press, London and New York, 1968).

12. P.-G. de Gennes and C. Taupin, J. Phys. Chem. **86**, 2294 (1982).

13. C. R. Safinya, D. Roux, G. S. Smith, S. K. Sinha, P. Dimon, N. A. Clark and A.-M. Bellocq, Phys. Rev. Lett. **57**, 2718 (1986).

14. G. Porte, R. Gomati, O. El Haitamy, J. Appell and J . Marignan, J. Phys. Chem. **90**, 5746 (1986).

15. P. Bassereau, J. Marignan and G. Porte, J. de Physique **48**, 673 (1987).

16. D. Roux and C. R. Safinya, J. de Physique **49**, 307 (1988).

17. W. Helfrich, Z. Naturforsch **33a**, 305 (1978).

18. a) P. Ekwall, L. Mandell and K. Fontell, J. Colloid Interface Sci. **33**, 215 (1970); b) K. Fontell, J. Colloid Interface Sci. **44**, 318 (1973).

19. D. J. Mitchell, G. T. J. Tiddy, L. Warring, Th. Bostock and M. P. McDonald, J. Chem. Soc. Faraday Trans. **79**, 975 (1983).

20. M. Kahlweit, R. Strey and D. Haase, J. Phys. Chem. **89**, 163 (1985).

21. D. Roux and A.-M. Bellocq, in *Physics of Amphiphiles: Micelles, Vesicles and Microemulsions*, V. Degiorgio and M. Corti, eds. (North-Holland, Amsterdam, 1985).

22. G. Guérin and A.-M. Bellocq, J. Phys. Chem. **92**, 2550 (1988).

23. C. R.Safinya, E. Sirota, D. Roux and G. S. Smith, Phys. Rev. Lett. **62**, 1134 (1989).

24. W. Helfrich, Z. Naturforsch. **28a**, 693 (1973).

25. a) I. E. Dzyaloshinski, E. M. Lifshitz and L. P. Pitaevskii, Adv. Phys. **10**, 165 (1959); b) J. N. Israelachvili, *Intermolecular and Surface Forces* (Academic Press, Orlando, 1985).

26. A. C. Cowley, N. L. Fuller, R. P. Rand and V. A. Parsegian, Biochem. **17**, 3163 (1978).

27. S. Engström and H. Wennerström, J. Phys. Chem. **82**, 2711 (1978).

28. a) V. A. Parsegian, N. L. Fuller and R. P. Rand, Proc. Natl. Acad. Sci. USA **76**, 2750 (1979); b) R. P. Rand, Ann. Rev. Biophys. Bioeng. **10**, 277 (1981).

29. W. Helfrich and R. M. Servuss, Il Nuovo Cimento **3**, 137 (1984).

30. P.-G. de Gennes, *Scaling Concepts in Polymer Physics* (Cornell University Press, Ithaca and London, 1979).

31. F. Brochard and P.-G. de Gennes, Pramana Suppl. **1**, 1 (1975).

32. F. Nallet, D. Roux and J. Prost, J. de Physique **50**, 3147 (1989).

33. F. Nallet, D. Roux and S. T. Milner, J. de Physique **51**, 2333 (1990).

34. a) A. Caillé, C. R. Hebd. Acad. Scien. **B274**, 891 (1972); b) L. Gunther, Y. Imry and J. Lajzerowicz, Phys. Rev. **A22**, 1733 (1980).

35. a) G. Porte, P. Bassereau, J. Marignan and R. May, in *Physics of Amphiphilic Layers*, J. Meunier, D. Langevin and N. Boccara, eds. (Springer-Verlag, Berlin and Heidelberg, 1987); b) G. Porte, J. Marignan, P. Bassereau and R . May, Europhys. Lett. **7**, 713 (1988).

36. R. E. Peierls, Annales Inst. Henri-Poincaré **5**, 177 (1935); Helv. Phys. Acta Suppl. **7**, 81 (1934).

37. L. D. Landau, Phys. Z. Sowjetun. **2**, 26 (1937), or in *Collected Papers of L. D. Landau*, D. Ter Haar, ed. (Gordon and Breach, New York, 1965).

38. P. Dutta, S. K. Sinha, P. Vora, L. Passel and M. Bretz, in *Ordering in Two Dimensions*, S. K. Sinha, ed. (Elsevier North Holland, Amsterdam, 1982).

39. J. Als-Nielsen, J. D. Litster, R. J. Birgeneau, M. Kaplan, C. R. Safinya, A. Lindegaard-Andersen and S. Mathiesen, Phys. Rev. **B22**, 312 (1980).

40. a) N. Lei, C. R. Safinya and R. F. Bruinsma, Phys. Rev. Lett. (submitted, 1993); b) N. Lei, Y. Shen and C. R. Safinya, in *Scaling in Disordered Materials: Fractal Structure and Dynamics* (Material Research Society Publishers, Pittsburgh, EA-25, 1990); c) E. B. Sirota, G. S. Smith, C. R. Safinya, R. J. Plano and N. A. Clark, Science **242**, 1406 (1988).

41. P. C. Martin, O. Parodi and P. S. Pershan, Phys. Rev. **A6**, 2401 (1972).

42. The permeation is taken explicitly into account by: X. Wen and R. B. Meyer, J. de Physique **50**, 3043 (1989).

43. P.-G. de Gennes, J. de Physique Suppl. **30**, 65 (1969).

44. C. Y. Young and N. A. Clark, J. Chem. Phys. **74**, 4171 (1981).

45. W. Chan and P. S. Pershan, Phys. Rev. Lett. **39**, 1368 (1977).

46. S. T. Milner, private communication.

47. This property, as well as the continuous connection between the baro-clinic mode and the undulation mode, depend crucially on the assumption that $\tilde{\delta c}$ is a high frequency mode. This assumption breaks down when the frequency of the membrane peristaltic mode is smaller than the frequency of the undulation mode, i.e., when $\mu B < K/\eta$. In this case, as for one-component smectics, the undulation mode is continuously connected to the second sound mode.

48. For very small repeating distances ($d \approx \delta$) the membrane thickness d varies with dilution owing to a competition between in-plane surfactant interactions and out-of-plane electrostatic repulsion [1].

49. F. Pincus, J.-F. Joanny and D. Andelman, Europhys. Lett. **11**, 763 (1990).

50. L. Golubovic and T. C. Lubensky, Phys. Rev. **B39**, 12110 (1989).

51. T. C. Lubensky, J. Prost and S. Ramaswamy, J. de Physique **51**, 933 (1990).

52. a) J. Als-Nielsen, in *Symmetries and Broken Symmetries in Condensed Matter Physics*, N. Boccara, ed. (IDSET, Paris, 1981); b) E. Nachaniel, E. N. Keller, D. Davidov and C. Boeffel, Phys. Rev. A **43**, 2995 (1991).

53. R. Lipowsky and S. Leibler, Phys. Rev. Lett. **56**, 2561 (1986).

54. S. Leibler and R. Lipowsky, Phys. Rev. **B35**, 7004 (1987).

55. a) Y. Kantor and M. Kardar, Europhys. Lett. **9**, 53 (1989); b) J. Toner, Phys. Rev. Lett. **62**, 905 (1989).

56. a) M. B. Schneider, J. T. Jenkins and W. W. Webb, J . de Physique **45**, 1457 (1984); b) I. Bivas, P. Hanusse, P. Bothorel, J. Lalanne and O. Aguerre-Charriol, J. de Physique **46**, 855 (1987).

57. G. S. Smith, C. R. Safinya, D. Roux and N. A. Clark, Mol. Cryst. Liq. Cryst. **144**, 235 (1987).

58. E. A. Evans, Biophys. J. **14**, 923 (1974).

59. A. G. Petrov and I. Bivas, Prog. Surf. Sci. **16**, 389 (1984).

60. a) R. Cantor, Macromolecules **14**, 1186 (1981); b) S. T. Milner and T. A. Witten, J. de Physique **49**, 1951 (1988).

61. I. Szleifer, D. Kramer, A. Ben-Shaul, D. Roux and W . Gelbart, Phys. Rev. Lett. **60**, 1966 (1988).

62. F. Nallet, D. Roux and J. Prost, Progr. Colloid Polym. Sci. **79**, 313 (1989).

63. F. Nallet, D. Roux and J. Prost, Phys. Rev. Lett. **62**, 276 (1989).

7

The Structure of Microemulsions: Experiments

Loïc Auvray[1]

7.1 Introduction

Microemulsions were discovered about 40 years ago [1]. Since that time many experiments have been performed to determine their structures [2–16]. We aim in this chapter to discuss the main results in connection with the theoretical ideas to be presented in the following ones (see Chaps. 8 and 9). We pursue in particular a description of microemulsions as phases of flexible interfacial film of surfactant [7].

The first question that we consider, therefore, concerns the existence of surfactant films. We present in Sec. 7.2 the experimental results showing that the oil and water are not mixed at the molecular scale but remain separated by well-defined interfaces, where all the surfactant is adsorbed in the form of a dense, saturated layer. We review the available information on interfacial films of surfactant: composition, density (or equivalently area per surfactant molecule), thickness, fluidity.

We then confront the diversity of microemulsion structures, which depend on the chemical composition, temperature and concentrations of the constituents. To present and discuss these observations, we make first a traditional (but not entirely justified) distinction between dilute and concentrated systems.

Dilute microemulsions contain by definition a small proportion of dispersed phase. In most cases, this dispersed phase is observed in the form of isolated, spherical swollen micelles of oil in water or of water in oil. We consider these cases with the additional condition that the micelle concentration is sufficiently low that their mutual interactions are very small.

The first studies of such simple systems by light scattering or ultracentrifugation were very important in establishing the colloidal nature of the microemulsions. They reinforced the classical description of these systems as suspensions of monodisperse undeformable spherical droplets [9,10].

---

[1]Laboratoire Léon Brillouin, CEN Saclay, 91191 Gif sur Yvette, France

However, the modern hypothesis of a flexible surfactant film [7] leads naturally to the question of droplet deformation [17]. This question has only recently been investigated experimentally, and we present the corresponding results in Sec. 7.3.1.

Sec. 7.3.2 is devoted to the numerous studies of the interactions between droplets. In many instances, the interactions are purely repulsive, hard sphere like, but in certain cases they exhibit an attractive component, first observed by Vrij [18]. We discuss these different situations and analyze the possible origin of the attractive interactions. There is an especially interesting relationship between them, involving the curvature properties of the surfactant films and the formation of droplet aggregates.

Microemulsions often exist over a broad range of concentration. In the case of most surfactants or mixtures of surfactant and cosurfactant, it is possible to obtain microemulsions containing comparable proportions of oil and water. In many cases, there are even continuous paths in the phase diagrams that lead from water-rich to oil-rich microemulsions. This means that the geometry of the surfactant films may somehow adapt itself to the oil and water content. An understanding of how precisely this can be done took many years to develop: there is a subtle competition between the entropy of dispersion of the oil and water and the surface and curvature energy [19] of the film [20,7,21]. Finally, the structure of concentrated systems depends mainly on the properties of the film. Not only the flexibility but also the asymmetry of the amphiphile films (with respect to the affinities for oil and water) play an important role.

If a film is very asymmetric with a strongly preferred (or spontaneous) curvature [19], c_0, it will preferentially form droplets with radii close to c_0^{-1}. If a film is flexible and symmetric, or weakly asymmetric, its local curvature may fluctuate strongly. In concentrated systems, this leads to fluid random structures, where the average curvature of the film is small. As first proposed by Scriven [22,23], there are no longer isolated droplets but rather a bicontinuous arrangement of oil and water channels.

These two kinds of structure have been observed and characterized by different techniques [24]. The most direct and powerful are certainly the small angle scatterings of X-rays or neutrons [24] or freeze-etching electron microscopy [25,26]. We discuss first, however, the measurements of electrical conductivity and of molecular self-diffusion coefficients—see Sec. 7.4.2. They yield indirect but useful information on the topology of the structures: discrete or bicontinuous. By comparison with precise theoretical predictions, we then describe and analyze a selection of light, X-ray and neutron scattering experiments performed on systems corresponding to the two cases presented above: droplet structure or random bicontinuous structure (Secs. 7.4.3 and 7.4.4).

We warn the reader that our descriptions and discussions will be incomplete in many instances. Certain aspects of the microemulsions are not yet fully understood and are being currently actively studied: aggregation phenomena in droplet microemulsions, existence of locally lamellar or tubular structure [27], and the whole field of the *dynamics* of concentrated systems. These topics will not be treated here. In most cases we will also not explain in detail the experimental techniques. The applications to surfactant systems of the most classical techniques, e.g., E.P.R. with spin-labelled surfactants, N.M.R. with the spin-echo pulsed gradient field technique, and static and dynamic scattering techniques, are described in recent books [10,16,24]. An exception is made for the small angle neutron scattering: we recall in our appendix (Sec. 7.6) the expression for the scattering intensity and introduce there the important notion of contrast. We also do not enter in the details of the physical chemistry of the systems, which are often the most important in practice. Finally, the phase diagrams [28], the particularities of the different classes of surfactants (ionic, nonionic [29], zwitterionic) deserve separate discussions and studies beyond the brief introductions we provide.

7.2   The Interfacial Film of Surfactant

The existence of a film of surfactant in microemulsions has for a long time been a source of controversies between the supporters of the two opposite descriptions that could explain the apparent cosolubilization of oil and water, either in terms of colloidal dispersions or of molecular solutions [4].

These controversies are now resolved. Whatever the system examined, dilute, concentrated or even critical, the observations of the structures by scattering [30] and electron microscopy [25,26] reveal the existence of well-defined microscopic interfaces between oil and water. Many important data have been gathered on the films concerning their composition, density, and local state. They have established the crucial role of films in stabilizing microemulsions.

7.2.1   COMPOSITION

It is frequently not possible to obtain a microemulsion phase in given proportions and for a given surfactant unless one adds to the mixture a weak amphiphilic substance, often a short alcohol, called cosurfactant. The action of this cosurfactant is complex and probably multiple [7]. Its presence raises the question of the composition of the film.

In many instances, this composition can be deduced from relatively simple studies of the phase diagrams. Very often the proportion of alcohol that is necessary to obtain a microemulsion is a linear function of the quantities of oil, water and surfactant. This indicates that the partition coefficients of an alcohol between the different parts of a microemulsion are constant to a good approximation and independent of the overall structure [28]. In particular, it has been checked that microemulsions consisting of water-in-oil droplets can be diluted at constant droplet radius and at constant composition of the film and of the continuous oil phase [9].

Depending on the system, the number of alcohol molecules per surfactant molecule varies between two and five. The cosurfactant is thus a major constituent of the film. The alcohol to surfactant ratio decreases when the length of the alcohol increases; in the case of ionic surfactant, it also decreases upon addition of salt.

7.2.2   DENSITY

The density of surfactant in the film is often expressed by its inverse, the interfacial area per surfactant molecule a. This quantity is measured directly or indirectly by different experiments, as described below.

7.2.2.1   Observations of Droplet Systems

a has been evaluated first in dilute droplet microemulsions from the measurements of the radius R of the droplet core (by scattering techniques or ultra-centrifugation). If one assumes that the droplets are all spherical and identical and that all the surfactant molecules reside at the oil-water interface, a is related to the volume fraction ϕ of the dispersed phase and to the surfactant concentration c_s (number of molecules per unit volume) by the relation:

$$a = \phi/3c_s R \qquad (7.1)$$

The main result is that for a given surfactant or surfactant-cosurfactant mixture and for sufficiently large droplets ($R \gg d$, thickness of the film), a is independent of the size of the droplets. As first affirmed by Schulman [31], the interfacial film can be argued to be in a well defined state that does not depend on its curvature.

A characteristic value of a, obtained from sodium dodecylsulfate (SDS) microemulsions, is $a = 60\text{Å}^2$. Notice that this includes the contribution from the cosurfactant. More explicitly, with two alcohol molecules per SDS molecule, one gets 20Å^2 per aliphatic chain, the typical value observed in bulk mesophases of surfactant: the film is indeed dense and saturated.

7.2.2.2   Direct observations of the films

It is significant that a can be measured without any assumptions concerning the geometry of a microemulsion, by observing directly the interfacial film with X-rays or neutrons [30,32]. These scattering techniques provide an absolute confirmation of the existence of the interfacial film, even when the microemulsions are concentrated or near a critical point.

If a film exists, the scattering intensity $I(q)$ reaches a characteristic asymptotic limit as soon as the scattering vector q is larger than the curvature of the film. This limit is in particular proportional to the area per unit volume of the film, $S/V = c_s a$:

$$I(q) = c_s a \, f(q) \qquad (7.2)$$

Here the function $f(q)$ depends only on the contrasts between the oil, water and film parts of the microemulsion, defined as the differences between their respective scattering length densities n_o, n_w and n_f (cf., Sec. 1.1 from the first chapter and the present appendix).

The simplest case arises when the oil-film contrast vanishes. This occurs, for example, by using heavy water with hydrogenated oil and surfactant in a neutron scattering experiment. Then the scattering intensity follows Porod's law [33]:

$$I(q) = 2\pi(n_w - n_o)^2 c_s a q^{-4} \qquad (7.3)$$

It is also interesting to observe *only* the film, by matching the contrast between the oil and the water ($n_o = n_w$). Then, one expects for a thin film of thickness $d(qd \gg 1)$:

$$I(q) = 2\pi(n_f - n_o)^2 c_s a d^2 q^{-2} \qquad (7.4)$$

Note that with X-rays it is not always possible to be in one of the two pure cases considered above, because the oil, the polar head and the water often have different electronic densities. This leads to a more complex angular variation [30]:

$$I(q) = 2\pi c_s a (A q^{-4} + B q^{-2}) \qquad (7.5)$$

These three behaviors have been observed [30,32] in concentrated Winsor microemulsions made of brine, SDS, butanol and toluene, proving definitely the colloidal nature of these phases (see Fig. 7.1a on the next page, and Figs. 7.1b and 7.1c on the following page). The area per surfactant molecule is $a = 70 \pm 10 \text{Å}^2$. It is independent of the proportion of oil, water and surfactant. From the observations of expression (7.4), one also obtains the thickness of the film d, about 10Å.

7.2.3 LOCAL STATE OF THE FILM

The local state of the surfactant film is studied by different techniques [10,16,24]: Electron Paramagnetic Resonance (E.P.R.) with spin-labelled surfactants, Nuclear Magnetic Resonance (N.M.R.), and fluorescence. These techniques yield a dynamical picture of the film in the time range $10^{-10}s$ (E.P.R.)$-10^{-5}s$ (N.M.R.).

The E.P.R. studies have been particularly powerful for detecting modifications of the local state of the film induced by the addition of cosurfactant, the variations with temperature, or the inversion of the microemulsions.

One observes in particular from the E.P.R. spectra that the order parameters of the spin-labelled surfactants adsorbed on the probed surfactant films are smaller and the correlation times shorter in microemulsions with cosurfactant than in ordinary lamellar phases without cosurfactant. *It thus appears that the films are more fluid in microemulsions.*

It is interesting to relate quantitatively this "fluidity" of the film to its elasticity, in particular its flexibility. This was first made possible by studying the ordered lamellar phases (called birefringent microemulsions) that often exist very close to the microemulsions in the phase diagrams of the amphiphile-oil-water systems. Different techniques enable one to observe the curvature fluctuations of oriented films: E.P.R. with labelled surfactant (sensitive to the slow dynamics of the probe in a film) [34,35,36], dynamic light scattering [37], and more recently X-ray scattering [38]. They yield values of the rigidity coefficient of the film that are of order $k_B T(10^{-14} \text{erg})$, confirming that the observed films are very flexible. One also observes clearly that the addition of cosurfactant lowers the film rigidity [35,36], a feature discussed in some detail in Sec. 1.3.4.

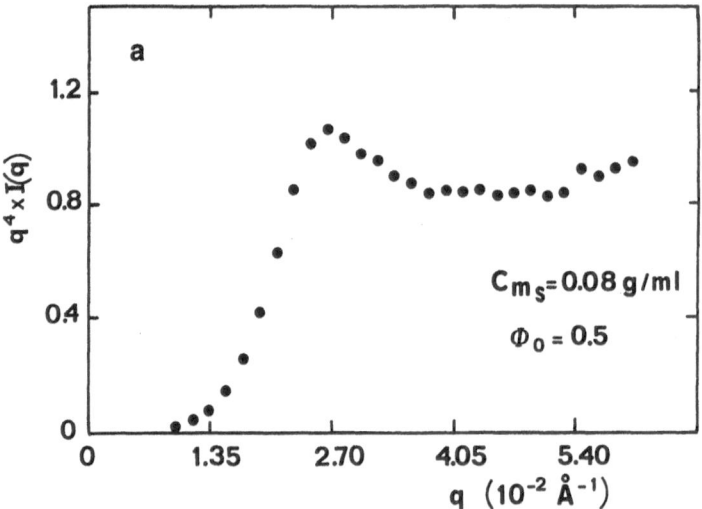

FIGURE 7.1. Asymptotic behaviors of the intensity (in arbitrary units) scattered by Winsor microemulsions (SDS, butanol, toluene, brine system) for different contrasts: (a) Oil-water contrast. The surfactant does not contribute to the scattering; the product $q^4 I(q)$ reaches a constant limit (neutron experiment, background substracted; c_m is the surfactant concentration, ϕ_0 the toluene volume fraction).

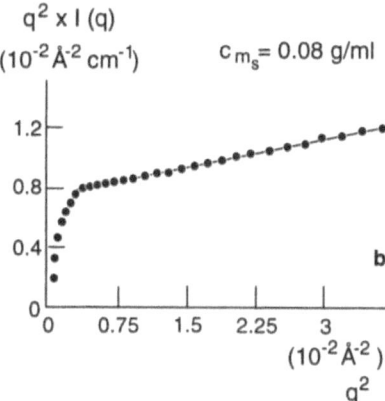

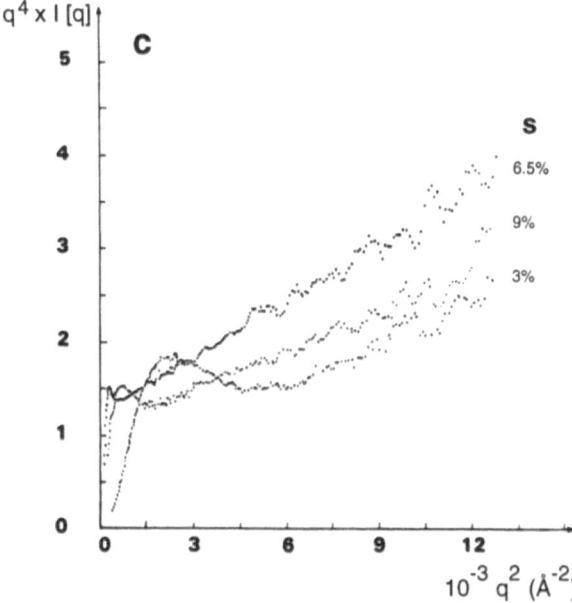

FIGURE 7.1. (b) Intensity scattered by the interfacial film (neutron experiment, background not subtracted). The representation $q^2 I(q)$ as a function of q^2 demonstrates the q^{-2} contribution of the film to the intensity (from [32]). (c) X-Ray experiment. The oil, the water and the polar heads of the surfactant contribute together to the intensity (shown here in arbitrary units); in the asymptotic regime, $q^4 I(q)$ varies linearly with q^2; both the slope and the intercept are proportional to the film area, i.e., to the surfactant concentration which varies with the brine salinity S (from [30]).

7.3   Dilute Microemulsions

Dilute microemulsions contain less than a few percent of dispersed phase. Depending on the initial proportions, and on the particular surfactant and cosurfactant, one obtains water- or oil-rich systems. In most cases they can be described as dispersions of small spherical droplets of oil in water or water in oil, the radius of the droplets being related to the area per polar head by expression (7.1). There are, however, known exceptions to the spherical shape in both ionic and non ionic systems [39]; they occur even when the aggregates of surfactant are only slightly swollen. We mention in particular the interesting case of cylindrical structures predicted theoretically [40] but only observed at present in unswollen micellar systems [41,42].

While the structure of dilute droplet microemulsions is apparently simple, these systems nevertheless raise several interesting questions concerning:

1. The polydispersity and shape fluctuations of the droplets related to the curvature energy of the film.

2. The origin of the interactions between the droplets.

7.3.1   POLYDISPERSITY AND SHAPE FLUCTUATIONS

If the film of the microemulsions is flexible, one imagines easily that the radius of the droplets is not strictly fixed but is distributed around an average value (polydispersity) and that the shape of the droplets fluctuates.

7.3.1.1   Polydispersity

Polydispersity effects have been noted qualitatively in several studies, particularly in contrast variation experiments on water-in-oil microemulsions [43]. There is, however, only one quantitative measurement, by Kotlarchyk et al. [44,45] on the AOT-water-heptane system: two parameters can be drawn from the scattering spectra—the average radius of the droplets R and the polydispersity p defined by $p = \Delta R/R$, ΔR being the root mean square deviation of the radii distribution. The observed value of p is about 30%, in agreement with the theoretical calculation. This value appears common to many systems, and is sufficiently high to explain why the characteristic oscillations of the form factor of "monodisperse" spheres are never observed in the scattering by microemulsions.

7.3.1.2   Shape deformations and fluctuations

The droplets of a microemulsion can be deformed by an external perturbation. This has been first observed [46] from the magnetic birefringence of a microemulsion with a strongly diamagnetic surfactant (octylbenzenesulfonate) and might also be detected by measurements of flow birefringence [7].

The shape of the droplets is also dynamically deformed by thermal fluctuations. In principle, these shape fluctuations contribute to the hydrodynamic radii measured by dynamical light scattering, which are often larger than the radii deduced from static light or neutron scattering, but no precise analysis has yet been made.

This field is, however, in rapid progress. Recently Huang, Milner, Farago and Richter [47] have studied the dynamics of dilute dispersions of water-AOT droplets in decane by neutron spin echo spectroscopy, probing the time range 1–15 nanoseconds at the spatial scale 10–1000Å. The neutron contrasts are chosen so that only the AOT film contributes to the scattering. By analyzing the effective diffusion coefficient of the droplets, as a function of the scattering vector q, the authors separate the contribution of the overall translation of the droplets from the contribution of the shape fluctuations. As the scattering vector goes to zero, the dominant relaxation phenomenon is the translational diffusion, but the shape fluctuations introduce a peak in the effective diffusion coefficient at a finite scattering vector q_m such that $q_m R \approx \pi$. The height of the peak is inversely proportional to the characteristic relaxation time τ of the shape fluctuations.

If the bending energy drives the fluctuation modes of the film, the characteristic relaxation time τ_b is related to the film rigidity k, the viscosity of the film η and the radius of the droplets R by the proportionality relation:

$$\tau^{-1} = \tau_b^{-1} \sim k/\eta R^3 \tag{7.6}$$

If, on the other hand, the fluctuations are driven by the surface tension of the film γ, the characteristic time is:

$$\tau^{-1} = \tau_s^{-1} \sim \gamma/\eta R \tag{7.7}$$

The dependence on the droplet radius is different in both cases: varying the radius, one distinguishes between the two possibilities. Experimentally, τ is proportional to R^3, and the effect of the bending energy is thus clearly demonstrated. This is a very important experimental result, which justifies all the recent theoretical efforts.

Experimentally, τ is found furthermore to be proportional and comparable to the relaxation time associated with the translational motion of the droplets: $\tau_t \sim \eta R^3/k_B T$. This means that the rigidity coefficient of the film k is of order $k_B T$ as expected theoretically and observed in ordered birefringent microemulsions by other methods. In the present case, $k_{AOT} = 5k_B T$, a value which depends slightly on the estimation of the spontaneous curvature of the AOT film, chosen to be $c_o = 0.5$ R.

7.3.2   INTERACTIONS BETWEEN DROPLETS

The study of the interactions between the droplets of a dilute microemulsion is a possible means to characterize the different factors which influence the microemulsion structure, stability and phase behavior in an elementary case. As presented in the following chapters, these factors are: the entropy of dispersion, the curvature energy of the film, and the van der Waals and electrostatic interactions. The interpretation of the data is, however, difficult and still controversial.

7.3.2.1   Measurements of the droplet virial coefficient

The interactions between droplets are classically studied as in other colloidal systems by light or neutron scattering. When the scattering vector goes to zero, the scattering intensity reaches a well-known thermodynamic limit, which is proportional to the osmotic compressibility of the dispersion, $\partial\phi/\partial\Pi$. Varying the droplet volume fraction ϕ at constant size (constant water-over-surfactant or oil-over- surfactant ratio) one extracts in the dilute limit the second virial coefficient B from the relation:

$$\partial\Pi/\partial\phi = (k_B T/\nu)(1 + B\phi) \tag{7.8}$$

ν here is the volume of one droplet. The droplets are assumed monodisperse. In the simple case, where one can define a potential of interaction between the droplets $V(r)$, B is expressed by the familiar relation:

$$B = -\frac{1}{V} \int d^3 r \left[\exp\left(-\frac{V(r)}{k_B T}\right) - 1\right] \tag{7.9}$$

As first shown by Vrij [8] in a complete study (see Sec. 7.4), the droplets of a microemulsion may behave essentially as hard spheres. The reference (hard sphere) value of the virial coefficient in this case is $B = +8$. However, small values of B, as low as -15, are observed in many systems [48,51], implying an attractive contribution to the interactions. These attractive interactions have been particularly studied in water-in-oil systems made with ionic surfactants (SDS [48–50], AOT [51]), but they also exist in certain oil-in-water and nonionic systems. In a general way, the strength of the attraction increases with droplet size. For neutral water-in-oil SDS droplets, it also increases with the volume of the oil molecules, but decreases as the length of the cosurfactant and the ionic strength of the water increases.

7.3.2.2   Attractive interactions and droplet association

The virial coefficient only yields average information on the interactions. One obtains more detailed information by observing the whole scattering spectra of the interacting droplets. This has first been done on the SDS-cyclohexane-pentanol-water system [48]. In this system, the interactions between water-in- oil droplets are attractive, and the scattering spectra exhibit strong deviations from the behavior of pure spherical droplets. These deviations can only be interpreted if one admits that a non-negligible fraction of the droplets (about 10%) are associated in dimers. It thus appears

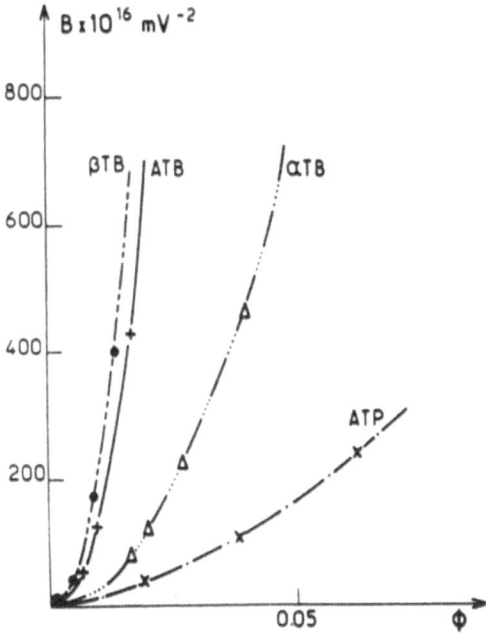

FIGURE 7.2. Kerr constant of dilute, quaternary, water-in-toulene microemulsions as a function of the droplet volume fractions. The strength of the attractive interactions in the systems studied increases in the order: ATP (toluene, pentanol, SDS system, $w/s = 1.25$), αTB (toluene, butanol, SDS, $w/s = 1$), ATB (toluene, butanol, SDS, $w/s = 1.25$) and βTB (toluene, butanol, SDS, $w/s = 1.75$). w/s is the water-to-surfactant weight ratio proportional to the radius of the droplets (from [49]).

that the attractive interactions may be associated with an aggregation between the droplets.

This association phenomenon is suitably studied by electric birefringence [49]. This technique detects any anisotropy induced by an electric field in the system, in particular the orientation of an anisotropic aggregate such as a dimer of droplets. If only dimers contribute to the electric birefringence, the Kerr constant is quadratic in the volume fraction of the droplets. This is indeed what is observed at small volume fraction in water-in-oil microemulsions with attractive interactions. The Kerr constant, i.e., the number of dimers, increases with the strength of the attraction between droplets, pointing up directly the relationship between the two phenomena (Fig. 7.2).

7.3.2.3 Origin of the attractive interactions

The rather general observation of attractive interactions and droplet associations may hide a diversity of phenomena. The starting idea, first emphasized by Vrij, is that the strength of the attraction, measured by the virial coefficient, is in general too large to be due only to the pure van der Waals interactions between the cores of the droplets separated by the interfacial film. Other mechanisms must be invoked to explain the observed effects. The proposed ones are very different:

a. *Mechanism of interpenetration*: Lemaire, Bothorel and Roux [52] assume that for water-in-oil microemulsions the tails of the surfactant molecules residing on two different droplets prefer to interpenetrate each other rather than to be in contact with the oil. The penetration depth depends on the length of the cosurfactant. The overlapping of the surfactant films and the increased proximity of the water cores result in an increase of the van der Waals attraction. This mechanism assumes that the droplets remain undeformed, although this condition may be relaxed; one is then led to consider the effects of sticking, *deformable*, droplets at the expense of bending energy [53].

b. *Mechanism of fusion induced by curvature*: Safran [17] first observed that a spherical droplet is unstable with respect to shape fluctuations which lower the bending energy when the spontaneous curvature of the film c_0 is adverse to the actual curvature of the film $1/R$, i.e., when $c_0 R \ll 1$. Similarly, under the same conditions, the fusion of two droplets in one dimer at constant surface area and volume lowers the curvature energy. A simplified calculation [54] of the virial coefficient leads to the expression:

$$B = 8 - \exp\left[(\alpha k/k_B T)(1 - \beta c_0 R)\right] \qquad (7.10)$$

α and β are two dimensionless constants of order unity, which depend on the precise shape of the dimer.

The first mechanism above interprets the relationship between the virial coefficient, the size of the droplets, and the alcohol and oil chain length in water-in-oil microemulsions. It does not explain, however, the existence of attractive interactions in direct oil-in-water systems nor the effect of the ionic strength.

The second mechanism applies both to direct and inverse droplets. For inverse droplets (by convention, of positive curvature), it predicts two main trends:

1. At constant droplet radius, B increases if c_0 increases, i.e., the droplets harden as the interfacial film becomes more lipophilic; this explains the effects of the alcohol chain length and of the water ionic strength.

2. At constant k and c_0, B decreases as the droplet radius increases, if c_0 is negative. This accounts for variation of B with the droplet radius observed in inverse SDS microemulsions without salt, if one admits that the SDS is actually very hydrophilic, favoring direct structures. However, it

does not explain the attractions observed in inverse microemulsions made with lipophilic surfactants like AOT [51] or BHDC [55].

Thus each mechanism alone cannot explain the whole set of observations already gathered. Experiments sensitive to the local distortion of the interface of colliding droplets would be necessary to classify more completely the phenomena and clarify the origin of the attractive inter actions.

7.4   Structure of Concentrated Microemulsions

7.4.1   DIFFERENT TYPES OF STRUCTURE

With many surfactants or surfactant-cosurfactant mixtures, one obtains concentrated microemulsions containing comparable proportions of oil and water. A great diversity of structures is observed that one may classify as follows.

7.4.1.1   Droplet structure

As in the dilute case, the microemulsion is a dispersion of well defined spherical droplets, either direct or inverse. This is the classical structure, first described by Schulman [1–3].

7.4.1.2   Intermediate structure

As shown in the preceding section, by starting from dilute droplets and increasing the volume fraction of the disperse phase, one often encounters aggregation phenomena associated with reorganizations of the interfacial film. The loose term "intermediate" encompasses all the structures occuring at intermediate concentration, which can be described as aggregates of stuck or fused droplets and which eventually lead to locally cylindrical or lamellar structures. These intermediate structures are still, by far, the least well known and studied.

7.4.1.3   Bicontinuous structure

In certain situations first imagined by Scriven, elementary objects, droplets, globules or aggregates can no longer be defined. The oil and water form two intertwined labyrinthic networks separated by the film of surfactant. This type of structure accounts for two essential observations:

1. The continuous inversion from water-in-oil to oil-in-water micelles observed in the phase diagram of many systems.

2. The existence of non-dilutable microemulsions, such as the Winsor middle phases [56].

As explained in the following chapters, the existence of these different types of structure is interpreted by considering the curvature elasticity of the surfactant film:

1. Disordered fluid microemulsions are associated with a flexible film of surfactant—otherwise ordered structures are preferred.

2. Strong electrostatic or steric constraints, leading to a large asymmetry between the polar and aliphatic part of the film and to a large spontaneous curvature, favor the formation of droplets.

3. If the film has a very small spontaneous curvature, it tends to adopt a local configuration with a vanishing curvature, either flat or saddle shaped. This can lead to ordered structures. However, if the film rigidity is of the order of $k_B T$, the ordered structures are not stable. The local curvature of the oil-water interface fluctuates strongly at scale larger than ξ_K, the persistence length of the film, defined by [7]:

$$\xi_K = l \exp(2\pi k / k_B T) \qquad (7.11)$$

l is a microscopic length comparable to the surfactant length.

The structure of the microemulsions (and their phase behavior) then depends strongly on the volume fractions of oil and water, ϕ_o and ϕ_w and on the concentration of surfactant c_s:

1. At small or intermediate volume fraction (ϕ_o or $\phi_w \ll 0.5$), one expects isolated droplets or random aggregates.

2. In the inversion domain ($\phi_o \approx \phi_w \approx 0.5$), one expects a random bicontinuous structure, the size scale of the dispersion being the persistence length of the film ξ_K.

The two extreme types of structure, the discrete structure of droplets and the bicontinuous structure, differ by their topology and geometry. They are experimentally distinguished by dynamical techniques probing the connectivity of the oil or water parts and by various scattering techniques.

7.4.2 DYNAMICAL CHARACTERIZATION OF THE DIFFERENT STRUCTURES

If a molecule or an ion is part of a big aggregate during a sufficiently long time, its mobility is governed by the motion of the entire aggregate. Its mobility is thus relatively small, compared to that of a free molecule. On the other hand, if the molecule belongs to a continuous phase, its motion is only slightly hindered. This difference in transport properties provides a simple means to characterize the dynamical connectivity of the microemulsions, which indirectly reflects their structure. The particular techniques differ by the nature of the molecular probe—i.e., whether it is a constituent of the microemulsion or an external probe—and by the time and length scales that are probed.

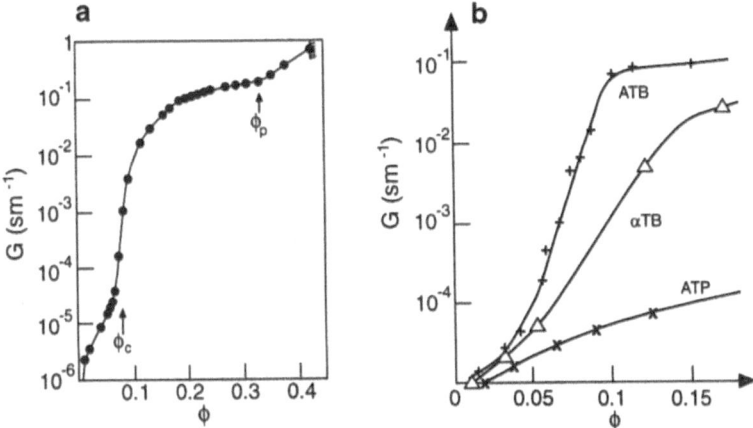

FIGURE 7.3. Electrical conductivity of quaternary water-in-oil droplet microemulsion as a function of the water volume fraction. (a) Water, pentanol, cyclohexane, SDS system ($w/s = 2.5$); ϕ_c is the percolation threshold for the conductivity—ϕ_p corresponds to the inversion of the droplets (from [57]). (b) Same systems as those quoted in Fig. 7.2; the variation of the conductivity clearly depends on the strength of the attractive interactions (from [49]).

7.4.2.1 Conductivity measurements

As the oil part of a microemulsion is electrically insulating, measuring the electrical conductivity is a simple means to study the connectivity of the water part [2]. The probe length scale is macroscopic, but the time scale is short, being on the order of $10^{-8}s$; rapid transient phenomena can contribute to the transport of ions, thereby complicating the interpretation.

The most interesting systems to study in this context are evidently water-in-oil microemulsions. Starting from the dilute droplets, one observes the variation of conductivity with droplet volume fraction ϕ [57]. This variation is correlated with the strength of the interactions [49,55] (Fig. 7.3). When the oil droplets behave as hard spheres, as in the SDS-toluene-pentanol-water system, the electrical conductivity increases slightly as ϕ increases, but always remains very small, even at $\phi = 0.4$. This is a strong argument in favor of a discrete structure of isolated globules.

Attractive interactions between droplets modify radically this behavior: at a certain volume fraction ϕ_p, of the order of 10%, the electrical conductivity increases by several orders of magnitude. ϕ_p depends on the strength of the attraction, decreasing as the attraction increases. This is characteristic of a percolation phenomenon [57] and confirms that the contacts between the water cores are favored by the attractive interactions.

7.4.2.2   Measurements of the self-diffusion coefficients

Self diffusion coefficients are measured by various techniques: Forced Rayleigh Scattering and Photobleaching Recovery use photochromic and fluorescent probes. Tracer and NMR techniques measure directly the self-diffusion coefficient of one or several constituents. The Fourier Transform Pulsed Gradient Field Spin Echo technique is particularly powerful: it covers a wide range of values $(D > 10^{-9} cm^2/s)$ and enables one to compare simultaneously the self-diffusion coefficient of every constituent in the microemulsion. One thus obtains more information on the structure. The time scale probed varies between 1 and 10^3ms, and the probed spatial scale is about 1μm, depending on the value of the field gradient. The technique and the results obtained for different ionic and non- ionic microemulsions are described in several reviews [58,59]. We only quote here three observations demonstrating the different possible cases.

The simplest situation, corresponding to water-in-oil droplets, is observed for the oil-rich phases of the AOT-xylene-water system [58]. For each sample, the self-diffusion coefficient of the oil is one or two orders of magnitude larger that the self- diffusion coefficient of water, D_w. Furthermore, D_w is equal both to the self-diffusion coefficient of the surfactant and to the self-diffusion coefficient of the whole droplet. This shows that the droplets remain isolated without significant exchange of water and surfactant between them.

A different case occurs when water is exchanged between colliding water-in-oil droplets: the water molecules will move faster than the surfactant molecules. As suggested by the conductivity experiments, the exchange of water is correlated with the interactions between the droplets. This correlation is clearly confirmed by the measurements made on toluene-in-water droplets stabilized by SDS, in the range of droplet volume fraction $0 < \phi < 0.3$. When the cosurfactant is pentanol, the interactions are repulsive, and there is no exchange of water between the droplets: $D_{water} \approx D_{SDS}$. When, on the other hand, the cosurfactant is butanol, the interactions are attractive and the self-diffusion coefficient is larger than both the self-diffusion coefficient of the surfactant and of the droplets, suggesting an exchange of water which is probably associated with a transient merging of the interacting droplets. The effect of the alcohol chain length on the structure of a more concentrated sample of the same system is described in [60]: changing the cosurfactant from butanol to decanol at constant composition decreases the water diffusion coefficient by two orders of magnitude, while D_{oil} remains constant. According to the measurements, isolated, well-defined, water-in-oil droplets exist only at high concentration if the cosurfactant is very long, i.e., if the spontaneous curvature of the

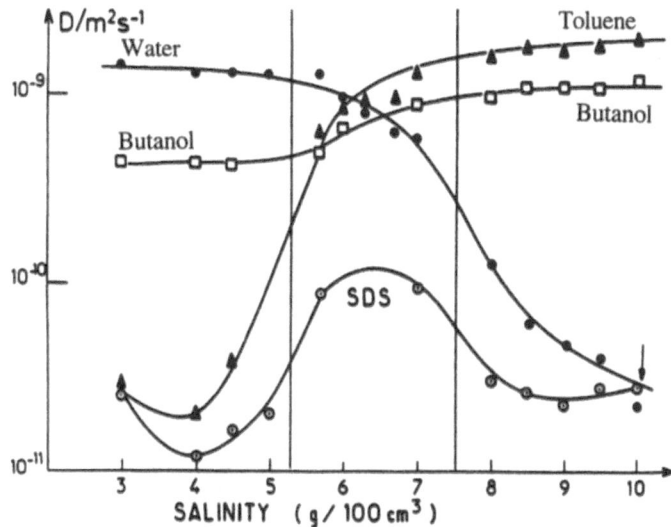

FIGURE 7.4. Self diffusion coefficients of the constituents of Winsor microemulsions as a function of the brine salinity (from [61]).

film is sufficiently high. For the shortest cosurfactant studied, butanol, the diffusion coefficients of oil and water are approximately equal and close to their free value, indicating a bicontinous structure.

The progressive evolution from a droplet structure to a bicontinuous structure is probably best appreciated by following the sequence of Winsor phases of a system of water, salt, oil, surfactant and cosurfactant as the ionic strength increases. This sequence realizes a progressive inversion of structure from oil-in-water to water-in-oil phases, and will be described more completely in Sec. 7.4.4.1.

Two NMR studies [61,62] have been performed independently on the same reference system, water-NaCl-toulene-SDS-butanol. As seen in Fig. 7.4 showing the self-diffusion coefficients of the constituents as a function of the water salinity, there are three distinct regimes:

1. At low salinity, D_{oil} is comparable to D_{SDS}, with both much smaller than D_{water}. This indicates oil-in-water droplets.

2. At high salinity, the inequalities are reversed—the diffusion of the water has been progressively hindered, and one observes $D_{oil} \approx D_{SDS} \ll D_{water}$. There are now water-in-oil droplets.

3. For the intermediate salinities, one notes that D_{oil} is of the order of D_{water} and that oil and water move almost freely. The diffusion of the surfactant is more rapid than in the two other regimes, characteristic of a

bicontinuous structure.

The simultaneous measurement of the diffusion coefficients of the constituents thus reflects sensitively the different structures of a microemulsion. These structures remain to be determined precisely in each case, but the information already obtained is highly useful, particularly for the exploration of systems which until now have been less studied, such as the non ionic systems [58].

7.4.3   SCATTERING STUDIES OF CONCENTRATED DROPLET MICROEMULSIONS

We describe in this section a few scattering experiments which establish the description of certain microemulsions as concentrated dispersions of well-defined, relatively monodisperse, spherical droplets. For the sake of simplicity, most of the structural studies have involved water-in-oil systems. With ionic surfactants, a water-in-oil inverse droplet is electrically neutral while a direct droplet is charged. In salt-free, direct, microemulsions, the long range electrostatic interaction therefore plays an important role. This role has, however, not been studied very much [64].

As discussed in Sec. 1.1 and shown in the appendix, the X-ray or neutron intensity scattered by a system of interacting spherical droplets is the product of three factors:

$$I(q) = CP(q)S(q) \qquad (7.12)$$

C is the number concentration of droplets. $P(q)$ is the form factor of the swollen spherical micelle. It depends only on its internal structure (namely on the core radius, R_c, and thickness of the surfactant film, d), and involves in particular all dependences on the contrast factors. $S(q)$ is the structure factor of the dispersion. It describes the correlations between the centers of mass of the droplets and depends only on the interparticle interactions but not on the contrasts.

7.4.3.1   Contrast variation experiments

The important method of contrast variation in neutron scattering was applied first to microemulsions by Taupin and Ober [65,57]. When the structure of a microemulsion is unknown, the observation of the separation of the scattering intensity into form and structure factor enables one to obtain a model-independent demonstration of the existence of isolated spherical droplets and to determine their internal structure.

a. *Existence of elementary droplets*: If a microemulsion is made of monodisperse spherical droplets, and therefore if the separation (7.12) applies, one can cancel the intensity scattered at vanishing angle, $I(q = 0)$, by a particular choice of the contrast. For water-in-oil droplets the expression

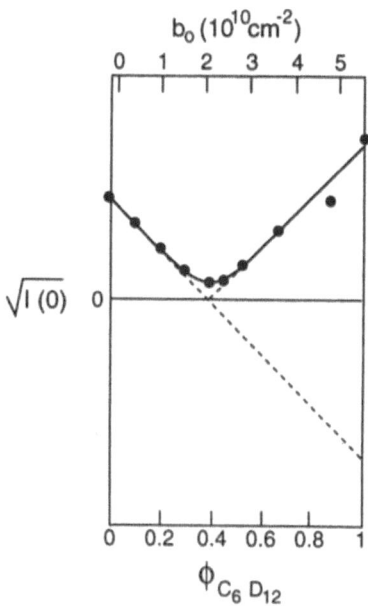

FIGURE 7.5. Square root of the intensity scattered at zero angle by a water-in-oil microemulsion as a function of the scattering length b_0 of the oil (a mixture of deuterated and hydrogenated cyclohexane) (from [65]).

for the forward intensity is:

$$I(q = 0) = [(n_w - n_o)V_w + (n_f - n_o)V_f]^2 S(q = 0) \qquad (7.13)$$

$n_w - n_o$ and $n_f - n_o$ are respectively the water-oil and film-oil contrasts; V_w is the volume of the water core (radius R_w); and V_f is the volume of the surfactant shell (thickness d). By changing the relative proportion of light and heavy water in the microemulsion, one varies n_w at fixed n_o and n_f. The square root $I^{1/2}(q = o)$ is a linear function of n_w. The extinction of the intensity occurs at the point $n_w^{w/o}$ such that:

$$\frac{n_w^{w/o} - n_o}{n_f - n_o} = -\frac{V_f}{V_w} \simeq -\frac{3d}{R_w} \qquad (7.14)$$

the last equality holding only if $d \ll R_w$. The observation of the extinction at $n_w^{w/o}$ proves that one is dealing with water-in-oil droplets and yields the ratio d/R_w (Fig. 7.5).

This method was used [65,57] first to determine the internal structure of dilute water in cyclohexane droplets stabilized by SDS and pentanol and also to show that this structure is conserved as one increases the droplet volume fraction up to 64% at fixed water-to-surfactant ratio and fixed composition of the continuous phase (cyclohexane + pentanol). More recently it has enabled one to confirm that the concentrated AOT-water-decane microemulsions have a structure of water-in-oil droplets even at equal volume fraction of oil and water [66].

b. *Inversion of structure*: An important feature of the contrast variation method is its sensitivity to the sign of the film curvature. For oil-in-water droplets (core radius R_o), one finds that the extinction at $n_w^{o/w}$ satisfies the relation:

$$\frac{n_w^{o/w} - n_o}{n_f - n_w^{o/w}} \simeq \frac{3d}{R_o} \qquad (7.15)$$

(assuming $d \ll R_o$). At fixed n_f and n_o the two values $n_w^{w/o}$ and $n_w^{o/w}$ cannot coincide. If, as usual, the surfactant is hydrogen rich, $n_f < n_o$, and one has the inequality

$$n_w^{o/w} < n_f < n_w^{w/o} \qquad (7.16)$$

One can thus know without further information if the droplets are direct or inverse. In this manner, Lagües, Ober and Taupin have obtained the first structural evidence of the inversion of a microemulsion [57].

7.4.3.2 Comparison with theoretical models of interacting spheres

The classical theory of liquids enables one to calculate the structure factor of interacting spheres as a function of the concentration and scattering vector, provided that one knows the potential of interaction between them. As first noted by Agterof, van Zomeren and Vrij [18], this theory can be applied directly to droplet microemulsions. The models used are tested by comparison with two kinds of scattering data:

1. The scattering intensity extrapolated to zero scattering angle. This intensity, obtained with light or neutrons, depends on the contrasts involved and is related to the osmotic compressibility of the dispersion of droplets.

2. The angular variation of the intensity, measured by neutron or X-ray scattering.

a. *Osmotic compressibility*: For a dispersion of monodisperse spherical particles of volume fraction ϕ, the scattering intensity extrapolated to the forward direction is related to the equation of state of the system by the relation:

$$I(q = 0, \phi) = CP(0)S(0) = \phi \nu P(0) k_B T \left(\frac{\partial \phi}{\partial \Pi} \right) \qquad (7.17)$$

k_B is the Boltzmann constant, ν the volume of the particle, T the temperature and $\partial \phi / \partial \Pi$ the osmotic compressibility of the dispersion.

As with simpler van der Waals fluids, the main contribution to the osmotic pressure of the microemulsion Π is expected to be a hard sphere repulsion, perturbed by an attractive term. Π is thus written semi-empirically as the sum of two terms:

$$\Pi = \Pi_{HS} + \Pi_{att} \qquad (7.18)$$

Π_{HS}, the pressure of a hard sphere fluid, is not known rigorously, but a very good approximation, given by Carnahan and Starling, is:

$$\Pi_{HS} = k_B T C (1 + \phi + \phi^2 - \phi^3)(1 - \phi)^{-3} \qquad (7.19)$$

The attractive perturbation is simply written to lowest order:

$$\Pi_{att} = -AC^2 \qquad (7.20)$$

A is a parameter; the volume fraction ϕ is related to the droplet concentration by the relation $\phi = C(4\pi R_{HS}^3/3)$; and R_{HS} is the hard sphere radius, which may differ from the actual external radius of a droplet.

The familiar expressions (7.19) and (7.20) were first proposed *for microemulsions* by Vrij and have been used by many authors. With two adjustable parameters, R_{HS} and A, they account remarkably well for the experimental results of the following systems, studied over a very large range of volume fraction ($0 < \phi < 0.5$ or 0.6): potassium oleate-hexanol-benzene-water [18]; SDS-pentanol-cyclohexane- water [48] and many others [9,10] (Fig. 7.6).

As expected, the hard sphere radius is larger than the radius of the water core; their difference—between 8 or 9Å for potassium oleate and SDS—is, however, usually smaller than the length of the extended surfactant chain, indicating a possible interpenetration of the droplets.

The attractive contribution is often weak, but non negligible; it has already been discussed in Sec. 7.3.2 in the context of dilute systems. The experiments made over the *whole* concentration range (and not only in the dilute regime) confirm that the attractive interactions depend on the alcohol chain length and water ionic strength and that they cannot be explained by simple van der Waals attractions between the water cores.

b. *Structure factors*: The measurements of the osmotic compressibility are macroscopic, and do not remove possible ambiguities concerning details of the structure. This is shown in particular by the experiments indicating attractive droplet aggregates in microemulsions satisfying Vrij's equation for the osmotic pressure [48]. A better description of the structure is attained by comparing the shape of the scattering spectra with the theoretical predictions. A very complete study of this kind is performed by the authors of [67] and [68] for two concentrated water-in-oil microemulsion systems: sodium dodecyl benzene, dodecane, water. A pure hard sphere model is used to interpret the data. The theoretical structure factor is evaluated by using an approximate semi-analytical expression due to Ashcroft and Lekner and a small polydispersity ($\Delta R/R \sim 0.2$) is taken into account to calculate the scattering intensity.

The repulsive interactions between the droplets influence the spectra significantly; one predicts in particular a peak in the scattering intensity as soon as the volume fraction is larger than 0.15. As apparent from Fig. 7.7, the shape of the experimental spectra, exhibiting a peak, is very well ac-

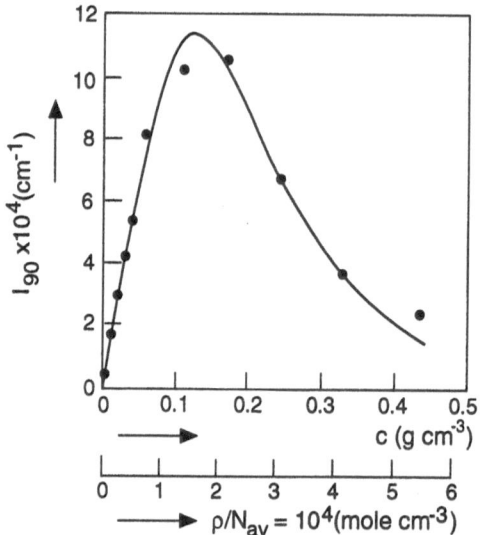

FIGURE 7.6. Light scattering experiment for a water-in-oil droplet microemulsion (potassium oleate, hexanol, benzene, water system). Normalized intensity scattered at right angle, as a function of the droplet concentration or number density. The circles are measured values, the curve is calculated from Eq. (7.18) (from [18]).

counted for by the theory. At constant water-to-surfactant ratio, the two free parameters of the models, the radius of the water core R_c and the hard sphere radius R_{HS}, are (as expected) independent of the droplet concentration and differ only by a small amount.

Detailed scattering studies like the one discussed above, and many others, thus yield unambiguous determination of the structure of droplet microemulsions.

7.4.4   STRUCTURE OF RANDOM MICROEMULSIONS WITH A FLEXIBLE INTERFACIAL FILM

In the droplet systems described above, the structure of the elementary micellar object does not depend on the concentration: there is no (or almost no) reorganization of the interfacial film. This is consistent with the fact that the curvature of the film is rather well determined by intrinsic

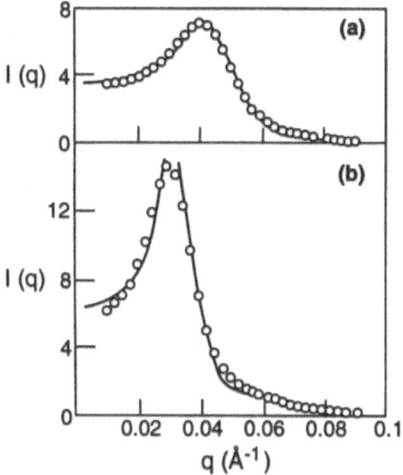

FIGURE 7.7. Comparison between neutron scattering data and fitted curves for two concentrated water-in-oil microemulsions (sodium dodecylbenzenesulfonate, hexanol, xylene, water system): (a) Water volume fraction $\phi = 0.356$, fitted core radius $R_c = 54.2A$. (b) $\phi_c = 0.436$, $R_c = 76.5A$ (from [67]).

factors. The opposite case, predicted by the theory, occurs when the film is flexible and without preferred curvature. The local curvature of the film may fluctuate strongly, so that the structure of the microemulsions may be considered as random at large scales. In practice one changes the preferred curvature of the film by varying different parameters: the ionic strength and the amount of cosurfactant for ionic systems, and the temperature for non ionic ones. The systems where these variations of curvature are particularly apparent are the Winsor phases. already encountered in Sec. 7.4.2.

7.4.4.1   Random microemulsions: The Winsor systems

a. *Composition and Phase Equilibria*: A mixture of comparable amounts of oil and brine, containing a small quantity of ionic surfactant ($\approx 5\%$) and cosurfactant, demixes very often into two or three phases, first observed by Winsor [56]. One of these phases is a microemulsion presenting several remarkable properties: extremely low interfacial tension ($\approx 10^{-3}$ dyne/cm), and maximal solubilization of oil and water for a given quantity of surfactant. Both properties could be used to advantage for enhanced oil recovery, and this fact has stimulated considerably the study of these systems [69]. (The fascinating interfacial properties of the Winsor microemulsions in particular are described in Chap. 10.)

The Winsor phase equilibria depend on the water ionic strength, in a most essential way. One observes in general three regimes:

1. At low salinity, an oil-in-water microemulsion ($\phi \ll 1$) coexists with an oil excess.

2. At high salinity, the equilibrium is inverted—a water-in-oil microemulsion ($\phi_w \ll 1$) coexists with an excess of oil.

3. In-between the equilibrium is intermediate—a microemulsion middle phase containing comparable proportions of oil and water coexisting with both a water phase and an oil phase.

A similar sequence of phases (presenting similar interfacial properties) is also observed in the case of non ionic surfactant, as the temperature increases [70]. A progressive inversion of structure is realized with both types of surfactant.

If one accepts the correspondence between curvature and sequence of phases, the vicinity of the point where the Winsor middle phase contains equal proportions of oil and water is particularly interesting, because the spontaneous curvature of the film then vanishes, or at least is very small.

Different systems made with ionic or non ionic surfactants [71] have been studied; they are either true Winsor phases or monophasic microemulsions close to the Winsor phases in the phase diagrams. The most studied one is probably that described initially in [63]. The samples contain by weight 46.3% of brine, 47.2% of toluene, 2% of SDS and 4% of butanol; the salinity, defined as a weight percentage of $NaCl$ in water, is varied between 3 and 10%. In this system, the optimal salinity corresponding to equal proportions of oil and water in the microemulsion phase is $S = 6.5\%$. Monophasic microemulsions made with the same constituents at this optimal salinity have also been obtained [30]. It was then possible to vary systematically the surfactant concentration and the oil volume fraction over a very large range ($0.2 < \phi_o < 0.9$), confirming the absence of preferred curvature.

b. *Models of structure*: Possible random structures of the microemulsions are described by simplified models: the system volume is divided into elementary cells of mean size ξ, randomly filled by oil and water, the surfactant being distributed at the oil-water interface generated by the filling procedure. These models [20,21,72,73] were first constructed to describe the thermodynamic properties of the microemulsions and account for the observed phase behavior (see following chapter); from the structural point of view they differ mainly by the way the random geometry is generated. In the first model, proposed by Talmon and Prager, the cells are generated by a Voronoi tessellation (Fig. 7.8a), with a broad size distribution. This model is adequate to evaluate the entropy and surface energy of the film, but does not incorporate the constraints imposed by the curvature energy.

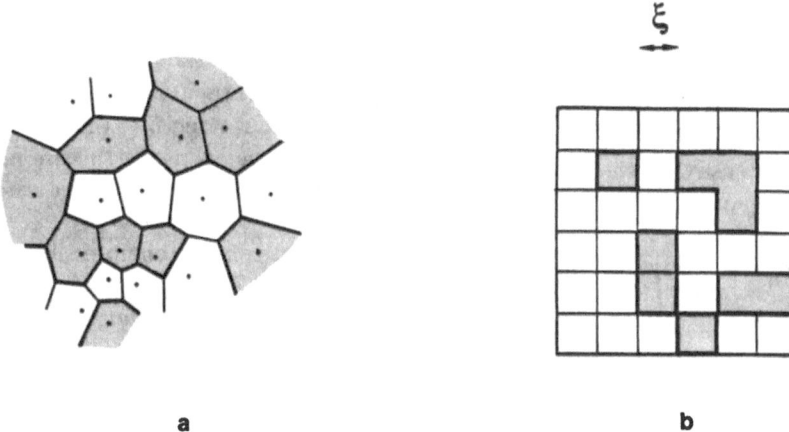

FIGURE 7.8. Structural models of random microemulsions: (a) Talmon-Prager model based on a geometry of cells generated by a Voronoi tesselation (from [20]). (b) de Gennes model.

Arguing that the film stiffness forbids the curvature fluctuations at spatial scales smaller than ξ_K, the persistence length of the surfactant layer, de Gennes, Levinson and Jouffroy assume a coarse-grained lattice of identical cubic cells of size $\xi = \xi_K$ (Fig. 7.8b). In both models, by contrast to Eq. (7.1) expressing the radius R of, for example, water-in-oil spheres, one finds $R = 3\phi_w/c_s a$. Furthermore, the mean size ξ of the elementary oil and water volumes, i.e., the length scale of the random structure, is related to the microemulsion composition by the constraint:

$$\xi = 6\phi_o\phi_w/c_s a \qquad (7.21)$$

ϕ_o and ϕ_w are the oil and water volume fractions ($\phi_o + \phi_w = 1$); c_s is the surfactant concentration; and a is the area per surfactant molecule in the film.

(7.21) points up the fact that in the random models, the area per unit volume of the film $c_s a$ is proportional to the number of contacts between oil and water cells $\phi_o\phi_w$; it interpolates continuously between the case of oil-in-water droplets ($\phi_w \ll 1$) and the case of oil-in-water droplets ($\phi_o \ll 1$). This means that the average mean curvature of the film $< C >$ is negative for $\phi_o < 0.5$ and positive for $\phi_o > 0.5$. It vanishes by symmetry when a microemulsion contains as much oil as water, i.e., at the inversion point $\phi_o = \phi_w = 0.5$.

c. *Predictions of the scattering intensity and experiments*: Contrary to the case of the spherical droplets, there is up to now no systematic theoretical prediction of the scattering by completely realistic random microemulsions. This has, however, not hindered the development of experimental studies, for several reasons:

1. Particular predictions based on the Talmon-Prager [20] or de Gennes [21] picture have been proposed. They provide useful elements of comparison for the shape of the spectra. We mention first the initial calculation of Kaler and Prager [74]. By neglecting all correlations between the elementary volumes of oil and water in the random structure, the scattering intensity by the oil and water is reduced to the average structure factor of an elementary cell of the Voronoi tessellation; it therefore decreases monotonously as the scattering vector increases.

2. Thermodynamic calculations of the partial osmotic compressibilities have been performed in different approximations [72,75]. They offer valuable points of reference for the intensities scattered at vanishing angle.

3. General scattering theory arguments as well as the practical possibility offered by the contrast variation method in neutron scattering (to separate the partial structure factors associated with individual constituents of a microemulsion) enable one to obtain conclusions on the structure, particularly on the film curvature, that are to a large extent model independent. We will emphasize these features in particular.

7.4.4.2 The size scale of the random microemulsions

a. *Procedures of estimation*: Because of the existence of a well defined microscopic interface between the oil and the water, it is possible to draw directly from the scattering spectra an estimation of the size scale of the microemulsion.

As the scattering vector decreases, from infinity, there is a range where q^{-1} becomes comparable to the average radius of curvature of the interface. In this range the scattering intensity departs from the asymptotic behavior discussed in Sec. 7.2. The observation of this deviation allows one to estimate a certain average radius of curvature of the interface. This idea is due to Kirste and Porod [76], who calculated the corresponding correction for a two-densities medium. In our case this corresponds to a distribution of scattering length density, where the contrast between the film and the oil is matched. The Kirste-Porod correction is expressed by the following formula, valid at large scattering vector:

$$I(q) = (n_w - n_o)^2 \frac{S}{V} q^{-4} \left[1 + q^{-2} \left\{ \frac{1}{4} - \langle (C_1 + C_2)^2 \rangle + \frac{1}{8} \langle (C_1 - C_2)^2 \rangle \right\} \right] \tag{7.22}$$

S/V is the surface area per unit volume of the medium ($c_s a$ for microemulsions), C_1 and C_2 are the local principal curvatures of the interface, and the brackets indicate that the average value is calculated over all the interface.

This expression shows that the q^{-6} curvature correction to Porod's law is always positive. As $q^4 I(q)$ vanishes at $q = 0$, this quantity gives rise to a bump before reaching the Porod plateau (Fig. 7.1a).

The Kirste-Porod correction enables one in principle to evaluate a certain mean square average of the film curvature. In practice, however, it is difficult to use it directly because the asymptotic development (7.22) diverges in the range of interest. When the q^{-6} term is sufficiently large to be measurable, the following term in the development (which may not be a power of q^{-1}) may also be large. Nevertheless, expression (7.22) justifies different phenomenological procedures, which have been used to estimate the size scale of the elementary water and oil domains in Winsor microemulsions.

i. The first estimation procedure consists in locating the cross-over between the asymptotic regime (where the film is seen as flat) and the regime of very small scattering vectors. This cross-over is well defined if there is only one characteristic length, either the persistence length of the film ξ_K for the random microemulsions or the radius R for the droplet microemulsions; it occurs at a characteristic scattering vector $\tilde{q}$ such that $\tilde{q}\xi_K \approx 1$.

Because the bump in the representation $q^4 I(q)$ as a function of q is related to the curvature of the interface, a simple and precise definition of $\tilde{q}$ is the abscissa value $\tilde{q}_2$ of the first maximum of $q^4 I(q)$. But this requires that the correct two-densities contrast condition associated with expression (7.22) be satisfied; if this condition is not fulfilled, the definition of $\tilde{q}$ is not so easy. With certain surfactants, like SDS, the electron density of the surfactant polar heads is higher than that of the other constituents; the X-ray scattering intensity is then dominated at large q by the contribution of the interfacial film ($I(q)$ decreases as q^{-2}), while at smaller q the volume contribution of the oil and water largely exceeds the surface contribution from the polar heads. The quantity $q^4 I(q)$ presents successively a maximum at $\tilde{q}_2$ and a minimum at $\tilde{q}_1$ before reaching the asymptotic interfacial regime in q^2 [30] (Fig. 7.1c and Fig. 7.9). Experimentally, the two characteristic scattering vectors $\tilde{q}_1$ and $\tilde{q}_2$ are proportional: $\tilde{q}_1 = 1.5\tilde{q}_2$. Both can be used to estimate the size scale of the dispersion. In practice $\tilde{q}_1$, the larger of the two, is less affected by collimation effects.

This procedure of size estimation has been tested on dilute and concentrated water-in-oil droplet microemulsions, the radius R of the droplets being known independently. In the neutron scattering experiments made under conditions of two-phases contrast ($n_o \neq n_w, n_o = n_f$), one obtains the relation:

$$R = 3\tilde{q}_2^{-1} \tag{7.23}$$

In the X-ray experiments, with the particular contrast imposed by the SDS molecules, one measures:

$$R = 4.6\tilde{q}_1^{-1} = 3\tilde{q}_2^{-1} \tag{7.24}$$

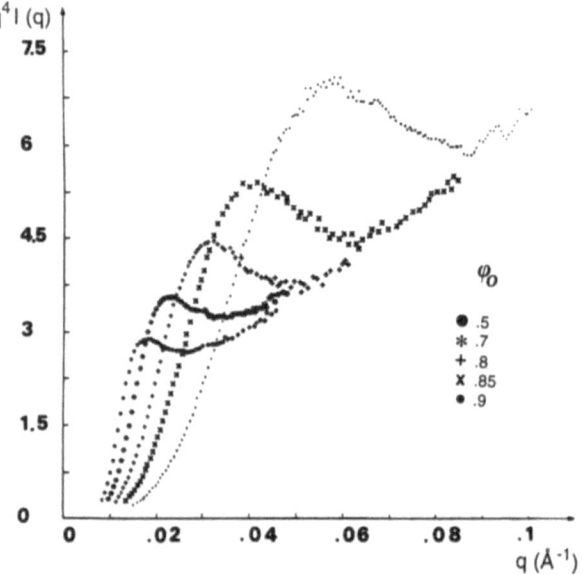

FIGURE 7.9. X-Ray scattering experiments on Winsor microemulsions at constant surfactant concentration for different oil volume fractions. $q^4 I(q)$ versus q. In this representation, the displacement of the bump with ϕ_0 shows directly that the average radius of curvature of the interfacial film decreases as ϕ_0 increases (from [30]).

Since this procedure of size estimation yields coherent results for spherical droplets, one can therefore assume its validity in other cases.

ii. Another (related) procedure is the following. When the scattering vector q is of the order of $\tilde{q}(0.15 < q/\tilde{q}_1 < 0.8)$, one notices [30,77,78] in the X-ray experiments that the scattering spectra are well approximated by a Guinier law according to which $I(q) \sim \exp[-(q^2 R^2_{g\,app})/3]$, although one is not in the very low-q Guinier range and the hypothesis of non-interacting objects is clearly inapplicable. An explanation of this feature is that the product $q^4 \exp(-q^2 R^2_{g\,app}/3)$ reproduces well the bump of the $q^4 I(q)$ versus q plot. The measurements of $R_{g\,app}$ yield another way to estimate the size scale of the microemulsions. One observes [30] $R_{g\,app} = 3.8\tilde{q}_1^{-1}$.

iii. Another approach is to Fourier transform the data. The expression (7.22) can be written in real space as a development in powers of r of the correlation functions of the oil or water [76]. The experimental determination of the behavior of these correlation functions at short distance, obtained by the Fourier inversion of the experimental intensity, yields a third tool for estimating the size of the elementary domain of a microemulsion [79,80,81,82].

b. *Results*: It is satisfying to note that all the procedures presented above yield similar results for the Winsor microemulsions and their monophasic equivalents. Three kinds of variations are performed:

i. Salinity scans along the demixing lines of the Winsor equilibria;

ii. Scans of oil volume fraction at constant surfactant concentration; and

iii. Scans of surfactant concentration at constant oil volume fraction.

In each case a size ξ_{exp} is measured. The main result is that in the inversion zone $(0.3 < \phi_o < 0.7)$, one observes experimentally the relation of proportionality:

$$\xi_{exp} \sim \frac{\phi_o \phi_w}{c_s a} \qquad (7.25)$$

The coefficient of proportionality depends on the practical definition of ξ_{exp}. In terms of a pseudo-radius of gyration, one gets, for instance, $R_g = 0.6\xi_K$ (a being defined by the relation (7.21)). The variations of the characteristic length of the structure of the Winsor microemulsions with the composition are thus in good agreement with the predictions of the models of random bicontinuous structure. This observation [77,78,30] was the first precise indication of the existence of such structures.

There are two other results of immediate interest here:

1. In the different scans performed there is a continuous evolution from the random bicontinuous structures in the inversion zone to discrete structures of droplets as the oil or water content increases; on the water-rich and oil-rich sides, one observes respectively $\xi_{exp} \sim \phi_o/c_s a$ if $\phi_o < 0.3$ and $\xi_{exp} \sim \phi_w/c_s a$ if $\phi_o > 0.7$.

2. The Winsor triphasic systems demix at constant ξ_K [77,78,83], i.e., at constant persistence length in the de Gennes model. This is in agreement with the theory if one admits that the variation of ionic strength does not modify the rigidity of the film, the partition coefficients of the alcohol remaining approximately constant. One may thus compare the rigidities associated with different surfactants [83].

7.4.4.3   Structure of the oil and water domains and rigidity of the interfacial film

We now restrict the discussion to Winsor microemulsions in the inversion zone $(0.3 < \phi_o < 0.7)$ and focus on the large scale $(r > \xi_K)$ arrangements of the oil and water. This corresponds to scattering spectra recorded under the usual conditions of contrast $n_o \neq n_w$ within the range of small scattering vectors, $q\xi_K < 3$, which is complementary to that investigated above.

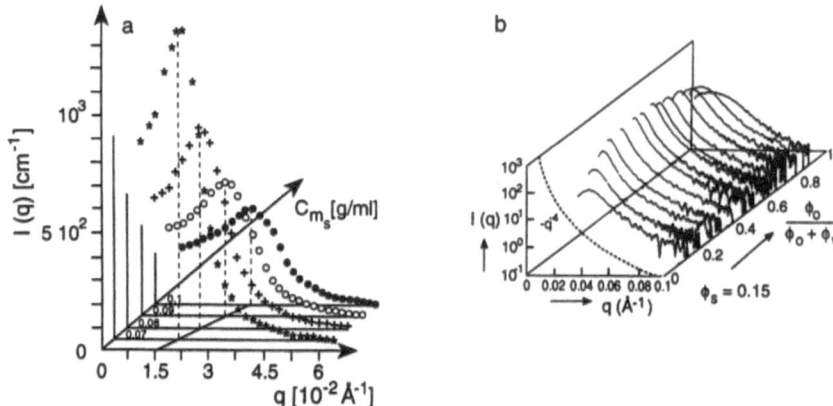

FIGURE 7.10. Intensity scattered at small angle by Winsor microemulsions in the zone of structure inversion. The scattering originates from the contrast between oil and water. (a) Water, toluene, butanol, SDS, NaCl system. Monophasic microemulsions containing as much oil as water, at different surfactant concentrations: neutron scattering experiments (from [32]). (b) Water, n-tetradecane, $C_{12}E_5$. Different water-to-oil ratios at constant surfactant volume fraction $\phi_s = 0.15$ (from [71]).

The central experimental result is the existence of a pronounced peak at a given scattering vector q^*, in both ionic and non ionic systems [30,32,71, 80-83] (Fig. 7.10a and 7.10b).

1. At fixed surfactant concentration, q^* is minimal at the inversion point $\phi_o = 0.5$ and decreases on each side of this point. Concomitantly, the scattering intensity is maximal at $\phi_o = 0.5$.

2. At fixed oil volume fraction ($\phi = 0.5$), q^* is proportional to the surfactant concentration. For the SDS system [32], $q^* = 122c_s$, with c_s in Å^{-3} and q in Å^{-1}. The intensity scattered at $q = 0$ decreases inversely with the cube of the surfactant concentration.

3. In the salinity scan of the Winsor middle phases, q^* is independent of the salinity [79-82].

These results show first that the Kaler-Prager calculation does not describe the scattering spectra, implying that the correlations between adjacent domains cannot be neglected. They also suggest that the position of the peak is related to the size ξ of the oil and water domains.

At $\phi_o = 0.5, \xi = 6\phi_o\phi_w/c_s a = 3/2c_s a$, using $a = 60\text{Å}^2$ (SDS system), one deduces the experimental relation $q\xi = 3$; the Bragg distance associated with the scattering peak is therefore:

$$2\pi q^{*-1} = 2\xi \tag{7.26}$$

The forward scattering intensity is also related to the size of the oil and water domains. It decreases as the size decreases. If the entropy of dispersion of the oil and water is the dominant term in the free energy of the microemulsions, the theory of random microemulsions predicts that $I(q = 0)$ has the form:

$$I(q = 0) \sim \phi_0 \phi_w \xi^3 \text{ with } \xi \simeq \frac{\phi_o \phi_w}{c_s a} \tag{7.27}$$

This explains quantitatively the dependence on surfactant concentration and qualitatively the variations with the volume fraction, which have not been tested precisely.

The existence of a peak in the spectra in the inversion zone has been puzzling for a long time: we know that it cannot be attributed to the existence of interacting objects, such as spherical droplets. One can, however, note that in this range of volume fraction ($\phi_o \approx 0.5$) the peak appears exactly at the position predicted by a model of droplets of radius $R = \xi = 3/2c_s a$. This coincidence indicates that the correlations between the oil (or water) domains are repulsive on a distance of order ξ in the Winsor microemulsions. As this can only occur if the size of the water and oil domains does not fluctuate very much, this is a strong argument in favor of the de Gennes picture, in opposition to the Talmon-Prager model [84]. Indeed, one can check that random bicontinuous structures with a well defined length scale, such as those arising in spinodal decomposition processes, exhibit naturally a peak in their scattering pattern [85].

The existence of a well defined length scale is theoretically related to the rigidity of the film. Recently, the de Gennes model has been extended to describe more generally the influence of the film rigidity on the microemulsion structure and phase behavior [75]. There are two essential modifications:

1. First, as in an earlier model of Widom [72], the persistence length of the film defining the size of the cubic cells of the lattice is treated as a variational parameter;

2. Then, one takes into account that the thermal fluctuations at small length scales renormalize the rigidity modulus of the film. This modulus thus depends on the size ξ of the elementary domains. A mean field analysis of the concentration fluctuations then enables one to calculate the scattering structure factor of this model microemulsion. Among many results concerning the phase equilibria, the model predicts a peak in the intensity scattered by the oil and water at a position $q^* = \pi/\xi$. This is in perfect agreement with the experiments. As expected, the physical origin of the peak resides in the curvature elasticity of the surfactant layer, which stabilizes the fluctuations with wave vectors corresponding to the lamellar,

tubular and cubic morphologies. The validity of this description is also confirmed by the generation of real space configurations, with the same pair correlations arising as those predicted by the model. The similarity between the generated structures and images obtained by freeze-etching electron microscopy of real Winsor microemulsions is striking [75,26]. (See Sec. 9.4.3 for a thorough discussion of this point, and Fig. 9.14 in particular.)

7.4.4.4   Fluctuations and curvature of the surfactant film

The intensity scattered by the oil and water do not contain all the information on the structure and topology of the microemulsions in the inversion zone. In particular, the interpretation of the scattering peak remains ambiguous. There are two reasons for expecting different partial structure factors associated with the interfacial film of surfactant:

1. The difference of geometry between droplet microemulsions and random structures is more pronounced when one observes directly the interfacial film, and

2. The thermodynamic models of microemulsions, in conjunction with general geometrical considerations, predict a direct relationship between the partial osmotic compressibility relative to the surfactant film and its algebraic average curvature.

a. The water film cross structure factor is the Fourier transform of the water film correlation function:

$$S_{wf}(q) = \int d^3r \langle \delta\phi_w(0)\delta\phi_f(r)\rangle e^{i\vec{q}\cdot\vec{r}} \qquad (7.28)$$

$\delta\phi_w$ and $\delta\phi_f$ are the local fluctuations of the volume fraction of water and film, including the cosurfactant; one thus neglects the fluctuations of composition of the film. Recall that the cross-structure factor cannot be measured directly, but must be extracted from a series of contrast variation experiments. In most cases this can only be done in neutron scattering studies.

i. *Macroscopic correlations of the water and film*: If $\delta\phi_w$ and $\delta\phi_f$ describe the fluctuations of water and film volume fraction in a macroscopic volume V, the value of the water-film cross-structure factor is given in the thermodynamic limit, at vanishing scattering vector, by the relation:

$$S_{wf}(0) = V\langle \delta\phi_w\delta\phi_f\rangle \qquad (7.29)$$

Because of Schulman's condition, the area per surfactant in the interfacial film does not fluctuate in a first approximation. A fluctuation of the film volume fraction in the volume V is therefore directly proportional to the fluctuation of the interfacial area S in V: $\delta\phi_f = d(\delta S/V)$ (d, thickness of the film). As first noticed by Widom [72] and developed in [32], the cross-correlations then have a simple geometrical interpretation.

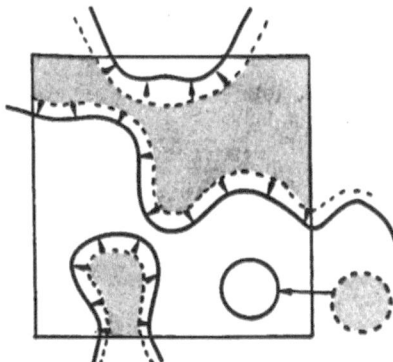

FIGURE 7.11. Fluctuations of interfacial area associated with variations of the oil and water volumes in a large volume V (from [84]).

Consider first the case of a droplet microemulsion, already discussed in Sec. 7.4.3. If the microemulsion is a dispersion of water-in-oil droplets, a water excess in a volume V, $\delta\phi_w > 0$, means a droplet excess in V implies a diminution of the number of oil droplets, hence a decrease of the surfactant volume fraction—$S_{wf}(0)$ is negative. This is the fundamental distinction explaining why the contrast variation experiments described in [57] can demonstrate the inversion of droplet microemulsions.

Let us consider now a random microemulsion and start from a given configuration of oil and water (broken line on Fig. 7.11). A small variation of the water volume $V\delta\phi_w$ in V displaces the oil-water interface. From the analytical definition of the mean curvature of the interface, one knows that to the first order in the displacement the variation δS of the interfacial area is proportional to the volume occupied by the film, that is to $\delta\phi_w$, and that the proportionality factor is the average mean curvature of the film $\langle C \rangle$. Up to a numerical coefficient and to the first order in d, the thickness of the film:

$$V\delta\phi_f = d\delta S \propto d\langle C \rangle V\delta\phi_w \qquad (7.30)$$

Therefore,

$$\begin{aligned} S_{wf}(0) &= V\langle\delta\phi_f\delta\phi_w\rangle \propto d\langle C\rangle V\langle\delta\phi_w^2\rangle \\ &= d\langle C\rangle S_{ww}(0) \end{aligned} \qquad (7.31)$$

$S_{ww}(0)$ is the water structure factor at vanishing scattering vector, proportional to the water partial osmotic compressibility.

In the mean field approximation the relation (7.21) between ξ and ϕ_w, valid at equilibrium, is still applicable for describing the large scale fluctuations in the volume V.

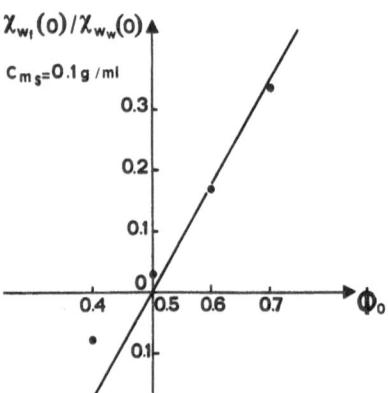

FIGURE 7.12. Variation of the ratio $S_{wf}(0)/S_{ww}(0)$ (noted $\chi_{wf}(0)/\chi_{ww}(0)$ on the figure) as a function of the oil volume fraction ϕ_0 at constant surfactant concentration for monophasic SDS Winsor microemulsions at the optimal salinity (from [84]).

From

$$\frac{S}{V} = c_s a = \frac{6\phi_w(1-\phi_w)}{\xi} \tag{7.32}$$

one deduces, assuming ξ constant:

$$\delta\frac{S}{V} \propto \frac{(1-2\phi_w)\delta\phi_w}{\xi} \tag{7.33}$$

Therefore the average curvature of the film, and $S_{wf}(0)$, are both proportional to $(1-2\phi_w)$. At the inversion point, when $\phi_o = \phi_w = 0.5$, the average curvature of the film vanishes and the fluctuations of the water and film are not correlated.

Figure 7.12 represents the experimental variation of the ratio $S_{wf}(0)/S_{ww}(0)$ as a function of ϕ_o at constant surfactant concentration for monophasic SDS Winsor microemulsions. This ratio increases when ϕ_o increases, and it is negative for $\phi_o < 0.5$. It vanishes and changes its sign at $\phi_o = 0.5 \pm 0.03$, in agreement with the prediction of the model of random microemulsions. This is a very strong and unambiguous indication that the curvature of the film vanishes on average at $\phi_o = 0.5$ in Winsor microemulsions.

In order to appreciate the sensitivity of the method of contrast variation, it is interesting to compare the experimental results with a model of water-

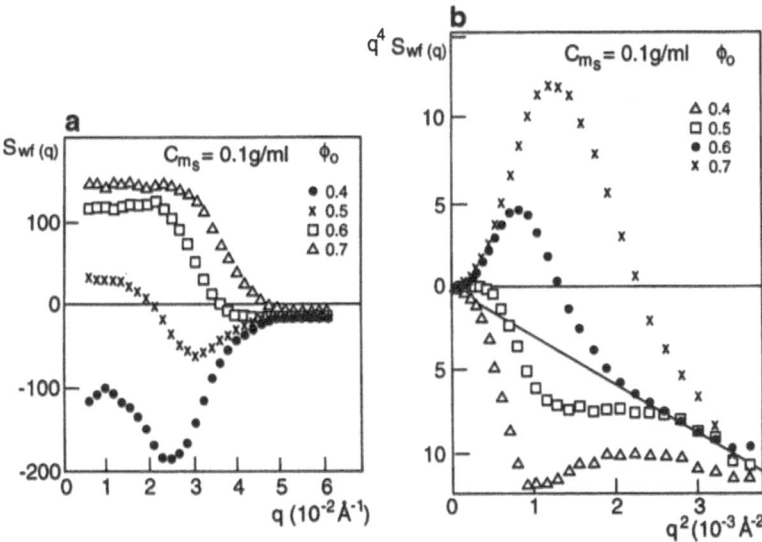

FIGURE 7.13. Water-film cross-structure factors of Winsor microemulsions for different oil volume fractions but constant surfactant concentration: (a) Representation $S_{wf}(q)$ versus q emphasizing the behavior at small angle. (b) Representation $q^4 S_{wf}(q)$ versus q^2 demonstrating the asymptotic behavior at large scattering vector (from [84]).

in-oil droplets. In this model, to first order in d:

$$\frac{S_{wf}(0)}{S_{ww}(0)} = \frac{3d}{R_w} \qquad (7.34)$$

R_w is the radius of the water core. With $d = 10$Å, $a = 60$Å^2 and $\phi_o = 0.5$, $3d/R_w = 0.25$. This large value is outside the range of the experimental uncertainty for the determination of $S_{wf}(0)$; a description of the $\phi_o = 0.5$ sample in terms of the droplet model is clearly not possible. Considering now the oil-rich sample $\phi_o = 0.7$, the droplet model yields $3d/R_w = 0.42$. This value is still larger than the observed $S_{wf}(0)/S_{ww}(0) = 0.33$. The curvature of the film in this sample is smaller than in an equivalent dispersion of droplets, implying that the structure of this water-in-oil microemulsion remains very connected.

ii. *Angular variations of the cross structure factor*: Figure 7.13 represents the variation of $S_{wf}(q)$ with oil volume fraction but constant surfactant concentration [84]. The samples are the same as those presented above in

Fig. 7.12. The curves show a plateau at the smallest scattering vector. The values $S_{wf}(0)$ discussed above are experimentally reached as soon as q is smaller than ξ^{-1}. This implies that the curvature of the film is completely averaged as soon as the spatial scales are larger than ξ and confirm that the structure is completely random at the larger scales $r > \xi$, in agreement with the de Gennes model and the real space simulations of [75].

For the largest observed scattering vector ($q\xi > 1, q < 0.06\text{Å}^{-1}$) the curves $S_{wf}(q)$ reach a common negative asymptotic limit. This limit is interesting because it is also related to the average curvature of the film. As explained in the appendix, the different partial structure factors $S_{ij}(q)$, relative to the oil, water and film, are not independent in the domain of small angle scattering. At each point the concentration fluctuations of the different constituents are related by a condition of incompressibility: $\delta\phi_f(r) + \delta\phi_w(r) + \delta\phi_o(r) = 0$. This implies in particular that:

$$S_{wf}(q) = -\frac{1}{2}S_{ff}(q) + \frac{1}{2}(S_{oo}(q) - S_{ww}(q)) \qquad (7.35)$$

We know from Porod's law that at large angle $S_{ww}(q)$ is proportional to the surface area S_{wf} of the water-film interface; similarly $S_{oo}(q)$ is proportional to the area S_{of} of the oil-film interface. We thus write, using also the asymptotic behavior of $S_{ff}(q)$:

$$S_{wf}(q) = -\pi c_s a d^2 q^{-2} + \pi V^{-1}(S_{of} - S_{wf})q^{-4} \qquad (7.36)$$

As the thickness of the film is small, the difference between the two areas S_{of} and S_{wf} is directly proportional to the average curvature of the film, and therefore:

$$S_{wf}(q) = \pi c_s a \left(d\langle C \rangle q^{-4} - d^2 q^{-2} \right) \qquad (7.37)$$

The product $q^4 S_{wf}(q)$ is plotted as a function of q^2 in Fig. 7.13b. In the experiment, q remains small and only the onset of the asymptotic regime is observed. The q^{-2} term of expression (7.37) is nevertheless well observed for $q > 0.05$ for the two samples $\phi_o = 0.5$ and 0.6, which have the largest ξ values. For the samples of largest curvature, positive ($\phi = 0.7$) and negative ($\phi_o = 0.4$), $S_{wf}(q)$ deviates respectively from above and below the asymptotic limit in q^{-2}. These experiments are in qualitative agreement with Eq. (7.37) and consistent with the measurements at vanishing scattering vector.

b. *The film structure factor*: With appropriate contrast, such that $n_0 = n_w$, the scattering intensity of the interfacial film is directly measured: one observes the fluctuations of a random interface. This is in principle interesting, but the present state of interpretation of the data is less satisfying than for the other structure factors.

i. *Concentration fluctuations of the film*: Considering the results of the preceding section, one may ask if the fluctuations of the film in a macroscopic volume V are correlated with the fluctuations of water and oil. The

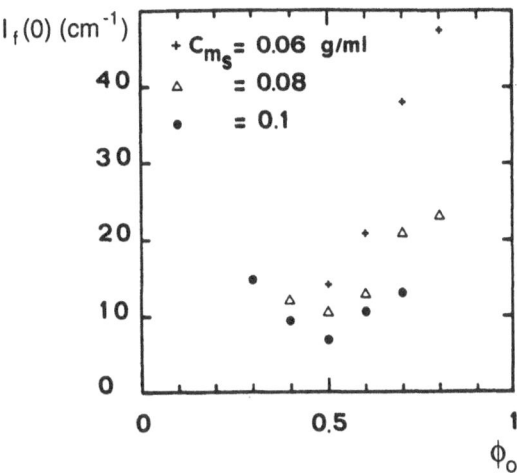

FIGURE 7.14. Intensity scattered at vanishing angle by the interfacial film of Winsor microemulsion as a function of the oil volume fraction for different surfactant concentrations (extrapolated values) (from [84]).

answer is evidently no. The fluctuations of the film at constant oil and water volume fraction do not contribute to $S_{wf}(0)$, but contribute to $S_{ff}(0)$. As a tentative approximation in order to understand the phenomena, we may, however, neglect this last kind of fluctuation. From the discussion of the preceding section, we obtain immediately an expression of the macroscopic mean square fluctuations of the film:

$$S_{ff}(0) \quad = \quad V\langle\delta\phi_f^2\rangle \propto \langle C\rangle^2 d^2 S_{ww}(0)$$

$$\propto \left[(1 - 2\phi_w)\left(\frac{d}{\xi}\right)\right]^2 S_{ww}(0) \qquad (7.38)$$

As already observed, $S_{ww}(0)$ has mainly an entropic origin in the model of random microemulsions: $S_{ww}(0) \propto (\phi_w(1 - \phi_w))^4/(c_s a)^3$. The above equation therefore leads to the very peculiar prediction that, at constant surfactant concentration, the intensity scattered by the film is minimal and in fact vanishes at the inversion point $\phi_o = 0.5$, where the intensity scattered by the water is maximal.

Figure 7.14 gathers the experimental results obtained on three series of samples at different surfactant concentrations [84]. In total contradiction to the predictions of any droplet model, the film intensity at zero angle is minimal at the inversion point in the three series, the minimum values being, however, not null. We thus conclude that the expression (7.38) and

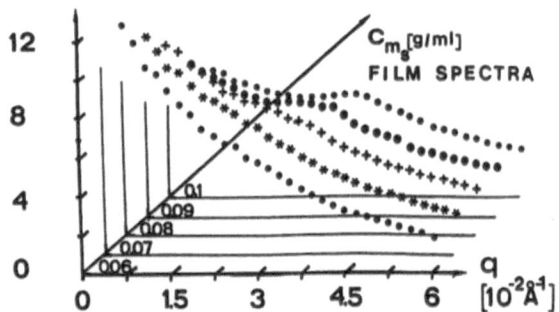

FIGURE 7.15. Intensity scattered by the interfacial film of Winsor microemulsions as a function of the scattering vector (water, toluene, butanol, SDS, NaCl system): monophasic microemulsions containing as much oil as water at different surfactant concentrations (from [32]).

the relation between curvature and macroscopic fluctuations of the film is sufficient to interpret qualitatively the shape of the observed variations, the contribution of the neglected modes of fluctuation being, however, clearly observable.

ii. *Angular dependence of the film intensity*: By comparison with the shape of the water structure factor, a surprising feature of the film structure factor is that it does not exhibit a peak, but is a monotonic decreasing function of the scattering vector, at least at sufficiently small surfactant concentration and sufficiently close to the inversion point [32,79–82] (Fig. 7.15). Although a particular model of random bicontinuous structure with a well defined length scale exhibits this characteristic [85], its origin is not understood. By continuity with the value of the intensity at $q = 0$, this absence of a peak is related to the existence of fluctuations of film concentration at constant oil and water volume fraction [84]. It would then be interesting to apply to the film scattering the formalism developed in [75] for the water and oil scattering, because one may conjecture that the film scattering at very small angle contains useful information on the interfacial rigidity.

7.5 Conclusions

Microemulsions are multicomponent disordered systems with a complex phase behavior. The determination of their structure is not always an easy task: it often necessitates the use of different techniques such as X-ray and neutron scattering, electron microscopy, N.M.R. and E.P.R. The versatility of these techniques has enabled in many instances the deduction of gross features and general tendencies that can be compared with those of theory. Above all, there are two facts to bear in mind:

1. Microemulsions are colloidal systems. Their colloidal structure is directly related to the existence of interfacial films of surfactant between oil and water. Films comprised of "short" surfactants (and cosurfactant) are characterized by local techniques (EPR with spin labels in particular) or scattering techniques. In the future, two kinds of films will certainly deserve more study:

a. Mixed films of surfactant and selected "impurities," in order to understand better the role of the cosurfactant,

b. The rather thick films of *polymeric* ("long") surfactants, which are fascinating model systems [7].

2. The microemulsion surfactant films are flexible. This is suggested indirectly by the absence of long-range order and the low viscosity of the phases, and is directly established by different experimental probes:

a. Observations of the fluctuations of the film driven by the curvature energy, either in the ordered birefringent microemulsions by EPR [35], light scattering [37] and X-rays scattering [38] or in droplet microemulsions by neutron spin-echo spectroscopy [47],

b. Structural studies of the Winsor systems:

i. Indicating a random bicontinuous dispersion of the oil and water;

ii. Showing that the average curvature of the film varies continuously with the volume fractions of oil and water and vanishes at the inversion point where both are equal; and

iii. Measuring length scales and correlation effects related to the film rigidity.

Indeed, these two basic facts do not suffice to complete our understanding of microemulsions structures: the film flexibility is not the only relevant factor.

In certain cases long range electrostatic interactions or the van der Waals forces play an important role; this has been investigated in droplet systems and ordered lamellar systems, and should be considered in more complicated structures.

The theory introduces also the spontaneous curvature of the film, c_o (without mentioning the saddle-splay rigidity $\bar{k}$), but no quantitative measurement method of this parameter exists at present. There are, however, clear experimental correlations between the existence of well-defined droplets, the observation of hard sphere repulsive interactions between them, and— depending on the continuous phase—the hydrophilic or lipophilic affinity of the film.

New structures, not discussed in this chapter, are being discovered. These "sponge phases" are locally lamellar [66] or tubular [27]. Although they are already described by certain geometrical models [27], much remains to be done to determine the constraints of curvature, packing and interaction, that lead to such structures.

7.6   Appendix: X-ray and Neutron Scattering

Expressions of the scattering intensity of X-rays and neutrons. Contrast.

Let us consider a sample illuminated by a monochromatic and collimated beam of X-rays or neutrons (wavelength λ, frequency ω). The waves associated with the incident particles are approximately plane waves of identical wave vector q_o ($q_o = 2\pi/\lambda$). Depending on the case, each nucleus or each electron of the sample interacts with the incident radiation (respectively neutrons or X-rays) and exchanges energy and momentum. Very often these energy transfers are very small in colloidal systems; interactions with the radiation are only weakly inelastic, and the scattering process is quasi-static. If multiple scattering is negligible, the intensity $I(q)$ scattered coherently per unit sample volume, per unit incident flux and detected in the solid angle $d\Omega$, is then written:

$$I(q) = \frac{1}{V} \sum_{i,j} \langle b_i \rangle \langle b_j \rangle \langle e^{i\mathbf{q}\cdot(\mathbf{r}_i - \mathbf{r}_j)} \rangle \qquad (7.39)$$

$I(q)$ has the dimension of a reciprocal length—it is a differential scattering cross-section per unit volume. $\mathbf{r}_i$ and $\mathbf{r}_j$ are the positions of the ith and jth scattering elements. V is the illuminated volume of the sample. The brackets implies that the thermodynamic average is taken. $\mathbf{q}$ is the scattering vector: its magnitude is $q = (4\pi/\lambda)\sin(\theta/2)$, θ being the scattering angle between the incident beam and the direction of observation. $\langle b_i \rangle$ is the coherent scattering length of the scattering element i that depends on its interaction with the incident particle:

1. In the scattering of X-rays, for the short wavelengths used and the small angles of observation considered, the electrons in the atoms behave as free electrons; the elementary process of interaction is the so-called Thomson scattering, and the scattering length b_e of one electron is therefore its classical radius $b_e = e^2/mc^2$ (e, electron charge, m, electron mass, c, velocity of light). The scattering length of an element of atomic number Z is then $b_Z = Zb_e$.

2. In the scattering of neutrons, b depends on the strong interactions between the neutrons and the nuclei and is not known theoretically. It is measured experimentally for each isotopic element and is eventually averaged over spin states and isotopic states of the elements, hence the notation $\langle b \rangle$. Let us recall that the scattering length of Hydrogen and Deuterium differ strongly—this the basis of the method of isotopic substitution and contrast variation.

It is clear from the preceding expression for the scattering intensity that at a given angle of observation the spatial resolution of a scattering experiment is of order q^{-1}. For an incident wavelength of order 1–10Å this implies in particular that small angle scattering is by definition not sensitive to the details of the molecular structure.

By defining the scattering length b_α of a molecule α as the sum of the scattering length of its constituents, it is then interesting to reexpress the scattering intensity as:

$$I(q) = \frac{1}{V} \sum_{\alpha,\beta} \langle b_\alpha \rangle \langle b_\beta \rangle \int d^3r d^3r' \langle \rho_\alpha(\mathbf{r}) \rho_\beta(\mathbf{r}') \rangle e^{i\mathbf{q}\cdot(\mathbf{r}-\mathbf{r}')} \qquad (7.40)$$

$\rho_\alpha(\mathbf{r})$ being the local concentration (number of molecules/cm^3) of the α molecules. For a non vanishing scattering vector, it is natural to introduce the local fluctuation of concentration $\delta\rho_\alpha(\mathbf{r})$ around the mean value ρ_α:

$$I(q) = \frac{1}{V} \sum_{\alpha,\beta} \langle b_\alpha \rangle \langle b_\beta \rangle \int d^3r d^3r' \langle \delta\rho_\alpha(\mathbf{r}) \delta\rho_\beta(\mathbf{r}') \rangle e^{i\mathbf{q}\cdot(\mathbf{r}-\mathbf{r}')} \qquad (7.41)$$

This expression relates the scattering intensity to the correlation functions of the concentration fluctuations of the molecules α and β.

For systems such as colloidal solutions which are practically incompressible at the macroscopic scale, it is convenient to introduce the volume fraction $\phi_\alpha(\mathbf{r})$ of constituent α, $\phi_\alpha(\mathbf{r}) = v_\alpha \rho_\alpha(\mathbf{r})$, v_α being the partial molecule volume of the molecule α. The condition of incompressibility,

$$\sum_\alpha \phi_\alpha(\mathbf{r}) = 1 \text{ or } \sum_\alpha \delta\phi_\alpha(\mathbf{r}) = 0, \qquad (7.42)$$

indeed reduces the number of independent correlation functions necessary to describe the scattering intensity. Relevant parameters of the scattering experiment are then the scattering length densities of the constituents, $n_\alpha = \langle b_\alpha \rangle / v_\alpha$.

Thus for the particular case of microemulsions, if one may consider that the chemical composition of the water, oil and film parts does not fluctuate, one may average over the composition (including cosurfactant particularly and other additives) and write:

$$\begin{aligned} I(q) &= (n_w - n_o)^2 S_{ww}(q) + 2(n_w - n_o)(n_f - n_o) S_{wf}(q) + \\ &\quad (n_f - n_o)^2 S_{ff}(q) \end{aligned} \qquad (7.43)$$

n_o, n_w and n_f are the mean scattering length densities of the oil, water and film parts of the microemulsions, respectively; ϕ_o, ϕ_w and ϕ_f are the corresponding volume fractions. One thus describes the scattering of a medium which may contain at least three (but often four or five) constituents by three independent functions, the partial scattering structure factors, $S_{ij}(q)$ with $i, j = o$ or w:

$$S_{ij}(q) = \int d^3r \langle \delta\phi_1(\mathbf{O})\delta\phi_j(\mathbf{r})\rangle e^{i\mathbf{q}\cdot\mathbf{r}} \qquad (7.44)$$

The factors $n_i - n_j$ are called the contrasts between constituents i and j. As these contrasts can be varied by isotopic substitution in the neutron scattering experiments, one may determine the three partial structure factors from a series of at least three experiments performed in different conditions of contrast.

The partial structure factors clearly depend on the geometry of the medium, which may impose certain relations between them. This is discussed in the text. A particular example is that of a suspension of identically spherical droplets, which leads to a simple expression for the scattering intensity. Let us identify the droplets by a greek symbol α or β and write i_α for the scattering elements of the droplet α.

Then:

$$I(q) = \frac{1}{V} \sum_{i_\alpha j_\beta} b_{i_\alpha} b_{j_\beta} e^{\mathbf{q}\cdot(\mathbf{r}_{i_\alpha} - \mathbf{r}_{j_\beta})} \qquad (7.45)$$

(We have already used the condition of incompressibility.) b_{i_α} is a scattering length with respect to the solvent: $b_{i_\alpha} = v_{i_\alpha}(n_{i_\alpha} - n_s)$, n_s being the scattering length density of the solvent. Let us introduce $\mathbf{R}_\alpha$ for the position of the center of the droplet α and write $\mathbf{u}_{i_\alpha} = \mathbf{r}_{i_\alpha} - \mathbf{R}_\alpha$. Then

$$\sum_{i_\alpha} b_{i_\alpha} e^{i\mathbf{q}\cdot\mathbf{r}_{i_\alpha}} = \sum_\alpha e^{i\mathbf{q}\cdot\mathbf{R}_\alpha} A_\alpha(q) \text{ with } A_\alpha(q) = \sum_{i_\alpha} b_{i_\alpha} e^{i\mathbf{q}\cdot\mathbf{u}_{i_\alpha}} \qquad (7.46)$$

For monodisperse particles with spherical symmetry, the factor $A_\alpha(q)$ is identical for each particle, i.e., independent of α. If N is the total number of particles in the volume V and $C = N/V$ the particle concentration, we arrive at the result:

$$I(q) = CP(q).S(q) \qquad (7.47)$$

with

$$P(q) = |A(q)|^2 \qquad (7.48)$$

and

$$S(q) = 1 + \frac{1}{N} \sum_{\alpha \neq \beta} e^{i\mathbf{q}\cdot(\mathbf{R}_\alpha - \mathbf{R}_\beta)} \qquad (7.49)$$

$P(q)$ and $S(q)$ are respectively called the form and structure factors.

References

1. T. P. Hoar and J. H. Schulman, Nature **102**, 152 (1943).

2. J. H. Schulman and D. P. Riley, J. Colloid Sci. **3**, 383 (1948).

3. J. H. Schulman and J. A. Friend, J. Colloid Sci. **4**, 497 (1949).

4. K. Shinoda and S. Friberg, Adv. Colloid Interface Sci. **4**, 281 (1975).

5. *Microemulsions, Theory and Practice*, L. M. Prince, ed. (Academic Press, New York, 1977).

6. *Micellization, Solubilisation and Microemulsions*, K. L. Mittal, ed. (Plenum Press, New York, 1977).

7. P. G. de Gennes and C. Taupin, J. Phys. Chem. **86**, 2294 (1982).

8. *Surfactants in Solutions*, vol. 1–3, K. L. Mittal and B. Lindman, eds. (Plenum Press, New York, 1984).

9. A. M. Bellocq, J. Biais, P. Bothorel, B. Clin, G. Fourche, P. Lalanne, B. Lamaire, B. Lamanceau and D. Roux, Adv. Colloid Interface Sci. **20**, 167 (1984).

10. *Colloides et Interfaces*, A. M. Cazabat and M. Veyssié, eds., Les Editions de Physique, 1984.

11. *Physics of Amphiphiles: Micelles, Vesicles and Microemulsions*, M. Corti and V. de Giorgio, eds. (North Holland, 1985).

12. *Surfactants in Solutions*, vol. 4–6, K. L. Mittal and P. Bothorel, eds. (Plenum Press, 1986).

13. *Physics of Complex and Supramolecular Fluids*, S. A. Safran and N. A. Clark, eds. (Wiley, New York, 1987).

14. *Microemulsion Systems*, H. L. Rosano and M. Clausse, eds., Surfactant Science Series, Vol. 24 (Marcel Dekker, New York, 1987).

15. *Physics of Amphiphilic Layers*, J. Meunier, D. Langevin and N. Boccara, eds. (Springer, Heidelberg, 1987).

16. *Microemulsions: Structure and Dynamics*, S. E. Friberg and P. Bothorel, eds. (CRC Press, 1987).

17. S. A. Safran, J. Chem. Phys. **78**, 2073 (1983).

18. W. G. M. Agterhof, J. A. J. van Zomeren and A. Vrij, Chem. Phys. Lett. **43**, 363 (1976).

19. W. Helfrich, Z. Naturforsch. C **28**, 693 (1973).

20. Y. Talmon and S. Prager, J. Chem. Phys. **69**, 2984 (1978).

21. P. G. de Gennes, J. Jouffroy and P. Levinson, J. de Physique **43**, 1241 (1982).

22. L. E. Scriven, Nature **63**, 123 (1976).

23. L. E. Scriven, in [5], vol. 2, p. 877.

24. *Surfactant Solutions: New Methods of Investigation*, R. Zana, ed., Surfactant Science Series, Vol. 22 (Marcel Dekker, New York, 1987).

25. J. Dubochet, M. Adrian, J. Teixeira, C. Alba, R. K. McFarlane and C. A. Angell, J. Phys. Chem. **88**, 6727 (1984).

26. W. Jahn and R. Strey, in [14], p. 353.

27. Th. Zemb, S. T. Hyde, P. J. Derian, I. S. Barnes and B. W. Ninham, J. Phys. Chem. **91**, 3814 (1987).

28. J. Biais, B. Clin and P. Lalanne, in [15], p. 1.

29. *Nonionic Surfactants: Physical Chemistry*, M. J. Schick, ed., Surfactant Science Series, Vol. 23 (Marcel Dekker, New York, 1987).

30. L. Auvray, J. P. Cotton, R. Ober and C. Taupin, J. de Physique **45**, 913 (1984).

31. J. H. Schulman, J. B. Montague, Ann. N. Y. Acad. Sci. **92**, 366 (1961).

32. L. Auvray, J. P. Cotton, R. Ober and C. Taupin, J. Phys. Chem. **88**, 4586 (1984).

33. G. Porod, Koll. Z. **124**, 83 (1951).

34. J. M. de Meglio, M. Dvolaitzky and C. Taupin, J. de Physique Lett. **44**, L-229 (1983).

35. J. M. di Meglio, M. Dvolaitzky and C. Taupin, J. Phys. Chem. **89**, 871 (1985).

36. J. M. di Meglio, M. Dvolaitzky and C. Taupin, in [14], p. 199.

37. J. M. di Meglio, M. Dvolaitzky, L. Léger and C. Taupin, Phys. Rev. Lett. **56**, 1686 (1985).

38. C. R. Safinya, D. Roux, G. S. Smith, S. K. Sinha, P. Dimon, N. A. Clark and A. M. Bellocq, Phys. Rev. Lett. **57**, 2718 (1986).

39. J. C. Ravey, M. Buzier and G. Dupont, in [14], p. 163.

40. S. Safran, L. Turkevich and P. Pincus, J. de Physique Lett. **45**, L-69 (1984).

41. G. Porte, J. Appell and Y. Poggi, J. Phys. Chem. **84**, 3105 (1980).

42. S. J. Candau, F. Hirsch and R. Zana, in [13], p. 569.

43. M. Dvolaitzsky, M. Lagúes, J. P. Lepesant, R. Ober, C. Sauterey and C. Taupin, J. Phys. Chem. **84**, 1532 (1980).

44. M. Kotlarchyk, S. H. Chen and J. S. Huang J. Phys. Chem. **86**, 3273 (1982).

45. M. Kotlarchyk, S. H. Chen and J. S. Huang J. Chem. Phys. **79**, 2461 (1983).

46. C. T. Meyer, Y. Poggi and G. Maret, J. de Physique **43**, 827 (1982).

47. J. S. Huang, S. T. Milner, B. Fargo and D. Richter, Phys. Rev. Lett. **59**, 2600 (1987).

48. R. Ober and C. Taupin, J. Phys. Chem. **84**, 2418 (1980).

49. A. M. Cazabat, in [10], p. 323.

50. S. Brunetti, D. Roux, A. M. Bellocq, G. Fourche and P. Bothorel, J. Phys. Chem. **87**, 1028 (1983).

51. J. S. Huang, S. A. Safran, W. M. Kim, M. Kotlarchyk and N. Quirke, Phys. Rev. Lett. **53**, 592 (1984).

52. B. Lamaire, P. Bothorel and D. Roux, J. Phys. Chem. **87**, 1023 (1983).

53. L. A. Turkevich, in [13], p. 241.

54. L. Auvray, J. de Physique Lett. **46**, L-163 (1985).

55. A. M. Cazabat, D. Chatenay, P. Guering, W. Urbach, D. Langevin and J. Meunier, in [14], p. 183.

56. P. A. Winsor, *Solvent Properties of Amphiphilic Compounds* (Butterworths, London, 1954).

57. M. Laguës, R. Ober and C. Taupin, J. de Physique Lett. **39**, 487 (1978).

58. B. Lindman and P. Stilbs, in [14], p. 129.

59. B. Lindman and P. Stilbs, in [16], p. 119.

60. B. Lindman, T. Ahnoläs, O. Söderman, H. Walderhaug, K. Rapacki and P. Stilbs, Faraday Disc. Chem. Soc. **76**, 317 (1983).

61. P. Guérin and B. Lindman, Langmuir **1**, 464 (1985).

62. M. T. Clarkson, D. Beaglehole and M. T. Callaghan, Phys. Rev. Lett. **54**, 1722 (1985).

63. A. M. Cazabat, D. Langevin, J. Meunier and A. Pouchelon, Adv. Colloid Interface Sci. **16**, 175 (1982).

64. P. Pincus, S. Safran, S. Alexander and D. Hone, in *Physics of Finely Divided Matter*, N. Boccara and M. Daoud, eds., Springer Proceedings in Physics, vol. 5, 1985.

65. M. Dvolaitzky, M. Guyot, M. Lagües, J. P. Lepesant, R. Ober, C. Sauterey and C. Taupin, J. Chem. Phys. **69**, 3279 (1978).

66. M. Kotlarchyk, quoted in S. H. Chen, Ann. Rev. Phys. Chem. **37**, 351 (1986).

67. D. J. Cebula, R. H. Ottewil, J. Ralston and P. N. Pusey, J. Chem. Soc. Faraday Trans. I **77**, 2585 (1981).

68. D. J. Cebula, O. Y. Myers and R. H. Ottewil, Coll. Polym. Sci. **250**, 96 (1982).

69. *Improved Oil Recovery by Surfactant and Polymer Flooding*, D. O. Shah and R. S. Schechter, eds. (Academic Press, New York, 1977).

70. K. Shinoda and H. Kunieda, J. Colloid Interface Sci. **42**, 381 (1973).

71. F. Lichterfold, T. Schmeling and R. Strey, J. Phys. Chem. **90**, 5762 (1986).

72. B. Widom, J. Chem. Phys. **81**, 1030 (1984).

73. D. Andelman, M. E. Cates, D. Roux and S. A. Safran, J. Chem. Phys. **87**, 7229 (1987).

74. E. W. Kaler and S. Prager, J. Colloid Interface Sci. **86**, 359 (1982).

75. S. T. Milner, S. A. Safran, D. Andelman, M. E. Cates and D. Roux, J. de Physique **49**, 1065 (1988).

76. R. Kirste and G. Porod, Koll. Z. u. Z. für Polym. **184**, 1 (1962).

77. E. W. Kaler, K. E. Bennett, H. T. Davis and L. E. Scriven, J. Chem. Phys. **79**, 5673 (1983).

78. E. W. Kaler, H. T. Davis and L. E. Scriven, J. Chem. Phys. **79**, 5685 (1983).

79. A. de Geyer and J. Tabony, Chem. Phys. Lett. **113**, 83 (1985).

80. A. de Geyer and J. Tabony, Chem. Phys. Lett. **124**, 357 (1986).

81. J. Tabony and A. de Geyer, in [12], p. 1287.

82. A. de Geyer and J. Tabony, in [15], p. 372.

83. D. Guest, L. Auvray and D. Langevin, J. de Physique Lett. **46**, L-1055 (1985).

84. L. Auvray, L. P. Cotton, R. Ober and C. Taupin, in [13], p. 449.

85. N. F. Berk, Phys. Rev. Lett. **58**, 2718 (1987).

86. M. E. Cates, D. Roux, D. Andelman, S. T. Milner and S. A. Safran, Europhys. Lett. **5**, 733 (1988).

8

Lattice Theories of Microemulsions

Gerhard Gompper[1]
Michael Schick[2]

8.1 Introduction: Aims of a Lattice Theory of Microemulsions

Ideally, one would like to have a microscopic basis for a theory of microemulsions. Such a theory would begin with the constituents of the system and produce from their known properties the observed behaviors of microemulsions. There are difficulties of many kinds in such an approach. Of course, the description of constituents must be simplified radically in order to extract general behaviors from particular systems. This simplification is common to the theoretical description of most physical systems. There is an additional difficulty, however, which arises in the theoretical description of a microemulsion; this is the lack of agreement on its defining behaviors. It is prudent, then, for us to state at the outset those properties which we consider to be characteristic of microemulsions, and therefore to be encompassed by any theory of them.

8.1.1 PHASE BEHAVIOR

Systems of oil, water and surfactant exhibit an interesting and characteristic pattern of phase behavior [1,2]. An oil-rich phase can be made to coexist with another phase which contains almost all of the water and most of the surfactant. By changing some external parameter, three phase coexistence can be brought about between an oil-rich phase, a water–rich phase, and a "middle phase" containing most of the surfactant and in which the water and oil are solubilized. By changing the external parameter further, two phase coexistence returns, but this time it is between one which is water-rich and another which contains almost all of the oil and most of

---

[1]Sektion Physik der Ludwig-Maximilians-Universität München, 8000 München 2, Germany

[2]Department of Physics, University of Washington, Seattle, Washington 98195

the surfactant. In systems containing a non–ionic amphiphile, the external parameter is usually temperature: in those containing an ionic surfactant, it is often the concentration of a fourth component, salt. The points at which three phase coexistence appear are critical endpoints, and the path in the phase diagram which joins them is a triple line. By changing a second parameter, such points can be made to coalesce at a tricritical point [2-4].

The description of phase behavior is one of the elementary tasks of statistical mechanics, and the description of this particular pattern seems to be a reasonable goal.

8.1.2   LOW SURFACE TENSIONS

One of the most interesting properties of microemulsions is that all surface tensions can be made to be low, perhaps three orders of magnitude lower than normal oil/water surface tensions [5]. Of course at either critical end point, at which the difference between the middle phase and one of the other two phases vanishes, the surface tension between the two merging phases must vanish. However the tension between the oil-rich and water-rich phases, the tension which is often of the most commercial interest, does not. In fact, it is usually higher at the critical end points than at some intermediate point of three phase coexistence at which all surface tensions are low and comparable.

The qualitative, if not quantitative, calculation of surface energies is well within the purview of statistical mechanics. Furthermore, the discipline makes connections between phase behavior and surface properties as the vanishing of surface tensions at tricritical points illustrates. Hence, we might hope that we can obtain some insight into the reason why such low surface tensions can be observed in some systems and, by extension, what should be done in general terms to bring them about in others.

8.1.3   STRUCTURE

The structure of the microemulsion is far from clear. There is certainly a consensus that it consists of coherent regions of oil and of water separated by sheets of surfactant. While we do not expect a completely detailed picture of this phase to emerge from a microscopic calculation, we do think that it should be able to produce a structure factor which resembles that of experiment: one with a peak at a non-zero value of momentum transfer [6]—see Chap. 7. This peak reflects the existence of a large length in the problem, typically of the order of 100 Å. Presumably, this is a typical thickness of coherent regions of oil or water. A reasonable theory should show how this length arises.

8.1.4 EFFICIENT SOLUBILIZATION OF OIL AND WATER

This property is also assumed to be related to that of structure. Certainly the larger the coherent regions of oil and of water, the lower the volume density of surfactant needed to separate them and the more efficient the solubilization. It is not too much to ask, however, to demonstrate that variations in the degree of efficiency are correlated with either some microscopic parameter of the amphiphile or macroscopic property of the microemulsion.

8.1.5 RELATION TO LYOTROPIC PHASES

The key observation here is that as the efficiency of the amphiphile in solubilizing oil and water is increased, lyotropic phases (particularly lamellar phases), increasingly compete with the microemulsion in phase space; that is, the range of amphiphile concentration over which the microemulsion can be found decreases as the lamellar phases are more easily produced. A realistic goal would be to shed some light on the relation between the lyotropic phases and the microemulsion.

It is our belief that a satisfactory account of all of these behaviors is currently within the purview of the kind of lattice theories which this chapter describes.

8.2 The Lattice Formulation

In order to proceed with a theory of microemulsions, we must simplify our description of them. The trick, of course, is to eliminate everything unimportant, maintain everything essential, and to know the difference between the two. Because microemulsions occur for a variety of ternary fluid mixtures, we conclude that the specific composition of oil or amphiphile is not essential for the phenomena and deal only with three generic components which will be labeled a, b, and c, or water, oil, and amphiphile. The reduction to three components is itself a simplification as the role of cosurfactants is ignored as inessential for the understanding of the phenomena.

The next simplification which many, but not all, of us who work in this field have chosen to employ, is to replace the continuum system by a discrete one. In particular, we choose to divide space into cells of such size that, at most, only one component of the multi-component system can be found there. This reduces the problem to some variety of a lattice gas. The reason we choose to do this is simple. There is a large literature on, and much experience with, discrete models in statistical mechanics. Such experience is a useful guide in the inevitable decisions as to which physical features of the system are crucial to a successful description of the system, and which features are not. By contrast, there is less experience with continuum models. One only has to compare the developments of the theories of the solid state with that of the liquid state to appreciate just how much more difficult the latter is. Nevertheless, some attempts have been made at a continuum theory of surfactants in solution [7]. We shall discuss only lattice descriptions.

8.2.1   THE WIDOM-WHEELER MODEL [8,9]

Consider a hypercubic lattice in d dimensions with a spin $\sigma_i = \pm 1$ at each site. Every nearest-neighbor pair of $(+1,+1)$ spins is to be thought of as a water molecule, every $(-1,-1)$ pair as an oil molecule, and every $(+1,-1)$ pair as an amphiphile. Each spin at any particular site belongs to $2d$ such pairs which can be thought of as located on the bonds connecting the lattice sites. By this construction, oil and water must always be separated by an amphiphile which is oriented correctly with respect to them. There is a chemical potential for each kind of molecule, so that the Hamiltonian is simply

$$\mathcal{H} = -\sum_{\langle ij \rangle} (\mu^a P_{ij}^a + \mu^b P_{ij}^b + \mu^c P_{ij}^c) \,, \tag{8.1}$$

subject to

$$\sum_{\alpha} P_{ij}^\alpha = 1 \,, \tag{8.2}$$

where the $P_{ij}^\alpha = 1$ if there is a molecule of species $\alpha = a, b$, or c (water, oil, or surfactant) on the bond ij. The sum is over all such bonds. In this version [8], all configurations with the same numbers of molecules of each type cost the same energy. The Hamiltonian of this model is easily mapped to that of a spin 1/2 Ising model via

$$
\begin{aligned}
P_{ij}^a &= (1 + \sigma_i + \sigma_j + \sigma_i \sigma_j)/4, \\
P_{ij}^b &= (1 - \sigma_i - \sigma_j + \sigma_i \sigma_j)/4, \\
P_{ij}^c &= (1 - \sigma_i \sigma_j)/2,
\end{aligned} \tag{8.3}
$$

The result is a simple Ising Hamiltonian

$$\mathcal{H} = -J \sum_{\langle ij \rangle} \sigma_i \sigma_j - H \sum_i \sigma_i. \tag{8.4}$$

with

$$
\begin{aligned}
J &= (\mu^a + \mu^b - 2\mu^c)/4, \\
H &= (\mu^a - \mu^b)d/2 \,.
\end{aligned} \tag{8.5}
$$

With J positive, there are only two low temperature phases, oil-rich and water-rich, which become identical at a critical point. There is no three-phase coexistence.

To remedy this, Widom [9] added a term which would distinguish different configurations of the same amount of oil and water, the motivation being that configurations with large curvature should cost more than those with less. Thus an energy $K(1 - \lambda)$ is introduced for every pair of amphiphiles whose +1 ends meet at the same site, and an energy $K(1 - \lambda)$ if their -1 ends meet. The factor of λ can be used to represent an intrinsic curvature which distinguishes between oil and water. By drawing a few configurations, the reader will see that additional energy is indeed introduced whenever a sheet bends one way or the other. However, the energy also appears in a perfectly flat bilayer, e.g., a sheet of (+1,-1,+1) spins, so that the energy does not represent only a curvature term.

Because all molecules consist of two spins, interactions between two molecules translate in spin language into interactions between two spins at second and third neighbor distances, and three spin interactions. The sign of the new interaction is very interesting. Because the configuration (+1,-1,+1) now costs more energy, the interaction introduced in the spin model between the first and third spins in this chain is antiferromagnetic. The same is true if the three spins form a right angle.

The simplest case to analyze is that for which there is no intrinsic difference between oil and water so that $\mu^a = \mu^b$ and $\lambda = 0$. Then one has a simple Ising Hamiltonian. In three dimensions, it is of the form

$$\mathcal{H} = -J \sum_{\langle ij \rangle} \sigma_i \sigma_j - 2M \sum_{\langle\langle ik \rangle\rangle} \sigma_i \sigma_k - M \sum_{\langle\langle\langle il \rangle\rangle\rangle} \sigma_i \sigma_l \qquad (8.6)$$

where the first sum is over all nearest neighbor pairs, the second sum is over all second neighbor pairs, the third sum over all fourth neighbor pairs, and

$$\begin{aligned} J &= [10K + (\mu^a + \mu^b) - 2\mu^c]/4, \\ M &= -K/4 \end{aligned} \qquad (8.7)$$

The competition between the ferromagnetic ($J > 0$) first term and antiferromagnetic ($M < 0$) second and third term is known to lead to interesting phases in other systems such as the axial next nearest neighbor Ising (ANNNI) model [10]. It is precisely the availability of this kind of information from other areas of statistical mechanics which makes a lattice formulation so attractive. Some results from the Widom model will be discussed later in Sec. 8.4.

8.2.2 THE THREE COMPONENT MODEL [11,12]

We begin with a Hamiltonian describing a hypercubic, d dimensional, lattice gas of three components, $a, b,$ and c, again representing water, oil, and surfactant. All components sit in the sites of the lattice, and interact via pairwise interactions. The Hamiltonian is

$$\mathcal{H} = -\sum_{\alpha\beta} E_{ij}^{\alpha\beta} \sum_{(ij)} P_i^\alpha P_j^\beta - \sum_\alpha \mu^\alpha \sum_i P_i^\alpha, \qquad (8.8)$$

subject to

$$\sum_\alpha P_i^\alpha = 1 \qquad (8.9)$$

where the indices α, β, take the values a, b, c, $P_i^\alpha = 1$ if site i is occupied by species α and is zero otherwise. The sum is over all possible pairs of particles. In practice, we consider only nearest and next-nearest-neighbor interactions. In order to distinguish one species, c, as the amphiphile, we introduce a term into the Hamiltonian which reduces the energy of configurations acb or bca of three particles in a row, while increasing that of configurations aca and bcb. The simplest form of this term, (i.e., one which introduces only one new parameter into the problem), is

$$-\sum_{(ijk)} L(P_i^a P_j^c P_k^a + P_i^b P_j^c P_k^b - P_i^a P_j^c P_k^b - P_i^b P_j^c P_k^a) \qquad (8.10)$$

where the sum is over all sets of three sites in a row. Again there is a competition between the pairwise terms, which in general favor like particles to sit at all distances from one another, and the three particle term which favors the amphiphile to sit between water and oil rather than to be surrounded by any one species. Note that in contrast to the Widom model, oil and water need not always be separated by amphiphile; the latter can choose to sit between the other components or not depending on its chemical potential and the energy gain L. In some regions of the phase space, we expect to find a majority of the surfactants at oil/water interfaces and a minority to be in less favorable configurations. However, in other regions of phase space, particularly with large concentrations of surfactant, we expect to find them in a surfactant-rich phase.

The three component model is easily recast into a spin-one model via the relations

$$\begin{aligned}
P_i^a &= (1+S_i)S_i/2, \\
P_i^b &= -(1-S_i)S_i/2, \\
P_i^c &= (1-S_i^2),
\end{aligned} \qquad (8.11)$$

where the S_i take the values 1, 0, -1. If we restrict the pairwise interactions to those between nearest neighbors only, then the spin Hamiltonian is, up to a constant,

$$\begin{aligned}
\mathcal{H} &= -\sum_{\langle ij \rangle} [JS_iS_j + KS_i^2S_j^2 + C(S_i^2S_j + S_iS_j^2)] - \sum_i (HS_i - \Delta S_i^2) \\
&\quad - \sum_{\langle ijk \rangle} LS_i(1 - S_j^2)S_k,
\end{aligned} \qquad (8.12)$$

where

$$4J = E^{aa} + E^{bb} - 2E^{ab}, \tag{8.13}$$

$$J + K + 2C = E^{aa} + E^{cc} - 2E^{ac} \tag{8.14}$$

$$J + K - 2C = E^{bb} + E^{cc} - 2E^{bc} \tag{8.15}$$

$$2H = \mu^a - \mu b - 2d(E^{ac} - E^{bc}), \tag{8.16}$$

$$2\Delta = 2\mu^c - \mu^a - \mu^b + 2d(E^{ac} + E^{bc} - 2E^{cc}), \tag{8.17}$$

Most calculations on three dimensional systems have been performed in the symmetric subspace in which oil and water are related by symmetry so that $C = H = 0$. Thus the Hamiltonian studied is

$$\mathcal{H} = -\sum_{\langle ij \rangle}(JS_iS_j + KS_i^2S_j^2) + \sum_i \Delta S_i^2 - \sum_{\langle ijk \rangle} LS_i(1 - S_j^2)S_k \tag{8.18}$$

In studying two-dimensional systems, a second-neighbor interaction $-\sum_{\langle\langle ij \rangle\rangle} K_2 S_i^2 S_j^2$ has been introduced in order to break a degeneracy between various lyotropic phases. Note that J distinguishes between oil and water, K between amphiphile and the other two components, Δ is the chemical potential of the amphiphile, and L is the strength of the amphiphilic interaction. Section 8.3 is devoted to the results obtained from this model.

8.2.3   THE ALEXANDER MODEL [13,14]

We consider a set of Ising spins $\sigma = \pm 1$ which occupy the sites of a d dimensional hypercubic lattice. The value $+1$ at a site indicates the presence of an oil molecule, -1 a water molecule. Additional variables $P_{ij}^c = 1, 0$ are placed on the bonds between sites. A value of $+1$ indicates the presence of an amphiphile between sites i and j, while 0 means no amphiphile. The Hamiltonian is

$$\mathcal{H} = -H\sum_i \sigma_i - J\sum_{\langle ij \rangle}\sigma_i\sigma_j - \mu^c\sum_{\langle ij \rangle}P_{ij}^c - J_1\sum_{\langle ij \rangle}\sigma_i\sigma_j P_{ij}^c. \tag{8.19}$$

Note that the last term above plays the same role as that introduced into the three-component model in Eq. (8.10), and the magnitude of the parameter J_1, like L in that equation, gives the strength of the amphiphile. As in the three-component model, oil and water are not necessarily separated by amphiphile, but are or are not depending on the strength of the amphiphile and the temperature. In contrast to the three-component model, the amphiphiles here interact only with the oil and water. Therefore it is easy to carry out the sum over the amphiphile variables in the partition function. This leaves a temperature-dependent interaction between the Ising spins, and so the phase diagram is qualitatively similar to that of the usual Ising model. In particular, there is no three phase co-existence. The situation is like that of the early version of the Widom-Wheeler model [8,9].

To remedy this, Chen et al. [14] followed a prescription similar to that of Widom. They introduced interactions between the amphiphiles, interactions representing curvature energies; i.e., an energy V if two amphiphiles were on neighboring links that met at $180°$, and an energy V_1 if the two links met at $90°$. Thus the Hamiltonian now reads

$$\mathcal{H} = H \sum_i \sigma_i - J \sum_{\langle ij \rangle} \sigma_i \sigma_j - \mu^c \sum_{\langle ij \rangle} P_{ij}^c - J_1 \sum_{\langle ij \rangle} \sigma_i \sigma_j P_{ij}^c$$
$$- V \sum_{\langle\langle ij \rangle\rangle} P_{ij}^c P_{ik}^c - V_1 \sum_{\langle\langle\langle ij \rangle\rangle\rangle} P_{ij}^c P_{ik}^c . \tag{8.20}$$

As in the case of the other two models, most work has been devoted to the symmetric subspace, $H = 0$.

The three models presented above are fairly similar in their treatment of the amphiphile. In none is much structure allowed in the interactions between amphiphile and oil and water. In the first, oil and water are always separated by an amphiphile which is correctly oriented; in the other two, the amphiphile has no orientation at all and is encouraged to find the most favorable location by a three-particle interaction. This latter is certainly not meant to represent a real three particle interaction, but only to imitate the real effect of the amphiphile in a simple way. The idea behind this simplified treatment of the amphiphile in these models is that the sole feature of the amphiphile which is important in the study of microemulsions is that it prefers to sit between oil and water.

8.2.4   OTHER LATTICE MODELS

The orientational degrees of freedom of the surfactant molecules can be taken into account by using vector variables τ_i in addition to the scalar occupational variables S_i. Such models [15,16,17,18] add more interesting details to the behavior but do not produce any qualitatively new behavior in the system, as long as the concentrations of oil and water are not too different [18]. Support for this view is provided by an analysis of a simple model for an interaction between an amphiphile and water, one which takes account of the directional properties of the former [15,17]. The directional dependence can be integrated out in the partition function leaving an effective interaction between water and a *structureless* amphiphile. Interactions between clusters of three and four particles are also generated. However, near the two-component, oil/amphiphile or water/amphiphile, side of the phase-diagram, the orientational degrees of freedom become essential, because they allow for a formation of amphiphilic bilayers, which are not possible in the simpler models introduced above [15,17,18].

Finally, we want to mention a model [19,20] in which the amphiphiles are described as short-chain polymers living on several lattice sites. A certain part of the polymer is hydrophilic, the rest hydrophobic. In such a model, the amphiphile has not only orientational degrees of freedom, but also internal ones, which may describe the bending modes of amphiphilic molecules.

With this background for lattice models of mixtures of oil, water, and surfactant, we shall turn to recent results on the model with which we are most familiar, the three component model. We then compare the results from this model with some from the others.

8.3   Some Results of the Three Component Model [21,22,23]

8.3.1   PHASE BEHAVIOR

The Hamiltonian of Eq. (8.12) with the parameter $L = 0$ is denoted the Blume-Emery-Griffiths model [24] which has been extensively studied, particularly within mean field theory [25]. There are, in general, three phases which are respectively rich in oil, or water, or surfactant. There is a large region of the phase space where these three phases can coexist. For our purposes, interesting phase behavior is obtained at fixed temperature by varying the parameter C which, from Eqs. (8.14-8.15) is related to the two-particle interactions by

$$4C = 2(E^{bc} - E^{ac}) - (E^{bb} - E^{aa}).   \tag{8.21}$$

From this expression, we see that this parameter is related to the difference between the interaction of the surfactant with water and with oil. When the surfactant is ionic, it is thought that adding salt to the water affects just this difference. When $C = 0$, one finds within mean field theory that there is a range of temperatures for which the system exhibits three phase coexistence. If the temperature is fixed at a value at which this is so and the value of C is made to increase, (surfactant prefers oil), a positive value is attained at which three phase coexistence ends at a critical end point. For larger values, only two phases coexist, one which is water-rich, the other a mixture of oil and surfactant. Similarly, if C is made to decrease, (surfactant prefers water), a critical end point is reached beyond which there is only two-phase coexistence between an oil-rich phase and a mixture of water and surfactant. Thus the sequence of two to three to two phase coexistence as a parameter, like salt concentration, is varied is given by this model. Further, this sequence remains when the amphiphilic interaction L is turned on [26]. From this we conclude that the unusual phase behavior observed with changing salt concentration which is characteristic of oil, water, ionic-surfactant systems has little or nothing to do with the strength of the amphiphile as measured by its ability to bring about micelles, lyotropic phases, etc.

To obtain this same phase behavior as a function of varying temperature, as is observed in systems with non-ionic surfactant, is a little more difficult. This is due to the fact that the model must generate a *lower* critical point as well as the usual upper critical point. At the former, phase separation occurs as the temperature is increased, a phenomena which is thought to be due to hydrogen bonding between the surfactant and water [27]. This can be accounted for by making the interaction energy E^{ac} temperature dependent [28]. When this is done, the two-three-two phase behavior now appears as a function of temperature [29]. The sequence is not very sensitive to the amphiphilic interaction L, so again we conclude that the phase behavior has little to do with the strength of the amphiphile. This conclusion is in agreement with experiment [2]. Additionally, the two critical end points can be made to vanish at a tricritical point as other interactions in the system are varied. Again this behavior is in agreement with experiment [2].

We note at this point that the result that the pattern of phase behavior is relatively insensitive to the strength of the amphiphile does not at all imply that the ability of the amphiphile to solubilize oil and water and to produce a fluid with structure is also independent of this strength. That this characteristic is in fact closely related to the ability of the amphiphile to bring about these properties will be seen in Secs. 8.3.2 and 8.3.4 below. The argument that amphiphiles should be distinguished not by the phase behavior they produce, but by the extent they solubilize oil and water and create structure has been stressed by Kahlweit, Strey, and Firman [4].

Let us now consider the phase diagram in the symmetric subspace governed by the Hamiltonian of (8.18), i.e., in which the amphiphile interacts equally strongly with water and oil which are present in equal amounts. In Fig. 8.1 is shown the phase diagram for parameter values for which a lamellar phase is present. The diagram, calculated via transfer-matrix and Müller-Hartmann Zittartz methods [30] for a two-dimensional system, is shown as a function of temperature, T, and surfactant chemical potential, Δ. For low temperatures and small values of Δ, there is two phase coexistence between oil-rich and water-rich phases. If the temperature is raised, a consolute point $T(\Delta)$ is reached. The nature of the transition there is continuous for sufficiently small Δ, and is first-order otherwise. A tricritical point, shown by an open circle, separates these two regimes. The significance of the dotted line, which is not a line of phase transitions, will be discussed in the next section. A lamellar phase exists over a region of surfactant chemical potential. The transition to it is first order. In the two-dimensional phase diagram, the lamellar phase never coexists with the oil-rich and water-rich phases; the disordered phase does so for temperatures below the tricritical point. This feature does not survive in three dimensions as the phase diagram of Fig. 8.2, obtained from mean-field

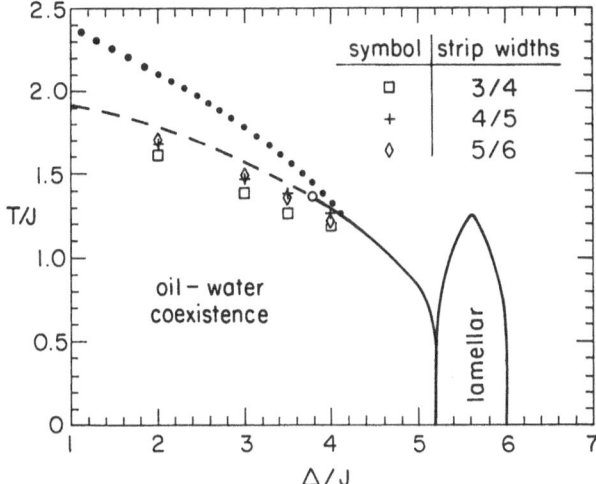

FIGURE 8.1. Phase diagram of the three-component model in two dimensions as a function of temperature and amphiphile chemical potential, Δ. The concentrations of oil and water are equal. Non-zero interaction parameters are $K/J = 2$, $K_2/J = -0.2$, and $L/J = -3$. Full lines and dashed lines, results of a Müller-Hartmann Zittartz approximation [21], indicate first-order and continuous transitions, respectively. The open circle denotes a tricritical point. The disorder line is shown dotted. Symbols show the results of transfer matrix calculations on strips of different widths. For details, see Ref. 21, 23.

theory, shows. The disordered phase coexists with oil-rich and water-rich phases down to a four phase point. Below this point, the lamellar phase coexists with the water and oil phases.

8.3.2 MICROEMULSION STRUCTURE

What are we to identify with the microemulsion? When the model was first studied [11], the microemulsion was identified with the lamellar phase. There were attractive reasons for doing so, but we believe that the identification is incorrect for several reasons. The most compelling of them is that if the identification were correct, then the model would provide no *real* lamellar phases, phases which are known to be present in the experimental systems. A second reason [31] is that the microemulsion so identified would be an anisotropic phase while experimentally it is not [32].

The microemulsion, then, is to be identified with a disordered phase [14]. But if this is so, what distinguishes it from an ordinary liquid? Presumably it is a question of structure. In order to determine the structure of the fluid phase which coexists with the oil and water phases, correlation functions were calculated using transfer matrix methods [33] in two dimensions,

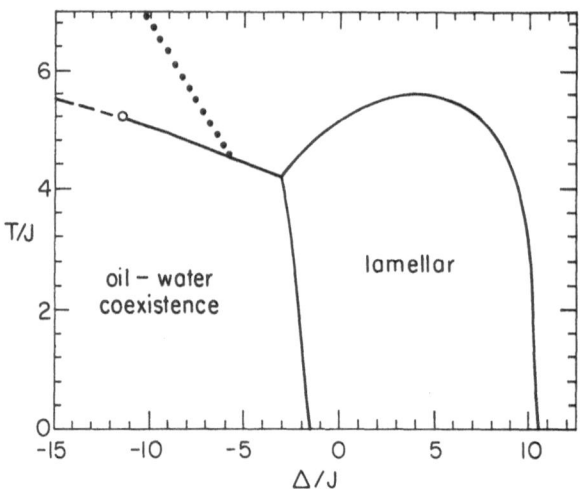

FIGURE 8.2. Phase diagram of the three-component model in three dimensions as obtained from mean field theory. The concentrations of oil and water are equal. Non-zero interaction parameters are $K/J = 0.5$, $L/J = -3.5$. Continuous and first-order transitions are denoted by dashed and solid lines, respectively. The open circle denotes a tricritical point. The dotted line is the locus of points at which the wavevector of the maximum of the water-water structure function approaches zero. The maximum is at non-zero wavevector to the right of this line.

and the Ornstein-Zernike approximation in three. In general the oil-oil, or water-water, correlation function has the form

$$\langle P_0^b P_r^b \rangle - \langle P_0^b \rangle \langle P_r^b \rangle \propto e^{-r/\xi} \cos(qr + \phi) \qquad (8.22)$$

for large separations r, where ϕ is an unimportant phase factor and we have ignored inverse powers of r. There are two length scales in this function: the correlation length ξ, which gives the typical size of correlated regions, and q^{-1}, which is the length over which significant changes in structure are found. In an ordinary fluid, the wavevector $q = 0$, and the correlation function simply decays exponentially. The vanishing of this wavevector indicates that significant changes of structure do not occur at finite lengths; the fluid is structureless. On the other hand, in a complex fluid with significant structure on a scale λ, q would be of the order of λ^{-1}.

In our calculation, we find that the fluid phase is characterized by a vanishing wavevector q in parts of the phase diagram, and by a non-vanishing q in others. The locus of points at which q just vanishes is called a disorder line [34]. It is shown in Fig. 8.1 as a dotted line; q is non-zero to the right of it. Note that over a large part of the triple line, the disordered phase is characterized by a non-zero value of q. The characteristic length q^{-1} is

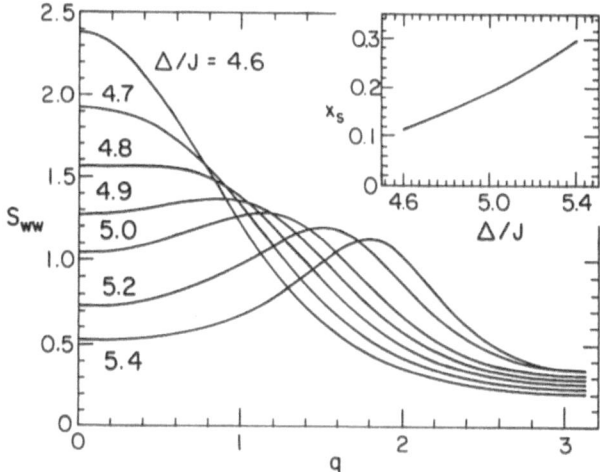

FIGURE 8.3. Unnormalized water-water structure functions vs. wavevector in the (11) direction calculated at $T/J = 1.1$ for the system of Fig. 8.1. The wavevector is in units of $\sqrt{2}/a$ with a the lattice constant of the model. The values of Δ span the region of existence of the disordered phase at this temperature. The variation in amphiphile concentration with chemical potential Δ is shown in the inset.

due to the same amphiphilic interaction which produces the relatively well-ordered lamellar phase at higher amphiphile concentrations, but which at smaller concentrations produces only a disordered phase of oil and water regions separated by amphiphile. *It is this phase, disordered but not structureless, which coexists with water-rich and oil-rich phases, that we identify with the microemulsion.*

It is not the correlation functions which are measured directly, but rather their Fourier transforms, or structure functions, which are determined in scattering experiments. Even though there are three components in the fluids, the individual water-water and surfactant-surfactant structure functions can be measured directly by neutron scattering with appropriate deuteration of one of the components, a process which exploits the very different neutron scattering lengths of hydrogen and deuterium (see introduction in Chap. 1 and Chap. 7 appendix). The water-surfactant structure function can then be inferred from a comparison of these results with the structure function of an undeuterated sample.

A calculated, unnormalized, water-water structure function in the two-dimensional disordered phase at $T/J = 1.1$ is shown in Fig. 8.3 on the range of Δ over which this phase exists. (The parameters are $K/J = 2$, $K_2/J = -0.2$, and $L/J = -3$). The variation of the surfactant concentration, x_s, with Δ in this range is shown in the inset. It is approximately 10% at

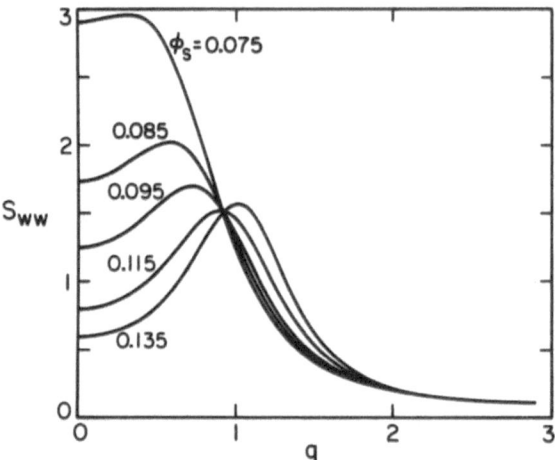

FIGURE 8.4. Unnormalized water-water structure functions vs. wavevector in the (111) direction calculated at $T/J = 4.45$ for the system of Fig. 8.2. The wavevector is in units of $\sqrt{3}/a$. The amphiphile concentrations span the region of existence of the disordered phase at this temperature.

three-phase coexistence at this temperature. The wavevector $\mathbf{q}$, which is in the (11) direction, is in units of the inverse lattice constant of our lattice model; in these units, the basic reciprocal lattice vector is of magnitude 2π. Figure 8.4 shows a similar result for the three dimensional system at $T/J = 4.45$ for the range of surfactant concentrations over which the phase exists. The surfactant concentration in the disordered phase at three-phase coexistence is 7.5%. (The parameters of this system are $K/J = 0.5, K_2/J = 0$, and $L/J = -3.5$.) The existence of a peak in the water-water structure function at a non-zero wavenumber, q_{max}, is a more experimentally useful (and more restrictive) criterion for the existence of a microemulsion than is a non-zero wavenumber λ^{-1} in the density-density correlation function. The dotted line in the phase diagram of Fig. 8.2 shows the locus $q_{max}(T, \Delta) \to 0$ which is given approximately by $x_s(T, \Delta) = -J/4L$. The microemulsion exists to the right of this line.

For comparison, the intensity of neutron scattering from a toluene, brine, sodium dodecyl sulfate microemulsion [35] is shown in the upper part of Fig. 8.5. This intensity is proportional to the water-water structure function. The similarities between our results and the experimental data are

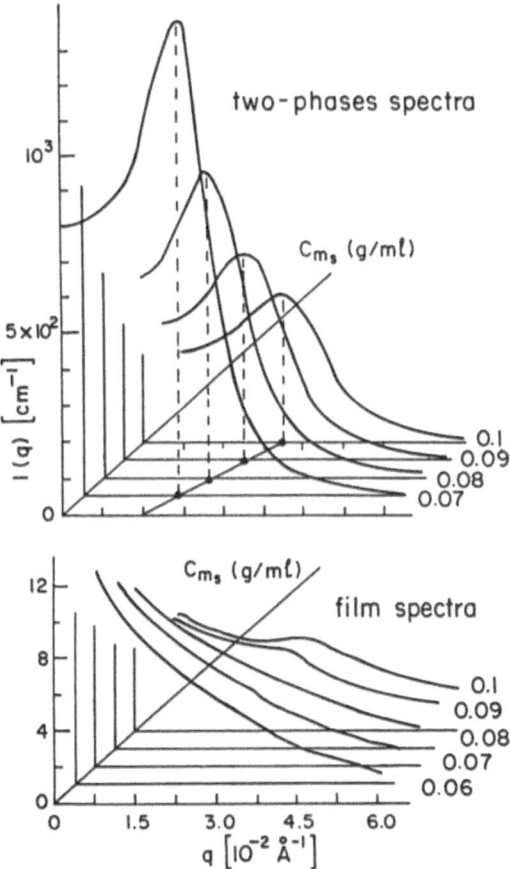

FIGURE 8.5. Neutron scattering intensities from the system brine, toluene, and sodium dodecyl sulfate as given in Ref. 35. The volume fractions of brine and oil are equal. The mass density of surfactant is denoted C_{m_s}. The upper spectra are proportional to the water-water structure function, the lower to the surfactant-surfactant structure function.

clear. In both cases, the peak in the water-water structure function moves out and decreases as the concentration of surfactant increases. Furthermore, if we take the size of our lattice constant to be 25Å, about the size of a typical surfactant molecule, then the peaks in our calculated structure function, Fig. 8.4, occur at $3 \times 10^{-2} Å^{-1}$ in agreement with the experimental data.

The intensity shown in the lower part of Fig. 8.5 is proportional to the surfactant-surfactant structure function. In general, it does not show the same features as the water-water structure function, but is characterized by a peak at zero wavenumber. A small shoulder at non-zero q may develop as the surfactant concentration is increased. For comparison, our calculated surfactant-surfactant structure function is shown in Fig. 8.6a for the same temperature and range of Δ employed in calculating the water-water structure function of Fig. 8.4. The function decreases monotonically with wavevector. If we break the oil water symmetry by taking C in Eq. (8.12) to be non-zero, then we find that a small shoulder develops in this structure function as the surfactant concentration is increased. An example is shown in Fig. 8.6b for a system at $T/J = 4.45$ with parameters $K/J = 0.5, K_2/J = 0, L/J = -3.5$, and $C/J = 0.3$. The difference between water and oil concentrations is fixed at 0.2 while the surfactant concentration is increased. The water-water structure function at these same values has a peak at non-zero wavenumber indicating that the disordered fluid is a microemulsion.

Finally, we show in Fig. 8.7a an example of the calculated water-surfactant structure function. The system parameters and temperature are the same as those of Fig 8.4. However in that figure, the oil/(oil+water) ratio was fixed at 0.5 and the surfactant concentration was varied. In Fig. 8.7a, the surfactant concentration is fixed at 0.1, and the oil/(oil + water) ratio is varied. For comparison, experimental results [36] are shown in Fig. 8.7b. Qualitative agreement is rather good.

8.3.3 SURFACE TENSION

The origin of the low surface tensions receives a pleasing explanation from this model and identification. Our results [37] for the oil/water interfacial tension at three-phase coexistence are shown in Fig. 8.8. To set the scale, the tension in the absence of amphiphile at $T/J = 4.45$ is approximately 2 in our units. That the oil/water tension should be small in the presence of the lamellar phase is easy to understand. At $T = 0$, the energy of a sheet of amphiphile between oil and water vanishes at some value of the chemical potential, leading to the formation of the lamellar phase. Entropy effects make the tension non-zero but small at finite temperatures [38]. This result extends to the oil/water tension in the presence of the microemulsion, because the tension is continuous at the four phase point. As a consequence, the oil/water tension in the presence of the microemulsion can be made

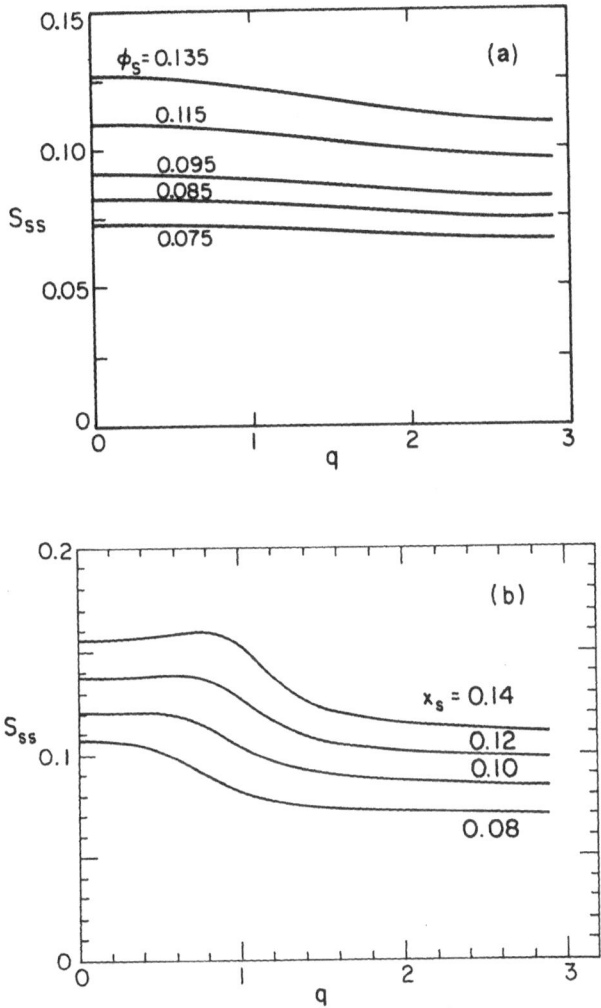

FIGURE 8.6. (a) Surfactant-surfactant structure function for several surfactant concentrations for the same parameters which apply to the water-water structure functions of Fig. 8.4; $T/J = 4.45, K/J = 0.5, L/J = -3.5, C/J = 0$, and equal concentrations of oil and water. The intensity scale is the same as in Fig. 8.4. (b) Surfactant-surfactant structure function for several surfactant concentrations for a slightly different system; $T/J = 4.45, K/J = 0.5, L/J = -3.5$, but $C/J = 0.3$. The difference in volume fractions of water and oil is fixed at $\delta x = 0.2$.

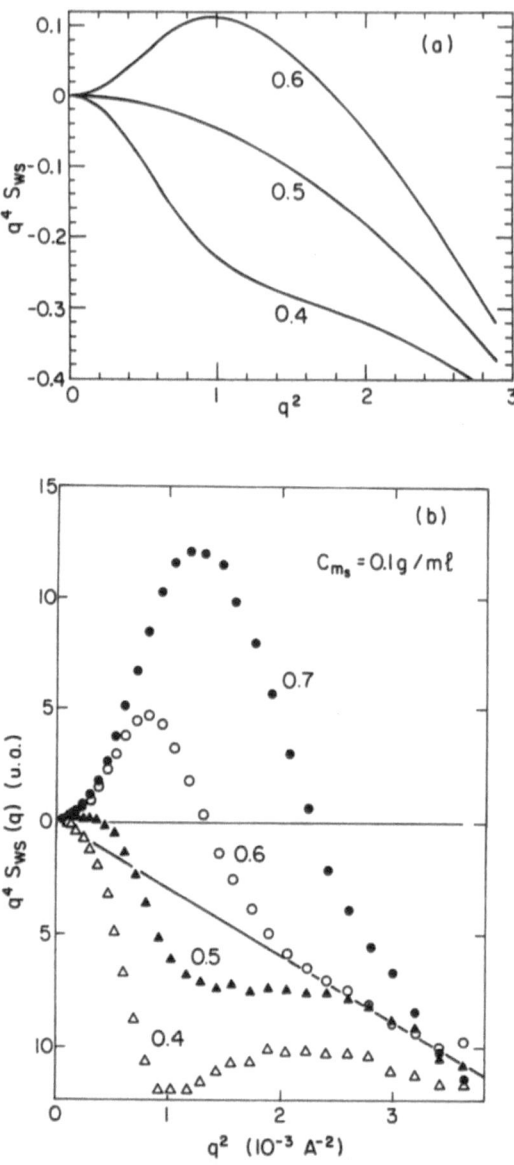

FIGURE 8.7. (a) Scaled water-surfactant structure function $q^4 S_{ws}$ vs. q^2 for three different volume ratios of oil to oil plus water. The system is the same as in Figs. 8.4 and 8.6a. The surfactant concentration is fixed at $x_s = 0.1$. (b) Scaled water-surfactant structure function as determined in Ref. 36 for four different volume ratios of oil to oil plus water. The experimental system is the same as that in Fig. 8.5.

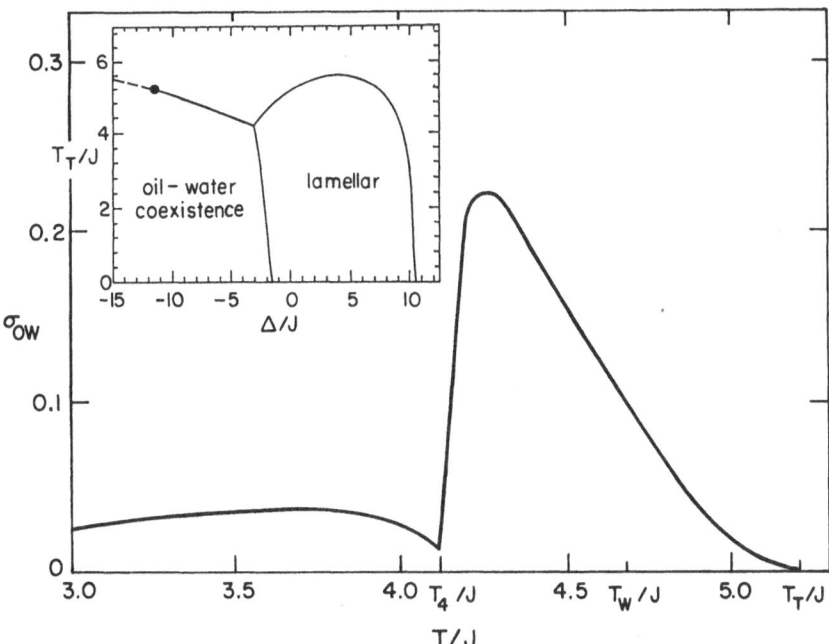

FIGURE 8.8. Variation of the oil/water interfacial tension at three-phase coexistence. The four-phase coexistence temperature T_4 and the wetting temperature T_w are indicated. Inset shows the phase diagram as in Fig. 8.2.

to be very low *if* the system can be brought close to its instability to the lamellar phase. The corollary that such systems which do not become unstable to a lamellar phase, but to another lyotropic one, will not exhibit such low tensions seems to be correct.

8.3.4 WETTING

Another interesting observation concerning the interfacial properties of this system is its wetting behavior. Wetting of the oil/water interface by the microemulsion does not occur in spite of the very low values of the water/microemulsion and oil/microemulsion interfacial tensions, values which would tend to favor such wetting [39,40]. A calculation [37] of the oil/water, the oil/microemulsion and the water/microemulsion density profiles, and a comparison of all calculated interfacial tensions, shows that the oil/water interface is not wetted by the disordered fluid provided that the disordered fluid has an oscillating correlation function, while it is wetted otherwise. At the disorder line at which the fluid loses its structure, an unusual critical wetting transition can occur, where the thickness of the microemulsion layer diverges.

The failure of the microemulsion to wet the oil/water interface is a direct consequence of the fact that the microemulsion *is* a structured fluid. To demonstrate this, we have calculated the free energy when there is a free oil/microemulsion interface at z, and a similar water microemulsion interface at $z + \ell$. The difference between the interfacial energy of this configuration at finite separation ℓ and infinite separation is the effective interface potential $V(\ell)$. The asymptotic form of the interface potential at large separations is

$$V(\ell) \propto \exp(-\ell/\xi) \cos(k\ell + \phi) , \qquad (8.23)$$

where ξ and k are the correlation length and characteristic wavevector of the bulk correlation function, respectively. Thus the oscillations of the effective interface potential reflect the oscillation of the bulk correlation function. As long as $V(\ell)$ has the above form with $k > 0$, the potential has an absolute minimum at finite ℓ, which corresponds to the equilibrium thickness of the microemulsion layer; the oil-water interface is therefore not wet [41]. As the disorder line is approached, the wavevector $k \to 0$, and the minimum of $V(\ell)$ moves to larger and larger values of ℓ diverging at the disorder line as $\delta^{-1/2}$, where δ measures the distance from the disorder line. This form, arising from the disorder line, should be compared with the logarithmic divergence usually seen in systems with short range interactions. The oil/water interfacial tension exhibits an essential singularity.

The effect of long range forces which favor wetting of the oil water interface are easily incorporated into our picture. Van der Waals forces, which couple to the average density of the fluid, produce an additional term in Eq. 8.23, one which varies as ℓ^{-p}, with $p = 2$. This causes the wetting transition to become *first order*. This transition does not take place at the disorder line, but on the microemulsion side of it.

As the first-order wetting transition is predicted to occur in the vicinity of the disorder line where the microemulsion is becoming less and less structured, one expects that the best chance for observing it would occur in a system containing a rather poor amphiphile. In fact, a wetting transition has recently been observed [42] in the system water/hexadecane/C_6E_2, the latter a poor amphiphile. The order of the transition has not yet been determined. The prediction of the theory is that it is first order.

8.3.5   SOLUBILIZATION OF OIL AND WATER

The efficiency of the surfactant as a solubilizer of oil and water in our model is limited only by the strength of the amphiphilic interaction. In particular, the concentration of surfactant required to cause the peak in the water-water structure function to move off of zero wavenumber is approximately $x_s = J/4|L|$ at equal oil, water concentrations, as noted earlier. For the system shown in Figs. 8.4, 6, and 7 for which $|L|/J = 3.5$, the minimum concentration of surfactant needed to create a microemulsion is about 7%. This is not the minimum concentration needed to solubilize oil and water; at some temperatures, less surfactant will produce a disordered phase in which oil and water are mixed, but the peak in the water-water structure function will be at zero wavenumber. It should be noted that a direct comparison of these concentrations with those measured experimentally is not straightforward because the assumption of the lattice model that all components of the system are the same size is certainly not correct.

8.3.6 SURFACTANTS VS. AMPHIPHILES

It is instructive to examine the behavior of the model as the strength of the amphiphilic interaction, L, is varied.

For a strong amphiphile, the disordered phase which coexists with oil and water is, over a large part of the triple line, characterized by structure of typical length q_{max}^{-1}. This is observable in the structure function which exhibits a peak at wavenumber $q_{max} \neq 0$. This disordered phase is the microemulsion. At higher temperatures, a disordered phase without structure exists and may co-exist with oil and water phases close to the tricritical point. There is no phase transition between these two disordered phases; in going from the microemulsion to the featureless disordered phase, the wavevector at which the peak in the structure function occurs decreases smoothly until $S(q)$ becomes a monotonically decaying function of wavevector which is characteristic of the featureless disordered phase.

As the strength of the amphiphile is decreased, the region of phase space in which the lamellar phase exists decreases as does the fraction of the triple line where the microemulsion exists with oil and water phases. Finally for very weak amphiphiles, one sees neither lamellar phases nor microemulsion, but only a surfactant-rich phase. (In real systems, this latter is probably another lyotropic phase which our model was not designed to describe.) Because the lamellar phase no longer exists, the oil/water surface tension no longer vanishes at zero temperature, so that there is no reason to expect that this surface tension is low except near the tricritical point.

We note again that the structure emerges with increasing amphiphile strength while the phase behavior is relatively unaffected, a picture which is in accord with experiment. The nature of the structure function provides the distinction between those ternary systems which exhibit microemulsion phases from those which merely exhibit the phase behavior of ternary systems. Amphiphiles with a strength sufficient to produce the microemulsion in our model are often denoted "surfactants" in the literature to distinguish them from the weaker ones for which the term "amphiphile" is reserved. This distinction exists in our model.

In sum, we believe that the calculation that we have described does accomplish the aims set out in the beginning of this chapter: it correctly describes the phase behavior observed in ternary systems; provides an explanation of the low surface tensions found in microemulsions, an explanation which is tied both to the phase behavior and the existence of lyotropic lamellar phases; correctly reproduces the relationship between solubilization of oil and water and strength of amphiphile; and provides a disordered phase with a long length scale and realistic structure factor, a phase identified with the microemulsion.

8.4   Some Results of Other Lattice Models

8.4.1   THE WIDOM-WHEELER MODEL

This model, with the Hamiltonian of Eq. (8.6), has been studied rather extensively, particularly by mean field theory [9,38] and Monte Carlo simulation [43,44] as well as other methods [45-47], and has been extended to encompass additional interactions [48,49]. The calculated phase diagrams are usually given in the space of dimensionless coupling constants J/k_BT and M/k_BT, where k_B is Boltzmann's constant. In order to compare more directly with the phase diagrams of the three component model, we show schematically, in Fig. 8.9, the phase diagram of the Widom-Wheeler model in space of temperature and surfactant chemical potential. There are three

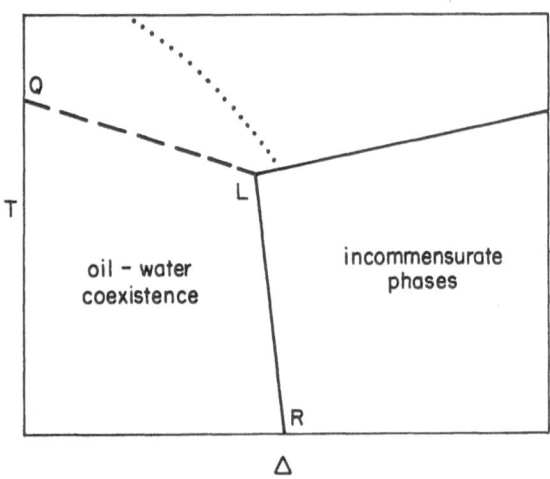

FIGURE 8.9. Schematic phase diagram of the Widom-Wheeler model in three dimensions in the temperature, surfactant chemical potential plane. The concentrations of oil and water are equal. Continuous transitions are shown by dashed lines, first-order transition by a solid line. Dotted line is a disorder line.

distinct regions. At low surfactant chemical potential, there is a coexistence region between oil-rich and water-rich phases. As the temperature is increased a consolute line, denoted QL in the figure, is reached. This is a line of continuous transitions to the disordered phase, which occupies the third region. If the surfactant chemical potential is increased, a triple line, LR, is encountered at which the oil-rich, water-rich and "incommensurate" phases coexist. This region actually consists of many different phases, each separated from those adjoining it in phase space by first order transitions. The particular phase which coexists with oil and water is periodic, the basic unit consisting of two sheets of oil followed by two sheets of water. In this model, the oil sheets are separated from the water by surfactant. The other phases differ from this either in the number of sheets of oil or of water, or by being periodic in more than one direction. This region is quite similar to the corresponding one in the well-studied axial next nearest neighbor Ising (ANNNI) model [10].

The region of incommensurate phases was identified as the microemulsion by Widom [9]. This is appealing in many ways. First of all, the existence of a very large number of phases is directly tied to the use of a lattice and would not survive in a more realistic continuum model. The destruction of these phases would result from the excitations of the interfaces between oil and water regions, excitations which make the interfaces rough, i.e., to fluctuate to such an extent that it is not possible to distinguish whether a phase has, for example, four or five sheets of oil alternating with three or four sheets of water. This destruction of phases is well known from the case of the two-dimensional ANNNI model in which the interfaces are rough even in this lattice model. Thus one expects to find in a continuum version a single phase in which regions of water and of oil are separated by surfactant. And, as stated before, this phase coexists with both oil-rich and water-rich phases. This identification also provides an explanation for the origin of low surface tensions. The oil/water surface tension vanishes at the point R and at the point L. Thus, one expects that it never gets very large anywhere along the triple line [38].

There are difficulties with this identification, however, the same that arise in the similar identification in the three component model. First, the phase is simply *too* structured. One expects that in a continuum version of this model, the phase which would coexist with oil and water would be a smectic liquid crystal phase, with correlations which decay with a power law, rather than exponentially. Such a distinction causes a difference in the structure factor, a difference which could be observed by very careful scattering experiments. More troubling is that the phase would be anisotropic, unlike the observed isotropic microemulsion. A phase transition would separate the microemulsion from the disordered phase. Finally, this identification leaves no other phases to be identified with the lyotropic phases which are usually found close to the microemulsion.

Recently, Dawson et al. [44] argued that the disordered phase should be identified as the microemulsion. (By extension, the incommensurate phases are identified as lamellar.) This phase is structured just as the disordered phase in the three component model, and exhibits [44,50] a structure factor with a peak at a non-zero wavevector. Although the identification of the disordered phase with the microemulsion is probably correct, it does give rise to a new set of difficulties. Most serious is that the disordered phase does not coexist with oil and water phases. Rather the line of transitions to this phase, the consolute line, is continuous. As a consequence, the oil/water interfacial tension is identically zero all along this line. These results are contrary to experiment. Also, the disordered phase which exists just above the consolute line is structureless (i.e., $q = 0$ in Eq. (8.22)), again in contrast to experiment. (The disorder line intersects the line of first-order transitions between disordered and lamellar phases. See Fig. 8.9.)

To summarize: the Widom-Wheeler model certainly captures the effect of the surfactant as evidenced by the lamellar phases it produces. The disordered phase has a structure factor characteristic of microemulsion in some parts of the phase diagram. But in the version presented here, this microemulsion does not coexist with oil and water phases, nor does it intrude between them and the lamellar phases. Thus the phase behavior is not given correctly by the model, with the additional consequence that it does not produce low surface tensions, but ones which are in fact zero. However, an extension of the range of interactions included in the model is sufficient to bring about a three phase coexistence between microemulsion and oil-rich and water-rich phases [49,51]. Because the surface tensions are intimately tied to the phase behavior, as we have argued, such an extension should produce non-zero, but low, surface tensions as well.

8.4.2 THE ALEXANDER MODEL

This model, with the Hamiltonian of Eq. (8.20), has been studied by mean field theory and Monte Carlo simulation [14]. The mean-field phase diagram from a two-dimensional system with a particular choice of parameters is shown in Fig. 8.10. The topology is probably very similar to the model's three dimensional phase diagram, with one exception noted below. Although the diagram is shown there in the space of temperature and the strength of the amphiphilic interaction, J_1, the topology is presumably the same in the space of temperature and surfactant chemical potential. There is a region of two phase coexistence between oil-rich and water-rich phases. As the temperature is increased, the consolute line is reached. For weak amphiphiles, this transition to the disordered phase is continuous, while for stronger amphiphiles it is first-order. A tricritical point separates the two regimes. As Chen et al. [14] identify the disordered phase with the microemulsion, they succeed in obtaining three phase coexistence between

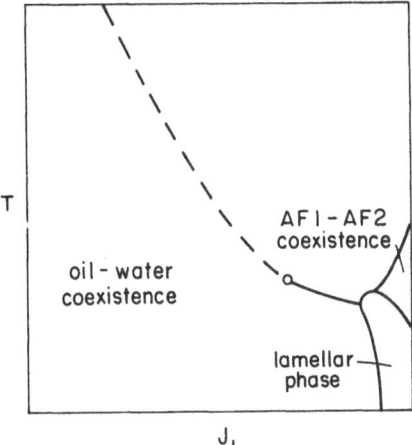

FIGURE 8.10. Phase diagram of the Alexander model modified by Chen *et al.* [14], as determined by mean field theory for two dimensions. The parameters in the Hamiltonian of Eq. (8.20) are $H = V_1 = 0, \mu^c = -0.2$. Continuous transition is shown by dashed line, first-order transitions by solid lines. The tricritical point is shown by an open circle.

microemulsion, oil, and water. As the amphiphile strength is increased at low temperatures, the third phase in coexistence with oil and water is no longer disordered, but is a layered phase consisting of single sheets of oil, surfactant, water, surfactant, oil, etc. Presumably, this should be identified with the lamellar phase. Another ordered phase with oil and water in a checkerboard pattern, the two possibilities in coexistence, also appears and is denoted AF1 and AF2. (In three dimensions, an ordered phase with different structure probably appears here.) The phase behavior leads to low surface tensions just as in the three component model; the oil/water surface tension vanishes at the high temperature end of the triple line, the tricritical point, and at the zero temperature end also. Thus it is not expected to be large anywhere.

One difficulty with this model in the version studied is that it produces no long period lamellar phases, i.e., ones with period longer than twice the basic lattice spacing. A consequence for the microemulsion is that it shows no structure at a long length scale. This is reflected in the structure function which has only a peak at zero wavevector, in contrast to experimental results. Thus the distinction is lost between the disordered phase produced by a strong amphiphile, which is a microemulsion, and that produced by a weak amphiphile, which is not. This problem is related to the short range of interactions taken; it is not difficult to conceive of only slightly longer ranged ones which would produce the long period lamellar phases and also a structured microemulsion.

To summarize: the Alexander model, as extended by Chen et al. and Stockfish and Wheeler captures the correct phase behavior and a lamellar phase which, as noted earlier, should lead to low surface tensions. Although with interactions which are too short-ranged it does not produce a spontaneously occurring long length in the phase identified as the microemulsion, it will certainly do so with only slightly longer-ranged ones.

The difference in the formulation of the three models we have discussed primarily is in the location of the particles. In the Widom-Wheeler model, all particles sit on the bonds of the lattice; in the three component model, all particles are located on the sites; in the Alexander model, oil and water sit on the sites, while the amphiphile is located on the bonds. Clearly, this distinction tells us nothing about microemulsions, but something about model building. The object of this activity is to construct a model sufficiently simple to focus on the essential physics, while sufficiently complex to describe the desired features. The differences in the formulations are then reflected in just how complex the interactions have to be to encompass the desired features. Although results from the initial formulations of these models may differ considerably, we have no doubt that refinements in each of them will produce very similar pictures of oil, water, surfactant systems.

8.5   Comparison with Phenomenological Models

As with the lattice models, there are several different phenomenological models of microemulsions which have been employed. The first kind are of the form of Landau-Ginzburg theories [37,52-54] which can, in principal, be derived from underlying microscopic theories when the order parameters are small. This is the case near the tricritical point, and phase diagrams and surface tensions have been well fit in this region [52]. The second kind pictures the microemulsion as a collection of surfactant membranes, and constructs a free energy from terms which are thought to be important, such as energies of curvature, entropy of mixing, etc. and then minimizes with respect to various parameters. These theories are summarized nicely by Andelman et al. [55] and by Safran in Chap. 9. The theory of Andelman et al. [55-57] has several successes. It produces three phase coexistence between oil-rich and water-rich phases and a disordered phase, one which is identified with the microemulsion. The characteristic length, ξ_K, in the latter phase is different from the atomic size characteristic of the other two phases. The water-water structure function shows a peak at the inverse of this length, a peak whose position moves out as the concentration of surfactant increases [58]. Lamellar phases are also described, and they compete with the microemulsion. The oil/water surface tension is expected to be of order $k_B T / \xi_K^2$, a small value if ξ_K is large.

There are two major ingredients in this cell theory. The first is the hypothesis that all surfactant resides between oil and water in a form much like a two dimensional liquid. This fixes the cell size in terms of the volume fraction of oil and of surfactant. The second is the form of the phenomenological free energy which contains two terms, an entropy of mixing, and an energy term which arises from the curvature of the surfactant film. The curvature energy is smallest when the cell size is large which occurs when oil and water are mixed. This energy competes with the entropy of mixing which, when combined with the constraint that the surfactant forms a liquid, favors the pure oil and water phases.

There are some similarities between the phenomenological and lattice theories, but more differences. As to the former, we note that the curvature term in the phenomenological model drives the oil and water to mix and form the microemulsion. Thus it is somewhat analogous to the amphiphilic interaction in the three component model which plays a similar role. The importance of the bending constant in the phenomenological theory is parallel to the importance of the strength of the amphiphilic interaction in the lattice theory.

There is a substantial difference, however, in the mechanism which produces the microemulsion and water-rich and oil-rich phases. In the phenomenological model, the curvature term favors the microemulsion and competes with the entropy of mixing which favors the oil-rich or water-rich phases. In the lattice models, the amphiphilic interaction *and* the entropy of mixing favor the microemulsion, and compete with oil-oil and water-water interactions which favor the pure phases. The role of the entropy of mixing is more intuitive in this scenario.

The phenomenological models assume well-developed sheets of amphiphile separating oil and water, i.e., that the amphiphiles are strong and that the temperature is not close to the tricritical point. The lattice theories are capable of describing both strong amphiphiles, denoted surfactants, as well as weak ones and are thus capable of describing the transition in phase behavior and structure as the amphiphile is varied.

The phenomenological models do not yet produce a tricritical point in the phase diagram. In contrast, some lattice models do, and the phase behavior produced by such models is becoming quite realistic.

Another major difference concerns the origin of the small wavevector q_{max} whose inverse is the scale of microemulsion structure. In the theory of Andelman et al. [55-57], q_{max} is essentially the inverse of the characteristic length ξ which is given in terms of the volume fractions of the three components. These fractions are determined by minimization of the Gibbs free energy. At the minimum identified with the microemulsion, ξ is of the order of a length ξ_K which appears in the curvature energy. This length is taken to be the persistence length [59], which is large. As its inverse is a small wavevector, it is not surprising that the characteristic wavevector of the structure function is small. This contrasts with lattice models in which the only length in the description is the cell size which is taken to be of molecular dimensions. Thus the natural wavevector in the model is quite large. Nonetheless, the maximum of the structure factor appears at a wavevector, q_{max}, which is only a fraction of this large value, a fraction determined by the interactions and chemical potentials. Taking the cell to be the size of a typical amphiphile, we find good agreement between the calculated and measured values of q_{max}.

In the phenomenological theory, surface tensions are of order $k_B T/\xi_K^2$, a relation which ties them to an input of the theory. Because the persistence length is large, these tensions are small. In the three component model, low surface tensions emerge from the model as a result of the phase behavior, i.e., the presence of the tricritical point and of the lyotropic phase.

The present lack of correspondence between phenomenological and lattice theories of microemulsions is not entirely unexpected. In constructing a theory of either kind, it is crucial that one have an idea as to what is essential and what is not. The disparity between the two kinds of theories simply indicates that, at this time, there is not agreement on what constitutes the essence of this interesting phase. As the field develops, this disparity will almost certainly give way to congruence.

Acknowledgments

We are deeply indebted to Eric Kaler for his interest and advice, and for sharing with us his intimate knowledge of the field of microemulsions. We also thank David Andelman, Gilson Carneiro, Manfred Kahlweit, Marcel den Nijs, Koos Rommelse, Reinhard Strey, and Sam Safran for profitable conversations. This work is supported by the NSF under grants no. DMR-8613598, and DMR-8916052.

References

1. B. M. Knickerbocker, C. V. Pesheck, H. T. Davis, and L. E. Scriven, J. Phys. Chem. **86**, 393 (1982).

2. M. Kahlweit, R. Strey, P. Firman, and D. Haase, Langmuir **1**, 281 (1985).

3. P. Firman, D. Haase, J. Jen, M. Kahlweit, and R. Strey, Langmuir **1**, 718 (1985).

4. M. Kahlweit, R. Strey, and D. Haase, J. Phys. Chem. **89**, 163 (1985); M. Kahlweit, R. Strey, and P. Firman, J. Phys. Chem. **90**, 671 (1986).

5. See Chap. 10 of this volume.

6. There is a large literature reporting such results. Some examples are as follows: L. Auvray, J.-P. Cotton, R. Ober, and C. Taupin, J. Phys. Chem. **88**, 4586 (1984); M. Kotlarchyk, S.-H. Chen, J. S. Huang, and M. W. Kim, Phys. Rev. Lett. **53**, 941 (1984); C. G. Vonk, J. F. Billman, and E. W. Kaler, J. Chem. Phys. **87**, 3195 (1987).

7. D. J. Lee, M. M. Telo da Gama, and K. E. Gubbins, J. Phys. Chem. **89**, 1514 (1985); M. M. Telo da Gama and J. H. Thurtell, J. Chem. Soc. Faraday Trans. 2 **82**, 1721 (1986).

8. J. C. Wheeler and B. Widom, J. Am. Chem. Soc. **90**, 3064 (1968); B. Widom, J. Phys. Chem. **88**, 6508 (1984).

9. B. Widom, J. Chem. Phys. **84**, 6943 (1986).

10. For reviews, see P. Bak, Rep. Prog. Phys. **45**, 587 (1982); M. E. Fisher and D. A. Huse, in *Melting, Localization, and Chaos*, edited by R. K. Kalia and P. Vashista (Elsevier, New York, 1982), p. 259; W. Selke, Physics Rep. **170**, 213 (1988).

11. M. Schick and W.-H. Shih, Phys. Rev. B **34**, 1797, (1986).

12. M. Schick and W.-H. Shih, Phys. Rev. Lett. **59**, 1205 (1987).

13. S. Alexander, J. Physique Lett. **39**, L1 (1978).

14. K. Chen, C. Ebner, C. Jayaprakash, and R. Pandit, J. Phys. C **20**, L361 (1987); Phys. Rev. A **38**, 6240 (1988). Another variant of this model has been studied by T. P. Stockfish and J. C. Wheeler, J. Phys. Chem. **92**, 3292 (1988).

15. J. W. Halley and A. J. Kolan, J. Chem. Phys. **88**, 3313 (1988).

16. A. Ciach, J. S. Høye, and G. Stell, J. Phys. A **21**, L777 (1988); J. Chem. Phys. **90**, 1214 (1989); A. Ciach and J. S. Høye, J. Chem. Phys. **90**, 1222 (1990).

17. G. Gompper and M. Schick, Chem. Phys. Lett. **163**, 475 (1989).

18. M. W. Matsen and D. E. Sullivan, Phys. Rev. A **41**, 2021 (1990); see also K. A. Dawson and A. Kurtovic, J. Chem. Phys. **92**, 5473 (1990).

19. R. G. Larson, L. E. Scriven, and H. T. Davis, J. Chem. Phys. **83**, 2411 (1985).

20. R. G. Larson, J. Chem. Phys. **91**, 2479 (1989).

21. G. Gompper and M. Schick, Phys. Rev. Lett. **62**, 1647 (1989).

22. G. Gompper and M. Schick, Phys. Rev. B **41**, 9148 (1990).

23. G. Gompper and M. Schick, Phys. Rev. A **42**, 2137 (1990).

24. M. Blume, V. Emery, and R. B. Griffiths, Phys. Rev. A **4**, 1071 (1971).

25. D. Mukamel and M. Blume, Phys. Rev. A **10**, 610 (1974); D. Furman, S. Dattagupta, and R. B. Griffiths, Phys. Rev. B **15**, 441 (1977).

26. M. Schick and W.-H. Shih (unpublished).

27. J. D. Hirschfelder, D. Stevenson, and H. Eyring, J. Chem. Phys. **5**, 896 (1937).

28. G. R. Andersen and J. C. Wheeler, J. Chem. Phys. **69**, 2082 (1978); J. S. Walker and C. A. Vause, Phys. Lett. **79A**, 421 (1980); R. E. Goldstein, J. Chem. Phys. **83**, 1246 (1985).

29. G. M. Carneiro and M. Schick, J. Chem. Phys. **89**, 4638 (1988).

30. For a review of transfer-matrix methods, see M. N. Barber in *Phase Transitions and Critical Phenomena* vol. 8, edited by C. Domb and J. L. Lebowitz (Academic, New York, 1984) p. 145. E. M. Müller-Hartmann and J. Zittartz, Z. Phys. B **27**, 261 (1977) introduced the approximation used to supplement the transfer-matrix results.

31. We are indebted to H. T. Davis for this observation.

32. A second study of the model, Ref. 12 above, focussed on a small region of phase space in which a fourth phase, also disordered, appears. Although there were reasons to believe that this phase represented the microemulsion, the fact that it does only appear in a small region of phase space and is uncorrelated with the lyotropic phases makes this identification much less compelling than that in the text.

33. N. C. Bartelt and T. L. Einstein, J. Phys. A **19**, 1429 (1986).

34. M. E. Fisher and B. Widom, J. Chem. Phys. **50**, 3756 (1969); J. Stephenson, J. Math. Phys. **11**, 420 (1970).

35. L. Auvray, J.-P. Cotton, R. Ober, and C. Taupin, J. Phys. Chem. **88**, 4586 (1984).

36. L. Auvray, J.-P. Cotton, R. Ober, and C. Taupin, Physica **136B**, 281 (1986).

37. G. Gompper and M. Schick, Phys. Rev. Lett. **65**, 1116 (1990).

38. K. A. Dawson, Phys. Rev. A **35**, 1766 (1987).

39. B. Widom, Langmuir **3**, 12 (1987).

40. M. Kahlweit, R. Strey, D. Haase, and P. Firman, Langmuir **4**, 785 (1988).

41. Similar considerations have been applied to the solid/liquid interface by A. A. Chernov and L. V. Mikheev, Phys. Rev. Lett. **60**, 2488 (1988).

42. M. Robert and J. F. Jeng, J. Phys. France **49**, 1821 (1988).

43. N. Jan and D. Stauffer, J. Phys. France **49**, 623 (1988).

44. K. A. Dawson, B. L. Walker, and A. Berera, Physica A **165**, 320 (1990).

45. Y. Levin and K. A. Dawson, Phys. Rev. A **42**, 1976 (1990).

46. A. Berera and K. A. Dawson, Phys. Rev. A **42**, 3618 (1990).

47. B. Kahng, A. Berera, and K. A. Dawson, Phys. Rev. A **42**, 6093 (1990).

48. T. Hofsäss and H. Kleinert, J. Chem. Phys. **88**, 1156 (1988).

49. A. Hansen, M. Schick, and D. Stauffer, Phys. Rev. A**44**, 3686 (1991).

50. B. Widom, J. Chem. Phys. **90**, 2437 (1989).

51. B. Widom, private communication.

52. H. Kleinert, J. Chem. Phys. **84**, 964 (1986).

53. K. Chen, C. Jayaprakash, R. Pandit, and W. Wenzel, Phys. Rev. Lett. **65**, 2736 (1990).

54. G. Gompper, R. Hołyst, and M. Schick, Phys. Rev. A**43**, 3157 (1991).

55. D. Andelman, M. E. Cates, D. Roux, and S. A. Safran, J. Chem. Phys. **87**, 7229 (1987).

56. See also D. A. Huse and S. Leibler, J. Phys. France **49**, 605 (1988); L. Golubović and T. C. Lubensky, Europhys. Lett. **10**, 513 (1989); Phys. Rev. A **41**, 4343 (1990).

57. S. A. Safran, D. Roux, M. E. Cates, and D. Andelman, Phys. Rev. Lett. **57**, 491 (1986); D. Andelman, S. A. Safran, D. Roux, and M. E. Cates, Langmuir **4**, 802 (1988).

58. S. T. Milner, S. A. Safran, D. Andelman, M. E. Cates, and D. Roux, J. de Physique **49**, 1065 (1988).

59. P. G. de Gennes and C. Taupin, J. Phys. Chem. **86**, 2294 (1982).

9

Fluctuating Interfaces and the Structure of Microemulsions

S. A. Safran[1]

9.1 Introduction

Mixtures of amphiphiles, water and oil can have structures [1, 2, 3] that have only microscopic correlations between the components (*e.g.*, a three component solution) as well as long-range ordered structures (*e.g.*, lyotropic liquid crystals). The term microemulsion, in its most general use, connotes a thermodynamically stable, fluid, oil-water-surfactant mixture. In practice, microemulsions are taken to consist of structures with intermediate-range correlations. The oil and water regions are fairly well separated and the surfactant molecules are organized as monolayers at the internal water-oil interfaces [4]. There are long-range correlations between the oil and water molecules which are segregated into domains of the order of hundreds of Angstroms. In addition, there are long range correlations among the surfactant molecules which self-assemble into a monolayer film at the set of internal water-oil interfaces. In this respect, microemulsions are differentiated from ordinary three-component solutions. However, the set of interfaces which comprise the microemulsion do not show long range order comparable to that found in lyotropic liquid crystals.

It is precisely for these reasons that the understanding of microemulsions is both interesting and difficult: interesting, because the structure can be idealized as a set of interfaces, and difficult, because there is no long-range order of these interfaces. In this review, the structures and phase behavior of microemulsions are discussed from the interfacial point of view. Although many workers [5] have implicitly used this "phenomenological" approach to analyze globular (mostly spherical) microemulsions, the application to random, or bicontinuous systems was initiated by Talmon and Prager [6] and further developed by de Gennes and co-workers [7], and Widom [6].

---

[1]Department of Materials and Interfaces, The Weizmann Institute, Rehovot 76100, Israel

More recently, Andelman, Cates, Milner, Roux, and Safran [9] have extended this approach to include the thermal fluctuations of the surfactant film and model both the thermodynamics and spatial correlations of bicontinuous microemulsions. One regards the oil and water as continuum liquids; the interfacial surfactant layer is treated as a flexible sheet. For systems where the surfactant film is a condensed, two-dimensional liquid, the dominant energy is the bending or curvature energy of the monolayer. While this energy is minimized by a droplet or domain with a given curvature (determined by the molecular details of the surfactant packing at the interface), the entropy tends to randomize the structure. Golubovic and Lubensky [10] have further analyzed this competition that gives rise to the rich [11] phase behavior and structures that are observed.

An alternative approach is based on the construction of microscopic lattice models in which a cell contains only a small number of molecules [12]. This point of view, which focuses on the microscopic interactions between the water, oil and surfactant molecules (see Chap. 8), is well suited to describe three-component solutions and their relation to microemulsions. The microscopic approach, though of undoubted fundamental interest, may be more difficult to implement than the phenomenological one. In particular, a microscopic model of microemulsions must produce structural organization on a length scale much larger than that of the molecular (or lattice) size. In contrast, the interfacial point of view presented here is tailored to focus on the intermediate scale microstructure and its relation to the phase behavior of microemulsions. The length scales of interest for the internal structures as probed by scattering experiments are in the range of hundreds of Angstroms for good microemulsions. These length scales are more easily treated by the continuum theory presented here; microscopic theories such as the lattice models [12] of Alexander, Widom, Schick, and others would have to include correlations between thousands of molecules to accurately model the interfaces at these length scales. In contrast, the interfacial models presented here assume that the strong correlations needed to self-assemble the surfactant molecules at the internal water-oil interfaces are always present.

The organization of this article is as follows: Sec. 9.2 begins with a brief review of some of the experimental observations and classifies microemulsions into their various structures. In addition, the theory of the curvature energy and an introduction to the necessary statistical mechanics is presented. The structure and phase behavior of spherical droplet microemulsions are discussed in Sec. 9.3.1 where attention is given to the problem of size (polydispersity) and shape fluctuations in both a static and dynamic sense. Sec. 9.3.2 deals with the analogy between worm-like, flexible, cylindrical microemulsions and polymers. In Sec. 9.4, the interesting and challenging problem of the connection between ordered, lamellar structures and random, bicontinuous microemulsions is explored. A unified, theoretical approach is presented for both the phase behavior and the microstructure of these systems and the scattering structure factor is calculated and compared with experiment.

9.2 Experimental Observations and Theoretical Models

This section summarizes the major properties of microemulsions which have to be addressed by any theoretical description. Attention is given to physical trends (*e.g.*, concentration dependence) as opposed to chemical trends (*e.g.*, counterion atomic number dependence) since the phenomenological theory most accurately predicts the former. A more complete discussion of the experimental situation can be found in Chap. 7. The phenomenological model is presented and the concept of the curvature or bending energy is explained. The effect of thermal fluctuations on both the interface structure and bending modulus is analyzed. The transitions between different microemulsion shapes are shown to arise from a competition between the bending energy and the conservation constraints.

9.2.1 EXPERIMENTAL OBSERVATIONS

In the present work, microemulsions are defined as thermodynamically stable, fluid, oil-water-surfactant mixtures. The surfactant volume fraction is typically small ($\sim 5\%$); many systems also contain cosurfactant (alcohol) and/or salt. The theoretical approach presented here considers only three component systems; the cosurfactant and the salt are taken to mix with the surfactant and water respectively and to only affect the numerical values of the theoretical parameters (interactions, moduli etc.) discussed here.

Microemulsions are differentiated from three-component solutions of oil, water and surfactant by the separation of the oil and water into coherent domains, typically tens or hundreds of Angstroms in size [2,3]. The surfactant molecules prefer the interfacial environment to either that of bulk water or bulk oil. The sizes or shapes of the microemulsion domains can be varied by changing the concentration, temperature T, or salinity [13]. When there is a strong tendency for the interface to bend towards either the water or the oil domains, the structure is generally that of globules (at least for small volume fractions of oil in water or water in oil). Both spherical [14] and cylindrical [15] structures have been proposed and examined in light of both structural and transport measurements. Analogies between these systems of interacting globules and colloidal suspensions have also been explored [5,16]. When the volume fractions of oil and water are comparable–and when there is no intrinsic tendency of the surfactant film to curve towards the water or the oil–one expects random, bicontinuous structures to form. Under appropriate conditions (particularly when the

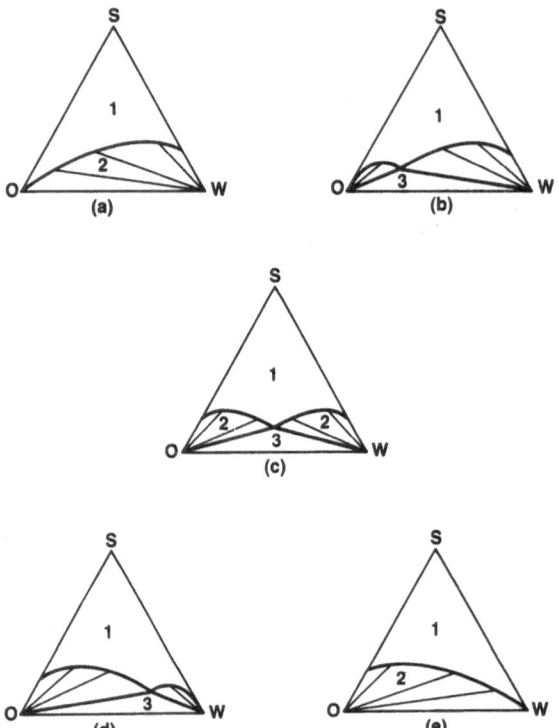

FIGURE 9.1. Evolution of phase diagram of microemulsions (schematic). The sequence of phase diagrams is obtained experimentally by varying temperature, salinity, oil chain length, or surfactant chemistry. Theoretically, many of these changes can be lumped into changes in the spontaneous curvature.

volume fraction of surfactant is higher than a few percent) various ordered structures, such as lamellae, cubic or cylindrical structures, may also arise.

The characteristic structure of the microemulsion associated with the spontaneous curvature of the interface (*i.e.*, the intrinsic tendency of the surfactant film to bend towards either the water or the oil regions) reflects itself in the equilibrium phase diagram as a function of concentration. The sequence of phase diagrams [17] shown in Fig. 9.1 is generally obtained by changing salinity which presumably affects the screening of the polar head groups of the surfactant molecules and hence the spontaneous curvature (see discussion below). Typically, one finds that globular systems (*e.g.*, water in oil) show a single phase region where the globular phase is stable and a two-phase region where the globules coexist with an excess phase (*e.g.*, water with some small amount of surfactant) as shown in Fig. 9.1a. As the salinity is varied and the surfactant film is "balanced" so that its spontaneous curvature is zero, another two-phase region consisting of a microemulsion in coexistence with a phase that is mostly oil, as well as a three-phase region consisting of a microemulsion in equilibrium with both oil and water, are seen (Fig. 9.1b). In Fig. 9.1c, the completely "balanced" system is shown; the phase diagram is oil-water symmetric. Figs. 9.1d and 9.1e show the evolution of the system as the spontaneous curvature (*e.g.*, controlled via salinity) is varied so that the surfactant film bends towards the oil.

At the balance point, the microemulsion which coexists with both the water and oil (*i.e.*, the "middle-phase") shows [18] ultralow interfacial tensions with both these phases; this results in a variety of technological applications, for example in chemically enhanced oil recovery [19]. The large length scale of the microemulsion is one factor that is responsible for the lowered interfacial tensions. By dimensional analysis, the macroscopic, interfacial tension, γ, varies as $\gamma \sim T/\xi^2$, where ξ is the structural length scale. Values of $\xi \sim 100 \ \AA$, can result in a reduction of γ by several orders of magnitude.

The theoretical model discussed in this article focuses on the following properties and observations:

(1) Shape transitions as the surfactant film properties or the concentrations are varied:

The spherical to lamellar transition is easily observed using optical birefringence or x-ray diffraction, since the lamellar phases have (quasi) long-range order [20]. The evolution from globular to bicontinuous phases is best evidenced in NMR experiments [21] by changes in the molecular, self-diffusion of the water, oil and surfactant as a function of salinity. The globular phases are characterized by a molecular self-diffusion constant of the internal (*e.g.*, water, for water-in-oil systems) species which is equal to the self-diffusion constant of the entire globule, which latter is measured by light scattering [21]. In the bicontinuous phase, the self-diffusion constant of both the oil and water approach their values in bulk liquids. Another experimental probe of the transition from globules to bicontinuity is the ionic conductivity of the water which shows a dramatic increase as the percolation transition is reached [22], presumably indicating a bicontinuous structure. However, the existence of a percolation transition can sometimes be consistent with a globular structure when the conduction is facilitated by the hopping of charges or charged surfactant molecules in clusters of globules [23]. Thus,2 conductivity measurements must be interpreted with some care.

(2) The static and dynamic fluctuations of globular microemulsions:

(a) Spherical globules: In systems of nominally spherical microemulsion droplets, analysis of the static neutron spectra indicate [24] that the systems are polydisperse with an rms polydispersity of about 25%. The shape fluctuations of the globules are difficult to extract from the static data. However, recent analysis of dynamic, spin-echo, neutron experiments have been analyzed in terms of the dynamical shape fluctuations of the nominally spherical droplets [25]. The scaling of the dynamical fluctuations with the droplet size indicates that the restoring force for these undulations is not the finite compressibility of the surfactant monolayer (surface tension of the droplet), but rather the bending modulus of the film [25].

(b) Cylindrical globules: For systems where the nominal globule shape is cylindrical, the major effect of the film fluctuations is to produce a random wandering of the cylinder axis [26]. This results in a worm-like behavior of the cylinder and allows an analogy with polymers to be made [26]. To date, there is evidence for cylindrical microemulsions from conductivity measurements as well as static scattering spectroscopy [15]. Measurements of the undulations of these cylinders and the expected polymer-like behavior have not yet been made. However, several groups have reported polymer-like behavior in systems of cylindrical micelles [27]. In these systems, the osmotic compressibility and cooperative diffusion constant show a dependence upon surfactant concentration that is similar (similar scaling exponents) as that found for semi-dilute polymer solutions.

(3) Transitions between ordered lamellar structures and bicontinuous microemulsions:

The phase diagrams for typical microemulsions show that at values of the salinity where the system is water-oil symmetric (no spontaneous curvature of the surfactant film), a lamellar phase exists [1] at large values of the surfactant volume fraction, ϕ_s. As ϕ_s is decreased, the structure undergoes a first-order transition to an isotropic, transparent, bicontinuous microemulsion. At even smaller values of ϕ_s, there is a three-phase coexistence between the microemulsion and two phases which are mostly oil and water respectively. Good microemulsions–those with large (hundreds of Angstroms) water and oil domains–show a large three-phase region where the microemulsion coexists with two phases that are mostly oil and most water respectively, all at low surfactant volume fractions. As Kahlweit [1] first pointed out, even a three-component mixture of water, oil, and surfactant is expected to have a three-phase equilibrium; it is the existence of this equilibrium *at small surfactant volume fractions* which leads to a long-length scale, which is the signature of the microemulsion as opposed to a multi-component solution with only short-ranged spatial correlations.

(4) Microstructure of bicontinuous microemulsions:

While the bicontinuous nature of balanced microemulsions has been established through probes such as conductivity [22] and NMR relaxation [21], the microstructure of these systems has been investigated by neutron [28] and x-ray scattering [29] as well as by electron microscopy (see Chap. 7). The scattering studies of middle-phase microemulsions which coexist with both oil and water show the following features for a wide variety of systems [28,29]:

(a) Scattering at large wavevectors, q, which obeys a Porod law (intensity proportional to q^{-4}) with corrections due to the finite thickness of the interface. The Porod law breaks down at a characteristic length scale related to the curvature of the interface.

(b) A peak in the intensity at a smaller value of $q = q_{max}$ with a half-width of the order of q_{max}.

(c) The ratio of the characteristic size obtained from the breakdown of the Porod law to that obtained from q_{max} is about 2.

9.2.2 THEORETICAL MODELS

Theoretical treatments of microemulsions can be divided into two categories: (i) microscopic models, which account for the properties of microemulsions in terms of discrete oil, water, surfactant molecules [12] (see Chap. 8) and (ii) interfacial models, which focus only on the properties of the surfactant film. The interfacial model, discussed here in detail, treats the surfactant monolayer separating water and oil domains as an incompressible, two-dimensional fluid, whose properties are controlled by curvature elasticity [30].

This section compares the results of the interfacial and microscopic models with the experimental observations discussed above. Although there is only one relevant energy in the problem, the predicted behavior of the system is rich and varied. Microscopic models which consider finite values for all the interaction terms described above (see Chap. 8) will have to be solved to a high degree of accuracy just to attain the highly correlated state which is the starting point of the interface model. In addition, these microscopic (e.g., lattice-gas) models will have to be treated in a very sophisticated manner to include the bending fluctuations which are best treated in the continuum manner described above. On the other hand, the continuum, interface model is inappropriate when the microemulsion and molecular length scales are comparable. In this limit, there is no well-developed interfacial structure; the system is best described as a three-component solution and there are no long-range correlations (except near critical points of second-order phase transitions); the microscopic models are then most appropriate for theoretical calculations.

9.2.2.1 Shape transitions as the surfactant film properties or the concentrations are varied

Most treatments of shape transitions have been based upon packing arguments in which the globule or interface shape is determined by the optimal packing of the surfactant heads and chains [31]. To date, lattice models of microemulsions have only been solved in mean-field-like approximations which are too crude to predict shape transitions. The approach presented here, in which the interfacial elasticity controls the energy of the surfac-

tant film, determines the shape of the interface by a minimization of the
free energy. Transitions from spheres to cylinders to lamellar structures
are predicted as a function of the spontaneous curvature and concentration. The entropy of dispersion and interglobular interactions can also be
included in a controlled manner. There have been many phenomenological treatments [14] of spherical microemulsions which have focused on the
colloid-like properties of systems with fixed spherical shapes. Interactions
between droplets can lead [16] to phase equilibria between droplet phases
with different concentrations of droplets [32]. These effects are beyond the
scope of the present work, although interaction effects can be included in
the free energy as discussed in [33].

9.2.2.2   The static and dynamic fluctuations of globular microemulsions

Microscopic models of microemulsions have not yet been developed which
can deal with the long-wavelength undulations which are responsible for the
shape changes of the spherical globules or the worm-like behavior of the
cylindrical systems. These effects are best treated by the effects of thermal
fluctuations upon the interface with the curvature elasticity as the restoring
force.

The dynamical fluctuations of the surfactant film can also be treated
within the interfacial model and detailed predictions for the fluctuation
timescales and amplitudes are obtained [34]. A comparison of theory and
experiment demonstrates that the restoring force for these fluctuations is
indeed the curvature elasticity and that the nearly spherical drops deform
at constant area and volume. In addition, fits of theory and experiment
yield the bending modulii of the system.

9.2.2.3   Transitions between ordered lamellar structures and bicontinuous
microemulsions

The transition between the lamellar phase and the microemulsion phase
has been discussed in terms of an increased flexibility of the surfactant
film. It has been observed that such a transition can be induced by the
addition of alcohol, which tends to decrease the bending modulus [30],
k, of the surfactant monolayer (see, for example, Sec. 1.3.4). In addition,
de Gennes and Taupin [7] have discussed the role of the film undulations
in nominally, lamellar systems. Stiff systems—with large values of k or
at low temperatures T–form lamellar structures, while floppy ones form
microemulsions. However, these findings do not explain the transition from
lamellar to microemulsion as a function of surfactant concentration. This
transition has been attributed [9] to the dependence of the effective bending
modulus on the length scale of the microemulsion and has been related to
the renormalization of the bending constant by thermal undulations as
discussed below [35]: see also Chap. 5. The interfacial model can treat
both the lamellar to microemulsion transition as well as the multi-phase
equilibria found [1,2] at small values of ϕ_s.

9.2.2.4   Microstructure of bicontinuous microemulsions

Previous theoretical treatments of the structure and scattering from bicontinuous phases have been of two sorts: (i) Extensions of the thermodynamic models, such as that of Talmon and Prager [36], which assume a random distribution of oil and water domains. These theories fail to predict a peak in the scattering due to the lack of correlations in the model. (ii) Geometrical theories for the structure factor which rely on either an assumed form [37] for the scattering or a highly constrained microstructure [38]. Although they fit the experimental data reasonably well, the former calculations have no thermodynamic origin; the $q \to 0$ limit of these theories does not originate from a free energy which is capable of predicting the equilibrium phase diagram. The disordered, open connected model (DOC) of microemulsion structure is envisioned [38] to be appropriate to systems with very large bending constants where film fluctuations can be ignored.

The theoretical treatment of bicontinuous microemulsion presented here extends the thermodynamic treatment to include fluctuations; the same theory predicts the phase diagram, the scattering peak, and the microstructure [39]. This theory predicts a peak in the scattering due to correlations in the water-oil domains. These correlations are present due to the existence of a finite bending modulus, so that the water-oil domains are never truly random. The theory also shows how the observed structure factor is consistent with the concept of bending modulus renormalization and that the predicted microstructure is in agreement with observation. Finally, the interfacial model relates the characteristic length scale of the microemulsion as well as the small value of ϕ_s in the middle-phase microemulsion to the natural persistence length of the surfactant film. This persistence length is determined by a balance between the entropic undulations and the curvature elasticity.

9.2.3   INTERFACIAL MODEL

9.2.3.1   Assumptions

The interface model which is the focus of the present work is a limiting case of a general model of three-component solutions, such as the microscopic "lattice" models of Ref.12 and Chap. 8. A general treatment of the statistical mechanics of water-oil-surfactant solutions would have to include several types of interactions between the components. The interface model simplifies the physics of these systems by considering a separation of energy scales in these interactions:

(1) Oil/water, water/surfactant-aliphatic-tail, and oil/surfactant-polar-head interactions are large ($>> k_B T$) and repulsive.

(2) Water/water, oil/oil, water/surfactant polar head, and oil/surfactant aliphatic tail interaction energies are large and attractive.

In the limit that these interaction energies are much larger than those associated with the surfactant film deformations, they act as constraints upon the allowed configurations of the system. In particular, there is no possible mixing of oil and water, and the surfactant is constrained to lie at the oil-water interface. The statistical mechanics of the system are then completely determined by the configurations of the surfactant film. Two types of interactions between surfactant molecules are then relevant.

(3) Elastic interactions which depend on the center of mass distance between surfactant molecules: These are assumed to be large ($>> k_B T$) and to constrain the surfactant film to act as an incompressible, two-dimensional fluid. The area per chain in microemulsions has been measured $\approx 20 \text{Å}^2$, consistent with the assumption of the surfactant film as an incompressible fluid [40].

(4) Orientational interactions between surfactant molecules at the water-oil interfaces: These interactions exist because the surfactant molecules are oriented at the water-oil interface; their polar heads are in the water region and their hydrocarbon tails are in the oil (consistent with assumptions (1) and (2) above). Since the angle between two surfactant molecules is a continuous variable which can have slow spatial variations, it is expected that this orientational or bending energy [30] of the interface will have the smallest magnitude of the energies considered above and will be the most strongly affected by thermal fluctuations.

9.2.3.2   Curvature elasticity

The interfacial model considers the bending or curvature energy of the surfactant film. Since the relevant structures have curvatures, c_1 and c_2, which are much smaller than $1/\ell$, where ℓ is a typical molecular dimension (*e.g.*, the surfactant size), the bending energy, F_b, can be written phenomenologically as an expansion in the product $c_1\ell, c_2\ell << 1$. Keeping terms up to quadratic order, noting that F_b must be symmetric in c_1 and c_2, and treating the film as an *isotropic* two-dimensional liquid, one finds [30,41]

$$F_b = \frac{1}{2}k \int (c_1 + c_2 - 2c_0)^2 dS + \bar{k} \int c_1 c_2 dS \qquad (9.1a)$$

This expression accounts for the energy cost for bending a surface (area element dS); deviations of the average curvature from the spontaneous curvature, c_0, raise the energy of the system by an amount proportional to k. This modulus is related to the splay constant of a single layer of a smectic liquid crystal, where the dominant energy is also due to the orientational interactions and hence the bending of the layer. The second term in Eq.(9.1) accounts for the energy cost for creating saddle-type deformations ($c_1 c_2 < 0$), and the modulus $\bar{k}$ is termed the saddle-splay modulus. The Gauss-Bonnet theorem implies that the term in the curvature energy proportional to $\bar{k}$ depends only on the topology, since

$$\int dS c_1 c_2 = 4\pi\chi, \tag{9.1b}$$

where χ is a topological constant (equal to unity for spheres), proportional to the difference between the number of disjoint surfaces and the number of handles.

The phenomenological bending energy of Eq. (9.1) accounts for all of the long-wavelength structures and phase equilibria of microemulsions. It is, nevertheless, instructive to consider [42] the microscopic origins of the modulii k and $\bar{k}$, and the spontaneous curvature, c_0 – see Sec.1.3.4. The numerical values of the phenomenolgical parameters are strongly influenced by changes in the chemistry, salinity and temperature; we therefore discuss the qualitative features of the bending modulii and the spontaneous curvature.

The spontaneous curvature describes the tendency of the surfactant film to bend towards either the water ($c_0 > 0$ by convention) or the oil. It is taken – in the absence of long-range interactions – to arise from the competition in the packing of the polar heads and hydrocarbon tails of the surfactant molecules. If the interactions between the polar heads (as mediated through the intervening water and salt molecules) favor a smaller packing area than that dictated by the tail-oil-tail interactions, the surfactant film will tend to curve so that the heads (and the water) are on the "inside" of the interface (an oil-external microemulsion). In the opposite case, the microemulsion will be water-external. One thus expects that with increasing salinity, the polar head repulsive interaction is screened, allowing the heads to pack more closely and favoring an increase in c_0. Naively, one might expect increasing temperature to favor an increased randomness in the tail packings, leading to a larger area per chain, and hence again an increase in c_0. A more accurate description of these trends would require a quantitative analysis of the screening and chain packing problems.

The bending modulii, k and $\bar{k}$ arise from the elastic constants determined by the head-head and tail-tail interactions. It is expected that these modulli increase with increasing surfactant chain length since k and $\bar{k}$ scale as the product of an elastic constant and a length squared. Models [42] of both short-chain and long-chain (polymer-like) surfactants indicate that in the case that the chain entropy dominates, k and $\bar{k}$ scale as the cube of the chain length. If the dominant interaction is the interchain attraction/repulsion, k and $\bar{k}$ are expected to increase with the square of the chain length (see Sec. 1.3.4).

It should be noted that the fact that the parameters k, $\bar{k}$, and c_0 can be temperature dependent, indicates that the bending energy discussed above should be considered a bending *free energy*. On short length scales, the phenomenological parameters are determined by the physical considerations discussed above. However, on long length scales, the undulations of the surface at smaller scales should affect the bending modulii and spontaneous curvature (see Chap. 5). This "renormalization", which is important for the coarse-grained configurations of the system, was first discussed by Helfrich [35] who showed that at large length scales, the bending modulus $k(\xi)$ is given by:

$$k(\xi) = k_0 \left[1 - \log(\xi/\ell)/\kappa + ...\right] \tag{9.2}$$

where $k_0 = k(\ell)$ is the bare bending modulus, ℓ is a microscopic cutoff (related to the distance between surfactant molecules), and

$$\kappa = 4\pi k_0/\alpha T \tag{9.3}$$

(We have used units where the Boltzmann constant $k_B = 1$.) Helfrich [35] determined α to be unity, while several other authors [35] find that $\alpha = 3$. In any case, for fluid films, the bending modulus is reduced by the short distance thermal undulations of the surface. Within this approximation $k(\xi) \to 0$ as ξ approaches the persistence length of the interface, ξ_K, defined as [7],

$$\xi_K = \ell \exp[4\pi k_0/\alpha T] = \ell e^\kappa \tag{9.4}$$

The expression shown in Eq.(9.2) is obtained from a perturbation expansion about a flat interface; as $k(\xi)$ approaches a value close to T, the higher order correction terms in Eq. (9.2) may be significant. The downward renormalization of k indicates that it becomes relatively easy to bend a sheet of size $\xi \approx \xi_K$, since such a sheet is spontaneously crumpled. Similar renormalizations may apply [35] to $\bar{k}$ and c_0, although the physical meaning of the renormalization of $\bar{k}$ is unclear, since it couples to a topological quantity. The effect of the renormalization upon c_0 is not significant for systems where the bare value of $k(\ell)$ is greater than several times $k_B T$. This is because systems with non-zero spontaneous curvature tend to form globules whose maximum size is of order $c_0^{-1} << \xi_k$; the system never gets large enough to sense the persistence length and hence the renormalization of k . The decrease in the bending modulus is most important is systems where $c_0 \approx 0$ as discussed below since these systems can be characterized by length scales $\xi \approx \xi_K$.

9.2.3.3 Microemulsion shapes

When the spontaneous radius of curvature, c_0^{-1}, is smaller than the persistence length, ξ_k, of the surfactant film, the fluctuations associated with the renormalization of k can be neglected. The bending modulus can be taken at its bare value, k_0, and the system can be studied by focusing on the minimization of the curvature energy. In the absence of any constraints, the minimum energy state for $k >> k_B T$ and negative values of $\bar{k}$ is one of monodisperse spheres of radius $\tilde{c}_0^{-1}$, where $\tilde{c}_0 = c_0 \left[1 + (\bar{k}/2k)\right]^{-1}$. However, such a configuration is only possible for particular values of the water, oil and surfactant concentrations because of the incompressibility assumption described earlier. These constraints require that the total surface area of all of the interfaces in the system is fixed by the surfactant concentration. Similarly, the total volume enclosed by the polar-head side of the surfactant film is fixed by the total concentration of water. Simple geometry indicates that the ideal curvature of $c = \tilde{c}_0$ for spheres can only be obtained when the surfactant/water ratio (for water internal globules) satisfies $\phi_s/\phi_w = 3\delta\tilde{c}_0$, where δ is a molecular size. Any other ratio of the concentrations can result in a system of monodisperse spheres, but with a curvature which is not equal to $\tilde{c}_0$. Therefore, even in the absence of thermal fluctuations, a large variety of structures are possible because of the competition between the tendency to minimize the bending energy (which prefers spheres of radius $\tilde{c}_0$) and the necessity to satisfy the incompressibility constraints. [It should be noted that for static properties these constraints apply globally to the entire system and not to each individual globule or section of film. In Sec. 9.3, the dynamic implications of these constraints are considered; they constrain the fluctuations of even a single droplet at the appropriate time scales.]

For systems where F_b dominates the free energy, the shape transitions due to the competition between the bending energy and the incompressibility can be simply calculated. Assuming a monodisperse set of globules, the incompressibility conditions are given by,

$$nA\delta = \phi_s \tag{9.5a}$$

and

$$nV = \phi_w \tag{9.5b}$$

where n is the density of globules, δ is the surfactant size, and A and V are the surface area and volume of the globules respectively. The bending energy for globules of different shapes can be calculated and compared to determine the microemulsion shape. Using the constraints given by Eq. (9.5) in the expression for F_b one finds the following expressions for the bending energies of spheres (F_b^s), and infinite cylindrical (F_b^c) and lamellar structures (F_b^l):

$$F_b^s = 6\tilde{k}\phi_w\tilde{c}_0^3(1 - r)^2 \tag{9.6a}$$

$$F_b^c = 6\tilde{k}\phi_w\tilde{c}_0^3 \left[\frac{9}{16}(1 + \bar{k}/2k)^{-1} - 3r/2 + r^2\right] \tag{9.6b}$$

$$F_b^l = 6\tilde{k}\phi_w\tilde{c}_0^3 r^2 \tag{9.6c}$$

In Eqs.(9.6), $\rho = 3\delta\phi_w/\phi_s$ is proportional to the volume to surface ratio of the structure, determined by the constraints, and $r = \tilde{c}_0\rho$ is the ratio of the two characteristic lengths in the problem: the volume to surface ratio, ρ (fixed by the concentrations) and the spontaneous radius of curvature, $\tilde{\rho}_0 = \tilde{c}_0^{-1}$ (fixed by the form of the bending energy). The modulus $\tilde{k} = k + \bar{k}/2$. Ref. [41] gives expressions for the bending energy for several other shapes as well; in the limit where the entropy of mixing and thermal undulations of the film can be neglected, the phase diagram is determined by the competition between the shapes discussed here.

For $r > 1$, the energy is minimized by a phase of spherical globules with $\rho = \tilde{\rho}_0$, coexisting with excess internal phase ("emulsification failure") or with a dilute, lamellar phase. This is because the lowest energy state of the system is one of spheres of radius $\rho = \tilde{\rho}_0$; if excess water is added, it is simply rejected into another phase. The effects of the entropy of mixing on this instability can be easily incorporated by adding a term to Eq. (9.1) which represents the entropy of a dilute "gas" of spherical droplets and by finding the value of at which the free energy is a minimum [43]. The result yields an expression for $\rho/\tilde{\rho}_0$ along the emulsification failure phase boundary which depends on the volume fraction of droplets:

$$\rho/\tilde{\rho}_0 = 1 + \frac{T}{8\pi k} \log(v_0 n) \qquad (9.7)$$

where v_0 is a molecular volume, and n is the number density of drops. As r is decreased, by a reduction of either the ratio of ϕ_w/ϕ_s (for water internal systems) or of c_0, first-order phase boundaries separate the regimes of different shapes. The regions where the different shapes (spheres, cylinders, lamellae) have the lowest bending energy are shown in Fig. 9.2 as a function of the ratio $x = -\bar{k}/2\tilde{k}$ and $r = \rho/\tilde{\rho}_0$. It should be noted that these regions include domains of two-phase coexistence of e.g., spheres and cylinders. The details of this coexistence may also depend on the entropic contributions to the free energy. However, it is important to note that a comparison of the bending energies of spheres and cylinders implies that for $\bar{k} = 0$ the two shapes are degenerate, albeit at different values of r. This implies that spheres and cylinders coexist in a two-phase equilibrium with no single phase of spheres when $\bar{k} = 0$ (see Fig. 9.2). Thus, a single phase of spheres is only stable when $\bar{k}$ is negative and non-zero; the size of this one-phase region is related to magnitude of $\bar{k}$. The cylindrical phase is stable in a region of r and $\bar{k}$ where it best accommodates the volume and surface constraints and still maintains an average radius of curvature close to $\tilde{\rho}_0$; for small saddle-splay elasticity, $\bar{k}$, there is little cost in generating an anisotropic structure.

For large values of the saddle-splay ($x > 1/3$), only shapes with identical orthogonal radii of curvature (i.e., spheres and lamellae) are present. Finally, at small values of $\tilde{c}_0$, for all values of x, the stable shape is lamellar;

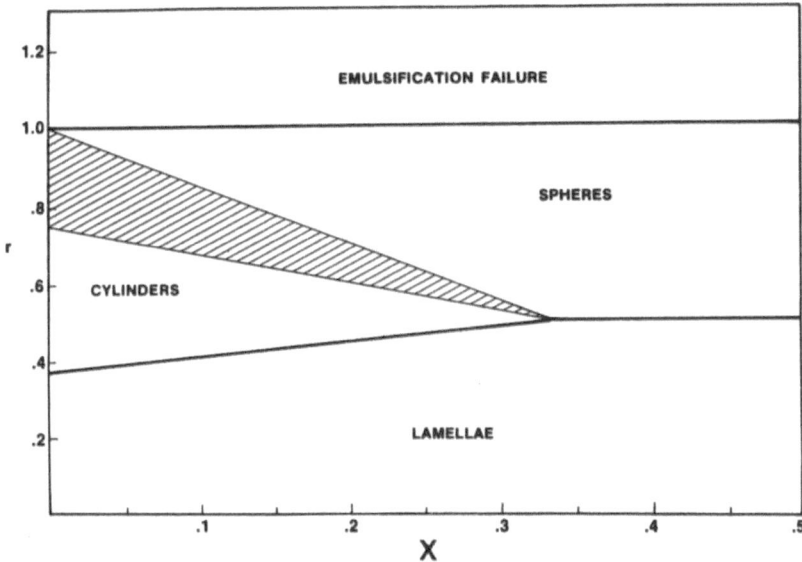

FIGURE 9.2. Phase stability of spherical, cylindrical, and lamellar phases. For simplicity, we show only the crossings with the lamellar free energy; the cross-hatched region is a two-phase coexistence of spheres plus cylinders. The parameter $r = 3\tilde{c}_0 \delta \phi_s / \phi_w$, where δ is the surfactant size, $\tilde{c}_0$ the spontaneous curvature defined in the text, and ϕ_s (ϕ_w) are the surfactant (water) volume fractions for water in oil systems; $x \sim -(\bar{k}/k)$.

there is no preferred globule since there is no energetic tendency to bend towards either the oil or the water.

The effects of including the entropy of mixing and excluded volume interactions between globules have been considered by Roux and Coulon [44]. For large values of k , they find a phase diagram similar to that described above; there is a two-phase coexistence region with spheres and cylinders in equilibrium at finite values of the volume fraction. Since entropy effects generally stabilize smaller objects, inclusion of the entropy of mixing would enlarge the region of stability of the spherical droplet phase compared with the cylindrical and lamellar structures.

9.3 Fluctuation of Microemulsions

The fluctuations of the system in both size and shape can modify the simple description of microemulsion shapes discussed above. The probability, P, that an arbitrary deformation of $e.g.$, spherical globules, will occur in thermal equilibrium is proportional to the Boltzmann factor

$$P \sim \exp(-\Delta F_b/T)$$

where ΔF_b is the difference in bending energy between the system of monodisperse spheres and the system of deformed droplets. For spherical globules, the main effect of these fluctuations on the ensemble of nominally spherical droplets is to induce a polydispersity in size and shape [45]. However, the integrity of the description of the system as one of approximately monodisperse spheres is maintained. This is not the case for the cylindrical [26] and lamellar [7] structures which have at least one very long length scale that can support long-wavelength undulations of the surfactant film. These long-wavelength undulations can have little bending energy cost and thus a large Boltzmann weight. For systems that are flexible (k/T not too large) and/or dilute enough, the zeroth order picture of infinite, rigid cylinders, or liquid crystalline, lamellar order may be radically modified by these fluctuations; the resulting structure - at long length scales - may be much more disordered.

9.3.1   Spherical Droplets

In this Section, the effects of the film fluctuations on a phase of spherical droplets is considered; both static and dynamic fluctuations are discussed.

9.3.1.1   Static Fluctuations

Although energetic considerations indicate that a phase of monodisperse, spherical droplets exists for values of $\tilde{c}_0$ close to unity [26,41], the ensemble of droplets may have individual members whose shape and size differ from that of the average drop size, $\rho = 3\delta\phi_w/\phi_s$. This polydispersity is intrinsic to the system in equilibrium and is unrelated to kinetic effects, sample preparation, etc. An estimate of this polydispersity has been discussed for microemulsions [45] and for vesicles [46], where the conservation constraints enter in a different manner. In both cases, one considers the deviation of a droplet from a spherical shape, and expands the bending energy to second-order in this deviation. The rms amplitude of the fluctuation is then calculated from the equipartition theorem.

The equation which defines the surface of the ith droplet with its center at the origin is

$$R_i(\vec{r}) = r - \bar{\rho}\,[1 + g_i(\Omega)] = 0, \tag{9.8}$$

where $\bar{\rho}$ is the average droplet radius which is independent of the solid angle, Ω, and the index i. The dimensionless fluctuation amplitudes $\{g_i\}$ are expanded in spherical harmonics as

$$g_i(\Omega) = \sum_{l,m,l\neq 1} a_{lm}^i Y_{lm}(\Omega). \tag{9.9}$$

The curvature energy, Eq.(9.1) is expanded to quadratic order in the $\{g_i\}$. The $l = 1$ mode is excluded since it represents a translation of the center of mass. The constraints of fixed total volume and area relate the droplet density, n, and the average drop size, $\bar{\rho}$, to the fluctuation amplitudes as

$$n = n_0 \left[1 - \frac{3}{4\pi} \sum_{l,m,l\neq1} <|a_{lm}|^2> (b_l - 1) \right] \qquad (9.10a)$$

$$\bar{\rho} = \rho \left[1 + \frac{1}{4\pi} \sum_{l,m,l\neq1} <|a_{lm}|^2> (b_l - 2) \right] \qquad (9.10b)$$

where $b_l = \frac{1}{2}l(l+1)$ and where $n_0 = (3\phi/4\pi)/\rho^3$ with $\rho = 3\delta\phi_w/\phi_s$ and δ being the typical size of a surfactant molecule. The resulting bending energy is thus given by

$$F_b = F_b^s + \sum_{l,m,l\neq1,i} \frac{1}{2}k\,\Delta F_l |a_{lm}^i|^2, \qquad (9.11a)$$

where

$$\Delta F_l = 4(3 - 4b_l + b_l^2) + V(b_l - 1) \qquad (9.11b),$$

with

$$V = [8\rho c_0 - 6\bar{k}/k] = [8\rho\tilde{c}_0(1 + \bar{k}/2k) - 6\bar{k}/k]. \qquad (9.11c)$$

In Eq. (9.11a), F_b^s is the bending energy of the monodisperse spheres discussed above. The equipartition theorem implies that the equilibrium value of fluctuation amplitudes are given by (note that ΔF_l is dimensionless)

$$<|a_{lm}|^2> = \frac{T}{k}[\Delta F_l]^{-1} \qquad (9.12a)$$

The fluctuations in the equilibrium ensemble are largest for small values of l. The $l = 0$ fluctuations represent polydispersity in size, while the $l = 2$ fluctuations result in a crimping of the spheres to an ellipsoidal shape. These two largest fluctuation amplitudes can be distinguished by their different dependence on the parameter V, which is most sensitive to the value of the saddle-splay modulus, $\bar{k}$. While both the $l = 0$ and $l \geq 2$ modes have amplitudes that are proportional to T/k, the $l = 0$ fluctuations are enhanced by a negative value of $\bar{k}$, while the $l \geq 2$ mode amplitudes are suppressed as $\bar{k}$ is made more negative. The physical origin of this effect is the fact that $\bar{k}$ couples to the number of droplets; negative values of $\bar{k}$ tend to increase the number of drops and hence the polydispersity. On the other hands, shape fluctuations tend to decrease the number of droplets, so $\bar{k}$ tends to inhibit these ($l \geq 2$) modes. At emulsification failure, i.e., when $\rho = \tilde{\rho}_0$, one has

$$< |a_{00}|^2 >= \frac{T/k}{2\left[2 + \bar{k}/k\right]} \qquad (9.12b)$$

$$< |a_{2m}|^2 >= \frac{T/k}{2\left[8 - 2\bar{k}/k\right]} \qquad (9.12c)$$

which show the trends discussed above. Again, there is no observed divergence in these modes because of the first-order shape transitions to cylinders and/or lamellae discussed above.

Neutron scattering measurements of the properties of dilute microemulsions, where interglobular interactions can be ignored, are sensitive to the polydispersity [24] (see Sec. 7.3.1.1). For a monodisperse system of spheres of radius R, the scattering intensity is given by the spherical form factor which vanishes when $QR = z_n$, where z_n are the zeros of the spherical Bessel function and Q is the scattering wavevector. For polydisperse systems, a minimum, rather than a zero, is predicted. Fits [24] of scattering data in systems which are well described as a dilute solution of spherical droplets, indicate that the rms value of the polydispersity is approximately 25%. A typical spectrum, which also shows a large amount of small angle scattering from droplet interactions in shown in Fig. 9.3. From Eq. (9.12), with $l = 0$ and $\tilde{c}_0\rho \approx 1$, the spectrum at high Q is consistent with a value of $k/T \approx 3$ and a value of $\bar{k} \approx -1.8k$.

9.3.1.2 Dynamic fluctuations

The previous section presented the results of a calculation of the equilibrium fluctuation amplitudes for size and shape fluctuations of nominally spherical microemulsion droplets. One way to think of these fluctuations is to imagine the equilibrium ensemble of drops, with an infinite number of droplets in the thermodynamic limit. Any given member of the ensemble will have a size and/or shape which differs from that of the average and is predicted by the expressions derived above. However, in addition to this description of the ensemble, it is possible to quantify the *dynamic* fluctuations of a single droplet using the interfacial model described here.

The bending modes of a single droplet accurately describe the dynamics of microemulsion systems on time scales which are short enough that any given droplet remains isolated from the ensemble. This time scale increases as the system is made more dilute, since the collision time is related to the droplet volume fraction, ϕ . The dynamic fluctuations of isolated vesicles was discussed by Schneider, Jenkins and Webb [47]; their work was extended to microemulsions by Milner and Safran [34]. Since a perfect sphere cannot fluctuate at constant area and volume, they considered a typical droplet of the ensemble which has a given volume, and an *excess area* (compared with the equivalent area of a perfect sphere) due to the static fluctuations discussed above. As above, the bending energy, area,

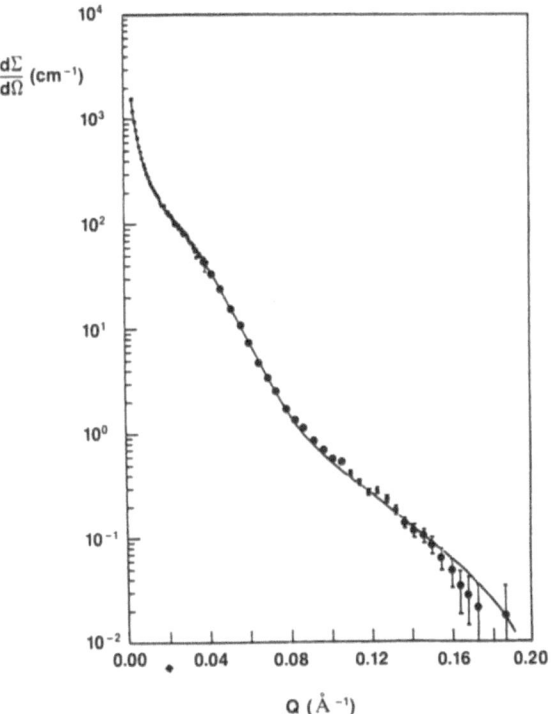

FIGURE 9.3. Neutron scattering spectrum (from [24]) measured for AOT/water/decane in a droplet phase. The wavevector is denoted by Q.

and volume of the droplet are expanded for small fluctuations, except that Eq. (9.10) is not used for the average size. A Lagrange multiplier is used to ensure that the mean excess area of the fluctuating droplet is equal to the value predicted by the equilibrium calculation. For small values of the bending modulus, all the modes contribute equally to the excess area, while for large values of k the $l = 2$ mode develops a large amplitude and governs the shape of the droplet which accommodates the (fixed) excess area.

The dynamical calculation is performed in the spherical geometry, which represents the lowest order in a systematic expansion of the deviations from the sphere. The incompressibility of the fluid is enforced by using a spatially varying pressure field, which cancels any compressional stresses set up by the fluid flow. The autocorrelation function for dynamical shape fluctuations, $u_{lm}(t)$ is,

$$< u_{lm}(t)u_{l'm'}(0) >= \delta_{ll'}\delta_{mm'} < |a_{lm}|^2 > e^{-\omega_{lm}t}. \qquad (9.13)$$

The time decay of the correlations is governed by ω_{lm} which is related to the bending energy cost of the fluctuation modes. For a droplet with average radius R (and V defined as in Eq. (9.11c))

$$\omega_{lm} = \frac{k}{\eta R^3} \left[\frac{(l+3)(l-2) + \frac{1}{2}V}{Z(l)} \right],$$ (9.14a)

where η is the viscosity, and $Z(l)$ is given by,

$$Z(l) = \frac{(2l+1)(2l^2 + 2l - 1)}{l(l+1)(l+2)(l-1)}.$$ (9.14b)

With these results for the dynamical modes of the droplet, the structure factor for scattering off of the thin shell of surfactant at the interface can be computed [34]. The resulting expression shows a decay of correlations due to the center of mass diffusion at small values of the wavevector $\vec{q}$. At larger values of $\vec{q}$, the decay of correlations responsible for the time dependence of the structure factor $S(\vec{q}, t)$ is related to the bending modes discussed above. In general, the wavevector does not select one or even a few of the spherical modes to dominate the sum, even though there is a rough correspondence ($q \approx 2\pi l/R$) between the index l and the wavevector. However, for $qR \approx \pi$, the sum is dominated by the $l = 2$ mode and an exponential fit of the initial time decay of $S(\vec{q}, t)$ should yield the $l = 2$ mode frequency as predicted by Eq.(9.14). More detailed fits of the experimental data are described in [25].

It is interesting to note that the dependence of the characteristic frequency, ω_c, (inversely proportional to the initial time decay of $S(\vec{q}, t)$) on the droplet radius serves to differentiate modes whose restoring force is the bending energy from those modes whose restoring force is the finite compressibility of the surfactant film ("surface tension" modes). For incompressible films, with only bending modes, Eq. (9.14) predicts that in the region, $qR \approx \pi$, ω_c is proportional to $1/R^3$. A similar analysis for compressible drops with no bending energy predicts ω_c proportional to $1/R$. The experimental results [25] for AOT/water/decane microemulsions with drop sizes, $27.5\text{Å} < R < 70\text{Å}$ shows that $\omega_c \sim 1/R^3$, consistent with the model of an incompressible surfactant monolayer with bending modes.

9.3.2   CYLINDRICAL GLOBULES

9.3.2.1   Worm-like tubes: polymer analogy

In the previous section it was shown that the bending fluctuations of spherical globules result in a measurable polydispersity in both size and shape (see Sec. 1.2). However, since the amplitude of these fluctuations is bounded (in the region where the spheres are stable), the description of the system as a phase of monodisperse spheres is still approximately correct. This is not the case for lamellar or cylindrical structures. Since these systems have at least one infinite dimension, they can support long-wavelength bending modes. These modes, which have a small free energy cost, can have divergent amplitudes, leading to a breakdown of the simple structural models which have been proposed. At short length scales, the structures remain locally cylindrical or lamellar. At long length scales, the undulations cause a random walk which leads to worm-like tubes in the cylindrical case (see Sec. 2.3).

The persistence length at which the cylinder is no longer rigid can be estimated from the bending energy [26,41]. Consider the bending energy of a cylinder with circular cross-section whose axis is permitted to wander through space. Fluctuations of the axis cross-section are similar to those of spheres and have bounded amplitudes. The position of the cylinder axis is given by $\vec{R} = X(z)\hat{x} + Y(z)\hat{y} + z\hat{z}$, and (for simplicity) we suppose that the cylinder is not permitted to double back upon itself. The radii of curvature are given by $R_1 = b$ and $R_2 = (bc\cos(\theta) - 1)/c\cos(\theta)$, where θ is the polar angle in the plane perpendicular to the cylinder axis. The cylinder radius b, is given by $b = 2\delta\phi/\phi_s$ and the curvature of the axis is $c \approx [(X')^2 + (Y')^2]^{1/2}$ where the prime denotes differentiation with respect to z. Using the expression for the bending energy, along with the conservation constraints, implies

$$F_b = F_b^c + \lambda \sum_q \left[|X_q|^2 + |Y_q|^2\right](1+x)q^4 b^2. \tag{9.15}$$

where F_b^c is the energy of the infinite cylinder given in Eq. (9.6b).

In Eq. (9.15), X_q and Y_q are the Fourier transforms of the functions X and Y, q is the wavevector in the $\hat{z}$ direction, $x = -\bar{k}/(2\tilde{k})$, where $\tilde{k} = k + \bar{k}/2$, and $\lambda \approx \phi(\tilde{c}_0 b)^3$. Using the equipartition theorem, one finds

$$\langle |X_q|^2 \rangle = \langle |Y_q|^2 \rangle = \frac{2T}{\pi(1+x)\tilde{k}b}q^{-4}. \tag{9.16}$$

This expression displays the divergence of the modes at small values of q.

The persistence length of the cylinder is the distance over which the axis of the cylinder is constant in direction. If $\hat{t}(z)$ is a unit vector along the cylinder axis, the axis fluctuation are measured by

$$\Delta(z) = \langle [\hat{t}(z) - \hat{t}(0)]^2 \rangle. \tag{9.17}$$

From the previous expression, one finds that

$$\Delta \approx 12Tz/(\pi^2 \tilde{k}b) \tag{9.18}$$

for large values of z. The persistence length, ξ_c, is defined as the distance z over which $\Delta \approx 1$ so that $\xi_c \approx (\pi^2 \tilde{k}/12T)b$. Note that the persistence length is controlled by both the bending modulus $\tilde{k}$, as well as by the volume fraction ratio of ϕ/ϕ_s which determines the cylinder radius, b. For length scales shorter than ξ_c the cylinders are rigid (rod-like), for longer length scales, the axes wander randomly in space.

The random walk of the cylinders suggest a polymer-like description for length scales larger than ξ_c. However, the polymerization index, N – *i.e.*, the number of persistence lengths per chain – must be determined from equilibrium considerations. The polymerization is *self-organizing*. The calculation of the size, N, follows the discussion of Ref. [26,48]. One considers finite cylinders capped by a spherical region. For $b/L \ll 1$, the energy of a finite cylinder of length L is

$$F_c = F_b^c + \alpha_0 \phi(b/L)k' \tag{9.19}$$

where $k' = (\pi\tilde{k}/2T)[14 - 10k - 16\tilde{c}_0 b]$ and α_0 is a constant of order unity. The entropy of mixing of the cylinders favors short globules ($N \approx 1$), while the energy of the spherical endcaps favors long chains ($N \to \infty$) which have fewer endcaps. A calculation of the statistics of the distribution of lengths yields an exponential distribution of lengths: $P(N) \sim \exp(-N/\bar{N})$, where the average degree of polymerization is $\bar{N}$ given by

$$\bar{N} \sim \phi^{1/2}\exp(k'/2). \tag{9.20}$$

This calculation, which is a mean-field treatment, neglects the formation of rings which can lead to a sharp polymerization transition [49]. However, since it is only the smallest rings that contribute in that case, the existence of a persistence length is important since it forbids rings smaller than ξ_c. The effects of the rings in the semidilute regime are expected to be greatly suppressed [49].

For small values of k', which depend on the concentrations and both bending moduli, the cylinders will be short with $N \approx 1$. In this limit, for high enough densities, the short rod-like aggregates may align in a nematic phase as discussed in [50]. In the limit $k'/T \gg 1$, the chains form flexible polymer-like tubes, with an average of $\bar{N}$ persistence lengths per tube.

It should be noted that both the persistence length, ξ_c, and the polymerization index, $\bar{N}$, depend on the volume fraction through ϕ and ϕ_s. Since the scaling laws for the radius of gyration, R_g, of polymer chains are functions of both the persistence length and $\bar{N}$, their dependence on the concentrations ϕ and ϕ_s can lead to unusual dependencies of R_g upon concentration [26].

9.3.2.2   Worm-like systems: experiments

While the theory discussed above presented the possibility of polymer-like behavior of cylindrical microemulsions, most of the experimental data on flexible, cylindrical systems comes from studies of micellar systems. Microemulsions formed from DDAB, brine, and oil have been discussed [38] in terms of cylindrical structures, but the evidence for polymer-like statics and dynamics is scant. The micellar systems generally consist of brine and surfactant, such as the CTAB system studied [26,27] by Porte *et al.*, Hirsch *et al.* and Langevin *et al.* In the micellar limit, ϕ refers to the surfactant volume fraction, and the cylinder radius, b, is fixed by the molecular geometry of the surfactant molecules. Nevertheless, a bending description can at least qualitatively describe the energetics of the interface and the concepts of the persistence length, polymerization index $\bar{N} \sim \phi^{1/2}$ *etc.*, should still apply.

The first evidence of cylindrical structures in the micellar systems came from magnetic birefringence measurements [27]. Since the birefringence saturated as the volume fraction of surfactant was increased, while the hydrodynamic radius as measured by quasielastic light scattering still increased with increasing ϕ_s, it was suggested that the micelles are flexible; the persistence length was measured [51] to be $\approx 150\text{\AA}$. The analogy to polymer solutions was established [27] by a quantitative measurement of the light scattering intensity at zero wavevector, $I(0)$, and the cooperative diffusion constant, D_c, as a function of ϕ_s. These quantities are independent of molecular weight in the semidilute regime and hence are a test of the polymer analogy but not the prediction of Eq. (9.20) for the polymerization index. The scaling laws for polymers in the semidilute regime with strong excluded volume and no rings, predict

$$I(0) \sim \phi^{-0.31} \tag{9.21}$$

$$D_c \sim \phi^{0.77} \tag{9.22}$$

The scattered intensity decreases upon an increase of ϕ due to the reduction of the osmotic compressibility associated with the formation of the transient network in the semidilute regime. (Light scattering indicates that the radius of gyration is $\approx 750\text{\AA}$ at the crossover between the dilute and semidilute regimes.) The cooperative diffusion constant increases with increasing ϕ since $D_c \sim 1/\xi$ where ξ is the mesh size which decreases as ϕ increases. This trend is opposite to that predicted for rigid rods in the semidilute regime. Thus the variation of D_c with ϕ offers a method to differentiate between elongated, flexible micelles and rigid rods. The experimental data for $I(0)$ and D_c are consistent with the polymer-like picture of flexible, worm-like chains. In addition, electron microscopy [52] of these systems yields pictures that strongly resemble flexible polymer networks. The observed cylinder diameter is about 30 Å, consistent with the molecular geometry.

While the light scattering data for these systems seems to be consistent with a polymer-like structure, measurements of the viscosity are not so simply interpreted [27]. For systems where the dynamics are dominated by reptation of the polymer chains along their contour lengths, the viscosity is predicted by mean-field theory to vary as:

$$\eta \sim N^3 \phi^4. \tag{9.23}$$

Viscosity measurements can therefore probe the concentration dependence of N. Using Eq. (9.20) for the average polymerization index, $\bar{N}$, would predict $\eta \sim \phi^{5.5}$. At high salt concentrations, where electrostatic effects are screened out, the experiments find that the exponent $\alpha \approx 3.5$. While the simple picture of self-assembling flexible cylinders predicts a polydisperse distribution of lengths, relaxation measurements indicate that the system can be described by a single exponential decay time. It is difficult to reconcile this result with reptation dynamics by a *polydisperse* system of flexible chains. Similar considerations apply to the results of photobleaching experiments which measure the self-diffusion constant of the polymer-like micelles [27].

The existence of a single relaxation time for the system indicates that the dominant dynamics are not sensitive to the reptation of the entire chains, but to a process that is fairly independent of chain length. Cates [49] has suggested that the breaking and recombination of chains must be considered along with the reptation if one wants to characterize these systems dynamically. After all, these kinetics are responsible for the equilibration of the chain lengths and lead to the concentration dependence of $\bar{N}$ predicted by Eq. (9.20). The calculation of the characteristic time scale considers that chains respond to stress by reptating along their contour length. Normally, the stress is completely relaxed when the chain has diffused along its entire contour length, implying that the characteristic reptation time, τ_r, should scale as $\tau_r \sim \bar{N}^2 D_r^{-1}$, where D_r is the curvilinear diffusion constant for reptation of the chain measured along its own arc length, in units of the persistence length, which varies as $(\bar{N}\phi^2)^{-1}$. This leads to $\tau_r \sim \bar{N}^3\phi^2$ which, combined with the usual polymer result $G_0 \sim \phi^2$ for the transient shear modulus, leads to $\eta \sim G_0\tau \sim \bar{N}^3\phi^4$, cf. Eq. (9.23). However, chains that can break and recombine behave differently. The motion of a chain end is by reptation, but each end can only move a curvilinear distance, λ, within its own lifetime, τ_b. This lifetime is determined by detailed balance to equal the mean lifetime of a chain of length $\bar{N}$ before breaking into two pieces. Thus, $\lambda^2 \sim D_r\tau_b$; the relaxation rate per monomer, τ^{-1}, is proportional to $\frac{\lambda}{\bar{N}\tau_b}$. [This arises because each chain end relaxes a fraction of order $\lambda/\bar{N}$ of the chain during its lifetime, τ_b.] Combining these expressions, one finds $\tau \sim (\tau_r\tau_b)^{\frac{1}{2}} \sim N\phi$. Again, invoking the result $G_0 \sim \phi^2$ for the plateau modulus, one predicts that the viscosity $\eta \sim \phi^{3.75}$. These predictions are in good agreement with the experimental results for worm-like micelles only at high salt concentration. At low salt concentrations, both the viscosity and self-diffusion exponents differ from the theory described above. However, the breaking process results in a stress relaxation function which is a single exponential, in agreement with experiment in both the high and low salt regimes. The discrepancy between the predicted and measured viscosity exponents at low salt concentrations can perhaps be understood if one accounts for the electrostatic interactions between the ionized surfactant molecules in these systems [53,54]. These interactions, which are important at low salt concentrations, can involve screening lengths that are comparable to the intermicellar spacing. The description of the chain no longer depends on locally determined, concentration-independent, elastic constants. The concentration dependence of the spontaneous curvature, c_0, has been discussed by Pincus and co-workers [53] for the case of oil-in-water systems of spherical microemulsions. At low salt concentrations, the end-cap energies which determine $\bar{N}$, are themselves ϕ dependent, and thus modify the scaling laws discussed above [54]. The low viscosities observed at *high* salt concentrations have been interpreted in terms of a branched, but very fluid, structure where the branch points are in thermal equilibrium [55].

9.4 Lamellar and Random/Bicontinuous Systems

In the previous section, the effect of thermal fluctuations on spherical and cylindrical microemulsions was discussed. While the spherical systems were only perturbed by undulations of the surfactant film, the cylindrical systems were fundamentally modified by fluctuations. At long length scales, the cylinders perform a random walk and behave like polymer solutions. In this section, the effects of thermal fluctuations on lamellar systems are analyzed. As shown above, the lamellar systems are stable in the region where the spontaneous curvature, c_0, is small. In the limit of $c_0 \to 0$, there is no simple globular description of the analogous, disordered microemulsion, and the resulting structure of the system is not obvious. A description of the structure and phase behavior of random microemulsions is presented in this section.

9.4.1 FLUCTUATIONS OF LAMELLAR SYSTEMS

The simple energetic considerations discussed at the end of Sec. 9.2 demonstrated that for small values of the spontaneous curvature, lamellar structures minimize the bending free energy. In many cases of experimental interest, microemulsions show lamellar phases with smectic order at high surfactant volume fractions [20,56]. At small values of ϕ_s, however, there is often a single phase which is isotropic [56]. For such systems (where the salinity has been used to "balance" the curvature of the film and establish $c_0 \to 0$ and where the volume fractions of water and oil are comparable), NMR and conductivity measurements indicate that both the oil and the water diffuse freely throught the system [21]. These structures were termed bicontinuous by Scriven [57] who presented examples of periodic, ordered, bicontinuous structures which may describe some systems at larger values of ϕ_s. In this review, the term random is used to denote microemulsions which are isotropic, non-periodic, and which do not have a particular globular component; they may or may not be bicontinuous, depending on the volume fractions of water and oil.

De Gennes and Taupin [7] first suggested that these random microemulsions are related to the ordered, lamellar phases which can occur at larger values of ϕ_s. Although the bicontinuous systems have a random structure, their characteristic length scale is related to the persistence length of an infinite surfactant monolayer [7]. The persistence length, ξ_K , is defined much in the same way as in Eq. (9.4): it is the typical distance over which the normal to the film for unidimensional structures is decorrelated by thermal fluctuations.

Jouffroy et al. [7] attempted to derive the phase diagram of microemulsions using this concept. However, only two-phase equilibria and not the characteristic three-phase coexistence of microemulsion, water, and oil were predicted. In a series of papers by Andelman, Cates, Roux, and Safran [9], a model was proposed which incorporated the physics of the fluctuations and also predicted the correct phase behavior. By allowing the length scale of the random microemulsion to be determined by the conservation of surfactant, an idea introduced by Talmon and Prager [6] (who did not include the bending energy in the expression for the free energy [58]) and extended by Widom [8] (who did not include the effects of thermal fluctuations on the free energy), it is possible [9,11] to derive phase diagrams which showed the regions of stability of the lamellar and bicontinuous phases as well as the appropriate multi-phase equilibria. As described below, the single phase, bicontinuous, microemulsion is stable when its length scale is of order the persistence length, but not identical to it.

The lamellar persistence length was introduced by De Gennes and Taupin by considering a single lamellar film, with $c_0 \to 0$. For small undulations about a flat sheet, the bending energy, Eq. (9.1), can be written:

$$U_b = \frac{1}{2}k \int dx \, dy [u_{xx} + u_{yy}]^2. \qquad (9.24)$$

The variable u is the deviation of the film from a flat monolayer whose normal points in the $\hat{z}$ direction. Since a single film with small undulations preserves its topology, the saddle-splay term can be neglected. The undulation modes are Fourier expanded as:

$$u_{\vec{q}} = \int dx \, dy \, e^{i\vec{q}\cdot\vec{R}} u(x,y) \qquad (9.25)$$

where the vectors $\vec{q}$ refer to the (x,y) plane.

The equipartition theorem determines the mean square amplitude of the fluctuation modes as:

$$\langle u_{\vec{q}} u_{-\vec{q}} \rangle = \frac{T}{2k} q^{-4}. \qquad (9.26)$$

The layer normal, $\hat{n}$, is given by $\hat{n}(x,y) = \vec{\nabla}u/|\vec{\nabla}u|$, so that the normal-normal correlation function is

$$|\Delta\hat{n}|^2 = \langle[\hat{n}(\vec{R}) - \hat{n}(0)]^2\rangle = \sum_{\vec{q}} 2q^2 \langle u_{\vec{q}}u_{-\vec{q}}\rangle[1 - \cos(qR)]. \qquad (9.27)$$

The normal will therefore be decorrelated (e.g., $|\Delta\hat{n}| \approx 1$) for two points on the film whose separation is equal to that of the persistence length, ξ_K. This determines ξ_K as

$$\xi_K = \ell\exp(2\pi k/\alpha T), \qquad (9.28)$$

where α is a number of order unity, discussed above. Thus, a single sheet will act as a rigid lamellae for length scales less than ξ_K but perform a surface "random walk" for length scales larger than ξ_K. The detailed nature of the "surface random walk", including the effects of self-avoidance, is a topic of current interest. In the context of smectic lamellar systems, this random walk is limited by the presence of the nearest-neighboring sheets. Helfrich [59] has shown that these steric interactions serve to reduce the free energy per unit area of a single, fluctuating surfactant sheet by an amount:

$$f_l = \chi T\frac{T}{k}d^{-2}. \qquad (9.29)$$

Here d is the interlamellar spacing and $\chi = (3\pi^2/128)$. While these steric repulsions stabilize the lamellar structure for small values of d, De Gennes and Taupin suggested that at large values of d, the system can more effectively minimize its free energy by forming a random, bicontinuous structure with a characteristic length of ξ_K. One test of this idea is whether this ansatz for the structure yields phase diagrams which are in agreement with experiment.

9.4.2   FILM FLUCTUATIONS AND THERMODYNAMICS OF RANDOM MICROEMULSIONS

The first phenomenological model of disordered microemulsions which idealized the system as a collection of interfaces was that of Talmon and Prager [6]. They considered a subdivision of space into random (Voronoi) polyhedra, which are filled at random with either oil or water, according to a probability proportional to the volume fraction of each component. The surfactant was presumed to form a monolayer at the interface between cells of water and oil; the area per surfactant molecule was taken as a fixed constant, $a = a_0$.

Jouffroy, Levinson, and de Gennes [7] attempted to include the effects of the bending energy and the thermal undulations in a model which was based on a simplified version of that of Talmon and Prager. They divided space into a lattice of cubes (rather than the Voronoi polyhedra); the cubes are filled at random with oil or water according to their volume fractions. As discussed in Sec. 9.4.1, they chose the lattice size to be always equal to the lamellar persistence length, ξ_K. Within a random-mixing approximation, where the number of interfaces is proportional to the product of the oil and water volume fractions, a fixed lattice size implies that the interfacial area per surfactant, a, depends on composition as discussed in Section 9.4.1, they chose the lattice size to always be equal to the lamellar persistence length, ξ_K. Within a random-mixing approximation, where the number of interfaces is proportional to the product of the oil and water volume fractions, a fixed lattice size implies that the interfacial area per surfactant, a, depends on composition as

$$a \sim \frac{\phi(1 - \phi)}{\phi_s \xi_k}. \tag{9.30}$$

Thus, the area per surfactant cannot be constant if the lattice constant is fixed and the model uses a free energy contribution per surfactant of

$$F_c = A(a/a_0 - 1)^2 \tag{9.31}$$

to account for the finite compressibility of the surfactant which has an optimum area per molecule of a_0. The bending energy was estimated under the assumption that the local radius of curvature of the interfacial film is comparable to ξ_K. While appealingly simple, this model does not predict the experimentally three-phase equilibrium. Instead, there is a two-phase region involving equilibrium between two microemulsions. The next development was the phenomenological theory of Widom [8]. He introduced a lattice of variable size, ξ which he treated as a variational parameter, but did not include the fluctuations of the surfactant film. For large bending energies, Widom's model predicts a strong dependence of the properties of the microemulsion on the water/oil interfacial tension, and only a very weak dependence on the surfactant monolayer properties. Schematic phase diagrams for the behavior of non-intersecting fluid films have also been studied [10], with a model that includes the saddle-splay interaction. This can stabilize cubic, saddle-shaped surfaces, first discussed by Scriven [57].

In a series of papers, the authors of [9] introduced a model for microemulsions which synthesized the approaches described above. They included the stabilization of the microemulsion by the entropy of mixing of the oil and water domains and used a variable value of the domain size, ξ. At the same time, they incorporated the effect of the thermal undulations of the surfactant film which give rise to the persistence length, ξ_K. The model, described in detail below, yields phase diagrams in agreement with experiments which vary temperature and/or salinity, which correspond to variations of either the spontaneous curvature or the bending constant. The middle-phase microemulsion turns out to have a length scale ξ which is indeed comparable to ξ_K; but this is a result of the theory, rather than an initial assumption.

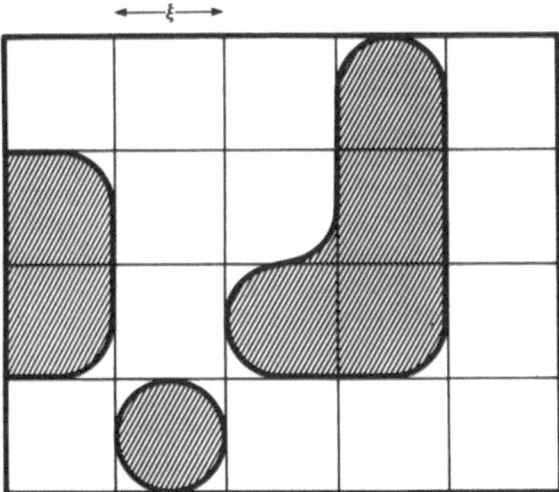

FIGURE 9.4. Schematic of random microemulsion model. The shaded region represents the water domains, the thick line delineates the surfactant film, and the remainder of the volume is filled with oil.

The microstructure and scattering structure factor are also in qualitative agreement with experiment.

In this model, space is divided into cubes of size ξ filled with either water or oil. The surfactant is constrained to stay at the water-oil interface (see Fig.9.4); it is divided equally between the oil and water domains. Using the random mixing approximation, the probability ϕ for a cube to contain water is

$$\phi = \phi_w + \phi_s/2 \tag{9.32}$$

where ϕ_w and ϕ_s are, respectively, the volume fractions of water and surfactant. The constraint for the surfactant to cover the water-oil interface at constant area per surfactant, a_0, determines the lattice size, ξ, within the random mixing approximation:

$$\xi = z \frac{v_s}{a_0} \frac{\phi(1 - \phi)}{\phi_s} \tag{9.33}$$

where $z = 6$ is the coordination number of the cubic lattice and v_s is the molecular volume of the surfactant. For convenience, the ratio v_s/a_0 is set equal to the molecular size, ℓ, discussed above. In this model, the lattice size, ξ, is neither fixed at a constant value (as in the model of [7]) nor a variational parameter (as in the model of [8]), but is determined by the constraint, Eq. (9.33).

The free energy per unit volume, f, has only two terms: the entropy of mixing of the water and oil domains giving rise to a contribution f_s, and the curvature energy, f_c. The entropy contribution is calculated using the random mixing approximation as

$$f_s = \frac{T}{\xi^3} [\phi \log(\phi) + (1 - \phi) \log(1 - \phi)] \equiv \frac{T}{\xi^3} S(\phi). \qquad (9.34)$$

In the limit $\bar{k} \to 0$, the curvature term, f_c, counts the number of bends in the system and fixes the bending energy of a cube of water (oil) of size ξ surrounded by oil (water) with that of a sphere of diameter ξ. Using the expression for the curvature energy in Eq. (9.1), and weighting the configurations of water and oil spheres by $\phi(1 - \phi)^2$ and $\phi^2(1 - \phi)$ respectively, the total curvature contribution to the free energy is

$$f_c = \frac{8\pi k(\xi)}{\xi^3} \phi(1 - \phi)[1 - 2\xi c_0(1 - 2\phi)]. \qquad (9.35)$$

In Eq. (9.35) a term linear in ϕ_s has been dropped. This term can be absorbed into the chemical potential of surfactant. Note that the length-scale dependent bending energy, $k(\xi)$, has been used; the result of the the theory is that the middle-phase microemulsion is stable when its length scale (determined from Eq. (9.33) is comparable to the persistence length. This indicates the significance of the renormalization of the bending modulus. The bending constant is a function of the volume fractions since the length scale, ξ, is a function of ϕ and ϕ_s.

9.4.2.1   Microemulsion free energy

In this model, the characteristic energy of the microemulsion is T and the characteristic length scale is the persistence length, $\xi_K = \ell e^\kappa$. It is therefore convenient to define the dimensionless free energy f_r, where

$$f_r \equiv (f_s + f_c)(\xi^3/T). \qquad (9.36)$$

The surfactant volume fraction is inversely proportional to the lattice size ξ. Scaling this length by the persistence length defines a reduced surfactant volume fraction as

$$\tilde{\phi}_s = \phi_s \exp(\kappa) \qquad (9.37)$$

where $\kappa = 4\pi k_0/\alpha T$—see (9.3), with α as the factor which determines the numerical value of the bending constant renormalization.

The function $f_r(\phi, \tilde{\phi}_s)$ determines the phase behavior of the system [9]. For $\tilde{\phi}_s$ greater than a maximum value of $\tilde{\phi}_s \approx 12/e = 4.4$, the reduced free energy has only one minimum at $\phi = \frac{1}{2}$. For values of $\tilde{\phi}_s$ smaller than this maximum value, the function has two minima and a maximum at $\phi = \frac{1}{2}$. As the value of $\tilde{\phi}_s$ further decreases, these minima approach the limiting values of $\phi = 0$ and $\phi = 1$ and the value of the free energy at these minima diverges to $-\infty$. This behavior was noticed by Widom and corresponds to the fact that the entropy of mixing favors small values of ξ. In order to obtain a physically meaningful result, he introduced a cutoff for the lattice size at $\xi = \ell$, where ℓ is a molecular length. In the model presented here, the finite volume of the surfactant molecule takes care of this problem. Indeed, for a given amount of surfactant, $(1 - \phi_s/2) > \phi > \phi_s/2$. These cutoffs limit the values of the free energy minima.

The origin of the three-phase equilibrium is related to these deep minima which generally occur at the boundaries of the phase diagram – $i.e.$, at small water or oil volume fractions. For the symmetric case, the three-phase equilibrium is found by plotting the free energy at the minimum value of ϕ, as a function of ϕ_s, as discussed in Ref. [9]. There it is shown that an equilibrium exists with coexistence of a phase with $\phi = \frac{1}{2}$ and a relatively large value of $\tilde{\phi}_s$ with two phases with small values of ϕ and $(1 - \phi)$ respectively. Both phases have very small values of $\tilde{\phi}_s$; they correspond to phases which are nearly all water or nearly all oil with some small amount of surfactant. The phase diagrams calculated below are somewhat sensitive to the free energies of the surfactant in these coexisting phases. In fact, the simple bending energy discussed above is not expected to accurately describe these cases of extreme water or oil asymmetry. In such phases, the system is more accurately described as a dilute solution of surfactant in either water or oil, whose free energy may not be simply related to the two elasticity parameters, k_0 and c_0, of a nearly flat monolayer. This is significant, since it is precisely the coexistence between a microemulsion and a very dilute phase of surfactant in oil or water that constitutes the phase equilibrium of interest. A generalization of the model to more accurately account for the free energy of the surfactant solution phases has been described in Ref. [9], but more work remains to be done. Qualitatively, however, it is expected that the model presented here gives a reasonable description at least in the case when the solubilities of the surfactant in the water and in the oil are both small and not too different from one another.

9.4.2.2 Microemulsion phase diagrams

The phase diagram is calculated via a common tangent plane construction. This is equivalent to the requirement that the chemical potentials of surfactant and water (or oil) be equal in all the coexisting phases and that the total volume of the system be conserved. Defining the potential $g(\phi, \tilde{\phi}_s)$ as

$$g(\phi, \tilde{\phi}_s) = f_r(\phi, \tilde{\phi}_s) - \mu_\phi \phi - \mu_s \phi_s \qquad (9.38)$$

the phase diagram is determined by minimizing $g(\phi, \tilde{\phi}_s)$ with respect to ϕ and $\tilde{\phi}_s$ with fixed values of the chemical potentials μ_ϕ and μ_s. The absolute minimum of g corresponds to the stable phase at a given chemical potential. Values of ϕ, ϕ_s determined by $\partial g/\partial \phi, \partial g/\partial \phi_s = 0$, must be compared with those corresponding to the possible free energy minima situated on the boundaries $[\phi = \phi_s/2, (1 - \phi_s/2)]$ of the phase diagram. Due to

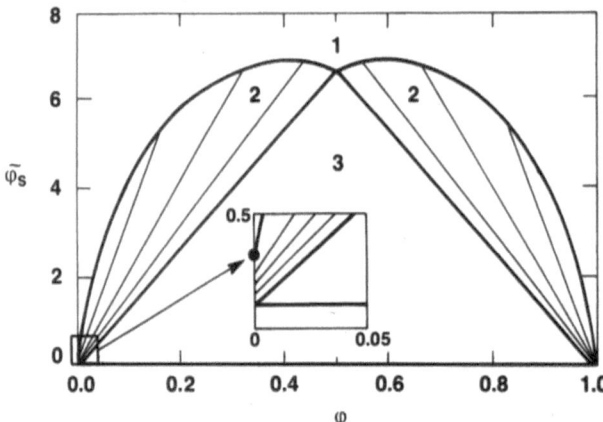

FIGURE 9.5. Theoretical phase diagram for zero spontaneous curvature ($c_0 = 0$). The numbers denote the numbers of coexisting phases and the tie-lines in the two-phase regions are shown. The inset shows the tie lines at small values of the volume fractions where the details are sensitive to the model.

the finite volume fraction of surfactant, these minima do not correspond to points where the first derivatives of the potential g are equal to zero. It is necessary, therefore, to compute these minima by finding

$$\frac{dg(\phi_s/2, \tilde{\phi}_s)}{d\tilde{\phi}_s} = 0 \tag{9.39}$$

for the water-surfactant side, and a similar expression (where $\phi_s/2 \to (1 - \phi_s/2)$) for the oil-surfactant side. There are thus three states that must be compared at fixed values of μ and μ_s: (i) the solution of

$$\left[\frac{\partial g}{\partial \phi}\right]_{\tilde{\phi}_s} = 0 \quad \text{and} \quad \left[\frac{\partial g}{\partial \tilde{\phi}_s}\right]_{\phi} = 0; \tag{9.40}$$

(ii) the solution of Eq. (9.39) to determine the minimum of g along the water-rich boundary of the phase diagram; and (iii) the solution of the equation corresponding to Eq. (9.39) for the case of the oil-rich boundary. The values of μ_s and μ_ϕ, where two or more minima have equal potentials, determine the chemical potentials for which two or more phases coexist. The values of ϕ and ϕ_s at those chemical potentials give the concentrations of the phases in equilibrium.

This procedure leads to the phase diagram shown in Fig. 9.5. It exhibits a one-phase region (the microemulsion phase), and three polyphasic regions: two two-phase regions where a microemulsion phase is in equilibrium with a phase which contains a small amount of surfactant in water or oil (upper- or lower-phase microemulsions, respectively), and a three-phase region where a middle-phase microemulsion is in equilibrium with both dilute phases.

Along the coexistence curve, the characteristic length scale is of order ξ_K and does not vary much except for volume fractions, ϕ, that are near either 0 or 1. In the middle phase, $\xi \approx 0.23\xi_K$; this relationship between the persistence length and the length scale of the middle phase (the volume fraction of surfactant in the middle phase also scales as $1/\xi_K$) arises naturally from our model which included the effect of the renormalization of the bending modulus. The phase diagram is a strong function of ξ_K and hence of k_0/T when plotted as a function of the unscaled volume fraction, ϕ_s. However, when plotted as a function of the scaled volume fraction $\tilde{\phi}_s$, the phase diagram is a slowly varying function of the bending modulus which only enters to determine the free energies of the dilute phases. It is also a slowly varying function of the effective cutoff at small values of $\tilde{\phi}_s$. [In general, the free energies of the surfactant in the oil and water phases enter as additional parameters, as discussed in [9]. In this dilute region, however, the simple model presented here is not expected to be accurate, as discussed above.]

The significance of the inclusion of the bending constant renormalization can be seen as follows: If the bending modulus, k is taken to be a constant, k_0, the free energy of the microemulsion with $\phi = \frac{1}{2}$ can be written $f = g_m\phi_s^3$, where g_m is a constant which depends only on the ratio k_0/T. As discussed in detail below, the free energy of the lamellar phase has been discussed by Helfrich [59] in terms of steric repulsion of the surfactant sheets and predicts a free energy per unit volume, f_l, which can also be written as $f_l = g_l\phi_s^3$, where g_l depends only on the ratio k_0/T. Thus, transitions from lamellar to microemulsion can only occur if changes in the bare bending constant force g_m to be less than g_l, and the observation of transitions as only the surfactant volume fraction is varied is not explained. On the other hand, the renormalization of k to depend on length scale implies that the microemulsion free energy is $f = g_m\phi^3 \log(\phi_s/\phi^*)$, where ϕ^* is inversely proportional to the persistence length, ξ_K. At some value of $\phi_s < \phi^*$, the logarithm forces the microemulsion to have a smaller free energy than the lamellar phase. Thus, a lamellar to microemulsion transition as a function of ϕ_s is naturally predicted by the model with the renormalized bending modulus.

The model also yields phase diagrams for the case of finite spontaneous curvature, $c_0 \neq 0$. Choosing positive values of c_0 focuses on the case of oil-continuous systems; negative values of c_0 pertain to water-continuous systems. When $c_0\xi_k \ll 1$, the phase diagram shows a slight asymmetry, but the three-phase equilibrium still exists. However, when $c_0\xi_k \gg 1$, the three-phase region vanishes as shown in Fig. 9.6. The phase diagram is then very asymmetric and most of the two-phase region consists of equilibria between a microemulsion phase and nearly pure water. The typical domain size now

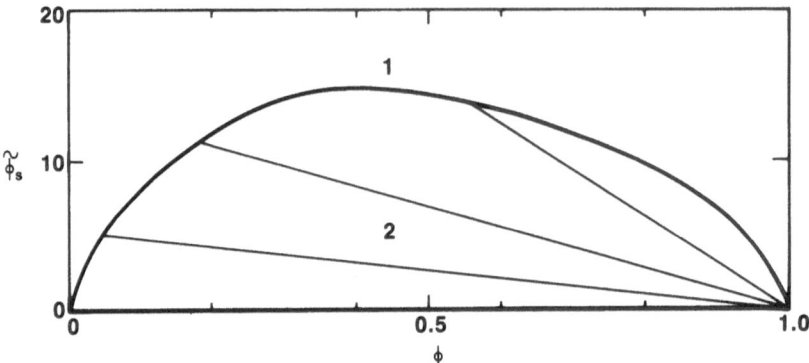

FIGURE 9.6. Phase diagram for finite spontaneous curvature ($c_0 \neq 0$), where $\xi_K c_0 \approx 10$. The tie-lines in the two-phase region indicate the coexistence of a microemulsion phase with nearly pure water. The three phase region no longer exists at this value of c_0. The saddle-splay constant $\bar{k} = 0$.

scales not with ξ_K, but with $1/c_0$, indicating that the structure is essentially one of globules whose maximum size is determined by the spontaneous curvature, with an additional contribution from the entropy of mixing. This is consistent with the emulsification failure instability discussed in Sec. 9.2 for the simple case of spherical globules.

The vanishing of the three-phase equilibrium as the spontaneous curvature is increased can also be seen in Fig. 9.7, where the phase diagram is shown as a function of $c_0 \xi_K$ and $\tilde{\phi}_s$, for the case of equal volume fractions of water and oil. In this case, the phase diagram has the shape of a "fish", in agreement with experiments on three-component, non-ionic systems (see Fig. 9.10a in following section) where the spontaneous curvature is varied by changing the temperature. There is no "head" to the theoretically derived fish (see inset for the experimental phase diagram) [1,17]. This is a result of the approximation, made in the model, that the dilute phases of surfactant in water and surfactant in oil can be represented by points on the edges of the phase triangle. This could be corrected by taking a more realistic description of the dilute phases as discussed in [9].

9.4.2.3   Microemulsion to lamellar phase transition

In this section, the stability of random microemulsions relative to a lamellar phase, which consists of surfactant monolayers dividing adjacent oil and water domains to smectic order, is considered. For the symmetric case ($c_0 = 0$), it shown that for a range of values of the bending modulus (k_0/T), the random microemulsion is more stable than the lamellar phase at small values of ϕ_s. The renormalization of the bending modulus to values of the

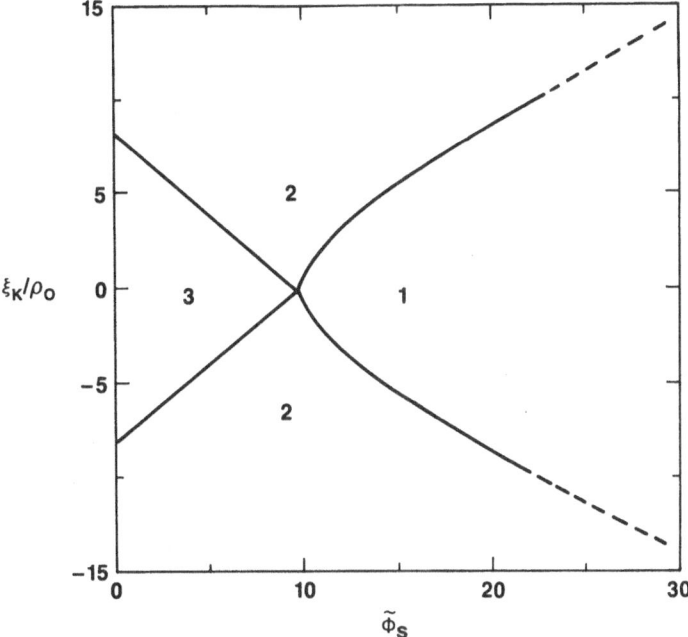

FIGURE 9.7. Phase diagram as a function of the spontaneous curvature ($\rho_0 = 1/c_0$) and the surfactant volume fraction, $\tilde{\phi}_s = \phi_s e^{\kappa}$. The water/oil ratio is fixed at unity and the bending and saddle-splay moduli at $\kappa = 6.6$ and $\bar{k} = 0$.

order of T, allows the energy cost of the bends in the random system to be compensated by the entropy gain from the random mixing of the water and oil. At large values of ϕ_s, the bending modulus approaches its bare value k_0. The energy cost of the curved interfaces in the random microemulsion is no longer compensated by the entropy of mixing, and the lamellar phase dominates. The same situation arises for large values of the bare bending modulus, as shown below.

Although the ordered, lamellar phase involves no entropy of mixing and no curvature energy, it has a non-zero free energy due to the steric repulsion of the surfactant sheets. This repulsion reduces the meandering entropy of the lamellae from its value in the limit of infinite separation of the surfactant sheets. The additional free energy per unit volume has been estimated by Helfrich for an ordered array of lamellae in a single solvent and is proportional to the product of the number of lamellae per unit length and the factor $T^2/(k_0 d^2)$, where d is the interlamellar separation. Helfrich's result can be heuristically generalized to describe the free energy of the lamellar microemulsion per unit volume, f_l as

$$f_l = \chi T \left[\frac{T}{k_0} \right] \left[\frac{1}{d_o^2} + \frac{1}{d_w^2} \right] \left[\frac{2}{d_w + d_o} \right]. \qquad (9.41)$$

In Eq. (9.41), $d_o \approx 2\ell\phi_o/\phi_s$ and $d_w \approx 2\ell\phi_w/\phi_s$, are the distances between the surfactant sheets separating oil and water domains, respectively. The surfactant volume fraction is given by $\phi_s = 2\ell/(d_o + d_w)$. The parameter χ express the uncertainty in the numerical coefficient of Eq. (9.41) which depends on the short distance cutoff in the plane of the sheets. In all of the present calculations, this cutoff has been set equal to the typical surfactant size. Since the molecules are anisotropic, there is some numerical error introduced by the simplification of a single microscopic length scale. The parameter $\chi = 0.15$ in these phase diagrams; if χ is taken to be Helfrich's value of $(3\pi^2/256) \approx 0.12$, the microemulsion is stable with respect to the lamellar phase even at high values of ϕ_s, for the chosen value of $\kappa \equiv 4\pi k_0/\alpha T = 5$. A unified theory accounting for the steric interactions in both the lamellar and microemulsion phases has been presented by Golubovic and Lubensky [10] who demonstrate that the system has potentially a rich and varied phase behavior.

The phase diagram is determined from a double tangent construction on the simpler model presented above, using a composite free energy surface that consists of both f and f_l. The results are shown in Fig. 9.8 for a value of $\kappa = 5$ where the transition between the lamellar phase and the random microemulsion phase can be seen as ϕ_s is increased. In Fig. 9.9, the phase diagram is shown as a function of κ and ϕ_s for equal volume fractions of water and oil, $\phi_w/\phi_o = 1$. For small values of κ (very flexible surfactant films) the lamellar phase is stable only for very high values of ϕ_s, and the microemulsion phase is stable for a large range of surfactant concentrations. When κ is increased, the surfactant concentration of the middle phase, ϕ_s^m, decreases as $\exp(-\kappa)$. At the same time, however, the lamellar phase is stable over a larger range of ϕ_s and the first order transition between the microemulsion and lamellar phase occurs at smaller values of the surfactant concentration. Since the slope of the boundary between the microemulsion and the region of three-phase equilibrium is larger than the slope of that between the microemulsion and the lamellar phase, the two lines meet (at $\kappa = 8.3$ in the present calculations which have used a value of $\alpha = 1$). For larger values of κ, there is no longer a stable microemulsion phase, and the region P corresponds to complex polyphasic equilibria between the lamellar, microemulsion, and water and oil phases. Thus, the price one pays for having a microemulsion phase with very small amounts of surfactant is to have a very narrow region of stability.

This behavior is in agreement with experimental results [1,17] on nonionic systems as shown in Fig. 9.10. Assume that the bending modulus increases monotonically with the surfactant length - i.e., thicker membranes resist bending more than thinner ones [42]. For small surfactant molecular lengths

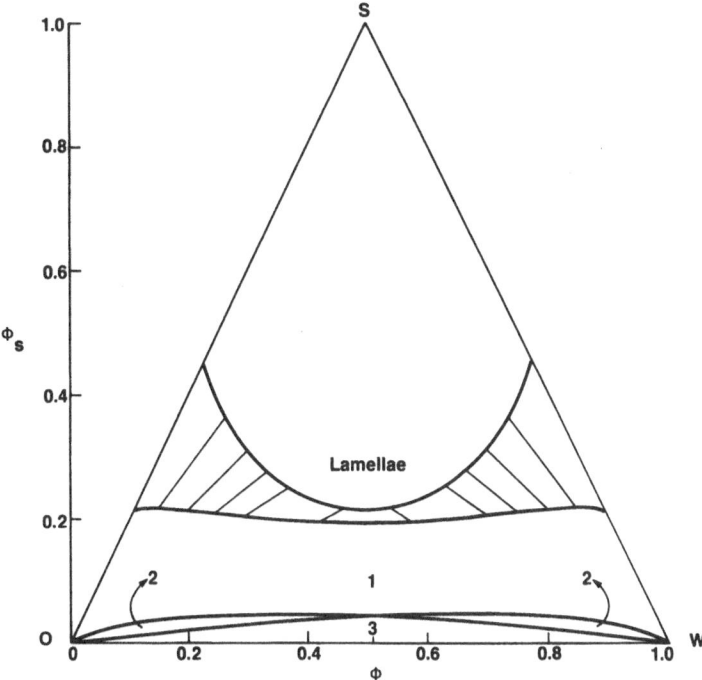

FIGURE 9.8. Microemulsion phase diagram for the case of no spontaneous curvature, showing the transition to the lamellar phase and the associated two-phase region.

(corresponding to small bending modulus, κ), the value of ϕ_s^m is large, but there is no lamellar phase, even for high surfactant concentration. On the contrary, for large surfactant length (corresponding to a higher value of the bending modulus, κ), the value of ϕ_s^m is much smaller and the lamellar phase boundary is nearer to the microemulsion phase. The dependence of these phase boundaries on the $\log(\phi_s)$ is in agreement with the model predictions $\phi_s^m \sim \exp(-\kappa)$ and suggests that the model correctly describes the competition between the lamellar and microemulsion phases.

9.4.3 SCATTERING FROM RANDOM MICROEMULSIONS

The model of random microemulsions described above relies upon a random mixing assumption to describe the occupancies of the lattice sites by oil and water. However, the contribution of the bending energy to the free energy implies that the structure of the oil/water domains will not be perfectly random. The bending energy will induce correlations between these regions which will serve to increase the amount of flat interface (for the

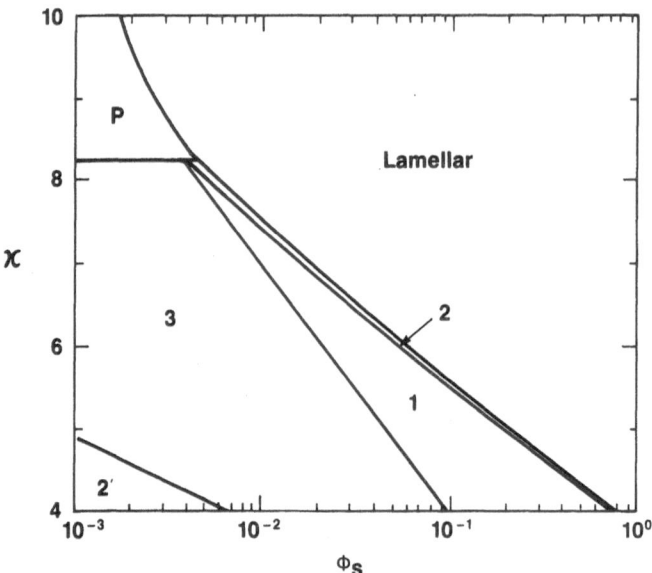

FIGURE 9.9. Phase diagram as a function of bending modulus, κ, and the surfactant volume fraction, ϕ_s. The oil/water ratio is fixed at unity and the spontaneous curvature is zero. The saddle-splay constant $\bar{k} = 0$.

case of $c_o = 0$), thus reducing the free energy. In Ref. [39] it was shown how these correlations give rise to a peak in the scattering structure factor, $S(\vec{q})$. Other theoretical treatments have also been presented to describe the observed peak. The extension of the Talmon-Prager model [36,28] predicts a monotonic decrease of $S(\vec{q})$ with increasing wavevector $\vec{q}$; the bending energy was not explicitly included. On the other hand, an ad-hoc Landau [37] expansion of the fluctuation contribution to the free energy predicts a structure factor of the form

$$S(q) = [A + Bq^2 + Cq^4]^{-1} \qquad (9.42)$$

which can describe the peak in the scattering if the coefficients B and C are taken to be negative and positive, respectively. However, the magnitudes of the coefficients A, B, and C are not predicted as a function of volume fraction *etc.*; the expression for $S(q)$ is unrelated to the free energy which governs the thermodynamic properties of the system. Furthermore, it is not at all clear that Eq. (9.42) correctly describes the small angle scattering at low $\vec{q}$, which often fits an Ornstein-Zernicke form with $B > 0$.

Other treatments of the scattering involve structural models, but with no thermodynamic basis. Kaler and co-workers [28,37] have modeled the scattering from a random microemulsion in terms of the scattering from

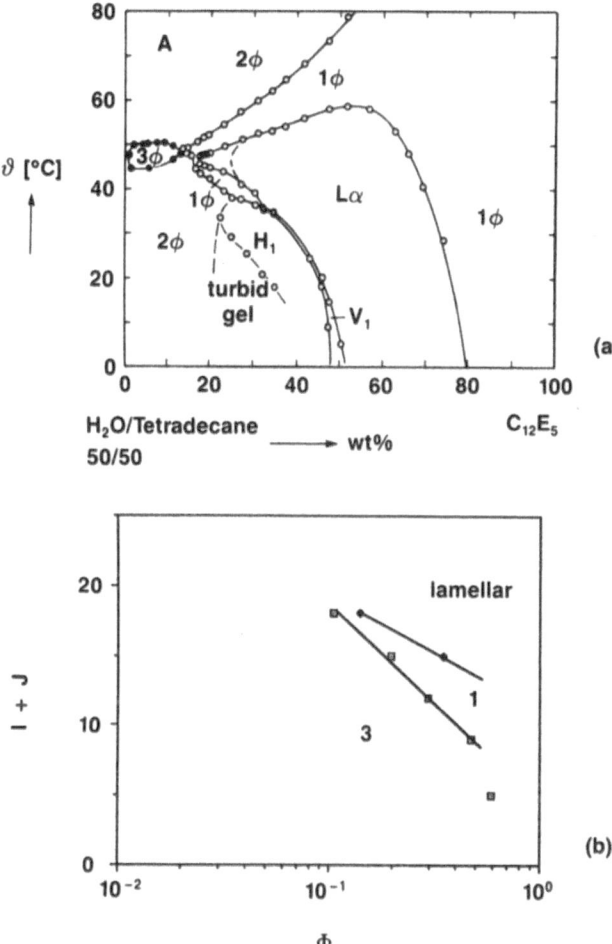

FIGURE 9.10. Experimental phase diagram for non-ionic surfactants from [1]. In (a) the temperature/surfactant phase diagram is shown for fixed water/oil ratio. In (b), the phase boundaries for a series of different surfactants $C_i E_j$ is shown as a function of the total surfactant *length* $i + j$, which is assumed to be directly related to the bending constant, ξ.

a lamellar system with a phenomenological factor introduced into the correlation function which accounts for the disorder. The correlation function is taken to be the product of the lamellar correlation function and an isotropic, exponentially decaying function of distance. Another model treats the scattering from a system with constant mean curvature (except for some defect regions) using a disordered open connected (DOC) model [38] consisting of spherical and cylindrical pieces, joined together in a random manner. This model is proposed for systems with large bending moduli, where the bending energy acts to constrain the film curvature to equal the spontaneous curvature, c_0. Cylindrical and spherical shapes are equally allowed, corresponding to the limit of no saddle-splay modulus as discussed above. However, this model does not account for microemulsions which consist of a single phase of spheres, nor does it describe systems where $c_0 = 0$.

In contrast to these structural models, the thermodynamic model discussed above can be generalized to include fluctuations in the oil/water occupancies and to predict the scattering peak and the microstructure. Instead of calculating the lattice size and the bending free energy from arguments about the average occupancy of the lattice by the oil and water, a local analysis is used. This local analysis reduces correctly to the random mixing results if departures from random mixing in the oil/water occupancies are ignored. The condition which conserves the total amount of surfactant can be written for an arbitrary configuration of water and oil domains:

$$\xi = \frac{1}{N} \sum_{ij} J_{ij} s_i (1 - s_j) / \phi_s \qquad (9.43)$$

where $s_i = 0, 1$ indicates that cell i is filled with oil or water respectively. The matrix J_{ij} equals one for nearest neighbor cells i and j, and is zero otherwise. In the random mixing approximation, this reduces to the relation between ξ and ϕ_s stated in Eq. (9.33).

Note that in Eq. (9.43), the sums on i and j run over the N cells of the lattice. However, for fixed total volume V, $N = V/\xi^3$. Thus, N depends on the configurations via Eq. (9.43). This has no effect on Eq. (9.43) itself since ξ is an intensive variable and has a value that is independent of N as long as N is large. Extensive quantities such as the thermodynamic potential, G, will have an added dependence on the configurations because of the relation between N and ξ. This is taken into account by writing the thermodynamic potential, G, as

$$G = \frac{V}{\xi^3} f - \Lambda M \qquad (9.44)$$

where f is the free energy per site, Λ is a Lagrange multiplier which fixes the average value of the water and oil volume fractions, and $M = \sum_i s_i$.

The structure factor can be calculated as the Fourier transform of the correlation function χ_{ij}, where

$$\chi_{ij} = \langle s_i s_j \rangle - \langle s_i \rangle \langle s_j \rangle. \tag{9.45}$$

The angle brackets denote a thermal average. In order to calculate χ_{ij}, a set of local fields, h_i, is introduced into the Hamiltonian, H, so that

$$H = H_b - \sum_i h_i s_i \tag{9.46}$$

where H_b is the bending Hamiltonian. The correlation function is calculated from the thermodynamic potential G, as

$$\chi_{ij} = \frac{1}{\beta} \frac{\partial^2 G}{\partial h_i \partial h_j} = -\frac{\partial m_i}{\partial h_j} \tag{9.47}$$

where $m_i \equiv \langle s_i \rangle$, and all the derivatives are evaluated at $\{h_i\} = 0$. The potential G is calculated from the partition function in the usual manner, while the partition function is calculated within a single-site approximation as detailed in Ref. [39].

To obtain the bending energy in this local analysis of an arbitrary configuration of water and oil domains, the different arrangements of cell edges are considered as shown in Fig. 9.11. In terms of the local cell occupancies $\{s_i\}$ on a cubic lattice, the bending Hamiltonian for the case of no spontaneous curvature is then:

$$H_b = (2\pi k(\xi)/3) \left[\sum_{ij} (-4J_{ij} s_i s_j + L_{ij} s_i s_j + 2J_{ij} s_i) + \right.$$

$$\left. \frac{1}{8}(C - 4) \sum_{ijkl} M_{ijkl} \{ s_i(1 - s_j)(1 - s_k)s_l + (1 - s_i)s_j s_k(1 - s_l) \} \right]. \tag{9.48}$$

The coupling tensor L_{ij} is equal to one if cells i and j are next nearest neighbors (i.e., have a common edge but no common face) and is equal to zero otherwise, while M_{ijkl} is equal to one if cells i, j, k, and l form the corners of an elementary plaquette with the pairs (i, l) and (j, k) being next nearest neighbors. Otherwise, M_{ijkl} is zero. In Eq. (9.48), C is a parameter which accounts for the energy of the diagonal configurations where the oil and water cells are next-nearest neighbors as shown in Fig. 9.11c and 9.11d. (For a more detailed analysis of these configurations and for an extension to the case of finite spontaneous curvature, see Ref. 39b.) The random mixing approximation considers only triplets of cells which share an edge, and is equivalent to choosing a value of $C = 4$ and setting each $s_i = \phi$. This results in the vanishing of the quartic terms in Eq. (9.48) and reduces to the

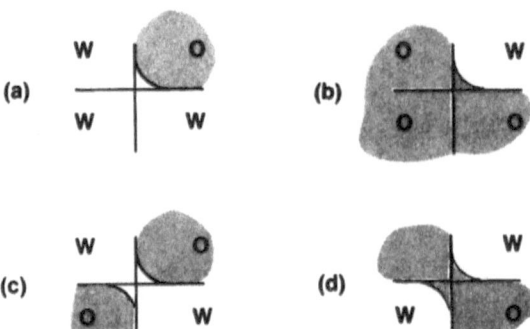

FIGURE 9.11. Geometries used to calculate bending energy in the lattice model. The letters O and W represent the oil and water regions respectively. The radius of the bends is approximately equal to ξ.

random mixing form of the bending energy used above. It may be argued that a more realistic choice is to choose $C = 2$, which would correspond to assigning two bends to the configurations shown in Figs. 9.11c and 9.11d. (The random mixing approximation effectively assigns four bends to these diagonal arrangements.) Values of $C < 2$ may correspond to attractive interaction between adjacent pieces of interface, while values of $C > 2$ may correspond to an effective repulsive interaction. Such a repulsion could arise, for example, from steric hindrance of the fluctuating interfaces or from electrostatic interactions.

With this model, the structure factor becomes [39]

$$S(\vec{q}) = N^{-1} \sum_j e^{i\vec{q}\cdot(R_i - R_j)} \chi_{ij} = \phi(1-\phi)\left[1 + \epsilon(\cos q_x\xi + \cos q_y\xi + \cos q_z\xi)\right.$$

$$\left. + \delta\left(\cos q_x\xi \cos q_y\xi + \cos q_y\xi \cos q_z\xi + \cos q_x\xi \cos q_z\xi\right)\right]^{-1} \qquad (9.49)$$

The quantities ϵ and δ are functions of C, the bending constant $k(\xi)$ (with ξ given by Eq. (9.33), the random mixing expression), and the volume fractions:

$$\epsilon = \frac{16\pi k(\xi)}{3T}\phi(1-\phi)\left[1 - (C-4)\phi(1-\phi)/2\right]$$

$$+ 2[\phi\log(\phi) + (1-\phi)\log(1-\phi)]$$

$$+ \frac{4\alpha}{3}\phi(1-\phi)[1 - (C-4)\phi(1-\phi)/2] \qquad (9.50)$$

$$\delta = \frac{8\pi k(\xi)}{3T}\phi(1-\phi)[(C-4) + 2 - 2(C-4)\phi(1-\phi)]. \qquad (9.51)$$

The terms included in ϵ represent the effects of all nearest neighbor interactions, while those included in δ are all from *next* nearest neighbors.

The unusually complicated forms of ϵ and δ are due in part to the quartic terms in the bending energy. More fundamentally, they arise from the dependence of the "lattice constant", $\xi(\{m_i\})$, on the configurations of the system. This represents a basic difference between our model of the random microemulsion and related Ising models where the lattice constant is independent of the state of the system. This unusual feature of the physics of the microemulsion is responsible for instabilities that are not present in the simple Ising systems.

The structure factor depends on the parameters: $k(\xi)/T, \alpha$, and C. However, the theory predicts the dependence of the structure factor on the volume fraction ϕ. Not all values of these parameters correspond to physically realizable microemulsions. It is only necessary to consider the parameters for which: i) $S(\vec{q})$ is finite and ii) the microemulsion is more stable than competing phases (e.g., lamellae, or a two-phase coexistence of oil-rich and water-rich phases). The results are shown in Fig. 9.12 where the instabilities (i.e., divergences) of the structure factor are shown as a function of $k(\xi)$ and the parameter C for the case of equal oil and water volume fractions. Also shown are estimates of the regions where the microemulsion

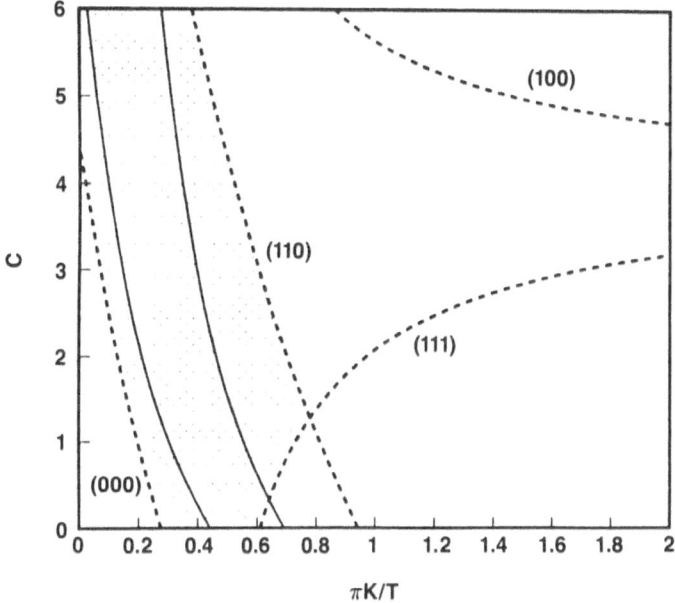

FIGURE 9.12. The linearly stable region (shaded area) and stability boundaries (dashed curves) for $S(q)$ with bending constant renormalization ($\alpha = 3, \phi = \frac{1}{2}$). The pair of solid curves are the estimated first-order phase boundaries which separate the microemulsion from the lamellar phase (right hand solid curve) and from the coexistence with oil-rich and water-rich phases (left hand solid curve).

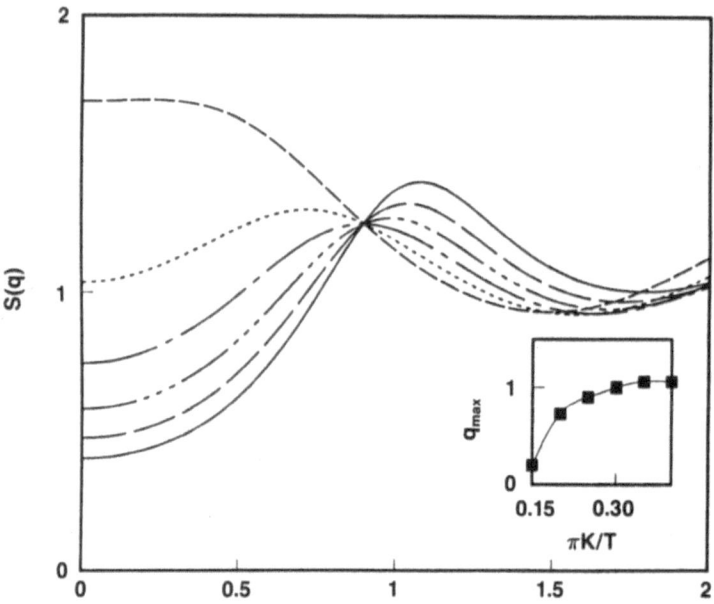

FIGURE 9.13. Rotationally averaged structure factor for random microemulsions for values of the bending constant $\pi k(\xi)/T = 0.15, 0.20, 0.25, 0.30, 0.35, 0.40$. The structure factor at $q = 0$ is a decreasing function of $k(\xi)$, which identifies the curves. The inset shows the position of the peak in the $S(q)$ at $q = q_{max}$. The results are for $\alpha = 3$, $\phi = \frac{1}{2}$ and $C = 3$.

is unstable to phase separation and to a transition to the lamellar phase. Outside the dotted region, the structure factor diverges. These divergences cannot be observed in the present model since they are usually preempted by first order transitions. The divergences, however, do indicate that the tendency to long-range correlations (and/or ordering) of several types does exist. For example, the (000) boundary indicates a phase separation instability, the (100) boundary a lamellar instability, and the (110) boundary tube-like correlations.

Within the region where the random microemulsion is stable, the structure factor shows a peak at a wavevector $q \approx \pi/\xi$ as shown in Fig. 9.13, for values of the bending constant that are not too small. The peak height indicates that the correlations in the oil/water occupancies span only a few cells, which implies that the calculation based on a small fluctuation expansion about random mixing is self-consistent. For moderate values of the bending constant, both the peak position as a function of concentration (recall that $\xi \sim \phi(1 - \phi/\phi_s)$) and the peak height are in semi-quantitative agreement with experiment [28,29]. At small values of k, both the peak position shifts to smaller values of q and the peak height decreases. The model is able to predict the complete behavior of the structure factor as a function of the concentrations and the bending modulus with no additional adjustable parameters. This is because the same Hamiltonian is responsible for the thermodynamics (*i.e.*, phase diagram) and fluctuation behavior (*i.e.*, scattering spectrum); all parameters necessary to describe the scattering and microstructure can be obtained, in principle, from thermodynamic measurements.

A real-space picture of the microstructure can also be obtained from the model as explained in [39]. This is shown in Fig. 9.14, along with an experimental micrograph [60]. It is interesting that despite the correlations which exist, the apparent structure shows no visible precursor to lamellar ordering, even though a transition from random to ordered lamellae occurs as ϕ_s is increased. In addition, the real-space picture generated by the model stand in comparison with the images as obtained in the freeze-fracture experiments.

9.4.4   SPONGE-LIKE L_3 PHASES IN TWO-COMPONENT SYSTEMS

The previous sections presented a comparison between random, bicontinuous, microemulsions and their related lamellar phases. A simple model of incompressible surfactant monolayers accounts for both the thermodynamic properties and the scattering and microstructure. The microemulsion was shown to be related to the properties of the fluctuating surfactant films through the dependence on the persistence length, ξ_K. As the lamellar phase is made more and more dilute it does not reach an infinitely "unbound" state as supposed in [61], but is unstable to a random microemulsion, albeit via a first-order transition.

This approach was extended in [62] to describe a sponge-like phase in a two-component system (surfactant and water or surfactant and oil). The sheets are composed of bilayers which are flexible and which are under some conditions ordered in a lamellar phase. The model was proposed to explain the observed isotropic (L_3) phase found in a small region of the phase diagram of some two-component systems where the surfactant forms bilayers [63]. The observed phase is strongly flow-birefringent and has a large structural length-scale of the order of hundreds of Angstroms. It was proposed that this L_3 phase consists of locally relatively flat pieces of surfactant bilayer, joined together randomly in such a way as to avoid all regions of strong curvature. Thus, a low bending energy and a relatively high entropy can be achieved at very low volume fractions of surfactant. The L_3 state is associated with an instability of the nearby lamellar phase; there is usually a weak, first-order transition between these two phases [64].

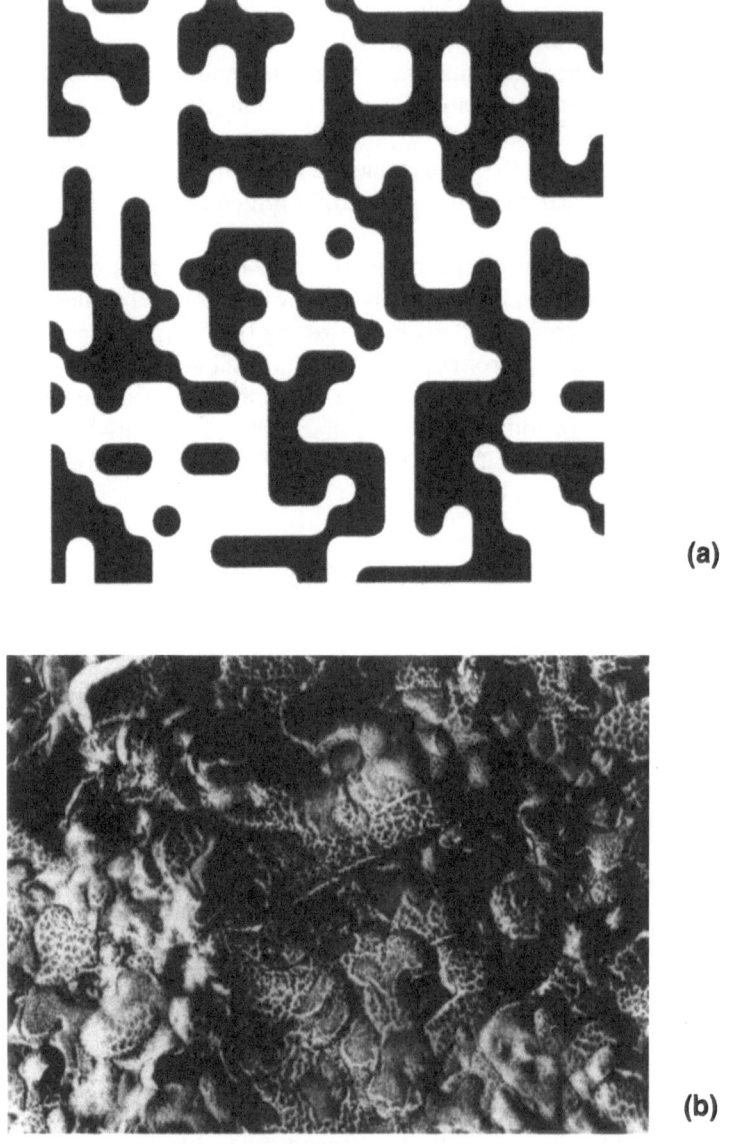

(a)

(b)

FIGURE 9.14. Real space representation of structure: (a) theory (see [39] for the details) and (b) experiment from [60].

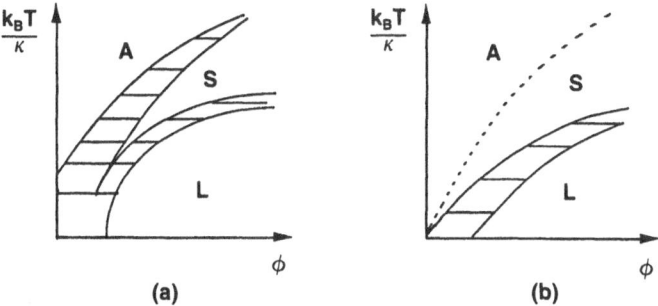

FIGURE 9.15. Typical phase diagram for sheet-like phases (from [65]). L is a lamellar phase, S, the symmetric sponge phase and A is an asymmetric sponge or (when very dilute) a phase of small vesicles. The phase transition between S and A can be first order as in (a) or second order - dashed line in (b).

The bilayer lamellar phase cannot be infinitely diluted; there is a first-order transition *not* to a phase of excess oil or water, *but* to an L_3 phase.

The phase diagram of the sheet-like phases in two-component systems can be analyzed with a model which corresponds to the bicontinuous, random, microemulsion as described above. For definitiveness, a surfactant-oil system is discussed. In contrast to the microemulsion where the volume fractions of oil and water are fixed, one considers a mixture of two, fictitious, different types of oil. The relative fractions ψ and $(1 - \psi)$ of the two oils are not fixed a priori, since in the real system these two "species" are identical. The variable ψ is thus a variational parameter of the theory. A bilayer-continuous, "maximally-random" interface structure is given by $\psi = \frac{1}{2}$, while a phase of dilute spherical vesicles is described by values of $\psi \ll \frac{1}{2}$. The calculation of the phase diagram is exactly the same as for the random microemulsion with zero spontaneous curvature, although the interpretation is different.

As shown in [62], for large values of the surfactant volume fraction, the minimum of the free energy over ψ occurs at $\psi = \frac{1}{2}$; the characteristic length scale is the persistence length ξ_K and the structure is the sheet-like, continuous interface. At small values of the surfactant volume fraction, the minimum of the free energy occurs for $\psi < \frac{1}{2}$, corresponding to a phase of vesicles. When the bending modulus is increased, transitions from the sheet-like to ordered lamellar phases occur, similar to the transitions between microemulsions and lamellar phases shown in Fig. 9.10. A typical phase diagram is shown in Fig. 9.15. The transition from the symmetric, sponge-like phase with $\psi = \frac{1}{2}$ to the vesicle phase (with $\psi < \frac{1}{2}$) can be either first or second order, depending on the bending modulus and the concentrations. This spontaneous symmetry breaking has been discussed in the context of microemulsions by Huse and Leibler [11] and Golubovic and Lubensky [10]. Its importance in the L_3 phase has been analyzed by Roux et al. [65].

Experimental phase diagrams indicate that the L_3 phase is indeed found to lie between an extremely dilute phase (D) and a swollen lamellar phase (L), separated from each other by a first-order coexistence region [64]. In oil continuous systems, it is often possible to make a stable L_3 phase even in the presence of small amounts of added water; this third component should reside in the middle of the bilayer as it does in the nearby lamellar phase [64]. For these systems, there are conductivity measurements which suggest the presence of water-continuous pathways, although the volume fraction of water is only a few percent. This is consistent with the idea of a random sheet-like phase exhibiting the continuity of the (water-swollen) surfactant bilayer.

An alternative explanation for the stabilization of the L_3 phase by the entropy of the random sheets has as been proposed by Porte [66] based on a study of surfactant/alcohol/water and surfactant/alcohol/oil systems. The transition from lamellar to L_3 occurs with the addition of alcohol in the former case and occurs with the reduction of alcohol in the latter. Porte argues that the L_3 phase is stabilized with a change in the spontaneous curvature and saddle-splay energy which tend to favor a periodic minimal surface for the L_3 phase. Of course, the lack of Bragg peaks and the large small-angle scattering argue against any ordered structure. However, it may still be possible that the L_3 phase is related to a saddle-splay instability which combines with entropic stabilization.

9.5   Outstanding Problems

In this Section, some of the outstanding problems relating to the theory and predictions of the interfacial description of microemulsions are discussed. These topics are currently under study or are areas for future investigations.

Most of the discussion above focused on the use of coarse-grained lattice models where the lattice size is an average domain size, to account for the curvature energy of a self-avoiding surfactant film and the entropy of randomly distributed water and oil domains. The predicted structure factor is not as simple as the Teubner-Strey formula [37a] used to fit scattering experiments, due to the lattice [39]; it is therefore not directly applicable to the analysis of the experiments. More importantly, effects related to the fluctuations of the domain sizes are not included.

Another class of models that has been used to study bicontinuous microemulsions uses a sum of sinusoidal waves with a distribution of wave vectors with random directions to describe the composition fluctuations. This description has a great advantage over the coarse-grained lattice models mentioned above since it is lattice independent and thus can account for local variations in the water or oil domain size, as well as for their correlations. The approach is phenomenological and was introduced by Berk [37b] and Teubner [71] in a geometrical context and by Gompper and Schick in a thermodynamic theory [72,73]; Pieruschka and Marčelja later motivated the theory using an entropy term taken from information theory [74]. However, none of these approaches have developed the model to the point that it can predict microemulsion structure and phase behavior as function of the surfactant concentrations and interfacial film properties.

Recently, a variational approximation has been used [75] to derive the fluctuation spectrum and scattering structure factor, the real space configurations, and the free energy of the random wave model from a unified, statistical mechanical approach. Starting with a Hamiltonian whose only input is the bending energy of the surfactant interface, it predicts how the structure factor and free energy vary as functions of the surfactant concentration, water/oil fractions, and bending moduli. This variational calculation indicates that in the microemulsion phase, the coefficients of the effective Ginzburg-Landau theory used by previous approaches [72,73] are not at all related to the microscopic water/oil interactions, but are strongly renormalized by the presence of the many, fluctuating interfaces and are thus sensitive functions of the concentration of surfactant and bending energy, which we predict. One important feature of this theory is the emergence of a new correlation length which characterizes the width in wavevector space of the structure factor and is in a large part of the phase diagram proportional to the bending energy. The linear dependence of the correlation length on the bending modulus seems to suggest an analogy to worm-like objects in contrast with the approaches described above, which related microemulsion properties to undulating membranes, where the persistence length depends exponentially on the bending energy. In addition, changes in the correlation length which occur as the surfactant concentration, ϕ_s, is decreased, result in a thermodynamic instability of the system at small wave vectors. How this random surface model relates to the models which focus on undulating membranes remains to be seen.

While the interfacial description presented here is well-suited to describe microemulsions where the characteristic length scale is much larger than a molecular size, it is not well-suited to describe systems which are closer to surfactant-water-oil solutions. On the other hand, the microscopic lattice models [12] are simple to deal with if correlations only involve several molecules, but are not optimized to describe long-wavelength undulations and structures. Experiments indicate, however, that there is a continuum of behavior which spans the small scale to large scale regime. Thus, a problem that is worthy of further study is the construction of a bridge between the continuum theory outline here and the more microscopic approach. Such a bridge will be able to treat micelles, microemulsions, vesicles, and ordered phases, on the same footing.

It should be noted that the continuum model presented here has not completely analyzed the implications of the bending energy on the phase diagram and structure. Most notably, the role of the saddle-splay in stabilizing random structures with zero mean curvature has not been discussed. Scriven [57] and later Huse and Leibler [11] have discussed the stability of an ordered version of such a state; a detailed treatment of its structure, phase behavior as a function of concentration, and the effects of the spontaneous curvature is only beginning to emerge [67].

Another such area is the dependence of the phase diagram for a random microemulsion on the chemical potential of the surfactant in the mostly water and most oil phases. The deep reason for this sensitivity is not apparent. Perhaps related to this is the absence of critical points in the balanced microemulsion. The relationship between the description of the surfactant as a monolayer (whose properties are determined by the curvature elasticity), and the surfactant monomers or micelles is also important for an understanding of the interfacial tensions between the microemulsion and the phases which consist of mostly oil or water. The implications of the model for the values of the observed interfacial tensions is an area of current study.

Another puzzle is the fact that the microemulsion model presented here only yields the three-phase equilibrium for a limited range of values [68] of the renormalization parameter α, defined in Eq. (9.3). For values of α that are larger than $\frac{3}{2}$, there are no free energy minima on the boundaries of the phase diagram (i.e., at values of $\phi = \phi_s/2$, $(1-\phi) = \phi_s/2$) and there is only a two-phase equilibrium between two microemulsions [11]. In Ref. [9], this problem was overcome by noting that the microemulsion free energy does not quantitatively account for the energy of the micelles present in the nearly all water and nearly all oil phases. An additional parameter, f^*, which represents the free energy of the micellar state, was introduced. For large enough values of $|f^*|$, the free energy minima at small values of ϕ and $(1-\phi)$ are restored. The phase diagram then also exhibits the three-phase equilibrium, in agreement with experiment. Again, the sensitivity to the micellar state is disturbing.

The sheet-like, random structure discussed in the previous section may apply to random microemulsions away from the point of equal oil and water volume fractions. The present microemulsion model, presented in Sec. 9.4.2, predicts that in the limit of small oil or water volume fractions, the structure is that of nearly isolated, dilute domains of oil in water or water in oil, respectively – even for the balanced case of $c_0 = 0$. In fact, the structure may consist of connected sheets of thin, surfactant coated oil or water layers in a water or oil continuum, as described in Sec. 9.4.2. It would be of great interest to construct a model which can describe both limits of $\phi \approx \frac{1}{2}$ (random cubes) and $\phi \ll \frac{1}{2}$ (sheet-like structure) in a unified manner. In addition, very few structural measurements have been performed on systems where both ϕ and c_0 are small. This is because these experiments are performed as salinity scans for systems where the overall volume fractions of oil and water are equal. Salinity scans for the case of small oil or water volume fractions may indicate whether the structure of the single phase–dilute microemulsion at the optimal salinity (where $c_0 = 0$)– is sheet-like or a dilute, droplet-like dispersion. Another question that deserves further study is the nature of the transition between the sheet-like and vesicle phases.

Finally, an area of future interest may involve the extension of the basic concepts of the interface and bending models presented here to other materials. In principle, the molecular analog of a surfactant molecule may exist for incompatible systems other than oil and water. Some work has already been done in the study of micellization and microemulsion formation in block copolymers [67,69,70]. The phenomenological parameters such as the bending modulus can be calculated as a function of chain length for these systems which are dominated by entropic interactions. The block co-polymers allow the solubilization of two, normally incompatible polymers [70]. Possible extensions of these ideas to *in*organic systems may be of both scientific and technological interest.

Acknowledgements

The work discussed represents the efforts of several productive and satisfying collaborations formed as a result of complex fluids research at Exxon Research and Engineering in the last ten years. The author and collaborators were/are staff members or visitors who benefitted from the Exxon research environment. The work on globular systems was performed in collaboration with L. Turkevich and P. Pincus. The theory of random microemulsions and the sheet-like phases was a collaboration that involved D. Andelman, M. Cates, and D. Roux. S. Milner developed the hydrodynamic theory for spherical droplets as well as the calculation of the scattering structure factor from random systems. The motivation for many of the theoretical ideas as well as their ultimate grounding in reality owes a large debt to the experiments and insights of J. Huang, M. W. Kim, D. Roux, and C. Safinya. Useful interactions with L. Auvray, J. Candau, S. Chen, P. G. de Gennes, J. M. DiMeglio, W. Helfrich, D. Litster, E. Kaler, D. Langevin, J. Meunier, P. Chandra, J. Samseth, R. Strey, Z. G. Wang and T. Zemb are also acknowledged. Finally, I am grateful to M. Cates for his comments on and suggestions for this manuscript.

References

1. M. Kahlweit, R. Strey, P. Firman, and D. Haase, Langmuir **1**, 281 (1985); Ang. Chem., Int. Ed. Engl. **24**, 654 (1985); J. Phys. Chem. **91**, 1553 (1987); J. Phys. Chem. **90**, 671 (1986); D. H. Smith, J. Colloid Interface Sci. **108**, 471 (1985).

2. For a general survey see (a) *Surfactants in Solution*, K. Mittal and B. Lindman, eds. (Plenum, New York, 1984), and ibid., (1987); (b) *Physics of Complex and Supermolecular Fluids*, S. A. Safran and N. A. Clark, eds. (Wiley, New York, 1987). (c) *Structure and Dynamics of Strongly Interacting Colloids and Supramolecular Aggregates in Solution*, S. H. Chen, J. S. Huang, and P. Tartaglia, eds., Vol. 369, NATO ASI Series, (Kluwer, Dordrecht, Boston, 1991).

3. For a recent survey of the physics of amphiphilic systems, see *Physics of Amphiphilic Layers*, J. Meunier, D. Langevin, and N. Boccara, eds. (Springer-Verlag, New York, 1987).

4. J. Meunier, J. de Physique Lett. **46**, L-1005 (1985).

5. A. Calje, W. G. M. Agerof, and A. Vrij, in *Micellization, Solubilization, and Microemulsions*, K. Mittal, ed. (Plenum, New York, 1977), p. 779; R. Ober and C. Taupin, J. Phys. Chem. **84**, 2418 (1980); A. M. Cazabat and D. Langevin, J. Chem. Phys. **74**, 3184 (1981); D. Roux, A. M. Bellocq, and P. Bothorel, in [2a], p. 1843; J. S. Huang, S. A. Safran, M. W. Kim, G. S. Grest, M. Kotlarchyk, and N. Quirke, Phys. Rev. Lett. **53**, 592, (1983); M. Kotlarchyk, S. H. Chen, J. S. Huang, and M. W. Kim, Phys. Rev. A **29**, 2054 (1984).

6. Y. Talmon and S. Prager, J. Chem.Phys. **69**, 2984 (1978); and **76**, 1535 (1982).

7. P. G. de Gennes and C. Taupin, J. Phys. Chem. **86**, 2294 (1982); J. Jouffroy, P. Levinson, and P. G. de Gennes, J. de Physique **43**. 1241 (1982).

8. B. Widom, J. Chem. Phys. **81**, 1030 (1984).

9. S. A. Safran, D. Roux, M. Cates and D. Andelman, Phys. Rev. Lett. **57**, 491 (1986); and in *Surfactants in Solution: Modern Aspects*, K. Mittal, ed. (Plenum, New York, in press); D. Andelman, M. Cates, D. Roux, and S. A. Safran, J. Chem. Phys. **87**, 7229 (1987); D. Andelman, S. A. Safran, D. Roux, and M. Cates, Langmuir **4**, 802 (1988).

10. L. Golubovic and T. C. Lubensky, Phys. Rev. A **41**, 4343 (1990).

11. D. Huse and S. Leibler, J. de Physique **49**, 605 (1988).

12. S. Alexander, J. de Physique Lett. **39**, 1 (1978); J. C. Wheeler and B. Widom, J. Am. Chem. Soc. **90**, 3064 (1968); B. Widom, J. Chem. Phys. **84**, 6943 (1986); M. Schick and W. H. Shih, Phys. Rev. **B34**, 1797 (1986) and Phys. Rev. Lett. **59**, 1205 (1987); K. Chen, C. Ebner, C. Jayaprakash, and R. Pandit, J. Phys. **C20**, L361 (1987); M. Gompper, M. Schick, and W. H. Shih, Phys. Rev. Lett. **59**, 1205 (1987); G. Gompper and M. Schick, Phys. Rev. Lett. **62**, 1647 (1989). Also, see the review by K. Dawson in Ref. 2c.

13. K. Shinoda and H. Saito, J. Colloid Interface Sci. **26**, 70 (1968); and M. L. Robbins in *Micellization, Solubilization, and Microemulsions*, K. Mittal, ed. (Plenum, New York, 1977), p. 713.

14. M. Kotlarchyk, S. H. Chen, J. S. Huang, and M. W. Kim, Phys. Rev. A **29**, 2054 (1984).

15. D. F. Evans, D. J. Mitchell, and B. W. Ninham, J. Phys. Chem. **90**, 2817 (1986); T. N. Zemb, S. T. Hyde, P. J. Derian, I. S. Barnes, and B. W. Ninham, J. Phys. Chem. **91**, 3814 (1987); and B. W. Ninham, I. S. Barnes, S. T. Hyde, P. J. Derian, and T. N. Zemb, Europhys. Lett. **4**, (1987).

16. M. W. Kim. W. D. Dozier, and R. Klein, J. Chem. Phys. **84**, 5919 (1986); and W. D. Dozier, M. W. Kim, and R. Klein, J. Chem. Phys. **87**, 1455 (1987).

17. M. L. Robbins in *Micellation, Solubilization, and Microemulsions*, K. Mittal, ed. (Plenum, New York, 1977), p. 713; M. Kahlweit and R. Strey, Angew. Chem. Int. Ed. Engl. **24**, 654 (1985); F. Lichterfeld, T. Schmeling, and R. Strey, J. Phys. Chem. **90**, 5762 (1986); and D. Andelman, S. A. Safran, D. Roux, and M. Cates, Langmuir **4**, 802 (1988).

18. A. M. Cazabat, D. Langevin, J. Meunier, and A. Pouchelon, J. Adv. Colloid Interface Sci. **16**, 175 (1982); and D. Guest and D. Langevin, J. Colloid Interface Sci. **112**, 208 (1986).

19. See, *Surface Phenomena in Enhanced Oil Recovery*, D. O. Shah, ed. (Plenum, New York, 1981); Proceedings of SPE/DOE Fifth Symposium on Enhanced Oil Recovery, Vols. 1-2, April, 1986, Society of Petroleum Engineering.

20. P. Ekwald, Adv. Liq. Crystals **1**, 1 (1975); A. M. Bellocq and D. Roux, in *Microemulsions*, S. Friberg and P. Bothorel, eds. (CRC PRess, New York, 1986); and D. H. Smith, J. Coll. Interface Sci. **102**, 435 (1984).

21. P. G. Nilsson and B. Lindman, J. Phys. Chem. **86**, 271 (1982); A. Khan, B. Lindstrom, K. Shinoda, and B. Lindman, J. Phys. Chem.

90, 5799 (1986); and K. Shinoda and B. Lindman, Langmuir **3**, 135 (1987).

22. E. W. Kaler, H. T. Davis, and L. E. Scriven, J. Chem. Phys. **79**, 5685 (1983); A. M. Cazabat, D. Chatenay, F. Guering, D. Langevin, J. Meunier, O. Sorba, J. Lang, R. Zana, and M. Pailette, in *Surfactants in Solution*, K. Mittal and B. Lindman, eds. (Plenum, New YOrk, 1984), p. 1553; A. M. Cazabat, D. Chatenay, P. Guering, and W. Urbach, in *Physics of Complex and Supermolecular Fluids*, S. A. Safran and N. A. Clark, eds. (Wiley, New York, 1987), p. 585; S. Bhattacharya, J. P. Stokes, M. W. Kim and J. S. Huang, Phys. Rev. Lett. **55** 1884 (1985).

23. M. Lagues, J. de Physique Lett. **40**, L331 (1979); S. A. Safran, I. Webman, and G. S. Grest, Phys. Rev. A **32**, 506 (1985); G. S. Grest, I. Webman, S. A. Safran, and A. L. R. Bug, Phys. Rev. A **33**, 2842 (1986).

24. S. H. Chen, T. L. Lin, and J. S. Huang, in *Physics of Complex and Supermolecular Fluids*, S. A. Safran and N. A. Clark, eds. (Wiley, New York, 1987), p. 285.

25. J. S. Huang, S. T. Milner, B. Farago, and D. Richter, Phys. Rev. Lett. **59**, 2600 (1987); and B. Farago, J. S. Huang, R. Richter, S. A. Safran, and S. T. Milner, Phys. Rev. Lett. **65**, 3348 (1990).

26. G. Porte, J. Appell, and Y. Poggi, J. Phys. Chem. **84**, 3105 (1980) and J. Phys. Lett. **44**, L689 (1983); H. Hoffman, G. Platz, H. Rehage, and W. Schorr, Adv. Colloid Interface Sci. **17**, 275 (1982); H. Hoffman and H. Rehage, Faraday Discuss. Chem. Soc. **76**, 363 (1983) and J. Phys. Chem. **92**, 5172 (1988); T. Shikata, H. Hirata, and T. Kotaka, Langmuir **4**, 354 (1988); G. Warr, L. Magid, E. Caponetti, and C. Martin, Langmuir **4**, 813 (1988); and T. Imae, J. Phys. Chem. **92**. 5721 (1988).

27. S. J. Candau, E. Hirsch, and R. Zana, J. de Physique **45** 1263 (1984), J. Colloid Interface Sci. **105**, 521 (1985), and in *Physics of Complex and Supermolecular Fluids*, S. A. Safran and N. Clark, eds. (Wiley, New York, 1987), p. 569; J. S. Candau, E. Hirsch, R. Zana, and M. Adam, J. Colloid Interface Sci. **122**, 430 (1988); S. J. Candau, E. Hirsch, R. Zana, and M. Delsanti, Langmuir **5**, 1525 (1989); and R. Messager, A. Ott, D. Chatenay, W. Urbach, and D. Langevin, Phys. Rev. Lett. **60**, 1410 (1988). See, also, the review by M. E. Cates and S. J. Candau, J. Phys. Conden. Mat. **2**, 6869 (1990).

28. L. Auvray, J. P. Cotton, R. Ober, and C. Taupin, J. Phys. Chem. **88**, 4586 (1984); A. DeGeyer and J. Tabony, Chem. Phys. Lett. **113**,

83 (1985) and **124**, 357 (1986); C. G. Vonk, J. F. Billman, and E. W. Kaler, J. Chem. Phys. **88**, 3970 (1988); and S. H. Chen, S. L. Chang, and R. Strey, Prog. Colloid Poly. Sci. **81**, 30 (1990) and J. Chem Phys. **28**, 679 (1990).

29. E. W. Kaler, K. E. Bennett, H. T. Davis and L. E. Scriven, J. Chem. Phys. **79**, 5673, 5685 (1983); L. Auvray, J. P. Cotton, R. Ober and C. Taupin, J. de Physique **45**, 913 (1985); L Auvray in *Physics of Complex and Supermolecular Fluids*, S. A. Safran and N. Clark, eds. (Wiley, New York, 1987), p. 449.

30. W. Helfrich, Z. Naturforsch. **28a**, 693 (1973).

31. J. N. Israelachvili, D. J. Mitchell, and B. W. Ninham, J. Chem. Soc. Faraday Trans. II **72**, 1525 (1976) and D. J. Mitchell, and B. W. Ninham J. Chem. Soc. Faraday Trans. II **77**, 601 (1981).

32. J. S. Huang and M. W. Kim, Phys. Rev. Lett. **47**, 1352 (1981); D. Roux and A. M. Bellocq, Proc. SIF Course XV, V. Degiorgio and M. Corti, eds. (North Holland, Amsterdam, 1985) p. 842; D. Roux and A. M. Bellocq, G. Fourche, and P. Bothorel, J. Chem. Phys. **83**, 1028 (1983).

33. S. A. Safran and L. A. Turkevich, Phys. Rev. Lett. **50**, 1930 (1983); B. Lemaire, P. Bothorel, and D. Roux, J. Phys. Chem. **87**, 1023 (1983); and C. Huh, J. Colloid Interface Sci. **97**, 201 (1984).

34. S. T. Milner and S. A. Safran, Phys. Rev. A **36**, 4371 (1987), and V. Lisy and A. V. Zatovsky, Phys. Lett. **166**, 253 (1992).

35. W. Helfrich, J. de Physique **46**, 1263 (1985) and ibid. **48**, 285 (1987); L. Peliti and S. Leibler, Phys. Rev. Lett. **54**, 1690 (1985); D. Foerster, Phys. Lett. **114A**, 115 (1986); H. Kleinert, Phys. Lett. **114A**, 263 (19860; and L. Golubovic and T. C. Lubensky, Phys. Rev. B **39**, 12110 (1989).

36. E. W. Kaler and S. Prager, J. Colloid Interface Sci.**86**, 359 (1982).

37. a) M. Teubner and R. Strey, J. Chem. Phys. **87**, 3195 (1987); and b) N. F. Berk, Phys. Rev. Lett. **58**, 2718 (1987).

38. B. W. Ninham, I. S. Barnes, S. T. Hyde, P. J. Derian, and T. N. Zemb, Europhys. Lett. **4**, 5651 (1987); most recently, polymer-like microemulsions have been reported by J. Eastoe, Langmuir **8**, 1503 (1992).

39. (a) S. T. Milner, S. A. Safran, D. Andelman, M. Cates, and D. Roux, J. de Physique **49**, 1065 (1988) and (b) P. Chandra and S. A. Safran, Europhys. Letts. **17**, 691 (1992).

482   S. A. Safran

40. Experiments have shown that the area per headgroup is approximately constant along the two-phase coexistence curves–see L. Auvray, J. P. Cotton, R. Ober and C. Taupin, J. de Physique **45**, 913 (1985).

41. S. A. Safran, L. A. Turkevich, and P. A. Pincus, J. de Physique Lett. **45**, L-19 (1984) and in *Surfactants in Solution*, K. Mittal and B. Lindman, eds. (Plenum, New York, 1984), p. 1177.

42. A. G. Petrov and I. Bivas, Prog. Surf. Sci. **16**, 389 (1984); E. A. Evans and R. Skalak, CRC Critical Reviews Bioeng. **3**, 181 (1979); A. G. Petrov and A. Derzhanski, J. Phys. Coll. **37**, C3-155 (1976) and Phys. Lett. **65A**, 374 (1987); S. A. Safran, L. A. Turkevich, and P. A. Pincus, in *Surfactants in Solution*, K. Mittal and B. Lindman, eds. (Plenum, New York, (1984), p. 1177; W. Helfrich in *Physics of Defects*, Les Houches School XXXV, R. Balian, ed. (North-Holland, Amsterdam, 1981); I. Szleifer, D. Kramer, A. Ben-Shaul, D. Roux, and W. M. Gelbart, Phys. Rev. Lett. **60**, 1966 (1988) and I. Szleifer, D. Kramer, A. Ben-Shaul, W. M. Gelbart and S. A. Safran, J. Chem. Phys. **92**, 6800 (1990). Estimates of the bending moduli in polymer systems have been discussed by R. S. Cantor, Macromolecules **14**, 1186 (1981); L. Leibler, Makromol. Chem. Macromol. Symp. **16**, 1 (1988); S. T. Milner and T. A. Witten, J. Phys. **49**, 1951 (1988); and Z. G. Wang and S. A. Safran, J. Chem. Phys. **94**, 679 (1991). Measurements of the local bending moduli have been discussed by J. M. di Meglio, M. Dvolaitzky, and C. Taupin, J. Chem. Phys. **89**, 871 (1985).

43. S. A. Safran in *Micellar solutions and Microemulsions*, S. H. Chen and R. Rajagopalan, eds. (Springer, New York, 1990).

44. W. Helfrich, Z. Naturforsch. **28a**, 693 (1973) and ibid. **33a**, 305 (1978). See also, D. Roux and C. Coulon, J. de Physique **47**, 1257 (1986).

45. S. A. Safran, J. Chem. Phys. **78**, 2073 (1983), and in *Surfactants in Solution*, Vol. 3, K. Mittal and B. Lindman, eds. (Plenum, New York, 1984), p. 1781. Reviews of droplet microemulsions can be found in [43] and N. Borkovic, Adv. in Colloid and Int. Sci. **31**, 195 (1992).

46. O. Y. Zhongcan and W. Helfrich, Phys. Rev. A **39**, 5280 (1989).

47. M. B. Schneider, J. T. Jenkins, and W. W. Webb, J. de Physique **45**, 1457 (1984).

48. J. N. Israelachvili, D. J. Mitchell, and B. W. Ninham, J. Chem. Soc. Faraday Trans. II **72**, 1525 (1976).

49. M. E. Cates, Macromolecules **20**, 2289 (1987); Europhys. Lett. **4**, 497 (1987); J. de Physique **49**, 1593 (1988).

50. W. M. Gelbart, W. E. McMullen, and A. Ben-Shaul, J. de Physique **46**, 1137 (1985).

51. G. Porte, J. Appell, P. Bassereau and J. Marignan, J. de Physique **50**, 1335 (1989), and J. de Physique II **1**, 1447 (1991).

52. Y. Talmon, unpublished.

53. For a discussion of the effects of long range electrostatic interactions and salt effects on the spontaneous curvature, see P. Pincus, S. A. Safran, A. Alexander, and D. Hone, in *Physics of Finely Divided Matter*, N. Boccara and M. Daoud, eds. (Springer-Verlag, New York, 1986), p.40.

54. S. A. Safran, P. A. Pincus, M. E. Cates and F. C. MacKintosh, J. de Physique **51**, 503 (1989); F. C. Mackintosh, S. A. Safran and P. Pincus, Europhys. Lett. **12**, 697 (1990).

55. G. Porte et al., J. Phys. Chem. **90**, 5746 (1986), S. J. Candau et al., J. de Physique (Colloque) in press and T. J. Drye and M. E. Cates, J. Chem. Phys. **96**, 1367 (1992).

56. C. Safinya, D. Roux, G. S. Smith, S. K. Sinha, P. Dimon, N. A. Clark, and A. M. Bellocq, Phys. Rev. Lett. **37**, 2718 (1986).

57. L. E. Scriven, in *Micellization, Solubilization, and Microemulsions*, K. L. Mittal, ed. (Plenum, New York, 1977), p. 877.

58. K. Bennet, C. Phelps, H. T. Davis, and L. E. Scriven, Soc. Pet. Eng. J. **21**, 747 (1981).

59. W. Helfrich, Z. Naturforsch. **33**, 305 (1978).

60. W. Jahn and R. Strey, J. Phys. Chem. **92**, 2294 (1988).

61. R. Lipowsky and S. Leibler, Phys. Rev. Lett. **56**, 2541 (1986).

62. M. E. Cates, D. Roux, D. Andelman, S. T. Milner, and S. A. Safran, Europhys. Lett. **5**, 733 (1988).

63. W. J. Benton and C. J. Miller, J. Phys. Chem. **87**, 4981 (1983); P. G. Nilsson and B. Lindman, J. Phys. Chem. **88**, 4764 (1984); and see references in [62].

64. A. M. Bellocq and D. Roux, in *Microemulsions*, S. Friberg and P. Bothorel, eds. (CRC Press, New York, 1987) p. 33, and D. Roux and C. M. Knobler, Phys. Rev. Lett. **60**, 373 (1988).

65. D. Roux, M. Cates, G. Olson, R. C. Ball, F. Nallet and A. M. Bellocq, Europhys. Lett. **11**, 229 (1990); D. Roux and M. Cates, in Proceedings of the 4th Nishinomyia-Yukawa Symposium, (Springer-Verlag, New York,1990).

66. G. Porte, J. Marignan, P. Bassereau, and R. May, J.de Physique **49**, 511 (1988) and J. Marignan, J. Appell, P. Bassereau, G. Porte and R. P. May, J. de Physique **50**, 3553 (1989).

67. Z. G. Wang and S. A. Safran, Europhy. Lett. **11**, 425 (1990).

68. The factor of $8\pi\alpha$ which enters the binding energy is based on a model of the bends as sections of spheres. The sensitivity of the phase behavior to the value of α can equally well be discussed in terms of a sensitivity to the factor of 8π which arises from the modeling of the bends as sections of spheres.

69. R. S. Cantor, Macromolecules **14**, 1186 (1981); and L. Leibler, Makromol. Chem. Macromol. Symp. **16**, 1 (1988).

70. Z. G. Wang and S. A. Safran, J. de Physique **51**,185 (1990) and J. Chem. Phys. **94**, 679 (1991).

71. M. Teubner, Europhys. Lett. **14**, 403 (1991).

72. G. Gompper and M. Schick, Phys. Rev. Lett. **65**, 1116 (1990).

73. G. Gompper, R. Holyst, and M. Schick, Phys. Rev. A **43**, 3157 (1991).

74. P. Pieruschka and S. Marčelja, J. de Physique **2**, 235 (1992).

75. P. Pieruschka and S. A. Safran, Europhys. Lett. in press (1993).

10

Interfacial Tension: Theory and Experiment

Dominique Langevin[1]
Jacques Meunier[2]

10.1 Introduction

10.1.1 MONOLAYER ADSORPTION

Surfactant molecules spontaneously adsorb at liquid interfaces: either liquid–air in the case of aqueous or organic surfactant solutions or liquid–liquid in the case of both aqueous and organic surfactant solutions. When the solubility of the surfactant in the aqueous and/or organic phases is very low, the surfactant molecules concentrate at the interface. They form a two-dimensional (2D) monolayer [1]. The energy per unit area of the interface is by definition the surface tension. Before adsorption the surface tension is γ_0, after adsorption γ. The change in free energy per unit area due to the monolayer is (by analogy with 3D systems) the surface pressure Π:

$$\gamma = \gamma_0 - \Pi \tag{10.1}$$

When the surfactant molecules in the monolayer are separated by large enough distances, the 2D system behaves like an ideal gas and:

$$\Pi a = k_B T \tag{10.2}$$

where a is the area per surfactant molecule, k_B the Boltzmann constant and T the absolute temperature. For more concentrated systems, other types of phases can be observed: "liquid expanded", "liquid condensed" and several crystalline phases, as well as equilibria between these phases in the monolayer. This subject will be discussed in detail in Chap. 12.

A surfactant monolayer is not fully characterized by its surface pressure. Like in 3D, its rheological behavior is described by both elastic and

---

[1]Laboratoire de Physique Statistique de l'ENS - 24 rue Lhomond, 75231 Paris Cedex 05, France
[2]Same as footnote 1.

viscous coefficients. Shear and dilational surface viscoelasticity are sufficient to describe in-plane perturbations of wavelengths much larger than the monolayer thickness. These properties can be experimentally measured with a variety of 2D rheometers. Out-of-plane perturbations are described by the bending elasticity already introduced in Chaps. 1 and 2.

When the surfactant molecules are more soluble in the aqueous and/or the organic phase, a surfactant monolayer is still formed at the interface, but exchanges between the monolayer and the bulk phases are now possible [2]. Eq. (10.1) remains valid, and shear and dilatational viscoelasticity can be defined as well. But, for instance, the low-frequency dilational elasticity vanishes because upon compression or dilatation of the monolayer, the equilibrium surface pressure can be attained by dissolution or adsorption of surfactant molecules. No clear evidence of phase transitions in these "soluble" monolayers has been reported up to now.

The interfacial tension at a liquid–air interface can never be very small. This can be simply understood qualitatively. The surface of an aqueous surfactant solution is hydrophobic, because the surfactant chains are oriented upwards and form an organic layer in direct contact with air; depending on their concentration, the surface pressure will be small or high. Its maximum value will correspond to a maximum chain coverage, i.e., to:

$$\gamma_{min} = \gamma_H = \gamma_0 - \Pi_{max} \tag{10.3}$$

where γ_H is a hydrocarbon liquid surface tension. Typically, $\gamma_H \approx 30 - 50mN/m$; thus with $\gamma_0 \approx 70mN/m$, one gets $\Pi_{max} \approx 20 - 40mN/m$.

On the contrary, the interfacial tension at an oil-water interface can be ultralow, possibly zero. In this case the polar parts of the surfactant molecules are in contact with water, and the hydrophobic chains in contact with oil: they are able to balance completely the interactions between oil and water. This happens at a particular value of the area per surfactant molecule, a^*, when the interface is said to be "saturated". The Schulman argument allows one to make precise the value of a^*: as presented by de Gennes and Taupin [3], this argument proceeds as follows. The surface free energy is written

$$f = \gamma_0 A + n_s G(a) \tag{10.4}$$

where n_s is the number of surfactant molecules, $A = n_s a$, the total surface area, and G the free energy per molecule describing surfactant-surfactant repulsions. The surface pressure of the film is (again in analogy with 3D):

$$\Pi(a) = \frac{-\partial G}{\partial a} \tag{10.5}$$

By minimizing f with respect to a, at fixed n_s, one obtains

$$0 = \frac{\partial f}{\partial a} = n_s(\gamma_0 - \Pi) = n_s\gamma \tag{10.6}$$

Whenever possible, then, the system will adjust its surface concentration so that

$$\Pi(a^*) = \gamma_0 \qquad (10.7)$$

requiring that there are enough surfactant molecules in the system, and that the surface pressure can reach large enough values before the monolayer collapses or before aggregates start to form in the bulk phases. In the latter case the excess surfactant molecules do not incorporate any more into the monolayer but serve rather to form new aggregates (micelles) in bulk.

10.1.2 SOLUBLE MONOLAYERS

10.1.2.1 Small surfactant concentration

The monolayer surface pressure is not easy to calculate. This problem is addressed in Chap. 12 for *insoluble* monolayers. When the surfactant molecules are soluble in the bulk phases, their chemical potential, μ_s, at the surface

$$\mu_s = \frac{\partial f}{\partial n_s} = G(a) + \Pi(a)a \qquad (10.8)$$

is equal to the surfactant chemical potential in the bulk phases. Using explicit expressions for $G(a)$ and the bulk free energy, one finds [2]

$$\mu_s = \mu_s^{oi} + k_B T \ln(f^i x^i) + \Pi a \qquad (10.9a)$$

$$\mu_s = \mu_s^{ob} + k_B T \ln(f^b x^b) \qquad (10.9b)$$

with i and b denoting "interface" and "bulk" respectively, f corresponding to the activity coefficients and x to the surfactant mole fractions. It follows that

$$\Pi = \frac{k_B T}{a} \ln\left(\frac{f^b x^b}{f^i x^i}\right) + \frac{\mu_s^{ob} - \mu_s^{oi}}{a} \qquad (10.10)$$

Equations similar to (10.9) can be written for the solvents. In this case, when the activities and mole fractions are equal to 1, the surface pressure is zero. If we characterize the solvent by zero subscripts, we have $\mu_0^{oi} = \mu_0^{ob}$ and,

$$\Pi = \frac{k_B T}{a_0} \ln\left(\frac{f_0^b x_0^b}{f_0^i x_0^i}\right) \qquad (10.11)$$

Introducing the surface concentrations Γ, we have $x_0^i = \Gamma_0/(\Gamma_0 + \Gamma_s)$; Γ_0 and Γ_s are related by:

$$a\Gamma_s + a_0\Gamma_0 = 1 \qquad (10.12)$$

At this stage, a convention has to be adopted to relate a and a_0. The usual one is $a = a_0 = 1/\Gamma_\infty$, with Γ_∞ denoting the saturation concentration. Since $f_0^b \approx 1$ and $x_0^b \approx 1$, we obtain from (10.11–12):

$$\Pi = -k_B T \Gamma_\infty \left[\ln\left(1 - \frac{\Gamma_s}{\Gamma_\infty}\right) + \ln f_0^i\right] \qquad (10.13)$$

Equation (10.13) gives the Langmuir equation of state when $f_0^i = 1$ and the Frumkin equation of state [2] when $f_0^i = (h/k_BT)(\Gamma_s/\Gamma_\infty)^2$, where h is a surface virial coefficient. We can also solve directly for the bulk surfactant concentration, c_s, and obtain adsorption isotherm expressions. More explicitly, equating (10.10) and (10.11) for Π, and writing

$$y = \exp\left(\frac{\mu_s^{oi} - \mu_s^{ob}}{k_BT}\right) \tag{10.14}$$

we find for $f_0^i = 1$ that

$$\frac{c_s}{y} = \frac{\Gamma_s/\Gamma_\infty}{1 - \Gamma_s/\Gamma_\infty} \tag{10.15}$$

which is the Langmuir adsorption isotherm. Similarly, for $f_0^i = (h/k_BT)(\Gamma_s/\Gamma_\infty)^2$, we obtain the Frumkin adsorption isotherm:

$$\frac{c_s}{y} \exp(\frac{2h}{k_BT}\frac{\Gamma_s}{\Gamma_\infty}) = \frac{\Gamma_s/\Gamma_\infty}{1 - \Gamma_s/\Gamma_\infty} \tag{10.16}$$

The Frumkin equations better describe the monolayers of soluble surfactants adsorbed at the free surface of water (see Fig. 10.1). This is because the surfactant chains are in contact with air and they strongly interact with each other. Soluble surfactants at an oil-water interface do not require the use of non-ideality corrections and are well described by the Langmuir equations.

10.1.2.2   Large surfactant concentration

When aggregates are formed in the bulk phases, a supplementary condition can be written by equating μ_s to the surfactant chemical potential in the aggregates:

$$\mu_s = \mu_s^{om} + k_BT \ln\left(f^m x^m\right) \tag{10.17}$$

with the superscript m meaning micelles. For this case, following Israelachvili [4], we get

$$\Pi = \frac{k_BT}{a}\ln\left(\frac{f^m x^m}{f^i x^i}\right) + \frac{\mu_s^{om} - \mu_s^{oi}}{a} \tag{10.18}$$

The micelles or microemulsion droplets form when the interface is almost completely saturated, so that $\Gamma_s \approx \Gamma_\infty$ and $x_i \approx 1$. Neglecting non ideality, i.e., putting $f^i = f^m = 1$, it follows from (10.18) that

$$\gamma = -\frac{k_BT}{a}\ln x^m - \frac{\mu_s^{om} - \mu_s^{oi}}{a} + \gamma_0 = \gamma_e + \gamma_c \tag{10.19}$$

Here, $\gamma_c = \gamma_0 + (\mu_s^{oi} - \mu_s^{om})/a$ represents the energy difference per unit area between surfactants in the surface of the micelle (or microemulsion droplet) and the plane-interfacial monolayer. This quantity has been calculated by

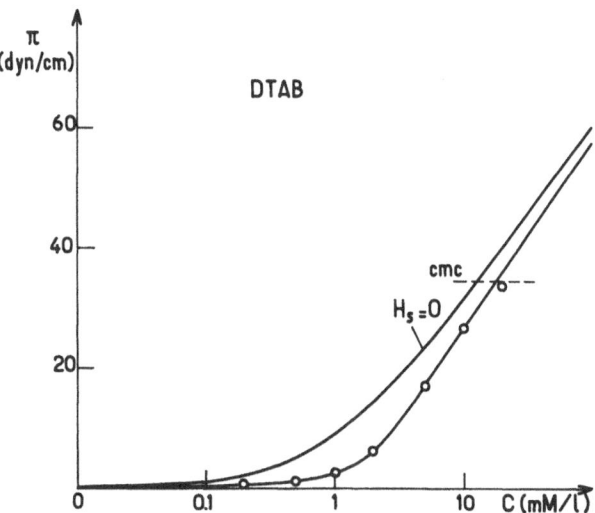

FIGURE 10.1. Surface pressure of aqueous solutions of dodecyl trimethyl ammonium bromide versus bulk surfactant concentration. The line with $H = 0$ is the fit with the Langmuir equation of state, the other line is the fit with the Frumkin equation of state.

many authors [5–8]. In the case of microemulsion droplets, γ_c can be related to the surfactant bending energy [9]

$$E_K = \int \left[k(c - c_o)^2/2 + \overline{k}c_1 c_2 \right] dA \qquad (10.20)$$

which should be added to the surface free energy f, cf. (10.4). In (10.20) k and $\overline{k}$ are, respectively, the "mean" and the "gaussian" bending elastic moduli, c_1 and c_2 the surface local principal curvatures (see earlier chapters); $c = c_1 + c_2$ is the mean curvature and $c_1 c_2$ is the gaussian curvature. De Gennes and Taupin have shown that [3]

$$\gamma_c = \frac{2k}{RR_0} \qquad (10.21)$$

where R is the droplet radius and R_0 the monolayer spontaneous radius of curvature. The physical interpretation of this equation is simple: when the plane interfacial area is increased, the minimum cost in energy is obtained if the area per polar head is kept constant. The new plane area must be covered by a surfactant monolayer which is taken from the bulk, i.e., from the monolayers around the droplets. The energy needed to unfold the monolayers is $2k/RR_0$ per unit area. This is accompanied by a change in the droplet number which is accounted for by the entropic term γ_e in (10.19). When using the relation between k, $\bar{k}$, R and R_0 (Chap. 9) and neglecting the entropy term γ_c, one finds

$$\gamma_c = \frac{2k + \bar{k}}{R^2} \tag{10.22}$$

The second term in (10.20) depends only on the *genus* of the surface: it is a constant when the shape of the surface is changed without changing its topology. But when the interfacial monolayer increases in area, this is accompanied by a change in the number of droplets, i.e., a change in the topology which introduces the term $\bar{k}/R^2$ in (10.22).

The first term in (10.19), γ_e, is less easy to calculate, because the exact form of the entropy of mixing of spheres is not known. Several approximations can be found in the literature [10–11]. They lead to

$$\gamma_e = \frac{k_B T}{R^2} \ln(\alpha\phi) \tag{10.23}$$

α being a numerical constant depending on the specific approximation employed, and ϕ the volume fraction of droplets.

When the structure of droplets is non-spherical, the calculation of γ_e is considerably more involved, and fewer models are available. In this connection we mention the work of Talmon and Prager concerning bicontinuous microemulsions [12]: according to this mean field calculation the interfacial profile indicates that the microemulsion becomes more oil-rich in the vicinity of the interface with the excess oil phase, as compared to the bulk. Similarly, the microemulsion is richer in water close to the interface with the excess water phase. However, this is in contrast to the experimental observations (see Sec. 10.2). More recent lattice (Ising type) theories of microemulsion systems [13] (see Chap. 8) lead to qualitative predictions which are in agreement with the observations. Quantitative comparisons are more difficult to perform.

10.1.2.3   Microscopic surface tension

When the aggregates in solution are large particles, like microemulsion droplets or vesicles, one can also define a surface tension γ_m for these entities. The free energy of the system contains a term equal to γ_m times the total area of the particles. For microemulsions, this term is assumed to be zero in certain theories [14]. This means that the area per surfactant molecules remains constant and equal to a^* while the microstructural entities are deformed. Several experimental investigations have been made recently to determine the values of γ_m in microemulsions, including electrical birefringence [15] and inelastic neutron scattering [16] measurements.

These experiments tend to show that γ_m is indeed zero, suggesting that the shape fluctuations of the droplets are governed by the bending elasticity. The above experiments allowed determination of the value of the elastic constant k and $\bar{k}$.

The macroscopic surface tension γ is, by definition, the energy variation required to increase the macroscopic surface area divided by the increase in surface area. It is then equal to the surfactant free energy difference between the aggregates in solution and the monolayer at the surface, as shown by Eq. (10.19). The measurement of γ therefore gives no information about γ_m.

The above introduction has shown the importance of the bending elasticity, which generally appears as a second-order term in the surface energy, but becomes important in systems of vanishing surface tension. The elastic bending constants also determine the value of the macroscopic surface tension; see (10.19)–(10.22). In the following sections we will describe experimental surface investigations that allow one to determine the bending elastic constant k. As this constant is known to be scale dependent (Chaps. 5–9), the interpretation of the measurements requires a renormalization of both the elastic constant and the surface tension. The surface tension that is measured in the experiments is the *renormalized* tension. It is therefore different from the bare tension defined previously for flat surfaces. For instance, although the bare tension of flaccid vesicles is non zero, the renormalized tension vanishes, because the vesicle can change its shape without changing its area [17]. In the following, we will deal mainly with experiments, so that γ will represent the renormalized tension except if stated otherwise. Results of measurements of surface tension in microemulsion systems will be presented together with data on the microstructure in the interfacial region.

10.2 Experimental Techniques and Data Analysis

Optical techniques such as ellipsometry, reflectivity measurements and dynamic surface light scattering are well suited for liquid interface studies [45]. They are non-perturbative and many liquids are reasonably optically transparent. X-ray or neutrons can be used instead of light, to extract additional information at short scales because of their short wavelength λ. X-rays are strongly absorbed by liquids, and hence their use (so far) is limited to the free surface of liquids. Neutron experiments are less sensitive to surface structure because of the weaker available fluxes and because of the smaller contrast of materials than with X-ray. But they have two advantages: neutron beams are generally only weakly absorbed, and the contrast can be varied by isotopic substitution.

In a reflectivity measurement (light, X-ray, neutron beam reflectivity or ellipsometry), the scattering vector is normal to the interface, allowing one to probe the refractive index variation along the normal. The index variation along the normal is determined by the structural details of the interface.

In the case of an interface between two structureless fluid phases, with or without an adsorbed monolayer, roughness arises from thermal fluctuations around the average, planar, shape of the interface. The distance of a surface point from the average plane $z = 0$ is $\zeta(\mathbf{r}, t)$ at time t, where $\mathbf{r}(x, y)$ are the coordinates of the point. One can write ζ as a sum of Fourier components:

$$\zeta(\mathbf{r}, t) = \sum_{\mathbf{q}} \zeta_{\mathbf{q}}(t) \exp(i\mathbf{q} \cdot \mathbf{r}) \qquad (10.24)$$

The $\zeta_{\mathbf{q}}$ are the amplitudes of the surface deformation modes.

In a surface light scattering measurement, the scattering wave vector has a component parallel to the interface which depends upon the scattering angle:

$$\mathbf{q} = \pm(\mathbf{k}_d - \mathbf{k}_r)_\sigma \qquad (10.25)$$

where $\mathbf{k}_r$ and $\mathbf{k}_d$ are, respectively, the wave vectors of the reflected light and of the light scattered around the reflected beam; the subscript σ describes projection onto the interfacial plane. Therefore the light scattered in the direction $\mathbf{k}_d$ gives information about the inhomogenieties in the interfacial plane of wave vector $\mathbf{q}$ (mode $\zeta_{\mathbf{q}}$) if their amplitudes are small. It is equivalent to say that the mode $\zeta_{\mathbf{q}}$ behaves like a small amplitude phase grating ($\zeta_{\mathbf{q}}\lambda \ll 1$) at the interface and scatters the light in the direction $\mathbf{k}_d$.

The surface light scattering technique gives information about the different modes $\zeta_{\mathbf{q}}$, while reflectivity measurements give information about the overall roughness.

Another optical technique, recently developed for the observation of two-dimensional phase transitions in monolayers, is the fluorescence microscopy which is described in Chap. 12.

10.2.1 ELLIPSOMETRY AND REFLECTIVITY

If E_j and E_{ij}^r are respectively the components of the incident and complex reflected electric fields, then

$$E_{ij}^r = r_{ij}(\theta)E_j \qquad (10.26)$$

with $r_{ij}(\theta)$ denoting the reflectivity coefficients. θ is the incidence angle and i, j are equal to s or p depending on whether the electric field is perpendicular or parallel to the incidence plane, respectively.

In ellipsometry one measures the ratio $(r_{pp} + r_{ps})/(r_{ss} + r_{sp})$. This ratio is positive for $\theta = 0$ and negative for $\theta = \pi/2$ because of a phase change of π between the two polarizations of the reflected field. For a Fresnel interface (i.e. an interface without thickness or roughness, for which the refractive index changes at $z = 0$ from n_1, the index of one phase, to n_2, the index of the other phase), this ratio vanishes at the Brewster angle defined by $\tan\theta_B = n_2/n_1$. For a real interface this ratio is complex; at the Brewster angle its real part vanishes but its imaginary part stays finite. The imaginary part is called the ellipticity $\bar{\rho}_B$ (at the Brewster angle) and is very sensitive to the interfacial structure and roughness because it measures a deviation from zero. In the case of thin interfaces (much thinner than the wavelength λ of the light), $\bar{\rho}_B$ is the sum of two terms:

i. A structural term $\bar{\rho}_B^L$ which depends upon $n(z)$, the refractive index profile across the interface [18] and upon a possible refractive index anisotropy [19]

ii. A roughness term [20–22] $\bar{\rho}_B^R \sim \sum_q q\langle\zeta_q^2\rangle$.

An ellipsometric measurement gives only the total ellipticity $\bar{\rho}_B$. In many cases, one of the two terms ($\bar{\rho}_B^L$ or $\bar{\rho}_B^R$) is smaller than the other and can be neglected or roughly estimated based on a theoretical model. In other cases it is possible to measure independently the two terms by varying a parameter of the system which affects only one of them, thus obtaining information about both the roughness and the structure of the interface.

In a reflectivity measurement the signal is the ratio between the reflected intensity and the incident intensity of the light; information about electric field phases is lost. This ratio R_j depends upon the polarization j of the incident light. If $R_j^F(\theta)$ denotes its value for a Fresnel interface, the ratio $R = R_j/R_j^F$ contains the information about the interface density profile. This ratio is equal to 1 for a Fresnel interface and is smaller than 1 for a real interface. But the difference is appreciable only when the wavelength of the incident beam is of the same order of magnitude as the thickness or the roughness of the interface. This is usually the case for X-ray and neutron beams. R is the product of a structural term F_L and a roughness term:

$$R(\theta) = F_L \exp\left(-Q_0^2 \langle\zeta^2\rangle\right) \qquad (10.27)$$

where $Q_0 = 4\pi\cos\theta/\lambda$. The X-ray wavelength being small ($\lambda \approx 1.5$ Å), one can separate the structural term and the roughness term in the R ratio by varying the incidence angle: $R(\theta)$ usually exhibits minima and maxima which reflect destructive and constructive interferences between the radiation reflected at different levels of the interfacial profile. The positions of these extrema give information about F_L while their amplitude variations with θ give information about the roughness term.

10.2.2 MEASUREMENT OF THERMAL ROUGHNESS

From the preceeding discussion we have seen that the reflectivity technique is sensitive to

$$\Sigma_1 = \langle \zeta^2 \rangle = \sum_q \langle \zeta_q^2 \rangle \qquad (10.28)$$

while ellipsometry measures

$$\Sigma_2 = \sum_q q \langle \zeta_q^2 \rangle \qquad (10.29)$$

It follows that the reflectivity is sensitive to all surface modes equally while the ellipsometry is more sensitive to the surface modes of high q. The calculation of Σ_1 and Σ_2 is a statistical mechanical problem. The probability of a configuration ζ is proportional to $\exp(-E(\zeta)/k_B T)$, where $E(\zeta)$ is the energy of the configuration ζ. E is the sum of three terms:

i. The gravitational energy:

$$E_G(\zeta) = \int_S dS \int_0^z (\Delta\rho) g z dz \qquad (10.30)$$

Here $\Delta\rho$ is the density difference between the two phases, g the gravitational acceleration, and S the projection of the area A of the interface onto the horizontal equilibrium plane.

ii. The capillary energy:

$$E_C(\zeta) = \int_A \gamma(dA - dS) \qquad (10.31)$$

where γ is the interfacial tension.

iii. The (relevant part of the) curvature energy:

$$E_K(\zeta) = \int_S \frac{1}{2} k c^2 dA \qquad (10.32)$$

where k is the bending elastic modulus and c the mean curvature of the interface. The gaussian curvature term in (10.20) is omitted because, for an insoluble surfactant, thermal fluctuations do not affect the film topology and do not change the integral of $c_1 c_2$ over the surface. For a soluble surfactant, it can be incorporated into the surface tension, Eq. (10.22). Similarly, the term containing the spontaneous curvature c_0, see (10.20), is omitted because the term linear in cc_0 in the curvature energy can be incorporated into the surface tension, (10.21), if the surfactant is soluble; its average value vanishes in the case of an insoluble surfactant. .

The calculation of $\langle \zeta_q^2 \rangle$ cannot be made without approximations. We will first consider the approximation of independent modes.

10.2.2.1   Approximation of independent modes

The roughness and the mean slope of liquid interfaces are generally small: $\nabla\zeta^2 \ll 1$. In most cases, an expansion of the interfacial energy in powers of $\nabla\zeta^2$ can be approximated by the first-order term. The energy of a configuration ζ is then the sum of independent mode energies, and the equipartition theorem gives:

$$\langle \zeta_q^2 \rangle = \frac{k_B T}{A(g\Delta\rho + \gamma q^2 + kq^4)} \tag{10.33}$$

This formula leads to:

$$\langle \zeta^2 \rangle = \frac{k_B T}{2\pi\gamma} \ln\left(\frac{q_R}{q_0}\right) \text{ where } q_R = \sqrt{\frac{\gamma}{k}} \tag{10.34}$$

$$\sum_q q\langle \zeta_q^2 \rangle = \frac{k_B T}{2\pi\gamma} q_e \text{ where } q_e = \frac{\pi}{2}\sqrt{\frac{\gamma}{k}} \tag{10.35}$$

In (10.34) q_0 is a large scale cut–off, due to gravity: $q_0 = (\gamma/g\Delta\rho)^{-1/2}$.

The experimental measurement of the roughness terms Σ_1 and Σ_2, which depend on k, can be used to determine the bending elasticity of monolayers. In the case of a reflectivity measurement, k appears in the logarithm (see (10.28) and (10.34)) and the dependence of Σ_1 on k is therefore weak. For a monolayer at a free surface of a liquid, the surface tension being large, reflectivity measurements are only sensitive to large k variations ($\approx 100 k_B T$). The roughness term in ellipsometry is proportional to $(\gamma k)^{-1/2}$. Therefore, ellipsometry is sensitive to small k values ($k \approx k_B T$). But the roughness term in the ellipticity $\bar{\rho}_B^R$ has to be larger than the structural term $\bar{\rho}_B^L$. This happens only when the surface tension is sufficiently small ($\gamma \approx 10^{-2} dyne/cm$), i.e., when the surface thermal fluctuations are large. It should also be noted that when the bending elastic modulus is of the order of $k_B T, \nabla\zeta^2$ is large and the approximation of independent modes fails; the coupling between modes must be taken into account, as outlined below.

10.2.2.2   Coupled mode theory

The first improvement of the (capillary) free energy, beyond the independent mode approximation, involves $(\nabla\zeta^2)^2$ terms, namely

$$E_C = \gamma \int_A \left\{ \frac{1}{2}\nabla\zeta^2 - \frac{1}{8}(\nabla\zeta^2)^2 \right\} dA \tag{10.36}$$

The second term in the integral accounts for coupling between modes. Using (10.24) one can write

$$E_C = \frac{\gamma}{2} \sum_q q^2 |\zeta_q|^2 \left(1 - 3 \sum_{q'>q} q'^2 |\zeta_q|^2 \right) \tag{10.37}$$

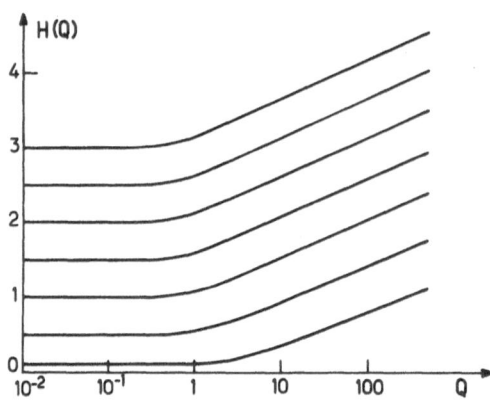

FIGURE 10.2. Solutions for the reduced bending elasticity H versus the reduced wave vector Q. H is a constant when the capillary energy is larger than the curvature energy.

In a surface tension measurement one does not measure the "bare" surface tension γ, but a smaller value because of thermal fluctuations [23]. For instance, in a static surface light scattering measurement, one observes a thermal fluctuation of wavevector q and deduces an experimental value for the surface tension from its average amplitude using (10.33). The coupling terms appearing in (10.37) which are q dependent are not taken into account, although they are included in the experimental value of the surface tension. The measured surface tension (and the bending elasticity) are therefore scale-dependent; γ and k depend upon the other thermal fluctuations of the interface—they are *renormalized*. A simple model calculation leads to [24]:

$$\gamma(q) = \gamma_\infty + b k_B T q^2 \qquad (10.38)$$

with $b = 3/8\pi$ and with γ_∞ denoting the surface tension measured at a macroscopic scale. The scale dependence appears only at a very short scale. The variation of k with the scale in the case $\gamma = 0$ was first calculated by Helfrich [25]. In the case $\gamma \neq 0$, $k(q)$ has no analytical expression and some solutions are given in Fig. 10.2 using the reduced parameters [24]

$$Q = \sqrt{\frac{k_B T}{\gamma_\infty}} \quad \text{and} \quad H(Q) = \frac{k(q)}{k_B T} \qquad (10.39)$$

At large scales, when the capillary energy dominates ($E_C \gg E_K$), the thermal fluctuations are small and the bending elastic modulus is constant

(k_∞). At short scales, the curvature energy dominates, the thermal fluctuations are large $(E_C \ll E_K)$ and using the second order expansion (10.37) one gets

$$k(q) = k_1 + bk_BT \ln\left[q^2 k_1/q_1^2 k(q)\right] \tag{10.40}$$

where k_1 is the bending elastic modulus observed at the scale q_1. The variation of k with observation scale is identical to the result obtained by Peliti and Leibler [26] when the k variation in the logarithm is neglected, i.e., when the role of the surface tension term is not taken into account.

It must be noted that the second term of $\gamma(q)$ varies as q^2. The corresponding surface energy varies as q^4 i.e., as a curvature term. This means that there is an apparent curvature energy, even for $k = 0$. The ellipsometric cut-off q_e never vanishes in this theory, and for $k = 0$ one obtains:

$$q_e = \frac{\pi}{2}\sqrt{\frac{\gamma}{bk_BT}} \tag{10.41}$$

This formula allows one to calculate the ellipticity $\bar{\rho}_B$ of a critical interface $(k = 0)$ using macroscopic parameters: γ_∞, n_1-n_2 and n_1+n_2. Surprisingly (in view of the crudeness of the theory used), good agreement is obtained between calculated and measured values (Fig. 10.3) without adding any intrinsic profile contribution $\bar{\rho}_B^L$ [24,29]. The agreement is still better with recent improvements of the theory [64].

10.2.3 DYNAMIC SURFACE LIGHT SCATTERING MEASUREMENT OF LOW SURFACE TENSIONS

A mode ζ_q at the interface behaves like a moving diffraction grating. The spectrum of the scattered light is changed by the Doppler effect and reflects the time evolution of the mode ζ_q. The time evolution and the wavelength of the observed modes are large compared to the molecular collision times and lengths: the evolution is governed by hydrodynamics [30].

At an interface between two liquids of densities ρ_1 and ρ_2, the modes ζ_q behave like capillary waves. The restoring forces come from surface tension (and gravity) and the damping forces from the bulk viscosities η_1 and η_2. (In light scattering experiments $\lambda_0 \ll 1$, and the curvature energy can be neglected.) For very low surface tensions as obtained in microemulsion systems, the ratio between the restoring force and the damping force is small:

$$\gamma(\rho_1 + \rho_2)/4(\eta_1 + \eta_2)^2 q \ll 1 \tag{10.42}$$

Thus, the thermal fluctuations observed by light scattering are overdamped. The spectrum of laser light scattered by a mode ζ_q has a lorentzian shape; it is centered at the frequency ω_0 of the incident light and has a half width [33]:

$$\Delta\omega_q = \frac{\gamma + g(\rho_1 - \rho_2)/q^2}{2(\eta_1 + \eta_2)}q \tag{10.43}$$

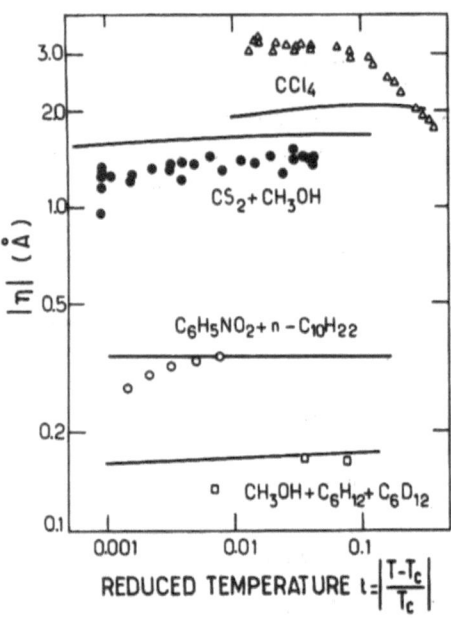

FIGURE 10.3. The experimental values for the ellipsometric parameter $\eta = (\lambda/\pi)((n_1^2 - n_2^2)/(n_1^2 + n_2^2)^{1/2})$ for several binary mixtures [28] and for the liquid-vapor interface of CCl_4 [27] close to their critical points. The lines are the theoretical values deduced from coupled mode theory. The origin of the discrepancy for CCl_4 is not clear; more experiments are needed in this case.

This spectrum is very narrow but can be analyzed with the heterodyne technique [32,33]. If the densities and the viscosities are measured independently, the slope of the curve $\Delta\omega_q(q)$ (at large enough q so that the gravity term is negligible) allows one to determine the surface tension with a high accuracy. The method is well adapted to the measurement of ultra-low surface tensions because the thermal fluctuations are large and the scattered intensity is large. The only perturbation to the interface is the laser beam, whose intensity can be significantly decreased if one of the liquid phases absorbs light strongly [34].

10.3   The Bending Elasticity of Monolayers: Experimental Results

The determination of the bending elasticity of monolayers based on measurements of interfacial density profiles is a very recent development. Consequently only few systems have been studied so far, mainly monolayers at the free surface of a liquid.

10.3.1   MONOLAYERS AT THE FREE SURFACE OF WATER

X-ray reflectivity measurements on the free surface of liquids are tedious and difficult. The refractive indices of the liquids are close to one and the reflected intensity is very weak as soon as the incident angle θ is smaller than the total reflection angle θ_L. For this reason the incident beam is always grazing: $0.001 < (\pi/2 - \theta) < 0.1$ radian, and the trough containing the liquid must be well isolated to prevent mechanical vibrations. But the refractive index values being close to 1, the theoretical calculations of the reflectivity at grazing incidence are much simpler for X-ray than for light. Another important point is that refractive indices can be easily calculated from the composition of the medium, and thus need not be measured. Therefore it is easy to have a model for the variation of the refractive index $n(z)$ in a monolayer.

The X-ray reflectivity curve of a lead-stearate monolayer [35] is shown in Fig. 10.4 for a surface pressure $\Pi = 25mN/m$ and an area per polar head of $19Å^2$. There is a chemical evolution of the monolayer and three curves are given for three different ages of this monolayer. The two minima M_1 and M_2 are due to the destructive interferences between the radiation reflected at different levels in the monolayer. Their coordinates give information about the position of the molecules in the monolayer (they are vertical, and the thickness of the monolayer is equal to the length of the molecules) and about the refractive indices of the polar heads and the tails. The reflectivity at the peaks (P_1) depends on the roughness of the monolayer. Here, one finds: $\langle \zeta^2 \rangle = 1.5Å^2$. This value is smaller than the one measured at the free surface of water [36] even though the surface tension is smaller ($47mN/m$) than for the water surface ($72mN/m$). This is consistent with a bending elastic modulus of $k = 25k_BT$; see (10.34). The uncertainty in the measured k value is large ($\pm 50\%$) because k appears in a logarithm.

The above technique can also provide interesting information about two-dimensional phase transitions in monolayers. For instance, the surface pressure isotherm of behenic acid monolayers at 20°C shows a kink for $\Pi = 20dynes/cm$. The corresponding phase transition has been studied by X-ray reflectivity [37]. A large decrease of the roughness was observed at this phase transition (Fig. 10.5) indicating that the bending elastic modulus of

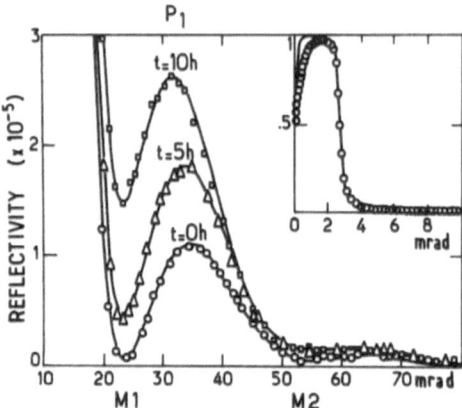

FIGURE 10.4. The x-ray reflectivity of a lead stearate monolayer at the free surface of water for different ages of the monolayers [35]. There is a chemical evolution of the monolayer. The minima M_1 and M_2 are due to destructive interferences between radiations reflected at different levels in the monolayer.

the monolayer changes from a small value ($k \approx k_B T$, that of a "liquid" phase) to a large value ($k \approx 200 k_B T$, that of a "solid" phase).

10.3.2   MONOLAYERS AT THE OIL-WATER INTERFACE

When the oil-water interfacial tension is small enough, the bending elasticity can be measured by ellipsometry. The main difficulty is the separation between the structural term $\bar{\rho}_B^L$ and the roughness term $\bar{\rho}_B^R$. This is possible with soluble surfactants, because the surface tension of the monolayer can be varied over a large range by changing slightly the salinity or the temperature, i.e., without varying too much the density of the monolayer.

One of the simplest cases is that of a pure surfactant, such as AOT, at the heptane-water interface [38,39]. The ionic strength is varied by adding sodium chloride. The concentration of AOT added to the brine-heptane system is just above the CMC. The monolayer at the interface is then saturated, and the bulk AOT concentration is small enough so that the refractive indices of the two coexisting phases are those of pure heptane and brine. Figure 10.6 shows the measured surface tension γ versus brine salinity S: a large variation of γ is observed with a minimum for $S^* = 0.225\%$. Figure 10.7 shows the measured ellipsometric parameter η versus S:

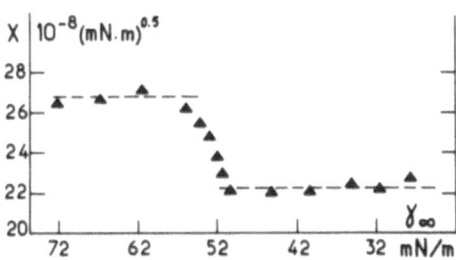

FIGURE 10.5. $X = (\gamma \langle \zeta^2 \rangle)^{1/2}$ versus the surface tension γ. For a constant k, X varies as $\log(\gamma)$. The large decrease of X for $X \approx 50$ dynes/cm is due to a large increase of k at this surface concentration.

$$\eta = \frac{\lambda}{\pi} \frac{(n_1^2 - n_2^2)}{\sqrt{n_1^2 + n_2^2}} \overline{\rho}_B \qquad (10.44)$$

A large increase of $|\eta|$ is observed around S^*. $|\eta|$ is proportional to $1/\sqrt{\gamma}$ indicating that $\eta^R \gg \eta^I$ (see Fig.10.8). (η^R and η^I denote the contributions to η from the roughness and the finite thickness of the interface, respectively.) The experimental points fall on a straight line as predicted by the independent mode theory. The coupled mode theory predicts deviations which are too small to be observed. In the independent mode approximation, we find $k = 1.10 k_B T$. Supposing that the coupling mode approximation is valid for these low k values and that $\gamma \ll k q^2$, we find $k = 1.65 k_B T$ at the molecular scale $q^{-1} = 5$Å.

More complex monolayers, which incorporate both surfactant and alcohol molecules, have also been studied. They can be obtained from Winsor microemulsion systems (see Sec. 10.4 and Chaps. 7–9). Winsor equilibria are observed upon mixing oil, brine and small amounts of surfactant and cosurfactant [40]. When increasing the brine salinity, one obtains:

i. For $S < S_1$, an oil-in-water microemulsion coexisting with an excess oil phase (Winsor I domain).

ii. For $S_1 < S < S_2$, a bicontinuous microemulsion (middle phase) coexisting with both excess oil and aqueous phases (Winsor III domain).

iii. For $S_2 < S$, a water-in-oil microemulsion coexisting with an excess aqueous phase (Winsor II domain).

An oil-brine interface can be obtained at each salinity from these systems with the following procedure (Fig. 10.9):

i. In the Winsor III domain, the oil-water interface is obtained upon removing the microemulsion phase. A small middle phase drop is added to

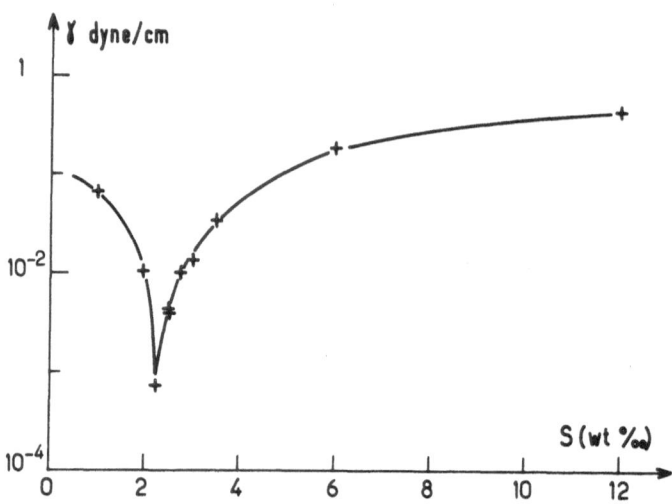

FIGURE 10.6. The interfacial tension γ of the heptane-brine interface covered with a monolayer of sodium bis-(2-ethylhexyl) sulfosuccinate (AOT), versus the brine salinity S. There is a deep minimum for $S = 2.25\%$.

ensure that the phases in equilibrium do not change. As we shall see later, such a drop does not wet the oil water interface. It usually stays at the "edge" of the interface and does not perturb the optical measurement.

ii. In the Winsor I (II) domain, the microemulsion is a dispersion of oil (water) droplets in a continuous aqueous (organic) phase, surrounded by a layer containing surfactant and cosurfactant molecules. This structure explains the fact that in most cases the microemulsion can be diluted by the continuous phase. This procedure allows one to change the droplet density without changing the droplet size [41,42]. The oil-water interface is obtained by mixing the organic phase and the continuous phase in the Winsor I domain, or the aqueous phase and the continuous phase in the Winsor II domain.

The aqueous and the organic phases obtained by the above procedure contain a small amount of surfactant (about the CMC). A monolayer is formed at the oil-water interface which is likely to be similar to the monolayer at the oil-water interfaces of the micro-structure in the microemulsion, especially in the Winsor III domain where these two monolayers have the same mean curvature. The study of the bending elasticity of the isolated monolayer at the oil-water interface allows one to test those microemulsion theories in which k appears as a key parameter (see Chaps. 7–9).

Various systems of this type have been studied. Two of them, using an *anionic* surfactant, are: i. Water + sodium chloride (47 wt %), toluene (47 wt %), butanol (4 wt %), and sodium dodecyl sulfate (SDS)

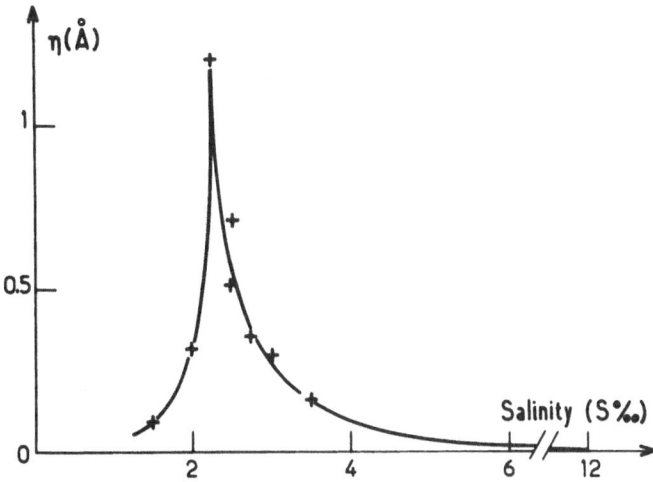

FIGURE 10.7. The ellipsometric parameter η of the heptane-brine interface covered with an AOT monolayer versus the salinity S. $\eta \sim 1/\gamma^{1/2}$ indicating that $\eta^R \gg \eta^L$.

(2 wt %) [43].

ii. Water + sodium chloride (56.83 wt %), dodecane (38.19 wt %), butanol (3.22 wt %), sodium hexadecyl benzene sulphonate (SHBS) (1.66 wt %) [39].

Two systems using a *cationic* surfactant are:

i. Water + sodium bromide (47 wt %), toluene (47 wt %), butanol (4 wt %), dodecyl trimethylammonium bromide (DTAB) (2 wt %) [39].

ii. Water + sodium bromide (47 wt %), dodecane (47 wt %), butanol (4 wt %), hexadecyl trimethylammonium bromide (CTAB)[39].

Table 10.1 gives the bending elastic moduli obtained with these different systems, for both the length-scale of the measurement, q_e^{-1}, using the independent mode approximation, and for $q^{-1} = 5$Å using the coupled modes theory.

The k values for these systems are very small and the scale dependence is large. Consequently there is a big difference between the values obtained on the scale of measurement and the renormalized values on the molecular scale. (Note, however, that the calculation of the molecular scale value was made by using a second-order expansion in terms of q^2 and is certainly questionable for such a weak bending elasticity.) For instance, the SHBS and the DTAB systems give the same k on the scale of the measurement, but different renormalized values on a molecular scale because the mea-

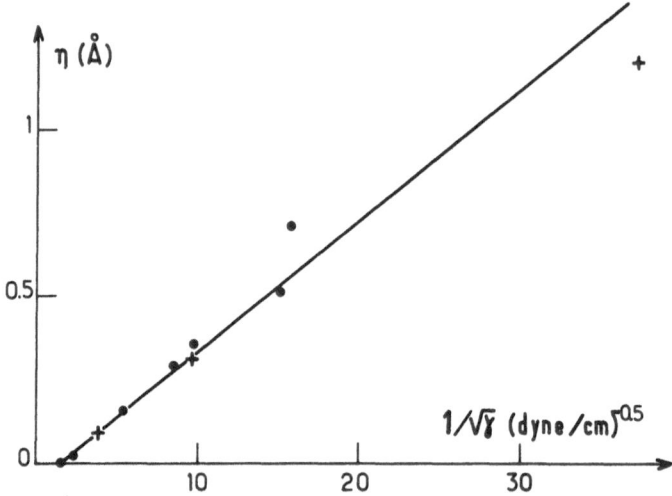

FIGURE 10.8. The ellipsometric parameter η of the heptane-brine interface with an AOT monolayer versus $1/\gamma^{1/2}$. The slope of the straight line gives the bending elasticity of the monolayer in the independent modes approximation. The limit for large tensions $\eta(0)$ is equal to η^L.

surements are not performed on the same scale (respectively $q_e^{-1} = 200\text{Å}$ and $q_e^{-1} = 40\text{Å}$).

The AOT bending elastic modulus was also measured by two other techniques:

i. By studying the shape fluctuations of microemulsion droplets by neutron scattering and neutron spin echo spectroscopy. The data analysis leads to different k values: about 0.5 and $5k_BT$ for the two length scales above [16]. But the existence of $\bar{k}$ has not been taken into account in the analysis of the scattering data. Further data treatment led to $k \sim 4k_BT$ and $\bar{k} \sim -2k$ [44].

ii. By measuring the Kerr constant, i.e., the optical anisotropy induced by an electric field, of a water-in-oil microemulsion at very small droplet volume fractions. This anisotropy is a measure of the ratio between the electrostatic forces which elongate the droplets and of the bending elastic forces which favor spherical droplets. The data analysis gives a small k value, $0.46k_BT$ [15], but the analysis requires many approximations whose effect on k is difficult to estimate. The droplet polydispersity has not been taken into account in the analysis, for example, but is expected to have a large effect because the Kerr constant is proportional to the sixth power of the droplet radius. Hence, the experiment is very sensitive to the droplets of large radius which are easier to elongate than the smaller ones; as a result, the apparent bending elasticity is smaller than the real one.

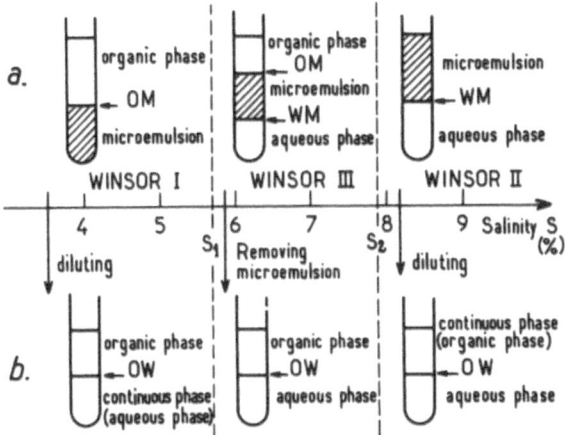

FIGURE 10.9. Winsor phase equilibria versus salinity (a) and the corresponding equilibria studied between aqueous and organic phases (b).

10.4 Experimental Study of Low Surface Tensions in Winsor Equilibrium

Two experimental techniques can be used to investigate the ultra low interfacial tensions in the Winsor equilibria: the dynamic surface light scattering described above and the spinning drop. In the latter, a horizontal capillary tube is filled with the phase of higher density and a small drop of the other phase is introduced. The tube spins around its axis and as the speed of

TABLE 10.1.

system	q_e^{-1}Å	"measured" k/k_BT	renormalized k/k_BT at $q = 2.10^7$ cm^{-1}
SDS	170	0.65	1.00
SHBS	200	0.40	0.86
DTAB	50	0.40	0.55
CTAB	50	0.40	0.55
AOT	200	1.10	1.65

rotation ω is increased, the drop moves towards the rotation axis until centrifugal forces and interfacial forces balance. For a drop whose length L is larger than four times its maximum radius ($L > 4R$), one finds [46]

$$\gamma = \Delta\rho\omega^2 R^2/4 \qquad (10.45)$$

The interfacial tension is deduced from measurements of the rotation speed, the maximum radius of the drop, and the density difference betwen phases. This technique can be applied to incompressible systems of small surface tension. In spite of the vibrations arising from the rotation, the monolayer at the interface seems to reach equilibrium. A comparison with surface light scattering [47] shows that the two methods give the same surface tension values.

A number of Winsor multiphase microemulsion systems have been studied in connection with tertiary oil recovery. The surface tension curves have similar shape in all cases. Consider, for example, the much-studied brine-toluene-SDS-butanol system (whose composition was given above). Fig. 10.10 shows the values of the interfacial tension measured at the oil-microemulsion interface (which is denoted 0/M) and at the water-microemulsion interface (W/M) versus the brine salinity S [48,50]. γ_{OM} decreases with the brine salinity while γ_{WM} increases. At "optimal salinity", S^*, the two surface tensions are equal: $\gamma_{OM} = \gamma_{WM} = \gamma^*$.

The interfacial tension was also measured at the oil-water interfaces obtained from the Winsor systems, as described in the previous section. In the Winsor I and II domains, the surface tension was found to be independent of the degree of dilution of the microemulsion i.e., independent of the presence of droplets in bulk [49]. This proves that the ultra low values of the surface tension at these interfaces arises from the large surface pressure Π in the interfacial monolayer (recall that the surface tension γ of an oil-water interface without an adsorbed monolayer is about $50nM/m$; in presence of the monolayer $\gamma_{OW} = \gamma - \Pi$; see (10.1)).

In the Winsor III domain there are two interfaces for each salinity S: an O/M interface and a W/M interface. The larger of the two surface tensions measured at each salinity (γ_{OM} for $S < S^*$ and γ_{WM} for $S > S^*$) is equal to the surface tension γ_{OW} at the O/W interface, obtained by removing the microemulsion phase [50]; see Fig. 10.11. As before, this surface tension is independent of the presence or absence of bulk structures in the microemulsion: it results from the pressure Π in the monolayer. The origin of the ultra low surface tension of the second interface (that of lower surface tension) is different [51]. Figure 10.12 shows the surface tension of the O/M and W/M interfaces versus the density difference between the coexisting phases, in a log-log plot. For γ_{OM} and $\gamma_{WM} < \gamma^*$ the interfacial tension values fall on a straight line whose slope is 4. This is characteristic of the vicinity of a critical consolute point. In the theory of critical phenomena,

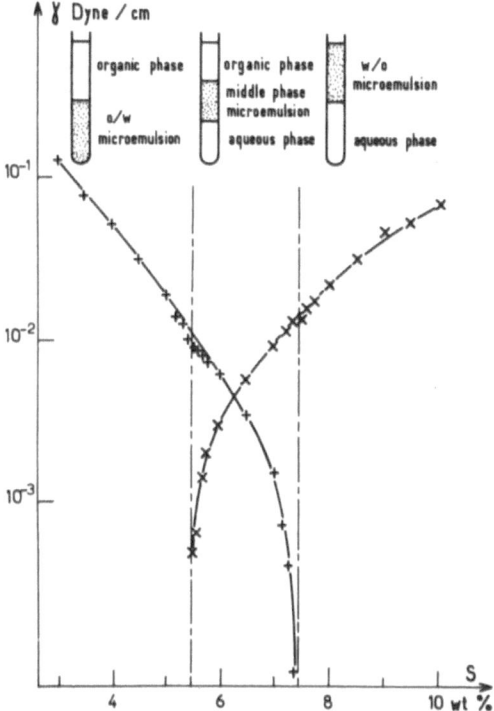

FIGURE 10.10. The interfacial tension γ at the O/M and W/M interfaces versus the salinity S in Winsor equilibria (toulene-brine-SDS-butanol mixture).

the relevant parameter is the distance ε to the critical point, defined by

$$\varepsilon = (\mu_i - \mu_i^C)/\mu_i \tag{10.46}$$

where μ_i is the chemical potential of component i, and μ_i^C is its value at the critical point. Scaling laws predict that

$$\gamma = \gamma_0 \varepsilon^\phi \tag{10.47}$$

where ϕ is a universal exponent. In Winsor systems, ε is not easy to determine, but it can be eliminated by using another measurable parameter like the density difference between the two phases $\Delta\rho$. In this case one writes

$$\Delta\rho = \Delta\rho_0 \varepsilon^\beta \tag{10.48}$$

where β is another critical exponent. Thus:

$$\gamma = \gamma_0 \left|\frac{\Delta\rho}{\Delta\rho_0}\right|^{\phi/\beta} \tag{10.49}$$

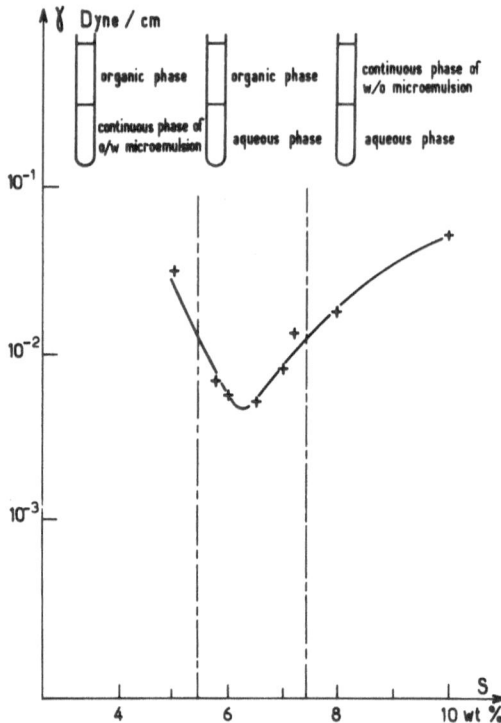

FIGURE 10.11. The interfacial tension γ at the O/W interfaces versus the salinity S in the systems of Fig. 10.9b (toulene-brine-SDS-butanol mixture).

Mean field theory predicts $\phi/\beta = 3$ while renormalization group theory predicts $\phi/\beta = 4$ in good agreement with experimental results. A study of the bulk properties of the coexisting phases confirms that the transitions at S_1 and S_2 are close to critical end points [51].

In conclusion, the origin of the lowest interfacial tensions ($\gamma < \gamma^*$) is the vicinity of critical points while that of the highest interfacial tensions ($\gamma > \gamma^*$) is the pressure in the interfacial monolayer. The transition between these two different behaviors, in this particular system, is close to γ^* as evidenced by Fig. 10.12.

In the Winsor III domain, it was found that:

$$\gamma_{OW} = \text{higher of } (\gamma_{OM}, \gamma_{WM}) \qquad (10.50)$$

The three surface tensions should satisfy the Antonov inequality:

$$\gamma_{OW} < \gamma_{OM} + \gamma_{WM} \qquad (10.51)$$

which is characteristic of a non-wetting situation. This can be checked by observing a middle phase microemulsion drop behavior at the oil-water

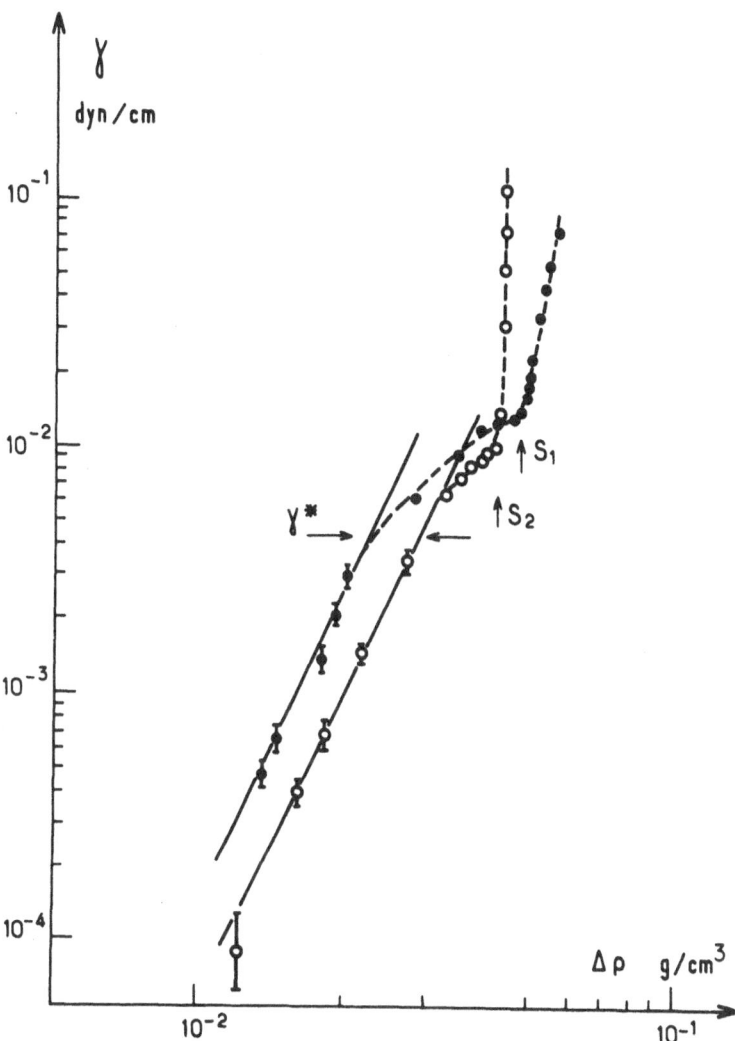

FIGURE 10.12. The interfacial tensions γ versus the distance to the critical point ϵ of the O/M and W/M in Winsor equilibria (toluene-brine-SDS-butanol mixture). The breaks in the curves for $\gamma = \gamma^*$ indicate the change in the physical origin of the low surface tensions.

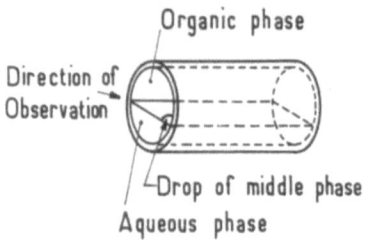

FIGURE 10.13. A drop of middle phase at the interface between the aqueous and oil phases in a toluene-brine-SDS-butanol mixture (Winsor equilibrium): the drop does not spread at the interface.

interface obtained (when the salinity is close to S^*) by removing the microemulsion phase in a Winsor III equilibrium. The drop does not spread at the interface [52,53], cf. Fig.10.13.

The direct observation of the drop behavior in the whole Winsor III domain is difficult because in the vicinity of the boundaries S_1 and S_2 the contact angles decrease quickly. However, ellipsometric measurements [43] at the O/W interfaces have proved that there is no wetting microemulsion layer at the interface even when a small microemulsion drop is added at the interface: this drop does not spread at the interface. This has been tested for all the Winsor systems mentioned above. It was conjectured that when approaching the critical point, a non-wetting transition should be observed [54-55]. But this transition has not yet been observed: probably, the mixtures studied do not have the exact critical composition and do not allow one to approach close enough the critical end points. The situation is different with short surfactant molecules for which the non-wetting/wetting transition was observed [56]. The non-wetting situation in three-phase equilibria observed in these mixtures is typical of long chain amphiphiles. It is not observed with short molecules (benzene-water-ethanol-ammonium sulfate) [57-58].

Equations (10.19), (10.22) and (10.23) allow one to calculate the interfacial tension γ between droplet microemulsions and excess phases and to compare it with the experimental value γ_{exp}. In most cases $\gamma_e \ll \gamma; \gamma$ is given by (10.22). The droplet radius R can be measured by neutron or X-ray scattering or by bulk light scattering in association with a dilution procedure [59-61]. Table 10.2 gives an example of the results for two systems and for $\bar{k} = 0$ at two different salinities. Although the calculated values have the correct order of magnitude and the correct variation with droplet size (salinity), the agreement between measured and calculated tensions is only qualitative [39,62].

TABLE 10.2.

System	S (wt %)	k/k_BT	$R(\mathring{A})$	γ_{exp}	γ
SDS	5 (γ_{OM})	0.65	130	$2.0\ 10^{-2}$	$3.4\ 10^{-2}$
	8 (γ_{WM})	0.65	180	$2.3\ 10^{-2}$	$1.8\ 10^{-2}$
AOT	0.175	1	69	$5.4\ 10^{-2}$	$2.0\ 10^{-1}$
	0.474	1	114	$5.6\ 10^{-2}$	$7.4\ 10^{-2}$

The origin of the discrepancy is not clear. It may be due to the failure of the theory which assumes that the monolayer thickness L is negligible ($L \ll R$); this condition is not well fullfilled in the above examples. It is also possible that the gaussian bending elastic modulus is of the same order of magnitude as k, cf. (10.22). These questions will be resolved when the origin of the discrepancies between the values of k obtained by different techniques are fully understood.

10.5 Structure of the Oil Microemulsion and the Water Microemulsion Interfaces in Winsor Equilibria

The O/M and W/M interfaces of the SDS system have been studied by ellipsometry [43]. Fig. 10.14 shows the experimental ellipsometric parameter η (see Eq. (10.44)) versus brine salinity. Interfacial tension measurements reported above have shown that:

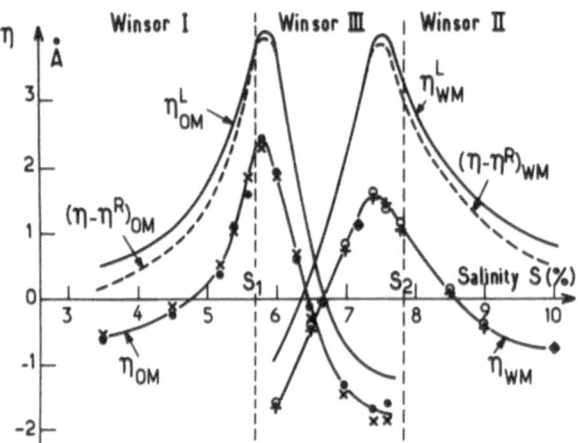

FIGURE 10.14. The ellipsometric parameter η for O/M and W/M interfaces in Winsor equilibria (toluene-brine-SDS-butanol mixture); $(\eta - \eta^R)$ (dashed lines), $\eta^L_{O/M}$ and $\eta^L_{W/M}$, deduced from the experimental values $\eta_{O/M}$ and $\eta_{W/M}$ as explained in the text.

i. For $\gamma > \gamma^*$ there is a flat surfactant monolayer at the O/M and W/M interfaces, the same as the one at the O/W interface for the same brine salinity.

ii. These O/M (or W/M) interfaces have the same interfacial tension as the O/W interface at the same salinity.

It is therefore possible to calculate the contributions to the ellipsometric parameter η from the thickness of the monolayer η^I and from the interfacial roughness η^R. But the sum of η^I and η^R is not equal to η: a third contribution $\eta^L = \eta - (\eta^I + \eta^R)$ must be added (Fig. 10.14). η^L, being positive in the Winsor I and II and in a large part of the Winsor III, cannot be a roughness contribution (always negative). A positive η is encountered only when the refractive index in the interfacial region takes values smaller or larger than n_1 and n_2, the refractive indices of the two phases. The following model can explain the experimental findings:

i. In the Winsor I (or II) domain, the microemulsions consist of oil (or water) droplets in a continuous phase. A droplet cannot get closer to the interface than its radius R. This "steric" repulsion produces a decrease in droplet density in the vicinity of the interface, equivalent to a layer of continuous phase. The thickness L of this layer can be deduced from η^L and compared with R. L and R are in good agreement (Fig. 10.15).

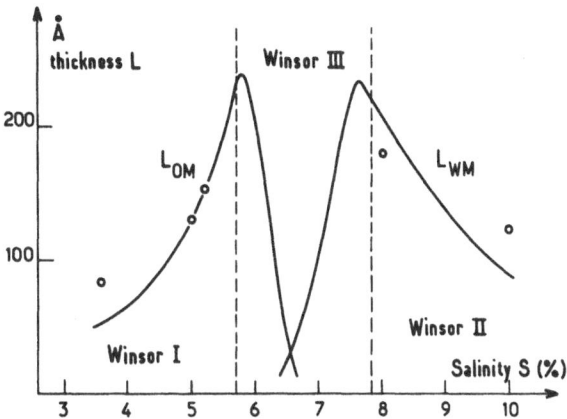

FIGURE 10.15. The thickness L of the depletion layer at the O/M and W/M interfaces versus the brine salinity S (full line) and the corresponding radii of the microemulsion droplets in the bulk for Winsor I and II domains (open circles).

ii. In the Winsor III domain, η^L reaches a maximum which agrees with a similar picture, i.e., the microstructural elements in the bulk stay at a distance L away from the monolayer. It is found that $L = 240\text{Å}$ which is close to the persistence length in the bulk ($\xi_K = 200 \pm 50\text{Å}$), i.e., the size of the microstructural elements. As the interfacial tension decreases, η^L and consequently the calculated L value decrease as well and are close to zero for $\gamma = \gamma^*$, probably because attractive forces between the interfacial monolayer and the elements increase quickly. For $\gamma < \gamma^*, \eta^L$ becomes negative, indicating that the refractive index in the interfacial layer has changed and lies between n_1 and n_2 (Fig. 10.16). This is explained by a model of open interfaces developed to account for the ultralow surface tensions in the vicinity of critical points [63]. Indeed, since the two coexisting phases become identical at the critical point, no monolayer should be present at the interface. As the interfacial tension decreases, the interface is expected to change from a closed to an open one, cf. Fig. 10.16.

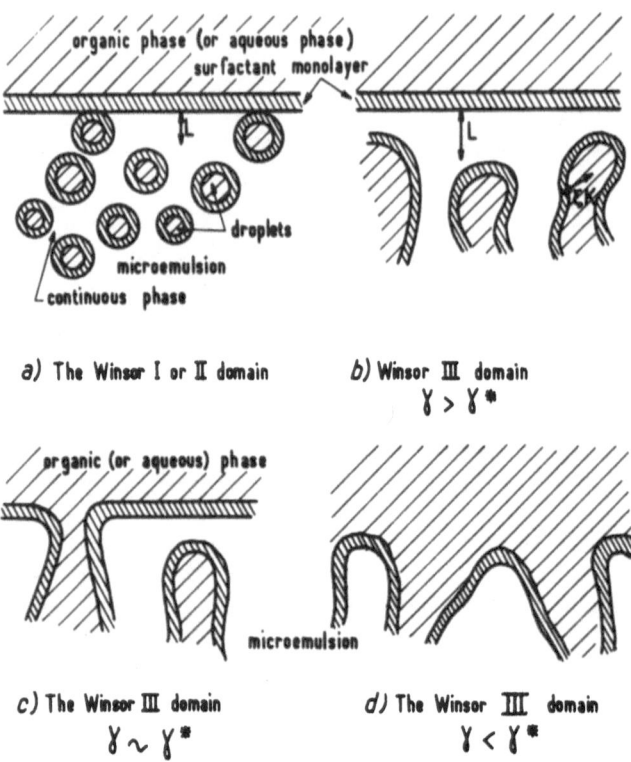

FIGURE 10.16. The structure of the O/M and the W/M interfaces in Winsor equilibria, versus brine salinity S. The interface is usually covered by a flat monolayer but can become gradually open close to the critical end points.

Acknowledgments

This research was supported by the Centre National de la Recherche Scientifique (France) and the Institut Français du Petrole.

References

1. G. Gaines, *Insoluble Monoloayers at Liquid-gas Interfaces* (Inter-science, 1966).

2. E. H. Lucassen-Reynders, *Anionic Surfactants* (Marcel Dekker, 1981).

3. P. G. De Gennes and C. Taupin, J. Phys. Chem. **86**, 2294 (1982).

4. J. Israelachvili, in *Surfactants and Solution*, vol. 4, K. Mittal and P. Bothorel, eds. (Plenum, 1986).

5. M. Robbins, in *Micellization, Solubilization and Microemulsion*, K. Mittal, ed. (Plenum, 1986).

6. C. A. Miller, R. Hwan, W. J. Benton and T. Fort, J. Colloid Interface Sci. **61**, 554 (1977).

7. C. Hu, J. Colloid Interface Sci. **71**, 408 (1979); **97**, 201 (1984).

8. B. W. Ninham and D. J. Mitchell, J. Phys. Chem. **87**, 2996 (1983).

9. W. Helfrich, Z. Naturforsch **28c**, 693 (1973).

10. E. Ruckenstein, in ref. 4.

11. G. J. Verhoeckx, P. L. de Bruyn and J. Th. Overbeek, J. Colloid Interface Sci. **119**, 409 (1987).

12. Y. Talmon and S. Prager, J. Phys. Chem. **76**, 1535 (1982).

13. B. Widom, C. Borzi and R. Lipowsky, J. Chem. Soc. Faraday Trans. 2 **82**, 1739 (1986).

14. S. A. Safran, D. Roux, M. E. Cates and D. Andelman, Phys. Rev. Lett. **57**, 491 (1986).

15. E. van der Linden, S. Geiger and D. Bedeaux, Physica **156**, 130 (1989).

16. J. S. Huang, S. T. Milner, B. Farago and D. Richter, Phys. Rev. Lett. **59**, 2600 (1987).

17. F. Brochard, P. G. de Gennes and P. Pfeuty, J. de Physique **37**, 1099 (1976).

18. P. Drude, Ann. Phys. **43**, 126 (1981).

19. R. M. A. Azzam and N. M. Bashara, in *Ellipsometry and Polarized Light* (North Holland, 1977).

20. P. Croce, J. Opt. (Paris) **8**, 127 (1977).

21. D. Beaglehole, Physica **B 100**, 163 (1980).

22. B. J. A. Zielinka, D. Bedeaux and J. Vlieger, Physica **A 107**, 91 (1981).

23. F. P. Buff, R. A. Lovett and F. H. Stillinger, Phys. Rev. Lett. **15**, 621 (1963).

24. J. Meunier, J. de Physique **48**, 1819 (1987); and in *Physics of Amphiphilic Layers*, J. Meunier, D. Langevin and N. Boccara, eds. (Springer, 1987).

25. W. Helfrich, J. de Physique **46**, 1263 (1985).

26. L. Peliti and S. Leibler, Phys. Rev. Lett. **54**, 1690 (1985).

27. D. Beaglehole, Phys. Rev. Lett. **58**, 1434 (1987).

28. J. W. Schmidt, Phys. Rev. A **38**, 567 (1988).

29. D. Beysens and J. Meunier, Phys. Rev. Lett. **61**, 2002 (1988).

30. L. D. Landau and G. Placzek, Physik z. Sowjetunion **5**, 172 (1934).

31. D. Langevin, J. Meunier and D. Chatenay, in *Surfactants in Solution*, vol. 3, p. 1991, K. L. Mittal and B. Lindman, eds. (Plenum, 1984).

32. M. A. Bouchiat, J. Meunier and J. Brossel, C. R. Acad. Sci. Paris **266**, 255 (1968).

33. S. Hard, S. Hamnerius and O. Nilson, J. Appl. Phys. **47**, 2433 (1976).

34. R. B. Dorshow and R. L. Swofford, J. Appl. Phys. **65**, 3756 (1989).

35. F. Rieutord, Thesis, Université Paris-Sud (1987).

36. A. Braslau, M. Deutsch, P. S. Pershan, A. H. Weiss, J. Als Nielson and J. Bohr, Phys. Rev. Lett. **54**, 114 (1985).

37. J. Daillant, L. Bosio, J. J. Benattar and J. Meunier, Europhys. Lett. **8**, 453 (1989).

38. J. Meunier and B. Jerome, in *Surfactants in Solution: Modern Aspects*, Proceedings of the 6th International Symposium on Surfactants in Solution (New Delhi, 1986).

39. B. P. Binks, J. Meunier, O. Abillon and D. Langevin, Langmuir **5**, 415 (1989).

40. P. A. Winsor, Trans. Faraday Soc. **44**, 376 (1948).

41. A. Graciaa, J. Lachaise, A. Martinez, M. Bourrel and D. Chambu, C. R. Acad. Sci. B **282**, 547 (1976).

42. A. M. Cazabat, J. de Physique Lett. **44**, 593 (1983).

43. J. Meunier, J. de Physique Lett. **46**, L-1005 (1985).

44. B. Farago, D. Richter, J. S. Huang, S. A. Safran, and S. T. Milner, Phys. Rev. Lett. **65**, 3348 (1990).

45. *Light Scattering by Liquid Surfaces and Complementary Techniques*, D. Langevin, ed. (Marcel Dekker, 1992).

46. B. Vannegut, Rev. Sci. Intrum. **13**, 6 (1942).

47. D. Chatenay, D. Langevin, J. Meunier, D. Bourdon, P. Lalanne and A. M. Bellocq, J. Dispersion Sci. and Technology **3**, 245 (1982).

48. A. Pouchelon, J. Meunier, D. Langevin and A. M. Cazabat, J. Phys. Lett. **41**, L-239 (1980).

49. A. Pouchelon, D. Chatenay, J. Meunier and D. Langevin, J. Colloid Interface Sci. **82**, 418 (1981).

50. D. Chatenay, D. Langevin, J. Meunier and A. Pouchelon, in *Scattering Techniques Applied to Supramolecular and Non-equilibrium Systems*, S. H. Chen, B. Chu and R. Nossal, eds. (Plenum, 1981).

51. A. Cazabat, D. Langevin, J. Meunier and A. Pouchelon, J. de Physique Lett. **43**, L-89 (1982).

52. D. Chatenay, O. Abillon, J. Meunier, D. Langevin and A. M. Cazabat, in *Macro- and Microemulsions*, D. Shah, ed., p. 125 (ACS Washington, 1985).

53. M. Kahlweit, R. Strey, D. Haase, H. Kunieda, T. Schmeling, B. Faulhaber, M. Borkovec, H. F. Eicke, G. Busse, F. Eggers, Th. Funk, H. Richmann, L. Magid, O. Söderman, P. Stilbs, J. Wrinkler, A. Dittrich and W. Jahn, J. Colloid Interface Sci. **118**, 436 (1987).

54. J. W. Cahn, J. Chem. Phys. **66**, 3667 (1977).

55. B. Widom, Langmuir **3**, 12 (1987).

56. M. Robert and J. F. Jeng, J. de Physique **49**, 1821 (1988).

57. J. C. Lang, P. K. Lim and B. Widom, J. Phys. Chem. **80**, 1719 (1976).

58. Y. Seeto, J. E. Puig, L. E. Scriven and H. T. Davis, J. Colloid Interface Sci. **96**, 360 (1983).

59. A. M. Cazabat, D. Langevin, J. Meunier and A. Pouchelon, Adv. Colloid Interface Sci. **16**, 175 (1982).

60. A. M. Cazabat, D. Langevin, J. Meunier and A. Pouchelon, J. de Physique Lett. **43**, L-89 (1982).

61. R. Aveyard, B. P. Binks, T. A. Lawless and J. Mead, Can. J. Chem.

62. D. Langevin, D. Guest and J. Meunier, Colloid and Surf. Sci. **19**, 159 (1986).

63. L. Auvray, J. de Physique Lett. **46**, L-163 (1985).

64. D. Bonn and G. H. Wegdam, J. Phys. I (France) **2**, 1755 (1992).

11

Critical Behavior of Surfactant Solutions

Anne-Marie Bellocq[1]

11.1 Introduction

Over the last 15 years or so, a great deal of progress has been made experimentally and theoretically in understanding critical phase separation in fluids. The general concepts of scaling and universality are well established and renormalization group calculations of critical exponents are generally in good agreement with experimental values. There exist many books and recent reviews of critical phenomena [1–12]. The best characterized systems are pure fluids near their liquid–vapor critical points and binary fluids near liquid–liquid consolute points [6–12]. Both types of critical points belong to the same universality class as the three-dimensional Ising model. For mixtures with three or more components, liquid-liquid phase separation seems also to be related to the Ising model [13-21].

More recently, the study of critical phenomena in fluids has been extended to binary micellar solutions [22–36] and multicomponent microemulsion systems [37–63]. The aim of these investigations in surfactant solutions was to point out differences (if they exist) between these "new" critical points and the liquid-gas critical points of a pure compound. The purpose of this chapter is not to treat the subject of critical phenomena in surfactant solutions exhaustively, but rather to point out certain topics specific to these systems. Surfactant mixtures are of great interest since they address several unique but generic problems. The behavior in such media with short–range organization appears in some cases to be far more complex than that of pure compounds or mixtures of molecular liquids. The complexity is related to the existence of microscopic interfaces. Therefore the main points to be understood are: i) What is the dependence of aggregate size and shape on the concentration and temperature in the critical domain, and what is the mechanism responsible for the phase separation?; ii) Why do the observed critical exponents not always follow the universal behavior predicted by the renormalization group theory of critical phenom-

---

[1]Centre de Recherche Paul Pascal (CNRS), Avenue Dr. A. Schweitzer, 33600 PESSAC, France.

ena?; and finally iii) Is the order of magnitude of the critical amplitudes comparable to that found in mixtures of small molecules?

11.2   Structure and Interactions

11.2.1   MICELLAR SOLUTIONS

Critical phenomena have been observed in both nonionic [22-34] and ionic [35,36] surfactant solutions. At low amphiphile concentrations micellar solutions generally exist as homogeneous isotropic liquid phases [64]. The phases are composed of micelles which are randomly dispersed in the solvent and do not exhibit long range positional or orientational ordering. For both ionic and nonionic aqueous micellar solutions, phase separation between two micellar isotropic phases and critical phenomena can be induced by changing temperature, salt concentration or by addition of additives. A typical phase diagram of a nonionic amphiphile in water is shown in Fig. 11.1. For these systems, the phase separation occurs by raising the temperature. Above the strongly asymmetric coexistence curve two micellar phases coexist; each of them contains micelles and solvent but they differ in surfactant concentration. At the critical point P_c, the minimum of the coexistence curve, the mixture separates into two phases of equal composition and volume.

The nonionic amphiphiles polyoxyethylene glycol monoethers $CH_3 - (CH_2)_{i-1} - (OCH_2CH_2)_j$ OH (called hereafter C_iE_j) have been exten-

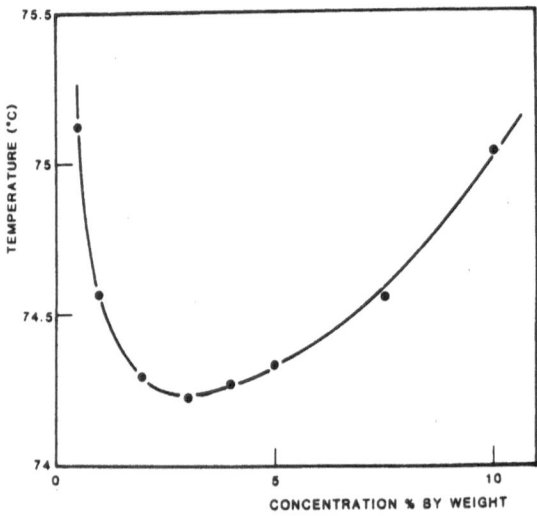

FIGURE 11.1. Lower consolute curve of the system $C_{12}E_8 - H_2O$ (from [30]).

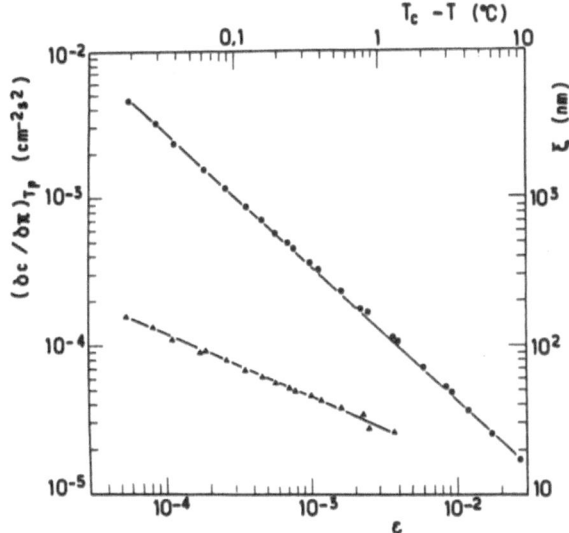

FIGURE 11.2. The osmotic compressibility (dots) and the correlation range (triangles) of $C_{12}E_8 - H_2O$ at the critical concentration in a temperature range very close to T_c (from [30]).

sively investigated by several techniques. Light scattering measurements performed with dilute aqueous solutions of C_iE_j have shown a considerable enhancement of the scattered intensity I_s as the temperature is increased. Two different interpretations of this phenomenon have been given. A review of both views is given in [27]. First the large increase in the scattering intensity near the cloud point was related to growth of the micelles [65-67]. The alternative explanation of the data was to assume that the micelles maintain constant size as the temperature raised but that critical concentration fluctuations arise, due to increasingly attractive intermicellar interactions as the critical temperature is approached. It is now recognized that the possible solution lies in a combination of these two views. However recent results suggest that the second effect is predominant in the concentration and temperature ranges near the critical point [23-34]. Indeed, static and dynamic light scattering data show several features which are typical of the behavior of critical mixtures [23-30, 33]. Both the osmotic compressibility and the correlation length of the concentration fluctuations diverge at T_c following power laws involving the distance from T_c (Fig. 11.2) (see next section). This is analogous to the phenomenon observed in binary simple liquid mixtures. Micellar solutions are of course different from binary liquid mixtures. Perhaps the most distinctive feature of solutions of amphiphilic molecules is that their microstructures, e.g., average aggregation numbers, depend sensitively on the bulk thermodynamic variables which specify the state of the system (see Chap. 1). Several groups have performed experiments in order to determine the temperature dependence of the micelle size using either neutron scattering [68-76] nuclear magnetic resonance [77-78] or fluorescence [79-80]. All these studies have led to contradictory results. Zulauf et al [68-70] have concluded from the analysis of the neutron scattered intensities that the micellar size at the critical concentration does not change with temperature whereas others infer from their data a substantial micellar growth with T. Indeed, an excellent fit to the small angle scattering curves can be obtained with a solution of uncorrelated cylinders [73]. This ambiguity makes it clear that static scattering does not provide enough information to distinguish between clustering of small micelles and growth of micelles. The most recent NMR self diffusion [76-78] and aggregation number measurements [79,80] seem to favor a moderate micellar growth. An increase of s from 100 to 400 is found for $C_{12}E_8$ as T approaches $T_c = 76°$ C. Similar variations have been obtained for $C_{10}E_8$, $C_{12}E_6$ and $C_{12}E_9$. In addition, within the experimental uncertainty, the substitution of H_2O with D_2O does not influence the aggregation number.

Hayter and Zulauf have examined the attractive interactions in the critical region [69]. They show that the assumption of a short-ranged temperature-dependent attractive pair potential between spherical micelles at constant size permits a quantitative analysis of the neutron scattering data for C_8E_5. They find that a potential with a range of only a fraction of nm is sufficient to generate spatial correlations over tens of nm as the attractive potential well deepens on approaching T_c. The primary effect of raising the temperature is to lower the degree of water structuring near the micelle surface, allowing increased Van der Waals attraction due to closer contact. Then the phase separation at the lower critical point in nonionic surfactant solutions is driven by the temperature dependence of the effective intermicellar interaction potential which is repulsive at low T but becomes attractive as T increases.

11.2.2 MICROEMULSIONS

Mixtures of water–oil–surfactant exhibit a complex range of phase behavior and offer an interesting area for observing critical phenomena [45]. Several detailed studies have established that ternary, quaternary and quinary microemulsion systems give rise to critical points and have suggested that some of them exhibit tricritical points [60,62,63]. Critical points have been observed in both the oil–rich and water–rich regions of the phase diagrams. For these multicomponent mixtures, phase separations between two critical micellar solutions are obtained either at constant temperature by varying composition or, as for binary micellar solutions, by raising temperature. This is illustrated in Fig. 11.3 which shows a section of the phase diagram of the quaternary system water-dodecane-pentanol-SDS obtained at

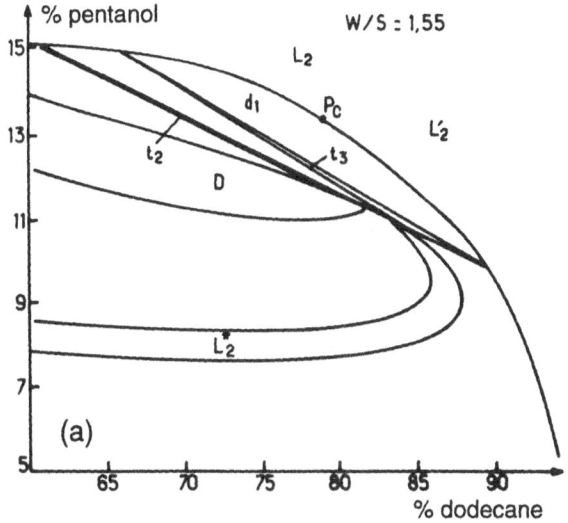

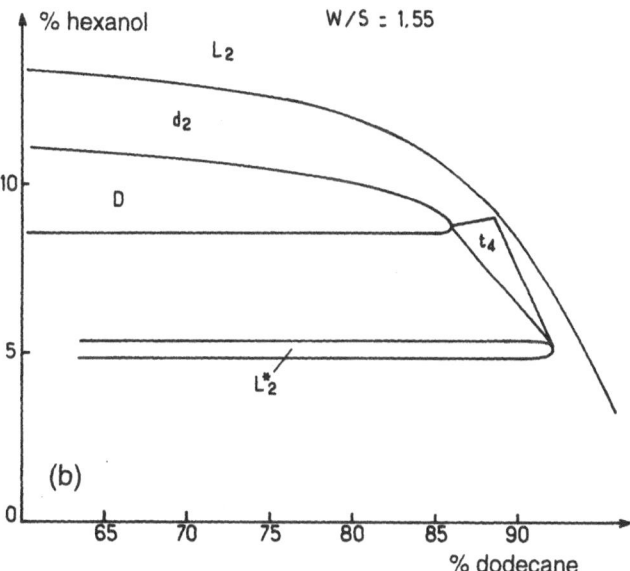

FIGURE 11.3. (a)Partial phase diagram of the system H_2O-dodecane-pentanol-SDS. $T = 21°C$. Magnification of the oil rich part of the section $X = 1.55$: d_1 is a two-phase region where the two microemulsions L_2 and L_2' are in equilibrium; t_2 and t_3 are three-phase regions; $t_2 = L_2^*DL_2$, $t_3 = L_2'L_2^*L_2$; P_c is a critical point. (Concentrations expressed in wt %) (from [30]). (b)Partial phase diagram of the system H_2O-dodecane-hexanolSDS, $T = 21°C, W/S = 1.55$: d_2 is a two-phase region where the microemulsion L_2 and the lamellar phase are in equilibrium. t_4 is a three-phase region, $t_4 = L_2DL_2^*$.

$T = 21°C$ [45]. In this pseudoternary diagram which holds constant the ratio X between the water and SDS concentrations (X=1.55), two isotropic phases L_2 and L_2^* and one lamellar phase D are identified. L_2 is the classical microemulsion phase; it consists of a dispersion of inverse micelles. In the two-phase region d_1, which bounds the phase L_2, two inverse micellar phases L_2 coexist. This region ends at the critical point P_c where both phases become identical. The three phases L_2, L_2^* and D are separated by a complex multiphase region, comprising five two–phase regions and two three-phase regions (t_2, t_3). Some of these equilibria also disappear at critical points.

In most cases, it appears possible to interpret the critical behavior of these mixtures as a liquid/gas-like critical point [81-84]. The structure of the medium can be described as a solution of interacting aggregates in a continuous phase comprised predominantly of water for oil-in-water microemulsions or oil for water-in-oil microemulsions [64]. Several light and neutron scattering studies on oil-rich ternary and quaternary microemulsions have clearly demonstrated that critical phenomena in these media are due to the existence of short-range attractive interactions between structural elements [39,43,48–57,83–87]. It has been established that the strength of attractions between w/o micelles is strongly dependent on the micellar size and on the chain lengths of the alcohol and oil molecules. In particular, attractions have been found to increase when the micellar radius increases or the alcohol chain length decreases and the molecular volume of oil increases [83-87]. By analogy with polymer solutions, the effect of oil can be easily understood as a change in the quality of the solvent for the chains of the film surrounding the micelles. In the language of polymer physics, an increase in the chain length of oil leads to a solvent of bad quality. A good solvent yields hard sphere interactions while a bad solvent yields an attractive interaction between the micelles. In such a solvent the aliphatic chains of the surfactant molecules prefer to mix together rather than be penetrated by the solvent. Lemaire et al [88] have established that this attraction (which derives from Van der Waals interactions) is due to the mutual interpenetration of the surfactants tails, the latter increasing with the micellar radius and with the difference in the surfactant and alcohol chain lengths.

As the potential between w/o droplets becomes strongly attractive a critical behavior is observed. This is illustrated in Fig. 11.4 where the curves of reduced compressibility versus the micellar volume fraction ϕ for three microemulsions formed with water-dodecane-SDS (Sodium dodecylsulfate) and heptanol, or hexanol, or pentanol, are reported [89]. The heptanol microemulsion is close to a hard sphere system. In the two other microemulsions, interactions are more attractive. In the case of pentanol both $\partial \pi / \partial \phi$ and $\partial^2 \pi / \partial \phi^2$ are close to zero, corresponding to a critical point. Then for

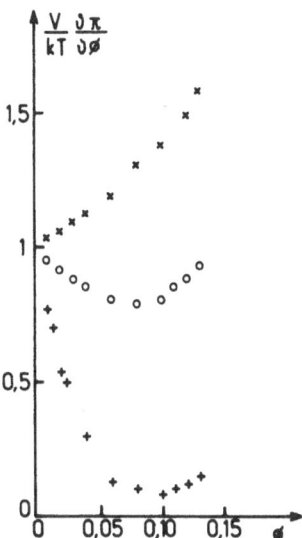

FIGURE 11.4. Normalized osmotic compressibility versus the volume fraction ϕ for three microemulsions formed with water-dodecane-SDS-pentanol(+), Hexanol (o) , heptanol(x) (from [30]).

this system it appears that attractions between droplets are strong enough to induce a phase separation between a micelle-rich phase and a micelle-poor phase.

Examination of the phase diagrams of the systems made with hexanol and pentanol reveals that they present two different types of phase behavior (Fig. 11.3). In the case of hexanol, the inverse micellar phase is bounded by a two–phase region where a microemulsion is in equilibrium with a swollen oily lamellar phase. By contrast, in the case of pentanol, the micellar phase domain is bounded by a region where two microemulsions are in equilibrium. In the first type of diagram, that of hexanol, no critical point is observed, whereas the second type of diagram is characterized by the occurence of a critical point. Some features of these phase diagrams can be understood from the light scattering results. The critical point and the two-phase region where two microemulsions coexist found in the phase diagram of the pentanol system have been interpreted as a liquid-gas transition due to intermicellar interactions [81,83,84]. A similar conclusion has been drawn for the ternary system AOT-water-decane [82,53]. A quantitative interpretation of the diagram of the quaternary system water–dodecane–pentanol–SDS has been made possible by using the intermicellar potential developed by Lemaire et al [88]. Interactions may explain the differences observed between the phase diagrams of the systems containing hexanol and pentanol. In the hexanol system, interactions remain moderately attractive; for this system the phase separation is not driven by interactions, but is due to another mechanism. (See Sec. 9.4.)

Likewise for micellar solutions, the study of changes in size and shape of the aggregates as the critical point is approached has led to contradictory results. Although a model of interacting spherical particles satisfactorily accounts for light and neutron scattering data obtained over a wide range of droplet volume fractions [85–94], results of electrical conductivity [95-97], transient electric birefringence [97,98], and ultrasonic absorption [99] provide evidence that the domain of validity of the droplet model in quaternary microemulsions depends strongly on the micellar potential. For large attractions, the description of the microemulsions in terms of individual droplets is not valid. For these systems, exchanges between droplets occur and the life time of the structure decreases as the attractions increase. The observation of an electrical percolation phenomenon suggests that in quaternary microemulsions large-size open structures are formed in the critical region by the merging of droplets during collisions; the droplet structure is then replaced by a bicontinuous structure. Self diffusion measurements done with photobleaching recovery techniques also provide evidence of the dynamical aspect of droplet aggregation in critical microemulsions [100]. Data suggest that at the concentrations higher than that of the percolation threshold, an infinite aggregate exists, but this is continuously renewed and has a short lifetime [100,101].

It seems that for the ternary microemulsion AOT–water–decane, the correlation established first for quaternary mixtures between the intermicellar exchange rate k_e and intermicellar interactions is also observed. Fluorescence decay results show that both k_e and the aggregation number s rapidly increase as the critical temperature is approached [102]. In contrast, Chen et al conclude from the analysis of neutron scattering data that the elementary droplets remain unchanged as they cluster near the critical point [103]. A similar conclusion has been drawn by Tabony et al. for a water rich critical microemulsion [104].

11.3   Critical Phenomena

Over the last few years a large number of papers have been published on the experimental study of critical phenomena in mixtures involving surfactants. Two-, three-, four- and even five-component mixtures were investigated [22-63]. In most cases, the aim of the experiments has consisted of measuring the critical exponents and the amplitude ratios since their values are characteristic of the universality class of the critical points considered. The main motivation of these studies was to know if the universal character of critical phenomena observed in pure fluids and usual binary mixtures also exists in mixtures as complex as microemulsions. The large experimental interest in the critical behavior of surfactant solutions has been due to a great extent to the very intricate results obtained. Indeed, whereas some experimental results are in agreement with those found in pure or usual binary fluids, other data indicate a more complex behavior. In some surfactant solutions the values of the critical exponents correspond to either negative or positive deviations from the Ising indices. An answer to the problem addressed has only been obtained very recently. As we will see later on in this chapter, the critical behavior in surfactant solutions is very similar to that of small molecules. The most remarkable feature when contrasted with the critical behavior of conventional mixtures is the order of magnitude of the critical amplitudes.

One of the main difficulties encountered in the study of multicomponent systems is to choose an appropriate path to approach the critical point. Usually the critical point is approached by varying temperature, keeping the composition of the mixture constant. However as pointed out by Griffiths and Wheeler [16] the critical behavior must be measured along special paths. Therefore before presenting the results obtained with surfactant solutions I briefly recall some general considerations on critical phenomena and the critical behavior in multicomponent systems.

11.3.1 GENERAL CONSIDERATIONS

The liquid–gas transition in a pure fluid and the liquid–liquid transition in a partially miscible liquid mixture are examples of phase transitions that have critical points. These systems exhibit striking macroscopic phenomena near their critical point.

The first major feature of critical phenomena is that sufficiently close to a critical point all equilibrium properties X vary as simple power laws of the deviations of the field variables from their critical values (such as $\varepsilon = (T - T_c)/T_c, (H - H_c)/H_c$ where T is the absolute temperature and H is the field conjugate to the order parameter M):

$$X = X_0 \varepsilon^\lambda \qquad \text{as} \qquad \varepsilon \to 0 \qquad (11.1)$$

In this equation, λ is a critical *exponent* and the coefficient X_0 defines the *amplitude* of the divergence. The definitions of most important critical exponents $\alpha, \beta, \gamma, \nu, \eta$, and μ are listed in Table 11.1.

TABLE 11.1. a) Critical power laws and exponent values (renormalization group R. G.)

QUANTITY	POWER LAW	THERMODYNAMIC PATH		EXPONENT R. G. d=3
Specific heat	$\tilde{C}_v = (\frac{A^+}{\alpha}) \mid \epsilon \mid^{-\alpha}$	$\rho = \rho_c$	$\epsilon > 0$	$\alpha = 0.109 \pm 0.002$
	$\tilde{C}_v = (\frac{A^+}{\alpha}) \mid \epsilon \mid^{-\alpha'}$	$\rho = \rho_c$	$\epsilon < 0$	$(\alpha = \alpha')$
Coexistence curve	$\Delta\tilde{\rho} = \pm B \mid \epsilon \mid^{\beta}$		$\epsilon < 0$	$\beta = 0.325 \pm 0.001$
Order parameter susceptibility	$\chi = C_0^+ \mid \epsilon \mid^{-\gamma}$	$\rho = \rho_c$	$\epsilon > 0$	$\gamma = 1.240 \pm 0.002$
	$\chi = C_0^- \mid \epsilon \mid^{-\gamma'}$	$\rho = \rho_c$	$\epsilon < 0$	$(\gamma' = \gamma)$
Correlation length for the order parameter fluctuations	$\xi = \xi_0^+ \mid \epsilon \mid^{-\nu}$	$\rho = \rho_c$	$\epsilon > 0$	$\nu = 0.630 \pm 0.001$
	$\xi = \xi_0^- \mid \epsilon \mid^{-\nu}$		$\epsilon < 0$	$(\nu' = \nu)$
Critical isotherm	$\mu = DM^\delta$	$T = T_c$	$\epsilon = 0$	$\delta = 4.81$
	$\xi^c = \xi_0^c \mu^{-\gamma}/\beta\delta$	$T = T_c$	$\epsilon = 0$	
Critical correlation function	$\tilde{\chi}(q) = \frac{C_1}{q^{2-\eta}}$		$\epsilon = 0$	$\eta = 0.0315 \pm 0.0025$
Interfacial tension	$\gamma = \gamma_0 t^\mu$		$\epsilon < 0$	$\mu = 1.260$

b) Amplitude relationships (R. G. values).

$A^+/A^- = 0.48$

$C_0^+/c_0^- = 4.5$

$\xi_0^+/\xi_0^- = 1.91$

$R_c^+ = \frac{A^+C^+}{B^2} = 0.666$

$R_\xi^+ = \xi_0^+(A^+)^{1/3} = 0.27$

$R_\chi^+ = C^+DB^{\delta-1} = 1.7$

$Q_2 = (\frac{C_0^+}{\tilde{C}_c})(\frac{\xi_0^c}{\xi_0^+}) = 1.21$

For the liquid–gas critical point in a pure fluid, the order parameter M and its field conjugate are defined by $(\rho - \rho_c)/\rho_c$ and $(\mu - \mu_c)/\mu_c$, respec-tively, where ρ is the density and μ the chemical potential. An alternative choice is $M \to (v - v_c)/v_c$ and $H \to (p - p_c)/p_c$ where v is the specific vol-ume and p the pressure. For the critical solution point of a binary mixture one choice of M and H can be $M \to (x - x_c)/x_c$ and $H \to (\Delta - \Delta_c)/\Delta_c$ where x is the concentration and Δ the difference in the chemical potential of the two components.

The second major important feature of critical phenomena is the concept of *scaling: the critical exponents are not independent*. There exist relations among the exponents called scaling relations, e.g.,

$$\alpha = \alpha\prime = 2 - \beta(\delta + 1); \gamma = \gamma\prime = \beta(\delta - 1); \alpha + 2\beta + \gamma = 2 \qquad (11.2)$$

Finally the third important feature of critical phenomena highlighted by modern theories is the concept of universality. According to this notion, the critical exponents associated with the singularities are identical for all the systems within a given universality class. A universality class is characterized by only two parameters, the dimensionality d of the system and the dimensionality n of its order parameter. Pure fluids and fluid mixtures near normal critical points belong in the universality class of the 3–dimensional Ising model. The renormalization group methods have yielded a detailed and accurate description of the critical thermodynamic behavior of such Ising-like systems. Another consequence of the concept of universality is the existence of universal combinations of the critical amplitudes called two–scale factor universality. Some universal relationships between amplitudes for the correlation length (ξ_0^+, ξ_0^-) and ξ_0^c on the critical isotherm, the susceptibility (C_0^+, C_0^-) and (D) on the critical isotherm, the order parameter (B), the specific heat (A^+, A^-) and the interfacial tension γ_0, are given in Table 11.1 [12].

11.3.2 CRITICAL BEHAVIOR IN FLUIDS AND MULTICOMPONENT MIXTURES

Table 11.1 indicates that the critical exponents are defined along precise paths. As an example, the definition of γ for a pure fluid applies to the path of density equal to the critical density (Fig. 11.5). As pointed out by Griffiths and Wheeler [16], the characteristic of this special path is that it is asymptotically parallel to the coexistence line of the space of fields. Along any other path, the exponents become renormalized. For systems defined with more than two independent variables it is then useful in the description of thermodynamic anomalies near critical points to distinguish two classes of intensive thermodynamic variables: field and density. Field variables are those variables such as the temperature (T), the pressure (P) or the chemical potential of the species (μ_i) which have equal values in the coexisting phases. Density variables are those variables which are not equal in coexisting phases. Examples of densities are concentration of the species (x_i), mass density, etc. Griffiths and Wheeler predict that the critical behavior for mixtures will be essentially the same as in pure fluids provided that the critical exponents are measured along equivalent directions in the space of independent variables of both systems. For a system made of n components, a situation equivalent to the case of the pure fluid will be obtained by holding n-1 fields constant (Fig. 11.6).

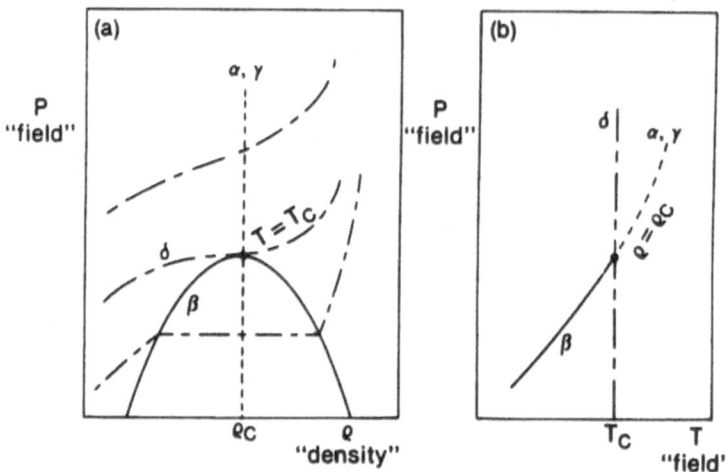

FIGURE 11.5. Pressure versus density isotherms, the critical isotherm $(T = T_c)$, the critical isochore (ρ_c), the critical point (•) and the coexistence curve (—) in (a) $P - \rho$ space and (b) P-T space. The paths along which the critical exponents α, β, γ and δ are defined have been indicated (from [10]).

One of the main advantages to drawing the phase diagram of a multicomponent system in a field space is that the section obtained in holding one or more fields constant has the same qualitative features as a system with one less thermodynamic degree of freedom. In a density space, or mixed field and density space, this is typically not the case. For systems with three or more independent variables there is in addition to the directions along and intersecting the coexistence surface a new way of approaching a critical point, namely along the critical line for $n = 2$ or parallel to the critical space for $n > 2$. Griffiths and Wheeler have predicted that the form of a divergence is determined by the orientation of a path L of approaching a critical point Y with respect to the coexistence surface on one hand and to the critical surface on another hand. These authors have considered three paths of interest.

Path I:

If L is asymptotically parallel to the coexistence surface but not parallel to the critical surface, the divergence of the compressibility is of the form

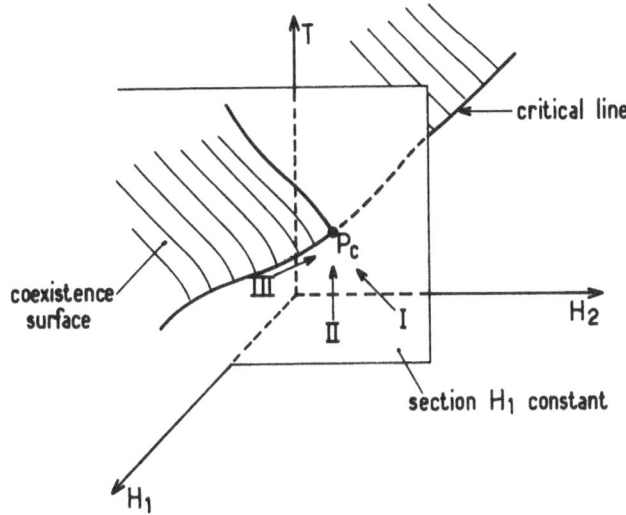

FIGURE 11.6. Phase diagram of a multicomponent system in the space of fields. Temperature T; chemical potentials H_1 and H_2. Paths of approach to a critical point P_c.

$|H - H_c|^{-\gamma}$, H being a field variable.

Path II

If L is not asymptotically parallel to the coexistence surface the divergence is of the form $|H - H_c|^{1-1/\delta}$.

Path III

The third case, where L is asymptotically parallel to the critical surface is more complicated since the form of the divergence will depend on the rate of approach to the critical surface. Given this rate of approach it is possible to calculate the form of the divergence. This has been carried out for double critical points for which one expects double values of the exponents.

For the liquid-liquid critical point of a binary mixture the path equivalent to the critical isochore for a pure fluid (path I) consists in keeping the concentration at its critical value, and by controlling the pressure to vary the temperature. In the cases of mixtures containing more than two components it is necessary to control (in addition to pressure) one or more chemical potentials. For a three component mixture, along a path of type I two fields (the pressure and one chemical potential) and one density must be kept constant, the variable of approach to P_c being temperature. In practice instead of holding two fields and one density constant, one fixes two densities at their critical values and controls the pressure to vary the temperature. The possibility for the second density, which is kept fixed, to fluctuate leads to a renormalization of the critical exponents which has been considered by Fisher [14]. The renormalized values of the exponents are ten per cent higher than the Ising values. They are multiplied by the coefficient $1/(1-\alpha)$ with $\alpha = 0.11$ (α being the exponent of the specific heat).

The possibility to approach the critical point of a ternary mixture at fixed pressure and temperature (Plait Point) has been studied by Widom and Wheeler [13,15] who have shown that in this case the critical behavior is identical in character to that of a binary liquid mixture of fixed temperature but variable pressure. Therefore as the plait point is approached along the line which is the extension into the one-phase region of the rectilinear diameter of the binodal curve, the osmotic compressibility diverges as the $-\gamma/(1-\alpha)$ power of the distance from the plait point:

$$\kappa_T \sim (\rho_2^c - \rho_2)^{-\gamma/(1-\alpha)} \qquad \text{fixed} P, T \qquad (11.3)$$

On the other hand, as the plait point is approached along the tangent to the binodal curve at that point, κ_T diverges as

$$|\rho_2 - \rho_2^c|^{-\gamma/\beta}, \qquad i.e., |\rho_2 - \rho_2^c|^{-(\delta-1)} \qquad (11.4)$$

For more than three components, it is necessary to reduce the number of variables until a situation equivalent to a known case of either binary or ternary mixtures is reached. That requires the knowledge of one or more chemical potentials. Obviously the main difficulty is then to find appropriate field variables different from temperature and pressure which can be experimentally controlled.

As already mentioned the critical exponents measured for binary molecular mixtures are in good agreement with those found for the liquid-gas critical point of pure fluids. In the case of ternary mixtures, experimental evidence for the predicted Fisher renormalization phenomenon has been given for the compressibility exponent γ [18–20] and for the exponent β[17]. Likewise, effects of renormalization of the exponent α have been clearly demonstrated for some ternary liquid mixtures [21].

11.3.3   EXPERIMENTAL RESULTS ON CRITICAL BEHAVIOR OF MICELLAR AND MICROEMULSION SYSTEMS

Several detailed experimental works on the critical phenomena of solutions containing surfactants have been reported. In most cases the critical behavior has been measured by its critical exponents ν and γ which characterize respectively the divergences in the correlation length and the osmotic compressibility. It has also been measured by its critical exponents β and μ. One common feature to all the critical points investigated so far is that they are *lower* critical points. The phase separation occurs as temperature is *raised*. In most of the surfactant systems investigated, the critical point was approached by raising temperature, but in some cases density variables were also used to approach the critical point. Table 11.2 summarizes the main results obtained for several systems.

TABLE 11.2. Critical exponents for surfactant-containing mixtures.

Exponents	(Ref)	(Variable)	ν	γ	β	μ
Ising Model	(1)		0.64	1.24	0.32	1.26
Renormalized Ising model	(14)		0.71	1.39	0.36	
Binary Mixtures						
C_1E_4, Water	(31)	T			0.36	
C_6E_3, Water	(30)	T	0.63	1.25		
C_8E_4, Water	(30)	T	0.57	1.15		
$C_{12}E_6$, Water	(30)	T	0.53	0.97		
$C_{12}E_6$, Water	(112)	T	0.60	1.20		
$C_{12}E_8$, Water	(30)	T	0.44	0.92		
$C_{12}E_8$, Water	(111)	T	0.63	1.20		
Ternary Mixtures						
AOT,Water,Decane	(37)	$T_c=36°C$	0.75	1.22		
AOT,Water,Decane	(56)	$T_c=43°C$	0.72	1.61		
AOT,Water,Decane	(43)	$T_c=26°C$	0.76	1.30		
AOT,Water,Decane	(43)	X	0.61	1.26	0.4	
AOT,Water,Decane	(52)	P	0.70	1.50		
AOT,Water,Decane	(57)	T	0.67	1.65		
CPBr,Water,$NaClO_3$	(35)	T	0.6			
SDS,Water,pentanol	(115)	T	1.27			
Quaternary and Quinary mixtures						
SDS,Water,Butanol,NaCl	(36)	T	0.62			1.24
SDS,Water,Pentanol, dodecane	(16,23) (118)		See	Table	11.6	
CTAB, H_2O,NaBr, Butanol,Octane	(38)	T	1.13	2.24	2.4	

+ See also Table 11.5

These latter are classified following the number of components. Values of the theoretical exponents are also given. The first point to be emphasized is that in most of the systems, values close to the Ising exponents (Fisher renormalized or not) are found independently of the number of components. However, for three particular series of mixtures, the measurements have led to very different values. As a paradox, it seems that the case of the quinary mixture is the easiest to explain. For that mixture, Dorshow and al. [38] have obtained values nearly twice those of the Ising model. These results were recently explained in terms of a tangential approach to a double critical point [105]. Further investigations of the phase diagram are required in order to verify if the path followed corresponds to this special approach. For the two other series of mixtures, even if a problem of path may exist in the case of the quaternary mixture, the intriguingly low values of the exponents are probably due to a more complex behavior that we will examine in the following.

In addition to these particular cases, a general question can be addressed. Can surfactant solutions because of their aggregate structure be viewed as a pseudobinary mixture or do they behave analogously to an ordinary molecular multicomponent mixture? A priori, an answer to this alternative should be given by the values of the exponents. For the first proposal one expects Ising values, while in the second case one should find renormalized exponents found for usual ternary mixtures. Unfortunately examination of the data does not allow a definitive answer. The expected difference in the values is only ten per cent and the experimental data are not accurate enough. In the following let us examine in more detail a few examples.

11.3.3.1 Micellar solutions

A comprehensive light scattering study of nonionic micellar solutions including temperature, concentration and chain length dependences has been performed by Corti and Degiorgio [22-30].

Static and dynamic light scattering measurements allow the determination of the correlation length ξ and the osmotic compressibility κ_T. Near a critical point, the inverse of the total intensity of the light scattered varies linearly with q^2 according to the law of Ornstein and Zernicke [1].

$$I(q) = \frac{I(0)}{1 + q^2\xi^2} \tag{11.5}$$

$I(0)$ is proportional to the osmotic compressibility (q is the wave vector of light $q = \frac{4\pi n}{\lambda} sin\theta/2$, n is the refractive index of the solution, λ the wavelength of light and θ the scattering angle).

The dynamic behavior in the vicinity of a critical point is more complex. The mode-mode coupling theories developed by Kawasaki predict the following behavior for the decay rate $< \Gamma >$ [106]:

$$< \Gamma >= \frac{k_B T}{16\eta_s} q^3 f(\frac{1}{q\xi}) \tag{11.6}$$

where η_s is the macroscopic shear viscosity and

$$f(\frac{1}{q\xi}) = \frac{2}{\pi}[\frac{1}{q\xi} + \frac{1}{q^3\xi^3} + (1 - \frac{1}{q^4\xi^4})tan^{-1}q\xi] \tag{11.7}$$

The hydrodynamic mean diffusion coefficient $\tilde{D}$ is derived as the limit of $< \Gamma > /q^2$ as q goes to zero:

$$lim_{q\to 0}\frac{< \Gamma >}{q^2} = \tilde{D}(1 + \frac{3}{5}q^2\xi^2 + \cdots) \tag{11.8}$$

with

$$\tilde{D} = h\frac{k_B T}{6\pi\eta_s\xi} \qquad \text{where } h \simeq 1 \tag{11.9}$$

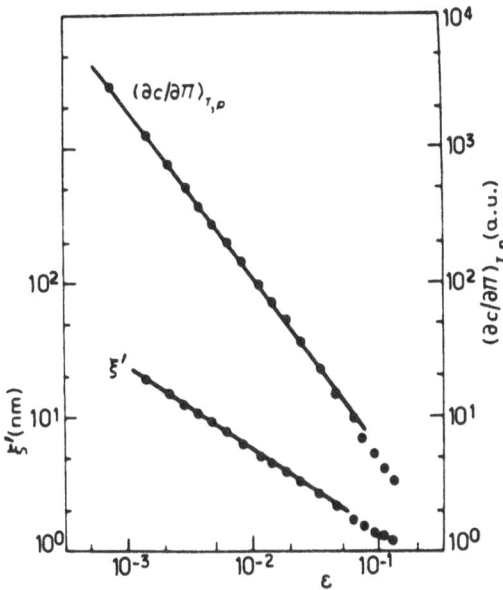

FIGURE 11.7. Variations of the osmotic compressibility and the correlation range as a function of the reduced temperature for the C_6E_3 aqueous solution at the critical temperature (from [26]).

The experimental data obtained by Corti and Degiorgio show typical features of critical phenomena. Their experiments have shown that the approach to T_c at constant concentration c_s is accompanied by power law divergences of the osmotic compressibility κ_T and of the correlation length ξ over a large range of reduced temperature, as happens for critical binary mixtures of usual components. As an example Fig.11.7 shows the variation of κ_T and $\xi' = \xi/h$ for the mixture $C_6E_3 - H_2O$ as a function of the reduced temperature ε.

Comparison of ξ' and ξ deduced respectively from dynamic and static measurements gives h=1.1. The critical exponents measured for this compound, $\nu = 0.63 \pm 0.03$ and $\gamma = 1.25 \pm 0.03$, are in good agreement with those predicted from RG. theory. Aqueous solutions of other compounds from the series C_iE_j show a qualitatively similar critical behavior. However the values of the exponents ν and γ and of the amplitude ξ_0 depend on the amphiphile, ν and γ going from their universal values to values smaller than the meanfield predictions. As ν and γ decrease, ξ_0 increases. The critical behavior appears also sensitive to the solvent. For the mixture $H_2O - C_{12}E_8$ the critical exponents are markedly influenced by isotopic substitution in the solvent and also by the addition of salt. Replacement of pure H_2O by pure D_2O results in an increase of the exponents ν and γ from $\nu = 0.44$ to $\nu = 0.58, \gamma = 0.88$ to $\gamma = 1.20$ and a decrease of the amplitude factor ξ_0 from 23Å to 9.2Å (Table 11.3). New accurate measurements near the critical point of the mixture $C_{12}E_8 - H_2O$ in the reduced temperature range $6.10^{-5} < \varepsilon < 2.10^{-2}$ confirm the anomalous values of ν and γ. Both κ_T and ξ follow a power law behavior as a function of ε over three decades without showing any crossover effect or correction to scaling contribution.

TABLE 11.3. Critical temperature T_c and concentration c_c and values of the parameters of the power-law fit to the osmotic compressibility and correlation range data. The last column refers to an experimental run performed much closer to T_c than the previous one (from [30])

SYSTEM	C_6E_3 H_2O	C_8E_4 H_2O	$C_{12}E_6$ H_2O	$C_{12}E_8$ H_2O	$C_{12}E_8$ D_2O	$C_{12}E_8$ H_2O 0.1M CsCl	$C_{12}E_8$ H_2O
$T_c(°C)$	44.7^a	40.3^a	50.4^a	75.5^a	71.04^a	72.64^a	74.205^c
$c_c(\%)$	13	7	1.25	3.2	2.5	3	3
γ	1.25^b	1.15^b	0.97^d	0.92^b	1.18^b	1.10^b	0.88^b
$B.10^7$ $(cm^{-2}s^2)$	0.2	0.8	13	5.8	2.1	2.3	8.29
ν	0.63^b	0.57^b	0.53^d	0.44^e	0.58^e	0.53^d	0.43^b
$\xi_0(nm)$	0.34^b	0.54^b	2.0^f	1.75^f	0.92^g	1.11^h	2.3^f

[a] Error of ± 0.1
[b] Error of ± 0.03
[c] Error of ± 0.003
[d] Error of ± 0.05

[e] Error of ± 0.04
[f] Error of ± 0.5
[g] Error of ± 0.25
[h] Error of ± 0.3

Corti and Degiorgio obtained data for different homologs C_iE_j in H_2O which seem to indicate that the deviation of ν and γ from Ising values is larger when the micellar size is larger. However the results obtained with $C_{12}E_8$ in D_2O or salted water show that the micelle size is not the only significant parameter but that the solvent–micelle interaction is equally important.

These observations have elicited intense theoretical interest. Several explanations for the apparently non universal exponents measured in nonionic surfactants systems have been given [107-110]. Fisher [108] proposes that the data are consistent with a crossover from classical to Ising behavior controlled by the reduced range of interaction Λ. This important parameter Λ is the ratio ξ_0/R where ξ_0 is the critical amplitude and R is the micellar radius. The interaction range is identified with the amplitude ξ_0. Based on the Ginzburg criterion for the validity of mean field theory, a minor change in Λ can move the crossover temperature t_x by several decades: $t_x \sim (\xi_0/R)^{-6}$ and hence leads to an apparent decrease in the exponents

γ and ν. This explanation, based on the idea that the asymptotic critical domain depends on the sample via ξ_0 and R, supposes that for a given sample Λ is constant as the temperature is varied. Another explanation has been given by Robledo et al [109] who suggest that the peculiarities of the crossover may be due to temperature dependent interactions. They point out that the interactions may change sign at a temperature T_D slightly below the critical temperature, T_D being analogous to the theta point of a polymer solution where the second virial coefficient vanishes. This leads to an unusual compression of the critical region for the mixture and to effective exponents.

Many aspects of these interpretations demand further study. What is the relationship between the micelle size and the range of interactions as measured by ξ_0? What physical features control ξ_0 and why is ξ_0 so sensitive to the isotropic substitution $H_2O \rightarrow D_2O$?

Very recently Dietler and Cannell [111] have reported light scattering measurements of the osmotic susceptibility χ and the correlation length ξ of binary mixtures of the nonionic surfactant $C_{12}E_8$ with H_2O and D_2O. They find 3D-Ising exponent values for γ and ν, regardless of which solvent is used. Their results being in striking contrast to those previously obtained by Corti and Degiorgio on different samples, they repeated experiments on a portion of the $C_{12}E_8$ sample used by Corti and Degiorgio. They found $\nu = 0.44$, i.e., a value identical to the (Corti and Degiorgio) value previously published. Therefore they conclude that the problem lies in the sample, but the origin of the differences between the two samples is not elucidated. Wilcoxon and Kaler [112] have studied $C_{12}E_6$ in H_2O obtaining $\gamma = 1.2 \pm 0.1$ and $\nu = 0.60 \pm 0.03$ whereas Corti and Degiorgio have obtained $\gamma = 0.97 \pm 0.05$ and $\nu = 0.53 \pm 0.05$ for this system.

11.3.3.2   Microemulsions

As previously mentioned, one of the main difficulties encountered in the study of multicomponent system is to find an appropriate path to approach the critical point. The detailed experimental study of the phase diagrams of both ternary (AOT–water–decane) and quaternary (SDS–pentanol–water–dodecane) systems has provided evidence that in the oil–rich part of the diagram the ratio X of the water and surfactant concentrations takes the same value in the coexisting phases [42-47]. The association of X with a field variable offers the possibility to study the critical behavior of these mixtures in following a path of approach to the critical point different from that corresponding to a variation of temperature. Therefore, for both systems, it has been tempting to follow paths equivalent to those known for ternary mixtures of simple molecules [42-47].

The ternary system AOT–water–decane presents in composition/temperature space a line of critical points [43]. The critical behavior at different points of this line was investigated by light and neutron scattering methods along two different paths [37, 43, 50–57] (see Fig. 11.8 and Table 11.4).

Table 11.4. Compositions in weight of the samples studied as a function of temperature (path 1) and ν_t, γ_t and ξ_o results.

SAMPLE	(1)	(2)	(3)	(4)
$T_c(C^o)$	26.30	30.02	36.01	43.3
Decane	87.27	86.96	89.96	89.07
Water	7.72	7.56	6.16	7.07
AOT	5.01	5.48	3.88	3.86
ν_t	0.76±0.05	0.71±0.05	0.75±0.05	0.72±0.04
γ_t	1.30±0.08	1.25±0.05	1.22±0.05	1.61±0.09
$\xi_o(\text{Å})$	7.2±2	9.3±2	12.2±2	11.0±2.1

(1,2) Light data from ref. (43).
(3) Light data from ref. (37).
(4) Neutron data from ref. (53).

Path I, considered as classical, consists in fixing the concentrations at their critical values and to raising temperature. The second one, path II, involves increasing X at constant temperature and constant decane concentration ($\phi_{decane} = 0.873$). Results show clearly that along both paths the osmotic compressibility κ_T and the correlation length ξ of the concentration fluctuations follow power laws as a function of either the reduced temperature $\varepsilon_T = (T_c - T)/T_c$ or the reduced variable $\varepsilon_X = (X_c - X)/X_c$. X_c is the value of X at the critical point considered.

$$Path I : I(0) = I_0\varepsilon_T^{-\gamma_t}, \eta = \eta_0\varepsilon_T^{-\nu_t} \qquad (11.10)$$

$$Path II : I(0) = I_0\varepsilon_X^{-\gamma_t}, \eta = \eta_0\varepsilon_X^{-\nu_t} \qquad (11.11)$$

Unfortunately, because of the low experimental accuracy it is not possible to conclude whether this ternary microemulsion behaves as a pseudobinary mixture or not.

The large value of ξ_0 suggests that the scale of the scattering units is not at the molecular level but corresponds to an aggregate. The value of ξ_0 increases with temperature. Such variation likely reflects the increase of attractions with temperature [56]. In pure fluids, interactions are mainly a function of temperature. In microemulsion systems several experimental studies have shown that interactions between w/o droplets are also dependent on many other parameters such as the micellar size or the chemical nature of the components. Therefore in these systems the phase transition may be controlled not only by temperature but also by an parameter that affects the interactions. Huang and Kim [113] had shown that variables such as the oil carbon number, oil composition, or salinity can be used to

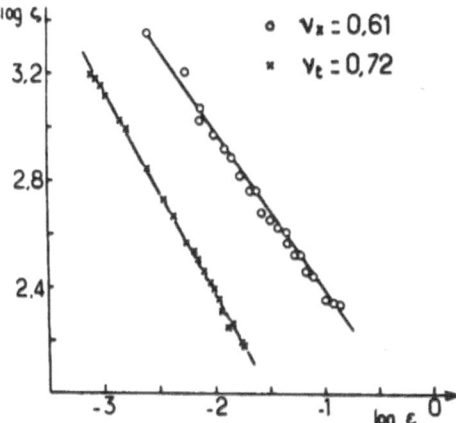

FIGURE 11.8. Variations of the correlation length ξ at the approach of a critical point for the mixture H_2O-AOT-decane as a function of the reduced temperature ($\varepsilon_T = (T - T_c)/T_c(X)$ (Path I, $\nu_t = 0.72$) and of the reduced variable $\varepsilon_x = (X - X_c)/X_c(O)$ (Path II, $\nu_x = 0.61$).

approach the critical point. The droplet size which is directly related to the water-over-surfactant ratio X is also a "critical value." Results shown in Fig. 11.8 identify a parallel between the two variables temperature and X that is consistent with the fact that phase separation is due to interactions.

The critical behavior of the four-component mixture consisting of water-dodecane–pentanol–SDS is more complex. At fixed temperature, T = 21°C, this quaternary mixture presents a line of critical points which extends in the phase diagram between a critical point P_c^A, belonging to the ternary mixture water–SDS–pentanol and a critical end point P_c^E located in the oil–rich part of the diagram (Fig. 11.9). The X values (expressed in weight) corresponding to P_c^E and P_c^A are, respectively, 0.95 and 6.6. Bellocq et al [114] have measured several critical exponents at different points of the critical line in order to test several scaling laws.

In the first place, the critical behavior at the point P_c^A has been investigated [115], using as variables either temperature (path I) or concentration at constant T (paths II and III). The paths II and III are respectively perpendicular at the direction of the tie–lines and tangent to the coexistence curve at the critical point (Fig. 11.10). These paths are equivalent to those described by Widom and Wheeler [13,15]. The results obtained by light scattering, turbidity and density measurements yield values in good

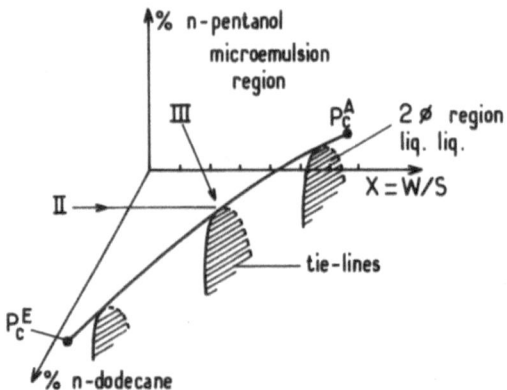

FIGURE 11.9. Schematic representation of the phase diagram of the quaternary mixture H_2O–dodecane-pentanol-SDS.

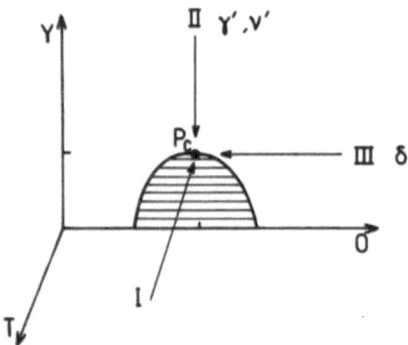

FIGURE 11.10. Schematic representation of the phase diagram of the ternary mixture water-pentanol-SDS and paths of approach to the critical point P_c. O and Y are related to concentrations.

agreement with those expected for a ternary mixture. The critical indices measured along the three different paths indicate an Ising–like behavior (Table 11.5).

Table 11.5. Ternary system : water-SDS-pentanol. The direction Y is perpendicular to the tie-lines. The direction O is tangent to the coexistence curve at the critical point.

PATH	VARIABLE	EXPONENT	
		THEORETICAL	EXPERIMENTAL
I	T	$\gamma\prime = \frac{\gamma}{1-\alpha} = 1.39$	1.27 ± 0.1
II	Concentration Y	$\gamma\prime = \frac{\gamma}{1-\alpha} = 1.39$	1.45 ± 0.1
		$\nu\prime = \frac{\nu}{1-\alpha} = 0.71$	0.75 ± 0.05
		$\beta\prime = \frac{\beta}{1-\alpha} = 0.36$	0.38 ± 0.03
III	Concentration 0	$\frac{\gamma}{\beta} = 3.81$	3.68 ± 0.2

However the question as to whether the Fisher renormalization is observed is open and requires a more accurate determination of the exponents.

As in the ternary system AOT/H_2O/decane, temperature and the X ratio were used as variables to approach six different critical points of the *quaternary* mixture water–dodecane–pentanol–SDS [44-47]. The six critical samples prepared are defined by $X = 1.03; 1.207; 1.372; 1.55; 3.45$ and 5.17. Both ξ and $I(0)$ increase as the critical point is approached. Figures 11.11a, 11.11b show the log–log plots of $I(0)$ and ξ versus the reduced temperature $\varepsilon_T = (T_c - T)/T_c$. $I(0)$ and ξ are found to follow power laws:

$$I(0) = I_0 \varepsilon^{-\gamma} \text{ and } \xi = \xi_0 \varepsilon^{\nu} \qquad (11.12)$$

For a given critical point, the values of ν and γ obtained as described above are very close to those found in independent experiments by varying X at constant temperature. These values are found to vary continuously from the Ising values to significantly smaller ones as the critical end point P_c^E is approached. For the six critical points studied the relation $\gamma = 2\nu$ is verified. The sharp decrease of the values of the exponents in the vicinity of P_c^E is accompanied by a rapid increase of the correlation amplitude $\xi_0 (30 - 300 \text{Å})$ (Fig. 11.12). This behavior is similar to that described earlier for nonionic aqueous micellar solutions as the surfactant is changed.

In view of the anomalous critical behavior of the correlation length and the osmotic compressibility, it appeared of interest to characterize the behavior of other properties. Recently Gazeau et al have investigated how the interfacial tension between the coexisting phases on the one hand and the difference of density of these phases on the other hand, vanish at various points of the critical line P_c^1 [114]. The aim of the experiments was to determine the associated critical exponents μ and β and to check if the scaling

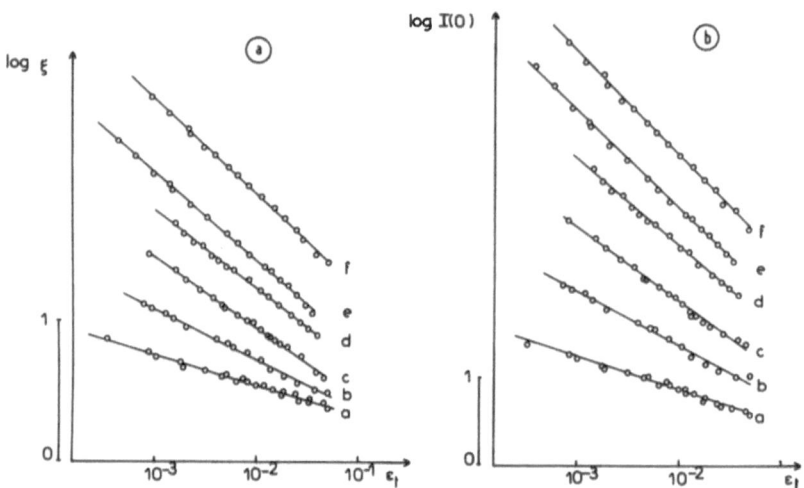

FIGURE 11.11. Quaternary mixture H_2O-dodecane-pentanol-SDS. Log-log plots of the correlation length ξ and the total intensity $I(0)$ vs. the reduced temperature for the six following critical mixtures. a : $X_c = 1.034$; b : $X_c = 1.207$; c : $X_c = 1.372$; d : $X_c = 1.552$; e : $X_c = 3.448$; f : $X_c = 5.172$. Each critical point has been approached in the single phase region (from [45]).

laws which relate ν, γ, β, and μ are valid all along the critical lines. Measurements were performed for the critical points defined by $X = 1.55$ and $X = 1.207$. The results indicate that the values of the critical exponents β and μ show a X dependence similar to that found for ν and γ. Furthermore, within the experimental accuracy, the obtained values of ν, γ, β and μ satisfy with reasonable agreement the theoretical predictions $\mu + \nu = \gamma + 2\beta = 3\nu$ (Table 11.6).

Table 11.6. Values of the critical exponents γ, ν, β, μ, measured for two critical samples defined by $X = 1.55$ and $X = 1.207$.

	γ	ν	β	μ
Ising	1.240±0.002	0.630±0.01	0.325±0.01	1.261±0.004
X=1.55	1.15±0.05	0.58±0.05	0.28±0.03	1.11±0.1
X=1.207	0.77±0.03	0.43±0.03	0.25±0.03	0.94±0.03

Other properties of these critical mixtures have also been investigated.

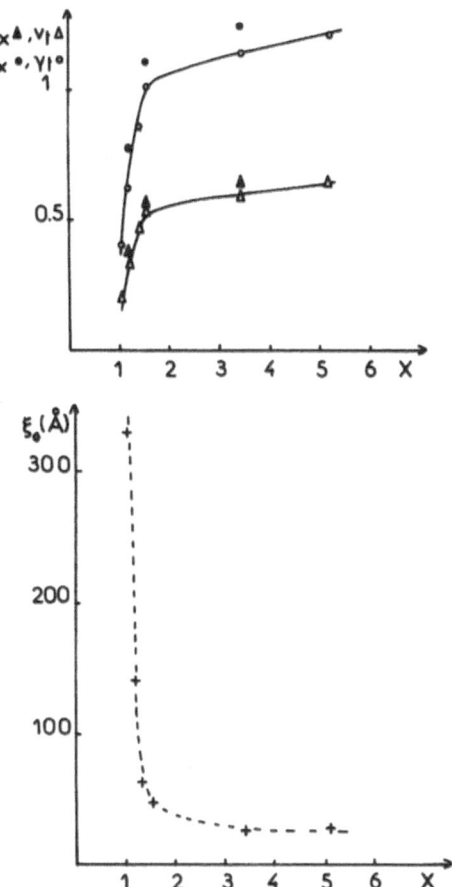

FIGURE 11.12. Variation of the critical exponents $\nu_t, \gamma_t, \gamma_x$ and of the scale factor ξ_o vs X (from [45]).

546   Anne-Marie Bellocq

In particular, the refractive index [116], the thermal conductivity Δ_{th} and the thermal diffusivity D_{th} [117] have been measured. Near the critical end point P_c^E, all these properties exhibit an anomalous behavior as the temperature approaches the critical temperature.

Since the smallest values of the exponents were measured near the critical end point P_c^E, new measurements have been performed in the plane X = 1.034 at three critical temperatures [118]. The new data suggest that the anomalous behavior can be interpreted as a crossover between two critical phenomena. This crossover results from the existence of a low temperature critical point P_c^2 in addition to the high temperature consolute critical point P_c^1 primarily considered as the one responsible for the critical behavior. Experimentally, the critical samples undergo a phase separation by raising temperature at T_c but also by lowering temperature at T_d. Both correlation length ξ and turbidity increase close to T_c and T_d and remain high in the interval T_c-T_d. This behavior suggests the existence of a critical point at low temperature. The high temperature critical point P_c^1 corresponds to a phase separation between two inverted micellar solutions L_2 and L_2' driven by intermicellar interactions; the low temperature one P_c^2 is related to the transition from the micellar phase L_2 to the sponge-like bicontinuous phase L_2^*. The structure of the L_2^* phase has been recently elucidated: it consists of a random bilayer continuous surface separating two oil–continuous regions [119]. Recent theories predict that the transition from the droplet structure to the sponge structure can be a second or a first order phase transition [120–121].

In order to account for the competition between two critical behaviors, light scattering data have been fitted assuming that the behavior at P_c^1 is Ising–like and that near P_c^2 it is mean-field-like [122]:

$$\xi_{eff}^2 = (\xi_1 \varepsilon_1^{-0.63})^2 + (\xi_2 \varepsilon_2^{-0.5})^{-2} \tag{11.13}$$

$$\chi_{eff} = \chi_1 \varepsilon_1^{-1.24} + \chi_2 \varepsilon_2^{-1} \tag{11.14}$$

where

$$\varepsilon_{1,2} = (T_c^{1,2} - T)/T_c^{1,2} \tag{11.15}$$

and $T_c^1, \xi_1, \chi_1, T_c^2, \xi_2, \chi_2$ are respectively critical temperatures and prefactors associated with the two critical points P_c^1 and P_c^2. Turbidity data have been fitted assuming [123]

$$\tau = \kappa \chi_{eff} \frac{2\alpha^2 + 2\alpha + 1}{\alpha^3} ln(1 + 2\alpha) - \frac{2(1 + \alpha)}{\alpha^2} \tag{11.16}$$

with

$$\kappa = \frac{\pi^2}{\lambda^4} (\rho \frac{\partial \varepsilon}{\partial \rho})_T^2 k_B T \text{ and } \alpha = 2(q_0 \xi_{eff})^2 \tag{11.17}$$

where λ is the vacuum wavelength and ε the dielectric constant of the medium. The solid lines of Figs. 11.13 and 11.14 are the best fits obtained.

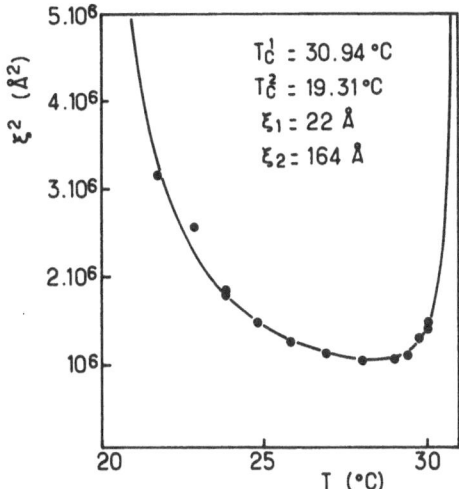

FIGURE 11.13. Temperature dependence of the square of the correlation length for a critical quaternary mixture H_2O-dodecane-pentanol-SDS defined by X = 1.034, $T_c^1 = 30.94°$ C. The full line is the fit according to Eq. (1a) from [118].

In the fits, T_c^1 (which is the only experimentally determined parameter) is fixed, while T_c^2, ξ_i and χ_i are free parameters. The amplitude $\xi_1(20Å)$ associated with the points of P_c^1 is in agreement with well known values for such micellar systems. ξ_2 is found to be definitively larger than ξ_1. Its value is of the order of the characteristic size of the L_2^* phase [119]. According to Eq. (11.13) with $\xi_2 >> \xi_1$, the size of the asymptotic critical domain (where the behavior characteristic of the point P_c^1 is expected) is very small and is dependent on the difference $(T_c^1 - T_c^2)$. In the cases investigated, Ising values are expected for $\xi < 10^{-4}$. From Eq. (11.13) it also appears that the competition between the two critical phenomena produces within some accuracy apparent power laws in the ξ range $10^{-3} - 10^{-1}$. This explains why the earlier measurements performed in this ξ range have led to effective exponents. For sections of the phase diagram corresponding to larger X values the temperature distance between T_c^1 and T_c^2 is more important. Thus the crossover is less and less pronounced and effective exponents are more and more Ising-like.

From the paper of Gazeau et al [118], it is apparent that a precise knowledge of the phase behavior of these sytems and its evolution with temperature is a necessary fundamental basis for understanding the light scattering data. Their interpretation is based on theoretical arguments which state that the "sponge" phase has a particular symmetry resulting from the equivalence of the two continuous media which form it. The "sponge"

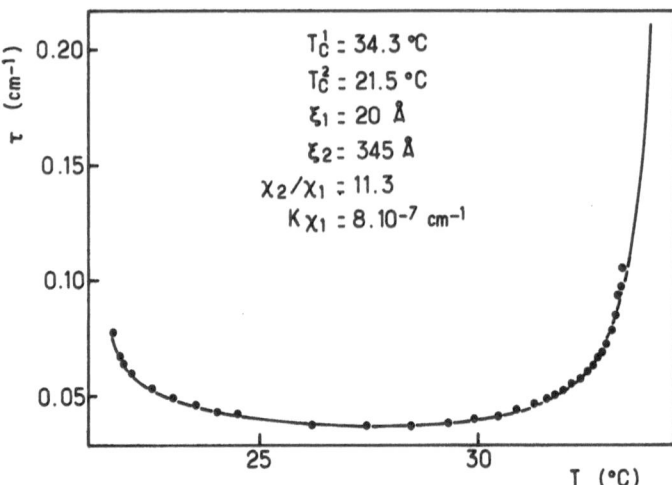

FIGURE 11.14. Turbidity versus temperature for a critical quaternary mixture H_2O-dodecane-pentanol-SDS defined by $X = 1.034, T_c^1 = 34.3°$ C. The full line is the fit according to Eq. (2) from [118].

to micellar transition breaks this symmetry and corresponds to a second order phase transition. Recent experimental work allows one to check this statement [124].

Investigations of the critical behavior of the so–called Winsor equilibria have been performed through studies of both their bulk and interfacial properties. Winsor equilibria are two– or three–phase systems where a microemulsion coexists with either an organic phase (Winsor I) or an aqueous phase (Winsor II) or simultaneously with both phases (Winsor III). The interfacial tension γ is expected to vanish at the critical point as $\gamma = \gamma_0 \xi^\mu$ where γ_0 is a scale factor and μ a critical exponent (Table 11.1). As previously mentioned, theory predicts relations between certain exponents and also that some combinations of scale factors are universal. In 3D Ising models one expects [125]

$$\mu = 4\beta \qquad (11.18)$$

and:

$$\frac{\gamma_0 \xi_0^{+2}}{kT_c} = 0.20 \qquad (11.19)$$

Experiments on Winsor equilibria lead to values of the exponents μ and β in agreement with the above relation but to a product $\gamma_0 \xi_0^2 (\approx 0.012 \text{ to } 0.025kT)$ which is much smaller than the theoretical value [126]. The origin of this discrepancy might be due to the fact that the critical points are not approached close enough.

In summary, most of the experimental results obtained so far on the consolute critical points of micellar solutions and microemulsions indicate that they belong to the 3D-Ising universality class. In particular it has been found that the values of the critical exponents ν and γ describing the divergences of the correlation length and the osmotic susceptibility, respectively, are in good agreement with those of simple fluids and binary mixtures of small molecules. Although the critical behavior in surfactant solutions is similar to that found in mixtures of simple molecules one important difference has been noticed. It concerns the order of magnitude of the amplitudes. Indeed the scale factors which control the amplitudes of the divergences are not universal. In the microemulsion systems the scaling factor for the interfacial tension is unusually small, over 100 times smaller than the corresponding numbers for simple binary mixtures [36]. In contrast, the amplitude of the correlation is typically one order of magnitude *larger* than in mixtures of small molecules. These results are consistent with the tensioactive properties of the surfactant on one hand and with the fact that the phase separation is due to interactions between large aggregates. Because of the two-scale factor universality principle, these unusual values of the scaling factors have several important consequences. Specially, the large value of ξ_0 means that some phenomena could not be observed while others will be (on the contrary) more easily measurable. For example the magnitude of the critical heat capacity in microemulsions should be smaller by a factor of several thousands than those observed in binary mixtures since $C_p^0 \xi_0^3$ =constant. Such a heat capacity would be too small to be detected. This explains why no critical singularity of C_p is observed in microemulsions [117]. In contrast, the large value of ξ_0 found in microemulsions has allowed one to follow the spinodal decomposition of these media by light scattering [127], something which is essentially impossible in binary mixtures. Likewise, giant optical non linearities were observed in critical microemulsions [128]. Two mechanisms, electrostriction and thermodiffusion have been shown to be at the origin of these non linearities which are known to become progressively larger as the medium is more compressible [129].

11.4   Conclusion

Important progress has been made during the last few years in understanding the critical phenomena of micellar and microemulsion systems. It is clear now that in these mixtures the phase separation between two micellar solutions is related to the existence of attractive forces between aggregates. The origin of these interactions depends on the system. More specifically, in nonionic amphiphile solutions the increase of Van der Waals attractions, as temperature is raised, results from a lowering of the degree of structuring of water molecules near the micellar surface. For water-in-oil microemulsions, the attraction is due to the mutual interpenetration of the surfactant tails and to the bad quality of the solvent. In these mixtures, in contrast to mixtures of simple liquids, interactions are not only a function of temperature but are also dependent on the various parameters which may affect the interpenetration of the surfactant layers. Thus, interactions and critical phenomena may be controlled by micellar size, number of carbons of the oil and surfactant molecules, or the alcohol chain length.

Except for some data on nonionic surfactant solutions, all the experimental results obtained on the critical behavior in micellar and microemulsions systems are sucessfully interpreted by the theory for simple molecules. The comparison of experimental results on the exponents shows a generally good agreement with the predictions of the renormalization group method. In particular the values of the critical exponents ν and γ are very close to those measured for small molecules. However due to the low experimental accuracy it cannot be concluded whether microemulsions behave as a pseudobinary mixture or a multicomponent mixture. It is very important to note that the amplitudes of the numerical scale factors which depend on structure are significantly different from those of small molecules. In particular ξ_0 is one order of magnitude larger than in usual mixtures. This is consistent with the fact that the interacting units are large aggregates. An analogous situation is found in critical polymer solutions [123]. Although up to now little consideration has been given to the amplitude ratios, one expects that the large value of ξ_0 has interesting consequences. This should lead to exceptionally low or high values for some thermodynamical constants and offer opportunities for studying phenomena not accessible in usual systems.

It is worthwhile to point out also that the precise knowledge of the phase diagrams has allowed us to understand the origin of the anomalous behaviors previously reported for quaternary and quinary microemulsions. Moreover, these multicomponent mixtures—which are known to present a rich polymorphism—are expected to exhibit a complex phase behavior and to offer an interesting area for observing a large variety of phase transitions. Future studies should focus, then, on the transition between micellar and sponge–like bicontinuous structures.

Acknowledgements

A part of the work described in this paper has been performed in collaboration with several scientists at the Centre de Recherche Paul Pascal. I would like to specially thank D. Roux, P. Honorat and D. Gazeau. It is also a pleasure to acknowledge fruitful discussions with the members of the French Greco "Microemulsions."

References

1. H. E. Stanley, *Introduction to Phase Transitions and Critical Phenomena* (Oxford University Press, Oxford,1971)

2. C. Domb and M. S. Green, *Phase Transitions and Critical Phenomena*, Vol. 1-6 (Academic Press, New York, 1972-1977).

3. J. M. H. Levelt Sengers, R. Hocken and J. V. Sengers, Phys. Today **30**, 42 (1977).

4. *Phase Transitions: Cargese 1980*, M. Levy and J. Zinn-Justin, eds. (Plenum Press, New York, 1982).

5. B. Chu, in *Dynamic Light Scatterings*, R. Pecora, ed. (Plenum Press, 1985).

6. S. C. Greer and M. R. Moldover, Ann. Rev. Phys. Chem. **32**, 233 (1981).

7. J. V. Sengers and J. M. H. Levelt Sengers, Ann. Rev. Phys. Chem.**37**, 189 (1986).

8. A. Kumar, H. R. Krishnamurthy and E. S. R. Gopal, Physics Report **98**, 57 (1983).

9. J. M. H. Levelt Senters, Pure Appl. Chem. **55**, 437 (1983).

10. J. M. H. Levelt Sengers, G. Morrison and R. F. Chang, Fluid Phase Equilibria **14**, 19 (1983).

11. D. Beysens and A. Bourgou, Phys. Rev. **A19**, 2407 (1979).

12. D. Beysens, A. Bourgou and P. Calmettes, Phys. Rev. **A26**, 3589 (1982).

13. B. Widom, J. Chem. **46** 3324 (1967).

14. M. E. Fisher, Phys. Rev.**176**, 257 (1971).

15. J. C. Wheeler and B. Widom, J. Phys. Chem. Soc. **90**, 3064 (1968).

16. R. B. Griffiths and J. C. Wheeler, Phys. Rev. **A2**, 1047 (1970).

17. J.A. Zollweg, J. Chem. Phys. **55**, 1430 (1971).

18. K. Obayashi and B. Chu, J. Chem. Phys. **68**, 5066 (1978).

19. L. E. Wold, G. J. Pruit and G. Morrisson, J. Phys. Chem. **77**, 1572 (1973).

20. W. I. Goldburg and P. N. Pusey, J. Phys. **33**, 105 (1972).

21. E. Bloemen, J. Thoen and W. Van Dael, J. Chem. Phys. **75**, 1488 (1987).

22. M. Corti and V. Degiorgio, Opt. Commun. **14**, 158 (1975).

23. M. Corti and V. Degiorgio, Phys. Rev. Lett. **45**, 13, 1045 (1980).

24. M. Corti and V. Degiorgio, J. Phys. Chem. **85**, 1442 (1981).

25. M. Corti, V. Degiorgio and M. Zulauf, Phys. Rev. Lett. **48**, 1617 (1982).

26. N. Corti, C. Minero and V. Degiorgio, J. Phys. Chem. **88**, 309 (1984).

27. V. Degiorgio, in *Physics of Amphiphiles : Micelles, Vesicles and Microemulsions* V. Degiorgio and M. Corti, eds. (North-Holland, Amsterdam, 1985), p. 303.

28. V. Degiorgio, R. Piazza, M. Corti and C. Minero, J. Phys. Chem. **82** 1025 (1985).

29. M. Corti and V. Degiorgio, Phys. Rev. Lett. **55**, 2005 (1985).

30. M. Corti, V. Degiorgi and L. Cantu, in *Physics of Complex and Supermolecular Fluids*, S. A. Safran and N. A. Clark, eds. (Wiley, New York, 1987) p. 463.

31. J. Lang and R. D. Morgan, J. Chem. Phys. **73**, 5849 (1980).

32. K. Hamono, N. Kuwahara, T. Koyama and S. Harada, Phys. Rev. A **32**, 3168 (1985).

33. Y. Dormoy, E. Hirsch, S. J. Candau and R. Zana, Progress in Coll. Polymer Sci. **73**, 81 (1987).

34. R. Strey and A. Pakusch, in *Surfactants in Solution*, K. L. Mittal and P. Bothorel, eds. (Plenum Press, New York, 1987) p. 456.

35. J. Appell and G. Porte, J. Phys. **44**, L689 (1983).

36. O. Abillon, D. Chatenay, d. Langevin and J. Meunier, J. Phys. Lett. **45**, L223 (1984).

37. J. S. Huang and M. W. Kim, Phys. Rev. Lett. **47**, 1462 (1981).

38. R. Dorshow, F. de Buzzaccarini, C. A. Bunton and D. F. Nicoli, Phys. Rev. Lett. **47**, 1336 (1981).

39. G. Fourche, A. M. Bellocq and S. Brunetti, J. Colloid Interface Sci. **89**, 417 (1982).

40. A. M. Bellocq, D. Bourbon and B. Lemanceau, J. Colloid Interface Sci. **79**, 419 (1981).

41. A. M. Bellocq, D. Bourbon, B. Lemanceau and G. Fourche, J. Colloid Interface Sci. **89**, 427 (1982).

42. D. Roux and A. M. Bellocq, Phys. Rev. Lett. **52**, 1895 (1984).

43. P. Honorat, D. Roux and A. M. Bellocq, J. de Physique Lett. **45** L961 (1984).

44. A. M. Bellocq, P. Honorat and D. Roux, J. de Physique **46**, 743 (1985).

45. A. M. Bellocq and D. Roux, in *Microemulsions, Structure and Dynamics*, S. Friberg and P. Bothorel, eds. (CRC Press, 1987) p. 33.

46. D. Roux and A. M. Bellocq, in *Surfactants in Solution*, K. L. Mittal and P. Bothorel, eds. (Plenum, New York, 1987) p. 1247.

47. A. M. Bellocq, P. Honorat and D. Roux, *ibid.* p. 1263.

48. A. M. Cazabat, D. Langevin, J. Meunier and A. Pouchelon, Adv. in Colloid Interface Sci. **16**, 175 (1982).

49. A. M. Cazabat, D. Langevin, J. Meunier and A. Pouchelon, J. de Physique Lett. **43**, L-89 (1982).

50. M. W. Kim and J. S. Huang, Phys. Rev. **B26**, 2703 (1982).

51. J. S. Huang and M. W. Kim, in *Physics of Amphiphiles : Micelles, Vesicles and Microemulsions*, V. Degiorgio and M. Corti, eds. (North-Holland, Amsterdam, 1985) p. 864.

52. W. M. Kim, J. Bock and J. S. Huang, Phys. Rev. Lett. **54**, 46 (1985).

53. S. H. Chen and M. Kotlarchyk, in *Physics of Amphiphiles : Micelles, Visicles and Microemulsions*, V. Degiorgio and M. Corti, eds. (North-Holland, Amsterdam, 1985) p. 768.

54. M. Kotlarchyk and S. H. Chen, J. Phys. Chem. **86** 3273 (1982).

55. M. Kotlarchyk, S. H. Chen, J. S. Huang and M. W. Kim, Phys. Rev. A **29** 2054 (1984).

56. M. Korlarchyk, S. H. Chen and J. S. Huang, Phys Rev. **A28**, 508 (1983).

57. C. Toprakcioglu, J. C. Dore, B. H. Robinson, A. Howe and P. J. Chieux, J. Chem. Soc. Faraday Trans. **80**, 413 (1984).

58. M. Kahlweit and R. Strey, Angew. Chem. **24**, 654 (1985).

59. M. Kahlweit, R. Strey and P. Firman, J. Phys. Chem. **90**, 671 (1986).

60. M. Kahlweit and R. Strey, J. Phys. Chem. **90**, 5239 (1986).

61. M. Kahlweit and R. Strey, J. Phys. Chem. **91**, 1553 (1987).

62. H. Kunieda and T. Arai, Bull. Chem. Soc. Jpn **57**, 281 (1984).

63. H. Kunieda, J. Colloid Interface Sci. **122**, 138 (1988).

64. For a general survey see *Surfactants in Solution*, Vol. 1-3, K. L. Mittal and B. Lindman, eds. (Plenum, 1984), Vol. 4-6, K. L. Mittal and P. Bothorel, eds. (Plenum, 1987).

65. R. R. Balmbra, J. S. Clunie, J. M. Corkill and J. F. Goodman, Trans. Faraday Soc. **58**, 1661 (1962); **60**, 979 (1964).

66. R. H. Ottewill, C. C. Storer and T. Walker, Trans. Faraday Soc. **63**, 2796 (1967).

67. D. Attwood, J. Phys. Chem. **72**, 339 (1968).

68. M. Zulauf and J. P. Rosenbush, J. Phys. Chem. **87**, 856 (1983).

69. J. B. Hayter and M. Zulauf, Colloid Polym. Sci. **260**, 1023 (1982).

70. M. Zulauf, K. Wechstrom, J. B. Hayter, V. Degiorgio and M. Corti, *J. Phys. Chem.* **89**, 3411 (1985).

71. L. Reatto and M. Tau, Chem. Phys. Lett. **108**, 292 (1984).

72. R. Triolo, L. J. Magid, J. S. Johnson and H. R. Child, J. Phys. Chem. **86**, 3689 (1982).

73. D. J. Cebula and R. H. Ottewill, Colloid Polym. Sci. **260**, 260 (1982).

74. J. C. Ravey, Colloid Interface Sci. **94**, 289 (1983).

75. L. J. Magid, R. Triolo and J. S. Johnson, J. Phys. Chem. **88**, 5730 (1984).

76. P. C. Nillsson, H. Wennerstrom and B. Lindman, J. Phys. Chem **87**, 1377 (1983).

77. D. J. Mitchell, G. J. T. Tiddy, L. Waring, T. Bostock and M. P. McDonald, J. Chem. Soc. Faraday Trans. I **79**. 975 (1983).

78. P. G. Neeson, B. R. Jennings and G. J. T. Tiddy, Chem. Phys. Lett. **95**, 533 (1983).

79. R. Zana and C. Weill, J. de Physique Lett. **46**, L954 (1985).

80. W. Binana-Limbelle and R. Zana, J. Colloid Interface Sci.**121**, 81 (1988).

81. D. Roux, A. M. Bellocq and M. S. Leblanc, Chem. Phys. Lett. **94**, 156 (1983).

82. S. A. Safran and L. A. Turkevich, Phys. Rev. Lett. **50**, 1930 (1983).

83. D. Roux and A. M. Bellocq, in *Physics of Amphiphiles : Micelles, Vesicles and Microemulsions*, V. Degiorgio and M. Corti, eds. (North Holland, Amsterdam, 1985) p. 842.

84. D. Roux and A. M. Bellocq, in *Macro- and Microemulsions. Theory and Practice*, E. O. Shah, ed. (ACS Symposium Series **272**, 105 (1985)).

85. D. Roux, A. M. Bellocq and P. Bothorel, in *Surfactants in Solution*, K. L. Mittal and B. Lindman, eds. (Plenum Press, 1984) p. 18443.

86. S. Bunetti, D. Roux, A. M. Bellocq , G. Fourche and P. Bothorel, J. Phys. Chem. **887**, 1023 (1983).

87. A. M. Cazabat and D. Langevin, J. Chem. Phys. **74**, 3148 (1981).

88. B. Lemaire, P. Bothorel and D. Roux, J. Phys. Chem. **87**, 1023 (1983).

89. A. M. Bellocq in *Physics of Complex and Supermolecular Fluids*, A. S. Safran and N. A. Clark, eds. (Wiley, New York, 1987).

90. A. M. Bellocq, J. Biais, P. Bothorel, B. Clin, G. Fourche, P. Lalanne, B. Lemaire, B. Lemanceau and D. Roux, Adv. Colloid Interface Sci. **20**, 167 (1984).

91. A. A. Calje, W. G. M. Agterof and A. Vrij, in *Micellization, Solubilization, Microemulsions* K. L. Mittal, ed. (Plenum Press, New York, 1976) p. 779.

92. M. Dvolaitzky, M. Guyot, M. Lagues, J. P. Le Pesant, R. Ober, C. Sauterey and C. Taupin, J. Chem. Phys. **69**, 3279 (1978).

93. R. Ober and C. Taupin, J. Phys. Chem. **84**, 2418 (1980).

94. A. M. Cazabat, in *Colloides et Interfaces* (Editions de Physique, 1984) p. 323.

95. M. Lages, R. Ober and C. Taupin, J. de Physique Lett. **39**, L487 (1987).

96. B. Lagourette, J. Peyrelasse, C. Boned and M. Clausse, Nature **281**, 60 (1979).

97. A. M. Cazabat, D. Langevin, J. Meuiner, O. Abillon and D. Chatenau, in *Macro- and Microemulsions* (ACS Symposium Series, 1985) p. 75.

98. P. Guering and A. M. Cazabat, J. de Physique Lett. **44**, L-601 (1983).

99. R. Zana, J. Lang, O. Sorba, A. M. Cazabat and D. Langevin, J. de Physique Lett. **43**, L-829 (1982).

100. D. Chatenay, W. Urbach, A. M. Cazabat and D. Langevin, Phys. Rev. Lett. **54**, 2253 (1985).

101. A. M. Cazabat, D. Chatenay, P. Guering and W. Urbach, *Physics of Complex and Supermoleculalr Fluids*, S. A. Safran and N. A. Clark, eds. (Wiley, New York, 1987) p. 585.

102. J. Lang, A. Malliaris and A. Jada, J. Phys. Chem. **92**, 1946 (1988).

103. S. H. Chen, T. L. Lin and J. S. Huang, in *Physics of Complex and Supermolecular Fluids*, S. A. Safran and N, A. Clark, eds. (Wiley, New York, 1987) p. 285.

104. J. Tabony, M. Drifford and A. de Geyer, Chem. Phys. Lett. **96**, 119 (1983).

105. C. M. Sorensen, Chem. Phys. Lett. **117**, 606 (1985).

106. K. Kawasaki, Ann. Phys.(N.Y.) **61**, 1 (1970).

107. Y. Shnidman, Phys. Rev. Lett. **56**, 2546 (1986).

108. M. E. Fisher, Phys. Rev. Lett. **57**, 1911 (1986).

109. G. Caflisch, M. Kaufman and J. R. Banavar, Phys. Rev. Lett. **56**, 2545 (1985).

110. C. Bagnuls and C. Bervillier, Phys. Rev. Lett. **58**, 435 (1987).

111. G. Dietler and D. S. Cannell, Phys. Rev. Lett. **60**, 1852 (1988).

112. J. P. Wilcoxon and E. W. Kaler, J. Chem. Phys. **86**, 4684 (1987).

113. J. S. Huang and M. W. Kim, in *Physics of Amphiphiles : Micelles, Vesicles and Microemulsions* V. Degiorgio and M. Corti, eds. (North Holland, Amsterdam, 1985) p.846.

114. D. Gazeau, Thesis, Bordeaux 1989, and A. M. Bellocq and D. Gazeau, J. Phys. Chem. **94**, 8933 (1990).

115. D. Roux and A. M Bellocq, in *Statistical Thermodynamics of Micelles and Microemulsions*, S. H. Chen and R. Rajagopalan, eds. (Springer, Heidelberg, 1990).

116. N. Rebbouh, J. Buchert and J. R. Lalanne, Europhys. Lett. **4**, 447 (1987).

117. P. Dorion, J. R. Lalanne, B. Pouligny, S. Imaizumi and C. W. Garland, J. Chem. Phys. **87**, 578 (1987).

118. D. Gazeau, E. Freysz and A. M. Bellocq, Europhys. Lett. **9**, 833 (1989).

119. D. Gazeau, A. M. Bellocq, D. Roux and T. Zemb, Europhys. Lett. **9**, 447 (1989).

120. D. A. Huse and S. Leibler, J. de Physique **49**, 605 (1988).

121. M. E. Cates, D. Roux, D. Andelman, S. T. Milner and S. A. Safran, Europhys. Lett. **5**, 733 (1988).

122. It has been considered that the system can be described by two different order parameters, weakly coupled.

123. V. G. Puglieli and N. C. Ford, Phys. Rev. Lett. **25**, 143 (1970).

124. C. Coulon, D. Roux and A. M. Bellocq, Phys. Rev. Lett. **66**, 1709 (1991).

125. H. Chaar, M. R. Moldover and J. W. Schmidt, J. Chem. Phys. **85**, 418 (1986).

126. D. Langevin in *Microemulsions : Structure and Dynamics*, S. E. Friberg and P. Bothorel, eds. (CRC Press, 1987) p. 173.

127. D. Roux, J. de Physique Paris, **47**, 733 (1986).

128. E. Redysz, W. Claeys, A. Ducasse and B. Pouligny, IEEE J. of Qu. Electronics **QE-22**, 1258 (1986).

129. B. Jean-Jean, E. Freysz, A. Ducasse and B. Pouligny, Europhys. Lett. **7**, 219 (1988).

12

Structures and Phase Transitions in Langmuir Monolayers

David Andelman[1]
Francoise Brochard[2]
Charles Knobler[3]
Francis Rondelez[4]

12.1 Introduction

Throughout the previous chapters, attention has been focused on the wide variety of structures and phase behaviors characterizing the self-assembly of amphiphilic molecules in aqueous solution (Chapters 1 to 6) and in oil/water mixtures (Chapters 6 to 11). In all cases the organization of amphiphilic (surfactant) molecules has involved one or more *three*-dimensional aggregates, *i.e.*, globular or cylindrical micelles, vesicles or undulating bilayers, and droplets or bicontinuous phases of microemulsions. As a consequence, elastic (curvature) free energy has played a crucial role in determining the relative stabilities of competing geometries, and the associated phase transitions have been naturally compared and contrasted with those familiar from the usual fluids, liquid crystals, and solids *in bulk*.

In the present, concluding, chapter we turn to an important class of *two*-dimensional (2D) phenomena, in particular those arising in the study of adsorbed amphiphilic monolayers on a liquid subphase. It has long been known that many amphiphilic molecules form monomolecular layers at the interface between water and air. If the amphiphile is only very slightly soluble in the bulk, it can be treated as a separate surface phase; such an

[1]Department of Physics, Tel Aviv University, Ramat Aviv, Tel Aviv, 69978 Israel

[2]Laboratoire de Physico-Chimie des Surfaces et Interfaces (CNRS-URA 1379), Institut Curie—Section de Physique et Chimie, 11 rue Pierre et Marie Curie, Paris 75231, Cedex 05, France

[3]Department of Chemistry and Biochemistry, University of California Los Angeles, Los Angeles, California 90024, USA

[4]Same as footnote 2

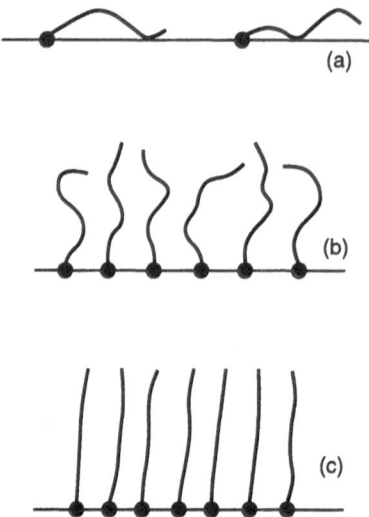

FIGURE 12.1. Schematic representation of the microscopic structure of a monolayer in different states. The dots represent the hydrophilic head and the wavy
lines the hydrophobic tails of the amphiphiles. (a) Gas: the molecules are widely
separated and the tails are in contact with the water surface; (b) Intermediate
density: the tails are upright but disordered; (c) Condensed state: the tails are
almost entirely in *trans* conformations and closely packed.

insoluble film is called a Langmuir monolayer. Langmuir monolayers can
be formed when there is the right balance between the solubility of the
hydrophilic head group of the amphiphile and the hydrophobic nature of
its tail. For substances like fatty acids, this balance can be achieved for
chain lengths that fall between 12 and 30 carbons. Insoluble monolayers of
phospholipids and simple esters are also easily formed.

The microscopic nature of a Langmuir monolayer is suggested by the
cartoons shown in Fig. 12.1. The head groups are immersed in the water
surface and the tails remain out of the water. At low density the tails
are likely to be disordered and will have conformations that bring them
in contact with the water surface. (Recall that hydrophobicity does not
imply a repulsive interaction between the tail and the water surface.) If the
monolayer density is increased by compression at constant temperature, the
chains begin to interact and are more likely to be found above the water
surface. As the packing becomes tighter the chains are increasingly in the
trans conformation (see Sec. 1.3.3). At the highest densities the chains and
head groups are completely ordered and the packing is similar to that found
in the three-dimensional solid.

The complex interplay between energetic and entropic factors in monolayers is similar to that found in other amphiphilic systems. What is new

here is the possibility of continuously changing the area per amphiphile, the constraint brought about by the pinning of the head groups to the surface and the interaction of both the chains and the head groups with the subphase. A further complication arises because the amphiphiles are necessarily dipolar. Thus, when the density is sufficiently high to orient the chains, the molecular dipoles are aligned and repel each other; if the chains are tilted with respect to the surface there will be both in-plane and out-of-plane dipolar interactions.

While the structures of Langmuir monolayers are far from being completely understood, there have been many recent experimental and theoretical advances, which are the subject of this chapter. We will begin (Sec. 12.2.1) with a review of results from what can be called classical studies, work that is not necessarily old but which has been carried out with the techniques that were developed by the early workers in the field. A variety of new experimental methods has led to a revitalization of monolayer research; these will be described briefly in Sec. 12.2.2. A critique of the current experimental situation then follows in Sec. 12.2.3. In Sec. 12.3 we discuss first (12.3.1) the equilibrium theories of successive fluid-fluid phase transitions and chain conformational statistics, and then the driving forces for spatially modulated states in Langmuir monolayers (12.3.2-12.3.3). *Dynamical* features of these systems are treated in Sec. 12.4, followed by a concluding discussion in 12.5.

12.2   The Experimental Situation

12.2.1   CLASSICAL STUDIES

Langmuir monolayers can be prepared by depositing a small amount of a solution of an amphiphile in a volatile solvent onto a clean water surface; the monolayer forms spontaneously as the solvent spreads and evaporates. The area available to the monolayer is controlled by a barrier that can be slid across the surface, as shown in Fig. 12.2. Detailed descriptions of the experimental methods outlined below are found in the books by Gaines [1] and Adamson [2].

The thermodynamic state of the monolayer can be described by an equation of state $\Pi = \Pi(a, T)$ where Π is a two-dimensional pressure, T is the temperature, and a is the area per molecule. The surface pressure is defined as the difference between γ_o, the surface tension of pure water and γ the surface tension in the presence of the monolayer (see Chap. 10):

$$\Pi = \gamma_o - \gamma \qquad (12.1)$$

In writing $\Pi(a, T)$ we have assumed that the properties of the subphase have been held constant. The properties of the monolayer, however, may be sensitive to the pH or ionic strength of the subphase or to the presence in the bulk of specific solutes such as divalent ions.

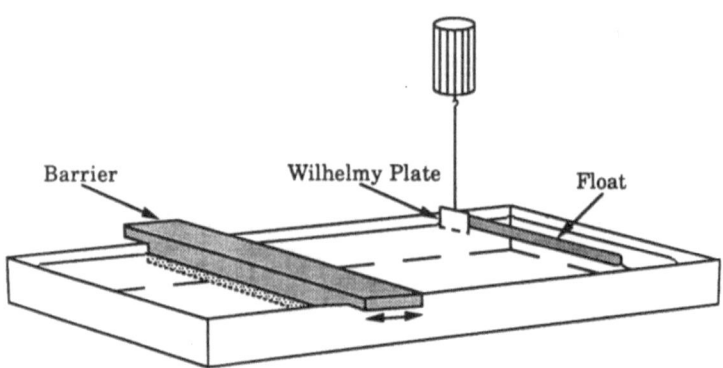

FIGURE 12.2. Schematic diagram of a Langmuir trough. The trough is typically constructed of a material such as Teflon that is not wetted by water; the water surface therefore rises above the edge. Monolayers can be compressed by sliding the barrier across the surface. The pressure can be determined by measuring the force on a plate (Wilhelmy plate) that passes through the surface or on a barrier that separates the monolayer-covered surface from pure water.

The most common measurement performed on Langmuir monolayers is the determination of surface pressure–area isotherms, and much of the information concerning the phase behavior of monolayers has been deduced from such studies. The schematic isotherm shown in Fig. 12.3 is typical of those observed for many pure amphiphiles. As in the case of $P - V$ isotherms, first-order transitions should be marked by horizontal portions of the isotherm and second-order transitions by changes in slope. Thus, for example, the long plateau at low pressure arises from a first-order transition between a gaseous phase and a fluid phase.

Unfortunately, the interpretation of isotherms is often not so straightforward because the slopes in first-order transition regions are rarely truly horizontal. Further, transitions between condensed phases may be accompanied by only very small changes in density so that the isotherms exhibit kinks rather than plateaus. Phase diagrams have therefore been established only by combining isotherm studies with measurements of other properties.

The character of a phase can be assessed, for example, from measurements of the surface potential ΔV, which can be related to p_z, the component of the dipole moment density of the film perpendicular to the interface, by the relation

$$\Delta V = p_z/\varepsilon^* \qquad (12.2)$$

where ε^* is the local dielectric constant seen by the dipoles. The observed potentials can be compared to those predicted for models of the monolayer in which the charge distribution along the amphiphile is estimated from

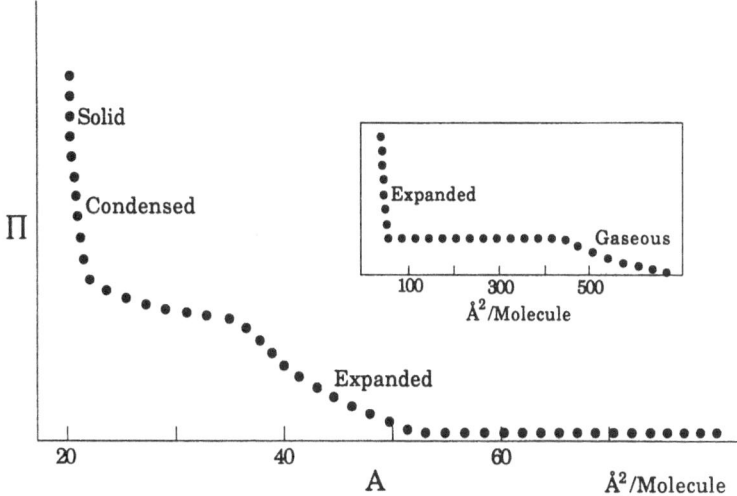

FIGURE 12.3. Schematic $\Pi - a$ isotherm. The long plateau corresponds to the G-LE two-phase region. The short plateau, which corresponds to the LE-LC two-phase region, is horizontal only in measurements made with extremely pure materials; it is not observed in all monolayers. The isotherm is simplified; other more subtle features may also be observed in precise measurements.

its chemical structure and for which a conformation and orientation are assumed. The interpretation of such data is not without its subtleties, but useful information can often be obtained without recourse to a model. For example, temporal variations in the surface potential indicate a first-order phase transition and the coexistence of domains of different densities within the monolayer. The fluctuations arise when domains move into or out of the region beneath the measuring electrode.

Although the location of phase boundaries and the characterization of the phases by these classical methods have often been ambiguous, there has been general (but not unanimous) agreement about the existence of at least four monolayer phases that are associated with easily detectable features in isotherms: (1) Gas (G); (2) Liquid [usually called *liquid expanded* (LE)]; (3) Solid (S); and (4) Another condensed phase that intrudes between the LE and S phases and is usually called *liquid condensed* (LC). However, other features such as kinks and changes in slope are often observed in isotherms. If each of these is taken as evidence of a phase transition, many more phases must exist.

Rather complex phase diagrams, similar to the general $\Pi - T$ diagram shown in Fig. 12.4, were constructed by Stenhagen [3,4] and Lundquist [5,6] from their detailed $\Pi - a$ measurements on fatty acids and fatty esters, respectively. Their proposals were looked upon with skepticism because of the difficulties in determining subtle features in isotherms, but recent

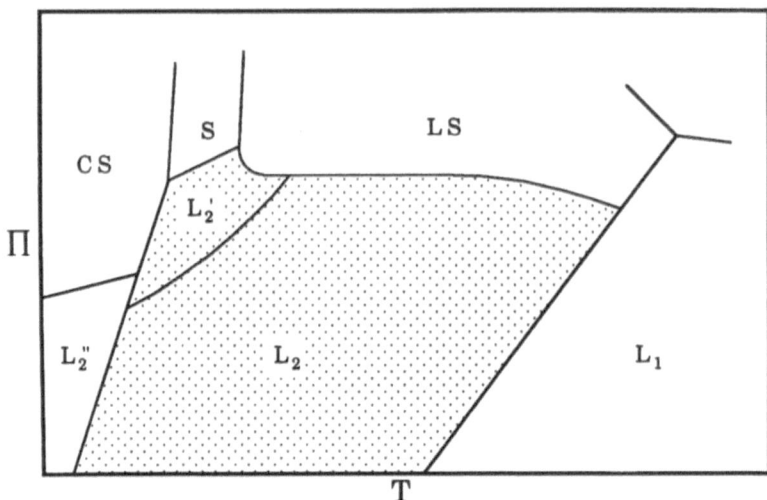

FIGURE 12.4. Generalized $\Pi - T$ diagram. The diagram is similar to the ones proposed by Stenhagen [3,4] and is applicable to long-chain fatty acids, esters, and alcohols. Adapted from Bibo *et al* [55].

studies by a variety of new methods have shown that many condensed phases *do* exist and that the Stenhagen and Lundquist diagrams are largely correct – see Section 12.2.3.1 below.

Much of the current research on Langmuir monolayers is directed toward determining the structures of the phases, understanding the relations between them and explaining them in terms of the molecular properties of the amphiphiles. These questions can be addressed only with great difficulty by the classical methods for studying monolayers, and a variety of new techniques have therefore been employed. We will give a brief overview of these methods in the next section; a more detailed discussion can be found in other reviews [7,8].

12.2.2   NEW EXPERIMENTAL METHODS

12.2.2.1   Fluorescence microscopy

The textures of Langmuir monolayers can be observed directly with the technique of fluorescence microscopy [9-11]. The amphiphile is mixed with a small amount ($< 1\ mol\%$) of a fluorescent probe, which is usually itself an amphiphile to which a chromophore has been chemically bonded. The monolayer is then prepared in a trough mounted on the stage of an optical microscope. The fluorescence is excited by illuminating the surface, generally from above, with light from a mercury arc or laser. The fluorescence image is then observed through the microscope; it is weak and is therefore usually detected with a high-sensitivity television camera and viewed on a monitor or recorded on videotape for subsequent analysis.

Figure 12.5 shows fluorescence images for monolayers of pentadecanoic acid at various stages of compression along a 25°C isotherm [12]. Midway through the G-LE plateau region, one sees dark circular bubbles of the G phase surrounded by the bright LE phase. The contrast between the phases is caused by the difference in density and also by a quenching of the fluorescence in the G phase. The proportion of light and dark regions varies with density according to the "lever rule"; at one phase boundary the image is completely dark while at the other it is completely bright.

The image remains uniformly bright as the monolayer is compressed through the LE one-phase region. Dark circular domains of the LC phase appear at the LE-LC phase boundary; the LC phase appears dark because the probe is only slightly soluble in it. As the compression proceeds in this two-phase region, the image becomes increasingly black as LE phase is converted into LC phase. If the probe concentration is sufficiently low, one can observe the complete transformation to the LC phase.

In some cases information about the tilt of amphiphile tails can be gained from fluorescence microscopy [13]. By tilt we mean a preferred orientation (direction) for the projection of the molecular axes onto the subphase surface. In such experiments a polarized laser that strikes the monolayer obliquely is used for excitation. If the electric field of the laser is properly oriented with respect to the surface normal, there can be a contrast in fluorescence between regions of differing tilt. This method has recently been used to examine domain structures in ester monolayers [14].

12.2.2.2 X-ray and neutron reflectometry and scattering

Monolayers have been investigated both by x-ray [15] and neutron [16] scattering. X-ray reflectivity measurements can be performed with conventional sources, but all of the X-ray diffraction studies have been carried out with synchrotron sources, which have very high intensity. The ease with which the neutron scattering cross section can be varied by isotopic substitution makes experiments with the much lower intensity neutron sources nevertheless attractive [17].

The information that can be obtained from scattering measurements on monolayers depends on the scattering vector Q, which is fixed by the source and detector geometry. Reflectance measurements are carried out by keeping Q perpendicular to the surface. In this geometry the variations of the scattered intensity with wave vector provide information about the density profile of the monolayer normal to the surface. Such measurements are analyzed by comparing the experimental scattering with that predicted for slab models in which the monolayer is envisioned as a series of layers of different thicknesses and scattering density, each slab representing some portion of the surface (subphase, head, tail). The scattering densities are computed

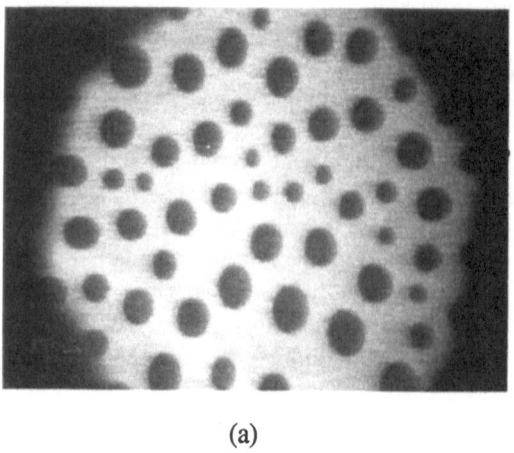

(a)

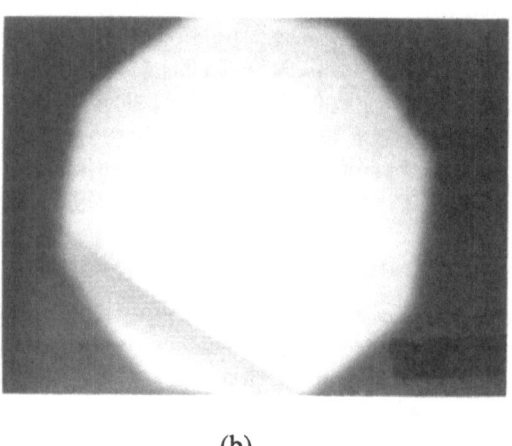

(b)

FIGURE 12.5. Fluorescence microscope images of a pentadecanoic acid mono-layer at various stages of compression along a 25°C isotherm. (a) G-LE coexis-tence at 61 Å^2/molecule; (b) LE phase at 37 Å^2 /molecule.

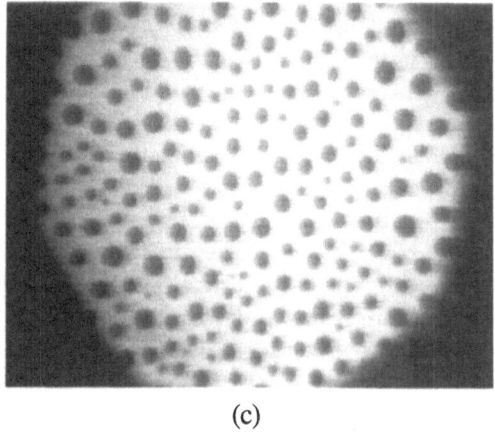

(c)

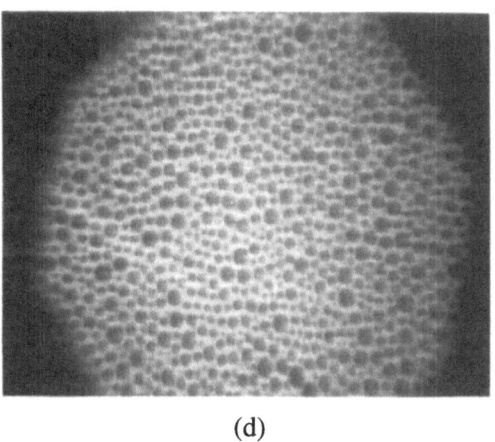

(d)

FIGURE 12.5. (c) LE-LC coexistence at 27Å^2/molecule; (d) LE-LC coexistence at 24Å^2/molecule.

from the molecular structures, and the thicknesses are determined by fitting the data; parameters are also included to account for surface roughness.

In the in-plane geometry, one studies the scattering parallel to the surface plane as a function of the scattering angle. This diffraction pattern is characteristic of the structure of the monolayer projected onto this plane. Because the source generally illuminates about $1\,\text{cm}^2$ of the surface, an area that includes many monolayer domains, the resulting pattern is powder averaged and does not show the isolated peaks characteristic of single crystals. The spacing of the diffraction planes can be calculated from the peak positions and the range of the translational correlations can be obtained from the peak widths. Measurements of the intensity of a Bragg reflection along a surface normal, "rod scans" [18], allow direct information to be obtained about chain tilt.

These diffraction experiments are always difficult to perform because of the small scattered intensities Even with the most powerful synchrotron sources, they require long-time exposures and care has to be exercised to maintain the monolayer chemical integrity under the X-ray irradiation and not to lose molecules by dissolution in the subphase. This explains why such experiments have generally been confined to the longest fatty acids (with eighteen carbon atoms or more) or phospholipids and at high surface densities. One notable exception is the study of fluorinated fatty acids with only twelve carbon atoms, and over a range of surface densities between 30 and 400Å^2 per molecule [19]. Advantage is taken of the larger scattering power of fluorine atoms compared to hydrogen, and of the rod-like conformation of the CF_2 chains.

12.2.2.3 Spectroscopic measurements

The monolayer represents an extremely small part of the entire experimental system: monolayer + substrate, so the most promising spectroscopic methods are those which somehow discriminate against bulk phases. One such technique is based on the nonlinear process of optical second harmonic generation (SHG). When an intense laser beam interacts with a medium that is not centrosymmetric, a second harmonic of the laser frequency is produced. Interfaces between two uniform media lack inversion symmetry and can therefore be used to generate second harmonics.

If the signal comes mainly from the monolayer, the nonlinear susceptibility that is responsible for the generation of the second harmonic is proportional to the product of the monolayer surface density and nonlinear polarizability averaged over the distribution of molecular orientations [20,21]. In general, the relation between the molecular orientation and the signal is complex, but with some simplifying assumptions it is possible to determine the average value of the polar angle between the surface normal and the axis of the part of the molecule whose nonlinear susceptibility is involved. In most cases this SHG-active part is localized in the *head group*, so one must be careful about inferring from these analyses the average orientation of the molecular *tails*; this point is especially problematic when the chain is flexible (as at larger areas in the LE phase).

In sum-frequency generation (SFG) spectroscopy [22], another nonlinear technique that can provide surface-specific information, light from a laser operating in the visible region is mixed on the surface with light from a laser operating in the infrared. When the frequency of the infrared laser matches an absorption frequency of the monolayer, there is a resonance enhancement of the signal. The technique therefore allows the infrared spectrum of the monolayer to be investigated. Direct measurement of the infrared spectrum of monolayers by external reflection infrared spectroscopy has also been carried out [23].

12.2.2.4   Langmuir-Blodgett films

A monolayer can be transferred to the surface of a solid substrate if the slide is passed from the water to the air side of an interface covered by the monolayer. If the solid is now passed back to the water side, a second layer is transferred. This process can be repeated to build up multilayer structures on solid supports; such structures are called Langmuir-Blodgett (LB) films. Much of the research on LB films has focused on the properties of multilayers, but monolayers have also been examined.

Many experiments that cannot be carried out on Langmuir monolayers can be performed with relative ease on LB monolayers. For example, the structures of LB films can be investigated by electron diffraction [24], a technique that can examine small domains and therefore does not produce a powder-averaged diffraction pattern. LB films can also be examined by electron microscopy [25] and by scanning tunneling [26] and atomic force microscopy [27]. The relevance of such experiments to Langmuir monolayers depends on the extent to which the properties of the monolayer are altered in the transfer process. This is not always clear, so comparisons between Langmuir monolayers and the LB films derived from them must be made with some caution.

12.2.2.5   Other methods

Several other techniques have been applied in a few instances to monolayer studies. Relaxation processes in monolayers, for example, can be studied by flow [1] or oscillating-disk viscometry [28, 29]; more detailed information on the 2D visco-elastic properties can be obtained by light scattering from thermally excited surface waves [30]. Ellipsometric measurements, which can be interpreted in terms of film thickness, have long been employed [1]; the development of modern instrumentation has sparked a renewed interest in such studies [31]. A recent development [32,33] is Brewster-angle microscopy, which allows imaging of a monolayer without the addition of fluorescent probes.

12.2.3   THE CURRENT SITUATION

The results of the classical studies – mainly isotherm and surface potential measurements, but also including investigations of quantities as diverse as the viscosity and the rates of chemical reactions in a monolayer – can be combined with those obtained with the newer techniques to provide a current consensus about monolayer phase diagrams and the microscopic nature of the phases.

12.2.3.1 Phase diagrams

Although there are certainly differences in detail, the number and type of
general monolayer phases – G, LE, LC, and S – and their relative locations
on a $\Pi - T$ diagram appear to be much the same for many phospholipids,
fatty acids, alcohols, and esters. One can envision a generalized phase diagram that is displaced in some regular fashion as the nature of the head
groups and tails are altered. There is, for example, a long-established rule
of thumb that the addition of one CH_2 group to the tail of a fatty acid displaces the phase boundaries to higher temperature by 8–10°C. The molecular areas at which phase transitions occur must also depend in a regular
fashion on the cross-sectional area and flexibility of the tails. For example, recent experiments on myristic, pentadecanoic, and palmitic acid have
shown that the part of the phase diagrams corresponding to the LE-LC
transition can be exactly superimposed by shifting the temperature scale
by 11.6°C [34]. Condensed phases for amphiphiles such as the phospholipids, that have two chains per head group, can be expected to differ in
detail from those for single-chain molecules because of packing constraints
produced by the mismatch between the chain and head-group areas and
differences in symmetry.

At high areas and low pressures all monolayers exist as gaseous phases.
In many cases, there is a transition to the LE phase on compression; it
is clear from a large variety of experiments - isotherms, surface potential,
fluorescence microscopy, scattering from surface waves - that this transition
is first order. It also appears likely that it ends at a critical point.

Highly precise measurements of G-LE isotherms for pentadecanoic
acid [35] appeared to show the existence of a G-LE critical point at 26.27°C,
but more recent isotherm studies [36] on the same system lead to the conclusion that the critical temperature must be in excess of 40°C. Fluorescence
experiments [12] are in accord with these latter studies but do not explain
why the two very careful isotherm experiments should give such disparate
results. A detailed analysis of the first isotherms had shown that the critical point had mean-field character rather than the non-classical Ising-like
behavior that might have been expected. The subsequent experiments call
this conclusion into question and the character of the critical point remains
unknown [37].

When the LE phase is compressed, a transition to a more condensed
phase, the LC phase, is often observed. With few exceptions [38,39],
horizontal isotherms are not observed in the LE-LC transition region and
there has therefore been considerable debate about the order of the transition. Fluorescence measurements clearly show the presence of two phases
throughout the transition region and the changes in the relative amounts of
the phases follow the lever rule, so the first-order character of the transition
is no longer in doubt.

The LE-LC transition is not found at low temperatures and the point at which it disappears, the LE-LC-G triple point, can be located from isotherm studies. The triple point can also be determined by fluorescence microscopy and in recent measurements on pentadecanoic acid [12] there was excellent agreement between the two methods. At temperatures below the triple point, there is a direct transition from the G phase to the LC phase without an intervening LE phase. This is the case, for example, for isotherms of stearic acid at room temperature, which lack the coexistence region associated with the LE-LC phase transition.

The existence of transitions between other condensed phases can be inferred from isotherm measurements [3,4,40] and has now been confirmed by diffraction measurements. X-ray studies of phospholipids [15,41-43], fatty acids [15,44-47] and a long-chain alcohol [48,49] have shown that there are a number of condensed monolayer phases. For example, five distinct phases that correspond to features in isotherms were distinguishable by in-plane scattering measurements on docosanoic acid. Following the notion established by Harkins and Stenhagen [3,4], they can be identified as the S, CS, LS, L_2, and L'_2 phases shown in Fig. 12.4, and will be discussed in Sec. 12.2.3.2 immediately below.

The fluorescence technique can also provide evidence of such transitions. When monolayers of methyl esters are formed in the LE-LC two-phase region at high temperatures, the LC domains are circular. If the monolayer is then cooled, the domains undergo a transition to a hexagonal shape at a well-defined temperature [14,50]. The shape transition is reversible and does not depend on the nature or concentration of the fluorescent probe. One can associate the domain shape change with a transition between two LC phases, a high-temperature phase that is isotropic, and a low-temperature phase that is anisotropic. The transition temperatures correlate well with those of the $LS - L'_2$ phase transitions deduced from Lundquist's isotherms [5].

Similarly, two different patterns have been observed for LC domains of long-chain fatty acids in the LE-LC two-phase region, depending on temperature [34]. At low temperatures the domains are circular whereas highly ramified, fractal-like structures are observed at high temperatures. The shape transition is extremely sharp and occurs within less than 0.1°C, evidence that the LE phase can transform into two different LC phases. The transition temperatures correlate well with the phase diagram proposed by Bibo and Peterson [40] on the basis of surface pressure isotherms. It should be emphasized however that in these cases, the LC domains are always circular at equilibrium, and the shape transitions are associated with nonequilibrium morphologies that occur under rapid compression or changes in temperature. The relaxation rate towards the equilibrium, circular, domain shape gives information about the monolayer viscosity. It appears that the viscosity of the high-temperature phase is three orders of magnitude larger than that of the low-temperature LC phase.

A transition, presumably LE-S, has also been observed by fluorescence microscopy in a highly-compressed monolayer of stearic acid, labeled with NBD (4-nitrobenz-2-oxa-1,3-diazole) which is itself a probe molecule [51]. In this case the solid domains take the form of extremely long and narrow needles. This peculiar shape has recently been explained on the basis of the dipole-dipole interactions associated with the conjugated rings of the NBD chromophore [52].

12.2.3.2   Structures of monolayer phases

The very low density of the gas phase makes it very difficult to study by any of the techniques that provide information about microscopic structure. The low surface pressures at which the gas phase is stable also present problems for the classical studies, because even very low levels of impurities can significantly affect the measurements. Isotherms at large areas per molecule can be fit to equations of state appropriate to a two-dimensional van der Waals gas. One finds then that the effective compact area per chain is considerably larger than that for a vertical chain, which is consistent with a structure in which the chains are disordered – both conformationally and orientationally – to a large extent. Such a structure makes sense on both energetic and entropic grounds.

The classical picture of the LE phase as one in which the chains begin to be raised off the surface is consistent with that obtained from the more recent studies. SHG measurements on pentadecanoic acid [18] have been shown to be sensitive to the orientation of the head group; if the chains are assumed to be rigid, then the measurements can be interpreted in terms of change in chain orientation. At 25°C and a molecular area of 45Å^2, the polar angle with respect to the surface normal is nearly 90° in the LE phase (*i.e.*, the chains are on average nearly horizontal) but it falls sharply upon compression to about 45° at the LE-LC phase boundary (molecular area of 32Å^2). The angle then changes linearly with area, as it would if there were a mixture of phases with different tilt angles, and reaches 30° at the end of the transition region. Similar results are found along other isotherms between 20 and 30°C, and it appears that the 45° angle at the start of the LE-LC transition is independent of temperature.

Sum-frequency spectra have also been measured [22] in the LE phase of pentadecanoic acid. A strong CH_2 asymmetric stretch signal at 2850cm^{-1} and a broad background in the range $2880 - 2930\text{cm}^{-1}$ that is also attributable to CH_2 stretches are observed. The intensities of these features increase when the monolayer is expanded further. If the chains were straight, there would be near inversion symmetry along them, and the CH_2 modes would not be observed. Thus the presence of the modes and the increase in intensity is indicative of an increase in *gauche* conformations in the tail as the area increases.

Fourier transform infrared spectroscopy (FTIR) studies of the LE phase in dipalmitoyl phosphatidylcholine (DPPC) [53] also provide information about changes in the *gauche/trans* ratio. For DPPC monolayers at a molecular area of 100Å^2, the CH_2 stretching frequency corresponds to that found in bulk suspensions; it decreases linearly with decreasing area, and in the LC phase reaches a value comparable to that in pure bulk DPPC in which the chains are all *trans*.

X-ray reflectivity measurements have been performed [15] on dimyristoyl phosphatidic acid (DMPA) at an area of $87 \pm 7\text{Å}^2$, which is in the LE one-phase region. The thickness in the LE phase is markedly less than in the LC phase. If one takes the chains in the LC phase to be vertical, then the thickness observed for the LE phase corresponds to a tilt of 50° if the chains were all *trans*.

The results of the early diffraction measurements on monolayers were difficult to correlate because they were performed on several different substances in different thermodynamic states. Some features of the structures could be generalized, however. The widths of the diffraction peaks showed that the positional correlation length in all but one of the phases (called the compact solid phase) was short, at most 100 lattice spacings. All of the structures can be seen to arise from an hexagonal packing of the head groups but may have a lower symmetry because of chain tilt or chain packing in phases in which the chains do not rotate. The chains are essentially fully *trans* in the condensed phase. On the other hand, electron diffraction experiments [24,25,54] have shown that there can be extremely long-range lattice orientational order since the diffraction spots are observed not to rotate in space as the sample is translated over several microns.

Peterson and his coworkers [46,55] have shown that the structures of the acids, esters, and alcohols can be understood in terms of four order parameters that have been used to characterize smectic liquid crystalline phases, and that each of the condensed monolayer phases can be seen as the 2-D analog of a known smectic phase [55]. These parameters are (1) Positional order (PO); (2) Bond or lattice orientational order (BO); (3) Tilt order (TO), which is the order of the molecular tilt azimuth with respect to the local orientational order; (4) Herringbone order (HO), which is the staggered ordering of the planes of the all-trans hydrocarbon chains. A distinction is made between quasi-long-range order, in which the order decays according to a power law and short-range order, in which the order falls off exponentially with distance. The order parameters for each of the smectic phases are shown in Table 12.1, and the correspondence between the monolayer and smectic phases is also given in Table 12.1.

Detailed isotherm measurements [55] on acid + ester mixtures suggest the existence of four "LC" phases, that is four phases that can coexist with the LE phase. All four of the phases (LS, L_2^*, L_1', and L_2) are hexatic phases in which there is quasi-long-range orientational order. They differ in their tilt order: the chains in the LS phase are untilted, so there is no TO; in the L_2^* phase the chains are tilted towards their next-nearest neighbor (NNN) in

TABLE 12.1. SMECTIC PHASES AND THEIR PROPOSED RELATION TO MONOLAYER PHASES. The smectic phases are characterized by four order parameters: PO (In-plane positional correlations); TO (Molecular tilt azimuth); BO (Bond or lattice orientation); and HO (Herringbone order or broken axial symmetry). The column T-O indicates the direction of the tilt azimuth with respect to the bond orientation: NN (toward nearest neighbors); NNN (toward next-nearest neighbors). The corresponding monolayer phases are specified in the Harkins-Stenhagen nomenclature. There are in addition a gaseous phase and the CS phase, which is taken to be a true 2-D crystal.

SMECTIC TYPE	PO	TO	BO	T-B	HO	MONOLAYER PHASE
A	S	S	S		S	LE
BC	L	S	L		S	
BH	S	S	L		S	LS
C	S	L	S		S	
E	L	S	L		L	S
F	S	L	L	NNN	S	L_2'
G	L	L	L	NNN	S	
H	L	L	L	NNN	L	S'
I	S	L	L	NN	S	L_2
J	L	L	L	NN	S	
K	L	L	L	NN	L	L_2''
L	S	L	L	I	S	L_1'

the locally hexagonal structure; the tilt is towards nearest neighbors (NN) in the L_2 phase; and toward a direction intermediate between NN and NNN in L_1'. There is no HO because the chains rotate freely and therefore have effective cylindrical symmetry.

All of these phases have been identified in diffraction studies. Proof of the existence of tilt order can be seen in the polarized fluorescence microscopy experiments by Xiu et al [14]. They showed that distinct regions of uniform tilt appeared in LC domains on cooling out of the LS phase. In some cases the tilt regions in a domain are organized into "star defects", Fig. 12.6, which had been observed in freely suspended films of hexatic liquid crystals [56]. It is evident from the figure that some of these defects are chiral, even though the ester and the probe are not. This identifies the phase as L_1', because it is the only intrinsically chiral hexatic (Table 12.1).

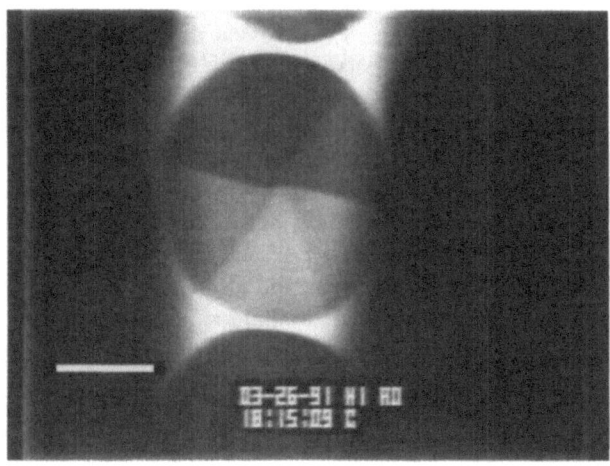

FIGURE 12.6. Regions of uniform tilt in an LC domain of methyl eicosanoate at 20°C. The tilt regions are evident only if the exciting radiation is polarized.

12.3   Equilibrium Theories of Monolayers

12.3.1   FLUID-FLUID PHASE TRANSITIONS AND CHAIN STATISTICS

At very low surface densities, when the average area per molecule a is much larger than the cross sectional area of an isolated chain, a^*, an adsorbed amphiphilic monolayer behaves as a 2D gas. Recall from Chap. 1 — Sec. 1.3.3 in particular — that a^* is the projected area of a single chain whose head is constrained at a planar interface: $a^* \simeq 40\text{Å}^2$, say, for a 12-carbon alkyl tail. For $a \geq a^*$, then, a molecule in a monolayer is still free to "express" its conformational entropy without interference from neighbors. As discussed above, a clear first-order phase transition from the gas-phase (G) to a "liquid-expanded" (LE) phase is observed upon compression of the monolayer. The area per molecule on the gas-phase side of the transition is typically $a_G \approx 300 - 1500\text{Å}^2$ whereas on the LE side it is on the order of the cross-sectional area of an *isolated* chain, *i.e.*, $a_{LE} \simeq a^*$. The second phase transition, to the "liquid-condensed" (LC) state, observed upon further compression of the monolayer, is associated with a change in area from $a_{LE} \approx 30 - 40\text{Å}^2$ to $a_{LC} \approx 22 - 25\text{Å}^2$, which is just barely larger than the cross-sectional area of a fully stretched (all-trans) chain. The main difficulty in understanding the LE-LC transition is in identifying the structure of the two phases and the driving force (order parameter) of the transition between them. Some insight can be gained by considering an analogy with liquid crystalline systems.

Rod-like (or disk-like) molecules forming liquid crystalline systems undergo an isotropic-nematic, (I→N) transition in three dimensions without the need for attractive forces between particles [57]. Here a discontinuous onset of orientational order (alignment) appears spontaneously upon compression (or cooling, in the case where attractive interactions are also operative [58]). Based on this analogy one naively expects that the LE→LC transition will be analogous to the I→N transition in liquid crystals. However, if the adsorbed molecules are treated as rigid rods that interact through their excluded volumes, it can be shown [59] that the increase in orientational order upon compression is continuous. Subsequently, it was shown by Chen *et al* [60] that because of the lower symmetry of the monolayer compared to that of a bulk liquid, a system of rigid, adsorbed rods will not undergo a first-order fluid-fluid phase transition upon compression unless there are attractive forces acting between the particles. For this latter case they showed that only a single transition occurs, essentially a gas→liquid condensation at which there is also (in addition to the jump in density) a discontinuous change from weak to stronger long-range orientational ordering.

This situation contrasts with the experimental situation described above in which the amphiphilic monolayer is observed to undergo two, successive transitions between fluid phases, *i.e.*, G→LE followed by LE→LC. As discussed in the preceding section, the LC "fluid" appears to be a hexatic phase, but we shall suppress this fact for the moment. Since amphiphilic molecules are not rigid rods, the existence of two fluid-fluid transitions can be related to the chain flexibility [60], *i.e.*, to the conformational degrees of freedom of the amphiphilic chains. In fact, it is possible for attracting *rigid* rods to undergo successive G→LE and LE→LC changes of state, but only if the particles are adsorbed at the surface with an energy that depends on orientation so as to favor "lying down" vs. "standing up" [61]. Nevertheless, a number of mean-field, renormalization group, and computer simulation studies of Ising- and Potts-like models have been carried out [62,63] in which the presence of a second phase transition is explicitly attributed to coupling of chain stretching and grafting density. To amplify this basic idea we briefly and qualitatively consider the various contributions to the free energy of the monolayer.

The four major terms in the free energy of the monolayer are:

$$F = F_{\text{tr}} + F_{\text{conf}} + F_{\text{att}} + F_{\text{head}} \tag{12.3}$$

The $F_{\text{tr}} = -TS_{\text{tr}}$ corresponds to the translational entropy of the molecules. To a first approximation, in the gas phase, F_{tr} per molecule is

$$F_{\text{tr}}/N \approx -k_B T \log(a - a^*) \tag{12.4}$$

where N is the number of molecules in the monolayer and $a - a^*$ represents the "free area" per chain, giving rise to a lateral pressure

$$\Pi_{tr} = -\frac{1}{N}\partial F_{tr}/\partial a = k_B T/(a - a^*) \qquad (12.5)$$

as in the familiar van der Waals equation. The second term, F_{conf}, represents the conformational free energy of the chain, which involves an energetic contribution associated with the internal energy of the chain (the energy of a "gauche" bond is of order $k_B T$) and an entropic term reflecting the large number Ω of possible chain conformations:

$$S_{conf} = N k_B \log \Omega \qquad (12.6)$$

For an isolated chain $(a \geq a^*)$ of $(n+2)$ C–C bonds, Ω can be estimated simply from, say, a 3-state rotational-isomer-state model: $\Omega \approx 3^n$. Of course as a decreases (below a^*) so does Ω, reaching $\Omega = 1$ if all chains are in the all-trans state. F_{att} denotes the attractive interaction energy between the chains. (Recall that the repulsive interactions are included in principle in F_{tr}). Like F_{tr} and F_{conf}, F_{att} is also a function of a. In the gas phase, i.e., when $a > a^*$, it has a van der Waals-like form:

$$\frac{F_{att}}{N} = \frac{E_{att}}{N} \approx \frac{-\text{const.}}{a} \qquad (12.7)$$

where the constant prefactor is temperature dependent. Finally, the last contribution to F includes the interaction between the head groups. This latter term, F_{head}, plays a central role in determining the character of the highly compressed (LC and, even more so, solid) phases. The contribution of the polar heads is discussed in the following section. Here, in discussing the G→LE and LE→LC transitions, we assume, as a first approximation, that F_{head} is a constant, independent of the thermodynamic variables a and T.

Now recall that in the G→LE transition both coexisting phases correspond to areas per molecule a that are larger than the area of an isolated, single, chain a^*. Thus, the conformational degrees of freedom are not severely perturbed in this transition, i.e., $F_{conf}(G) \approx F_{conf}(LE)$. In other words, G→LE involves mainly an interplay between F_{tr} and F_{att}. Of course these are the same two terms responsible for ordinary gas-liquid transitions. Indeed, in our case, we should not expect a different behavior as long as the conformational degrees of freedom are not affected. (The chains "look the same" on both sides of the transitions.) Here the chain "blobs" play the role of the inflexible molecules in an ordinary G-L transition: the "order parameter" of the transitions is still the surface density, $1/a$.

Once the LE phase is formed and $a \simeq a^*$, very little translational freedom is left to the chains, a condition that does not change much as the monolayer is further compressed to $a < a^*$. On the other hand this compression involves important, and opposing, changes in F_{conf} and F_{att}. Recall that

— by definition of a^* — as soon as a decreases below a^* the chains begin to feel significant constraints on their conformational degrees of freedom. Regarding F_{conf}, it can be shown (based on simple scaling arguments [64]) that to a first approximation the (dominant) entropy term varies with a according to $S_{\text{conf}} \sim 1/a^2$. It can also be argued that as the chains are compressed and thus stretched the average attraction energy between different chains varies as

$$F_{\text{att}} = E_{\text{att}} \sim -d/a^k \qquad (12.8)$$

where $k \geq 1$ and d is a (nearly temperature-independent) constant. The basic idea here is that chains can attract each other more strongly ($k > 1$) when they are stretched, since a greater number of monomer-monomer contacts becomes possible between them. (An actual estimate of k requires information on the density profile of the chains, as discussed below.) Accordingly, it is the interplay between the loss of conformational entropy and the gain in chain monomer contacts that induces the LE→LC transition. In this sense, the LE→LC transition is similar to the so-called "liquid crystalline-gel" transition in bilayers [65] in which the two major contributions to the free energy are also the chain segment density and the chain conformational free energy.

In the theory of Shin et al [66], the decomposition described above is treated systematically via introduction of explicit models for the flexible chains. In particular, they consider the free energy of the adsorbed amphiphilic monolayer to be a sum of the following contributions: mixing entropy associated with amphiphilic molecules and solvent (water) in the surface layer; interaction energy involving chain monomers on the surface, i.e., adsorption on the surface and attraction to other monomers (both intra- and inter-chain) which are adsorbed; and interaction energy involving the chain segments that lie above the water surface. The first contribution is taken to have the Flory–Huggins form [67], and the second to be described by a random mixing (Bragg–Williams [68]) approximation. For the third contribution, the interaction between the chain segments that lie above the surface, three very different models for chain behavior are considered, referring specifically to the extreme cases of flexible chains and stiff chains. The key result is that conclusions concerning the nature of fluid-fluid phase transitions are largely independent of details about the inter-chain monomer-monomer interactions above the surface or about the adsorption energy.

Even in the dilute gas phase, part of the hydrocarbon chain is found to be above the surface, consistent with the demands of conformational and/or orientational entropy. The G-LE transition is then seen to be a condensation of the in-surface portions of the chains, with essentially no change in the overall chain configuration. The LE-LC transition, on the other hand, involves a significant increase in the fraction of chain lying above the surface and — concomitantly — a condensation of the chain segments above the surface. Consistent with this interpretation, it is found that the in-surface monomer-monomer attraction produces the G-LE transition, but has essentially no effect in the LE-LC region. Conversely, a sufficiently negative above-surface interaction energy is necessary for the LE-LC transition to occur, whereas it is largely irrelevant to G-LE.

Cantor and McIlroy [69] had earlier found a similar set of conclusions, but from a somewhat different theoretical description. They consider flexible chains attached via their head groups to a planar interface and characterize each chain configuration by the profile of volume it occupies perpendicular to this plane. The lateral-excluded-area profile of each conformation is approximated as an average over all pairs of conformations, and the monolayer entropy is determined by assuming ideal, two-dimensional, mixing of the chains with "solvent" (air or oil) and with each other. The amphiphilic chains are also characterized by the position and orientation of their surface area available for nearest-neighbor contact, for purposes of calculating interaction energies. These latter are reduced to two parameters, one related to the chain-"solvent" interfacial tension (*i.e.*, either chain/air or chain/oil) and the other to oil-water interfacial tensions. A modified cubic lattice model of chain conformations is then used to calculate $\Pi - a$ isotherms: for strong enough attractions between monomers it is found that two, successive, fluid-fluid phase transitions occur, both being characterized by a critical point. Taking the limit of rigid chains, only a single G→LE transition appears, consistent with an earlier conclusion [60]. This last point has been recently verified experimentally [70]. If the monolayer is formed with perfluorinated fatty acids, which are rigid, only a single G → LC phase transition is observed at all experimentally accessible temperatures. However a second, LE → LC phase transition is restored if a flexible spacer of four methylene units is introduced between the rigid chain and the head group.

Detailed insight into orientational and conformational structure can be obtained in principle from computer simulation of amphiphilic monolayers. Extensive molecular dynamics computations of this kind have been carried out [71,72]. However, in this work the special structure of the subphase (water!) has been neglected and the simulations have been performed in the small area per molecule limit of the adsorbed monolayer. The methylene groups of the aliphatic tails interact with each other and with the surface via a Lennard-Jones potential. Valence bond bending (*e.g.*, deviations from tetrahedral C–C–C angles) and dihedral torsional (*i.e.*, *trans-gauche* energy difference) potentials are used to provide the correct conformational statistics for the chains. Consistent with the "rod-scan" diffraction measurements, the angle of tilt is found to vary sharply upon change in the area per molecule but quantitative agreement with experiment is still lacking.

More importantly, the machine computations do not yet provide any indication of a phase transition between "expanded" and "condensed" state of the adsorbed liquid. In order to treat properly the subtle effects associated with this transformation it will almost certainly be necessary to confront directly the molecular structure of the water "subphase". One expects in particular that the preferred orientations and interactions between head groups will depend sensitively on the "reconstruction" of water near its surface. While simulations of the free surface of water have been carried out by several groups [73], no consensus has emerged for describing the local structure at the interface. Furthermore, the presence of an adsorbed monolayer of amphiphiles will surely disrupt the surface to a significant degree: much work remains to be done before a microscopic understanding of this situation can be achieved. In general, because of the delicate interplay of competing forces involved, one must expect a wide range of different behaviors as one looks from one system to the next. For example, some will show a molecular tilt in the LC phase, whereas others will not; others will "skip" directly from the LE phase to the high-pressure solid (*i.e.*, without passing first through the LC phase) at room temperature.

12.3.2   PATTERN FORMATION AND DOMAIN SHAPES

A striking phenomenon observed in monolayers is the formation of patterns, some quite complex, which can be observed directly by fluorescence microscopy. The patterns range from foam structures that are seen at low densities corresponding to the G-LE coexistence region, to fractal structures, dendrites, hexagonal arrays, and spirals that have been observed at higher densities. What is the origin of these patterns and what controls their shapes? Are these patterns equilibrium structures or do they appear only when a system is driven out of equilibrium, for example either by temperature or pressure quenches? Can one learn anything about the microscopic structure of monolayer phases from the study of these patterns?

McConnell, Möhwald and their coworkers [74,75] have made extensive studies of pattern formation in phospholipid monolayers of DPPC. These experiments have been interpreted independently by several groups [76-80] in terms of a model in which the equilibrium shape of an isolated domain is determined by the competition between electrostatic forces and line tension. In the condensed phases to which this model is applied, the amphiphiles have similar orientations and there is a repulsive, long-range, dipole-dipole interaction.

The free energy of a domain can be written as $F = \lambda L + F_{el}$ where λ is the line tension, L is the perimeter of the domain, and F_{el} is the electrostatic energy resulting from the interaction of the molecular dipoles of the amphiphiles. Surface potential measurements demonstrate that there is a marked difference in the dipole density μ_z between the LE and LC phases. Miller and Möhwald [81] have also demonstrated this in elegant experiments in which individual domains are manipulated by electrodes. As a result of these opposing contributions, the free energy of a domain depends on both its shape and on its size. The model therefore predicts that transitions between one shape and another can occur at certain critical sizes, in accord with experiments in which circular domains undergo shape transitions as they grow [82].

In the expression for the free energy of a domain written above, the line tension is assumed to be isotropic. The formation of hexagonal LC domains in methyl esters, spiral structures in DPPC monolayers, chiral patterns in monolayers prepared from pure DPPC enantiomers and dendrites in chiral fatty amino acids demonstrate that this is not the case. This anisotropy can result from in-plane components of the molecular dipole, related either to chain tilt and/or anisotropy of the head group.

The key parameter in pattern formation is the ratio λ/μ_z^2. The addition of the "line-active" agent cholesterol to a monolayer reduces λ and leads to more extended structures. Changes of λ with temperature can also induce shape transitions. In the binary system DPPC + cholesterol, there is coexistence between two phases, one DPPC-rich, the other cholesterol-rich, that ends at an upper critical solution temperature [83,84]. As the critical point is approached from below, the line tension between the phases decreases and there is a sharp transition from circular to noncircular droplets [84].

Fluorescence micrographs show that large areas in monolayers are often organized into "superlattices" of regions with different density. These may be hexagonal arrays or lamellar structures. Some of these spatial patterns, such as foams, are clearly nonequilibrium structures because they evolve with time, but other patterns do not change during the period of typical experiments, times of a day or more.

One can take the point of view that these superlattices represent ordered regions of two distinct phases (e.g., LE and LC, or LE and G) that are metastable. The slow evolution to larger domains might be attributed to the lack of gravity as a driving force and the small differences in curvature free energy between relatively large but different size domains. From another point of view these superlattices can be taken to be one-phase regions in which there is a spatially modulated density. Just such an interpretation has been applied to the patterned phases observed in magnetic bubble domains and ferrofluids [85-87]. The morphology of superlattices in such iron garnet films are remarkably similar to those observed in monolayers.

Can one decide between these two pictures: metastable ordered two-phase regions or modulated one-phase regions? Helm and Möhwald [88] found that when dilauryl phosphatidylethanolamine (DLPE) was compressed into the LE-LC coexistence region the shapes of the domains were an equilibrium property but the size and number of the domains (and hence the superlattice spacing) depended on the nucleation kinetics. Studies [89] on another phospholipid, dimyristoyl phosphatidyl ethanolamine (DMPE), similarly argue for the two-phase interpretation. The circular LC domains pack into what appears to be a hexagonal superlattice, but the characteristic length varies continuously and uniformly with compression from one phase boundary to the other.

In contrast to these results, however, it has been [84] found that lamellar (labyrinth) and hexagonal phases exist in well-defined areas of the DPPC + cholesterol phase diagram and that it is possible to observe coexistence between these two modulated phases. Similarly, To *et al* [90] have observed a stable stripe pattern in the LE-G coexistence region of D-myristoylalanine, an enantiomeric fatty amino acid. The stripe spacing is uniquely defined by the relative amounts of the G and LE phases. The experimental situation is therefore ambiguous and it is useful, then, to examine theoretically the nature of possible modulated phases in monolayers.

12.3.3 DIPOLAR AND CHARGED LANGMUIR MONOLAYERS

For neutral amphiphiles, we consider for simplicity only the case where the average dipole moment points along the perpendicular $\vec{z}$-direction, $\mu_z \equiv \langle \mu \rangle$. A non-vanishing in-plane component of the dipole moment stabilizes elongated structures and has been investigated as well [52,76]. As is shown below, an in-plane density wave of the dipoles will reduce the overall long-range repulsive dipolar interaction. Let $\phi(\vec{\rho})$ denote the amphiphile density at position $\vec{\rho}$ in the surface. For an amphiphile density with wavevector $\vec{q}$, i.e., $\phi(\vec{\rho}) = \phi_o + \phi_q \cos(\vec{q}\cdot\vec{\rho})$, this energy lowering is given by:

$$\Delta F_{el} = -\frac{1}{2}\frac{\varepsilon_0}{\varepsilon(\varepsilon + \varepsilon_0)}|q|\mu_z^2\phi_q^2 \tag{12.9}$$

ε (ε_0) is the water (air) dielectric constant as shown in Fig. 12.7. In Eq. (12.9), the dipoles are assumed to be *totally* immersed in the water and the interaction is screened by approximately ε^2. However, since the polar moieties lie very close to the air-water interface, the *effective* dielectric constant, ε^*, felt by the dipoles is expected to be intermediate between the bulk water value, $\varepsilon = 80$, and that of vacuum, $\varepsilon_0 = 1$. In addition, if the aliphatic tail of the molecules also contributes to the molecular dipole

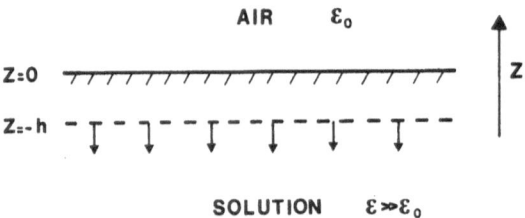

FIGURE 12.7. A flat interface at $z = 0$ separates air (dielectric constant ε_0) from an aqueous solution ($\varepsilon \gg \varepsilon_0$). Dipoles are confined to the plane $z = -h$ and the electrostatic energy is calculated in the limit $h \to 0$

moment, then some of the dipolar interactions are not screened. The linear dependence of ΔF_{el} in (12.9) on the modulation wavevector $|q|$ is a consequence of the long-range ($\sim 1/r^3$) character of the dipolar forces. In general, an algebraically decaying interaction, $F_{el} \sim 1/r^{d+\sigma}$, where d is the spatial dimension, will have a Fourier component with a $|q|^\sigma$ dependence. For dipoles, $d + \sigma = 3$, $d = 2$, hence, $\sigma = 1$.

If the amphiphiles are *charged* and an electrolyte is added to the water, the electric potential can be obtained from the Poisson-Boltzmann equation [78]. Here, we restrict ourselves to the linearized version of the Poisson-Boltzmann equation for the electrostatic potential V:

$$\nabla^2 V(\vec{r}) = \kappa^2 V(\vec{r}) \quad \text{for} \quad z < 0$$
$$\nabla^2 V(\vec{r}) = 0 \qquad\quad \text{for} \quad z > 0 \qquad (12.10)$$

where κ^{-1} is the Debye-Hückel screening length. We solve the electrostatic problem (12.10) for a modulation in the surface charge density $\sigma(\vec{\rho}) = e\phi_0 + e\phi_q \cos(\vec{q} \cdot \vec{\rho})$, with the appropriate boundary conditions at $z = 0$.

The electrostatic energy is then given by

$$F_{el} = \frac{1}{2} \int \sigma(\vec{\rho}) V(\vec{\rho}, z = 0)\, d^2\vec{\rho} \quad = \quad \frac{e^2 \phi_q^2}{2(\varepsilon \kappa_1 + \varepsilon_0 |q|)} \qquad (12.11)$$

where in (12.11) κ_1^{-1} is the effective screening length for the q-mode, $\kappa_1^2 = \kappa^2 + q^2$, and we omit the average electrostatic contribution depending only on ϕ_0. For concentrated ionic solutions, the Debye-Hückel screening length is small: $\kappa \gg q$. In this case, when (12.11) is expanded in powers of q/κ, equation (12.9) is recovered. The counter ions together with the surface charges form effective dipoles at the water surface with a dipole moment

$$\mu_z = \frac{e}{\kappa} \sqrt{1 + \varepsilon_0/\varepsilon} \quad \simeq \quad e\kappa^{-1} \quad \text{for} \quad \varepsilon \gg \varepsilon_0 \qquad (12.12)$$

By changing the ionic strength in the solution, it is possible to reach $\kappa^{-1} \simeq 10\text{Å}$, corresponding to strong dipoles. In the opposite limit of dilute electrolytes, $q \gg \kappa$, the electrostatic interactions are almost unscreened [91] and the interaction is Coulombic $\sim 1/r$.

As we shall see below, electrostatic interactions tend to stabilize phases with modulated density. Phase diagrams incorporating the possibility of modulated phases were calculated within the framework of the mean-field approximation in two cases: (*i*) close to a critical point where only the most dominant q-mode is considered, and (*ii*) at low-temperatures where entropy is neglected.

12.3.3.1 Landau theory close to a critical point

Close to the critical point, the free energy can be written phenomenologically [78-79] as a Landau expansion in the order parameter,

$$\psi(\vec{\rho}) = \phi(\vec{\rho}) - \phi_c \qquad (12.13)$$

where ϕ_c is the amphiphile density at the critical temperature, T_c. The expansion contains only even terms in ψ

$$\Delta F_0/k_B T = \frac{1}{2} d(T - T_0)\psi^2(\vec{\rho}) + \frac{1}{4} u\psi^4(\vec{\rho}) \qquad (12.14)$$

where the coefficients, $d > 0$ and $u > 0$, can be obtained from an expansion of the monolayer free-energy of mixing. Note that T_0 is the critical point *only* in the absence of the dipolar interactions. Since the electrostatic interactions favor spatial modulations, we also have to consider the energy gain and loss as an in-plane modulation of the two-dimensional concentration created. The *gain* will be in the electrostatic energy

$$F_{el}/k_B T = \frac{1}{2} \int \psi(\vec{\rho})g(\vec{\rho} - \vec{\rho}\,')\psi(\vec{\rho}\,')\,d^2\rho\,d^2\rho' \qquad (12.15)$$

where $g(\rho) = k_B T b^3/2\pi\rho^3$ and $b^3 = \mu_z^2\varepsilon_0/[\varepsilon(\varepsilon + \varepsilon_0)k_B T]$.

The energy *loss* is the interfacial line energy between domains in a 2D geometry. For small density variations, the line energy is written to lowest order in a gradient expansion as:

$$F_I/k_B T = \frac{1}{2} a^{*2} \int (\nabla\psi)^2 \, d^2\rho \qquad (12.16)$$

with a^* being the minimum (compact) area per molecule. In Fourier space, eqs. (12.15) and (12.16) can be expressed as

$$(F_{el} + F_I)/k_B T = \frac{1}{2} \sum_q (a^{*2}q^2 - b^3|q|)\psi_q^2 \qquad (12.17)$$

where $\psi(\vec{\rho}) = \psi_0 + \sum_q \psi_q \cos(\vec{q} \cdot \vec{\rho})$. The dominant q-mode is the one minimizing the ψ_q^2 coefficient in (12.17).

$$|\vec{q}\,|^* = b^3/2a^{*2} \qquad (12.18)$$

Close to a critical point, it is a good approximation [85,92] to consider only modulations with magnitude q^* in addition to the homogeneous ($q=0$) solutions: $\psi(\vec{\rho}) = \psi_0$ and $\psi(\vec{\rho}) = -\psi_0$ for the condensed and dilute phases, respectively. Two types of spatial modulation of the two-dimensional density are considered [78,85]:

(i) a stripe-like phase, $\psi_S(\vec{\rho}) = \psi_0 + \psi_q \cos q^* x$. (See Fig. 12.8.)

(ii) a hexagonal phase, $\psi_H(\vec{\rho}) = \psi_0 + \sum_{i=1}^3 \psi_q \cos(\vec{k}_i \cdot \vec{\rho}_i)$, with $|\vec{k}_i| = q^*$ and $\sum_{i=1}^3 \vec{k}_i = 0$. (See Fig. 12.9.)

The phase diagram can be calculated by comparing the homogeneous free energy F_0 with the stripe and hexagonal ones, F_S and F_H, respectively.

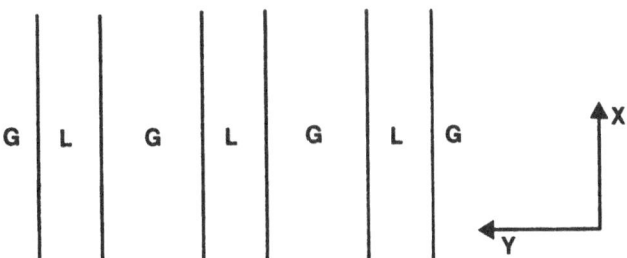

FIGURE 12.8. The stripe phase is shown schematically, where the stripes are chosen to be in the x direction. Domain walls (which are sharp only at low temperatures) separate denser liquid (L) from dilute gas (G). Close to the critical point, the density profile is sinusoidal and given by ψ_S in Sec. 12.3.3.1.

$$F_0 = \frac{\delta}{2}M_0^2 + \frac{1}{4}M_0^4 \qquad (12.19)$$

$$F_S = F_0 + M_q^2(\delta - 1 + 3M_0^2) + \frac{3}{2}M_q^4 \qquad (12.20)$$

$$F_H = F_0 + M_q^2(3\delta - 3 + 9M_0^2 + 12M_0M_q) + \frac{45}{2}M_q^4 \qquad (12.21)$$

where,

$$\eta^2 = b^6/a^{*3} \quad , \quad \delta = 4d(T - T_0)/\eta^2$$
$$M_0^2 = (4u/\eta^2)\psi_0^2 \quad , \quad M_q^2 = (u/\eta^2)\psi_q^2 \qquad (12.22)$$

In Fig. 12.10, the phase diagram in the reduced temperature, $\delta \sim T - T_0$, — reduced average concentration, $M_0 \sim \langle \phi \rangle - \phi_c$, plane is shown, whereas in Fig. 12.11, the same phase diagram is drawn in the chemical potential μ — reduced temperature δ plane. The chemical potential is the thermodynamical variable that is coupled to the average concentration M_0. The usual coexistence region between liquid and gas regions, $M_0^2 = \delta$, is largely modified. The critical point at $\delta_c = 0$ ($T_c = T_0$) is renormalized upwards to $\delta_c = 1$ ($T_c = T_0 + b^3/4da^{*3}$), and is the termination point of five distinct phases: gas (G), stripe (S), hexagonal (H), inverted hexagonal (IH), and liquid (L). The gas and liquid phases are isotropic dilute and dense phases, respectively. All the transition lines below the critical point ($M_0 = 0$, δ_c) are first-order. Consequently, there are four regions of two-phase coexistence between the phases. Although the single q-mode expansion is valid close to a critical point, it cannot be trusted for low temperatures. For example, this theory predicts a disappearance of the stripe and hexagonal phases for δ less than δ_c, as seen in Figs. 12.10 and 12.11. To overcome this difficulty we give in the next section a proof of the existence of modulated phases at low temperatures.

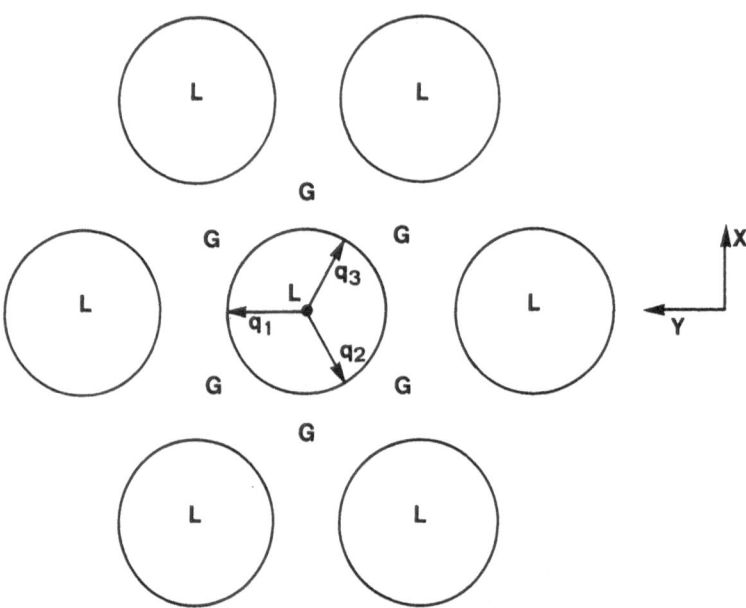

FIGURE 12.9. The hexagonal phase is shown schematically. Denser liquid "bubbles" (L) are separated via domain walls from a dilute gas background (G). Domain walls are sharp only at low temperatures. Close to T_c, the density profile is given by ψ_H in Sec. 12.3.3.1.

12.3.3.2   Modulated phases at low temperatures

At low temperatures (far from the critical point) it is natural to replace the Landau expansion by a direct calculation of the modulated phase free energy. Here we neglect contributions from the entropy of mixing and focus wholly on the electrostatic energy and line tension effects. We show that at low temperatures modulated phases are still expected to be stable over a range of concentrations. For simplicity, only the stripe phase with sharp domain walls is considered. The stripe phase is formed from a periodic arrangement of stripes of the dilute phase of size D_G and of the dense phase of size D_L. The electrostatic free energy of the stripe phase is

$$F_{\mathrm{el}}/k_B T = \frac{b^3}{\pi \ell}[x\phi_L^2 + (1-x)\phi_G^2] - \frac{b^3}{\pi D}(\phi_L - \phi_G)^2 \log\left(\frac{D}{\ell}\frac{\sin \pi x}{\pi}\right) \quad (12.23)$$

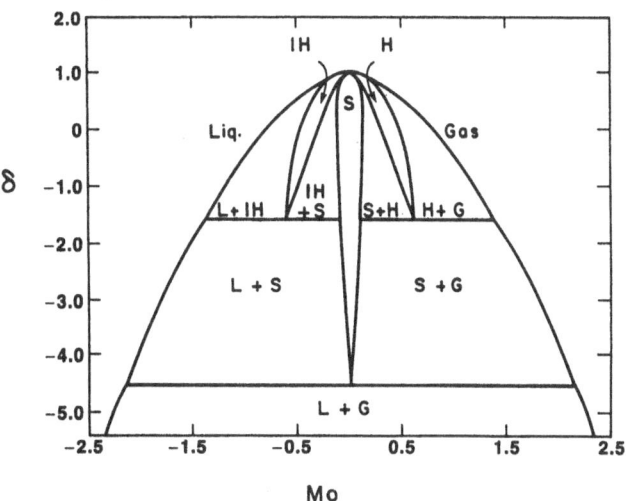

FIGURE 12.10. Phase diagram in the (M_0, δ) plane where $\delta \sim T - T_0$ is the reduced temperature and $M_0 \sim \langle \phi \rangle - \phi_c$ is the reduced concentration. The two isotropic phases, liquid (L) and gas (G), are separated by the hexagonal (H), stripe (S), and inverted hexagonal (IH) phases. Two-phases coexistence regions are also indicated. The phase diagram is obtained from Eqs. (12.19) – (12.22) and is valid only close to T_c.

where $x = D_L/D = D_L/(D_L + D_G)$ is the relative concentration of L and G, and $\ell \simeq \sqrt{a^*}$ is a microscopic cutoff. The first term represents the overall average contribution to the electrostatic energy and is independent of the periodicity D. The last term is an exact summation of the intra- and inter-stripe electrostatic interactions [93]. An additional contribution to the free energy difference ΔF between the stripe and the homogeneous phases with the same concentration comes from the line tension λ associated with every domain wall separating a G domain from an L one. The total free energy difference is thus

$$\Delta F = -\frac{k_B T b^3}{\pi D}(\phi_L - \phi_G)^2 \log\left(\frac{D}{\ell}\frac{\sin \pi x}{\pi}\right) + \frac{2\lambda}{D} \qquad (12.24)$$

The equilibrium periodicity D^* of the stripe structure is given by minimizing (12.24) with respect to D, giving

$$D^* = \frac{\ell \pi}{\sin \pi x} \exp \beta \qquad (12.25)$$

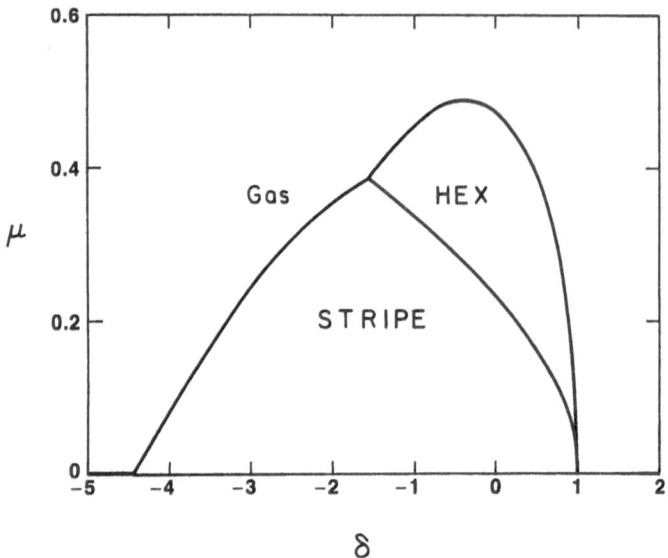

FIGURE 12.11. Phase diagram in the chemical potential μ – reduced temperature δ plane. The phase diagram is symmetric about $\mu = 0$ and is presented only for $\mu > 0$. First-order transition lines separate the isotropic dilute gas (G) from hexagonal (H) and stripe (S) phases. The phase diagram is obtained from Eqs. (12.19) – (12.22) and is valid only close to T_c.

where

$$\beta = \frac{2\pi\lambda}{k_B T b^3 (\phi_L - \phi_G)^2} + 1 \qquad (12.26)$$

The exponential dependence of the periodicity D^* on the ratio between the line tension λ and the dipole interaction coefficient $b^3(\phi_L - \phi_G)^2 \sim \mu_z^2$ makes it difficult to give an *a priori* estimate of D^* since neither λ nor b are known accurately from experiments. In principle, b can be changed in a controlled way by tuning the electrolyte strength for a charged monolayer, allowing a check of Eq. (12.25). In addition, it has been found that cholesterol [74-77, 81-84] reduces the line tension, causing thinning of the domains in qualitative agreement with the theoretical prediction [78,94] of Eq. (12.25). Recently, direct measurements of the stripe periodicity with the surface fraction of G and LE phases in a single-component system and in the absence of line-active cholesterol [90] have been performed. The results are consistent with Eq. (12.25).

12.4   Dynamical Properties of Amphiphilic Monolayers

12.4.1   LATERAL DIFFUSION IN MONOLAYERS

Since a monolayer of amphiphilic molecules spread at the water/air surface creates a 2D flat interface, it provides an example of a system undergoing a phase separation at zero gravity [95]. However, the dynamics of such a monolayer is not strictly two-dimensional since the motion of the amphiphilic molecules on the interface induces a *backflow* in the *water* subphase. This backflow, in turn, gives rise to hydrodynamic interactions between the amphiphilic molecules. Below we discuss the first stages of phase separation when viscous dissipation and electrostatic interactions are included in the analysis. It is important to note that since the dissipation takes place mostly in the water subphase due to the backflow effect, it has a three-dimensional character.

Imagine that we have a monolayer with an average surface concentration ϕ on the surface of a water subphase of thickness h. For simplicity, we assume that the water thickness h is smaller than the wavelength associated with concentration fluctuations of the monolayer. In this case, we can use the *lubrication approximation* [96] to describe the flow in the subphase, since the flow in the water is to a good approximation a local Poiseuille flow. A local fluctuation of the monolayer concentration, $\delta\phi(x)$, leads to a local modulation of the surface pressure $\Pi(x)$ and hence, modifies the surface tension, $\gamma(x) = \gamma_0 - \Pi(x)$ and produces a stress at the free boundary (water/air interface). For a thin water subphase, the flow is a simple shear flow as shown in Fig. 12.12a, *i.e.*, the x-component of the velocity increases with z according to

$$v_x(z) = v_S(1 + \frac{z}{h})$$ (12.27)

where v_S is the velocity of the amphiphiles at the surface $z = 0$, and $z = -h$ is the position of the water/solid interface.

At the free surface, the viscous stress must balance the gradient of surface tension

$$\sigma_{xz} = \eta\frac{\partial v_x}{\partial z} = \eta\frac{v_S}{h} = \frac{\partial\gamma}{\partial x} = -\frac{\partial\Pi}{\partial x}$$ (12.28)

and the surface flow of the amphiphile is

$$\vec{J_s} = \phi\vec{v_S} = \phi(-\vec{\nabla}\Pi)\frac{h}{\eta} = -D_c\vec{\nabla}\phi$$ (12.29)

Equation (12.29) defines the cooperative diffusion coefficient of the amphiphiles, D_c

$$D_c = \phi\frac{h}{\eta}\frac{\partial\Pi}{\partial\phi}$$ (12.30)

which is concentration dependent.

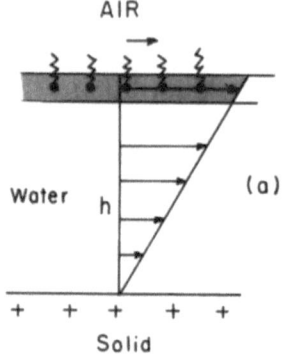

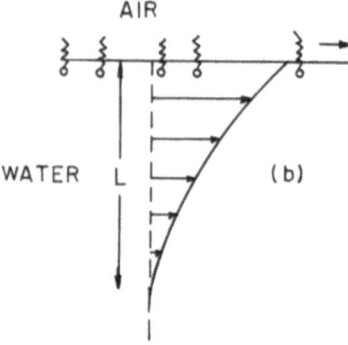

FIGURE 12.12. Backflow induced by motion in the monolayer: (a) for a thin liquid subphase of thickness h on a solid. The flow is a simple shear flow; (b) for a bulk liquid, the flow penetrates into the liquid with an exponential decay characterized by a small penetration length of thickness L.

The diffusion coefficient is calculated for two regimes:

The first is the very dilute regime corresponding to the gas state of the monolayer: the surface pressure obeys an ideal gas law, $\Pi = \phi k_B T$ and D_c is proportional to $\phi \sim 1/a$

$$D_c = \phi \frac{k_B T}{\eta} h = (\phi \ell h) D_0 \qquad (12.31)$$

where $D_0 = k_B T / \eta \ell$ is the molecular diffusion constant and ℓ is a molecular length. Equation (12.31) describes the cooperative part of the diffusion. In principle, we should add a term of order D_0 due to the diffusion of single amphiphilic molecules in the absence of any neighbors. However, since $\phi \ell h > 1$ in all practical cases, $D_c \gg D_0$, and we neglect the molecular diffusion term.

We note that (12.31) represents a *"hypodiffusive"* case [97] since D_c vanishes in the limit of small surface concentration, $\phi \to 0$, and the spreading of amphiphilic molecules at the free surface is described by a nonlinear diffusion equation

$$\frac{\partial \phi}{\partial t} = D_c(\phi) \nabla^2 \phi \qquad (12.32)$$

A sharp edge of the amphiphile concentration profile, $\phi(x)$, is expected [98] in this hypodiffusive case.

The second regime we consider is when the monolayer is in a liquid state in which case $\phi \partial \Pi / \partial \phi$ becomes of order $k_B T / \ell^2$. Equation (12.30) can then be written as

$$D_c = D_0 \frac{h}{\ell} \qquad (12.33)$$

The diffusion coefficient D_c is extremely large; it is the product of the molecular diffusion D_0 and a factor h/ℓ, where h is a macroscopic length and ℓ is a molecular length.

In conclusion, backflow leads to a strong enhancement of the amphiphile diffusion coefficient for a shallow water subphase. For a bulk water subphase, the same conclusion holds but the flow penetrates only up to a layer of thickness L (from the surface), which is frequency dependent [96], as can be seen on Fig. 12.12b. For a mode of wavevector q and frequency ω one has

$$L^{-1} = \sqrt{q^2 + \omega \rho / \eta} \qquad (12.34)$$

where η / ρ is the kinematic viscosity . For q vectors larger than $q_c \simeq \omega \rho / \eta$, inertial effects are negligible: $L^{-1} \simeq q$ and $D_c(q)$ is then given by

$$D_c(q) = \frac{D_0}{q\ell} \qquad (12.35)$$

12.4.2   Dynamics of Phase Separation in Monolayers

Now imagine a rapid change in either temperature or surface pressure such that the monolayer is rapidly quenched from a homogeneous one-phase region into a two-phase coexistence region or into a one-phase modulated region. If in the final state the original homogeneous phase is *metastable*, the initial stages of the growth dynamics are characterized by a nucleation process requiring an activation energy to initiate the phase separation. On the other hand, if the original homogeneous phase is *unstable* after the quench, the initial stage of the dynamics is characterized by a fast amplification of local concentration fluctuations [99-100]. For the monolayer geometry this amplification causes a divergence in the so-called longitudinal Lucassen modes [101-102]. This dynamical process is at the basis of *spinodal decomposition* and we will consider it now in the specific case of dipolar monolayers with electrostatic interactions.

The longitudinal modes of a monolayer on a bulk liquid have been studied in detail by Lucassen [101]. Contrary to usual liquid – liquid phase separation where the soft mode is diffusive, they are propagative in the range of small wavevectors because the inertial component is important. For simplicity, we discuss here [103] only the case of an amphiphilic monolayer covering a thin liquid subphase. As in the previous section, the hydrodynamics is treated only within the lubrication approximation which simplifies the dispersion relation.

When the monolayer in a homogeneous state is suddenly quenched into the coexistence region, it behaves as an elastic membrane with $u_q = u_0 e^{iqx} e^{i\omega t}$ being its displacement in the x-direction for a wavevector q. The fluctuation of the surface density ψ_q is related to u_q by a conservation equation $\psi_q = -\partial u_q / \partial x$. For any concentration fluctuation, there is an elastic restoring force f_{elastic}, which can be calculated from the free energy density (12.14-16)

$$(\Delta F_0 + F_{\text{el}} + F_I)/k_B T = \frac{1}{2}d(T - T_0)\psi^2(\vec{\rho}) + \frac{1}{4}u\psi^4(\vec{\rho}) + \frac{1}{2}\sum_q (a^{*2}q^2 - b^3|q|)\psi_q^2 \qquad (12.36)$$

where the order parameter ψ and the coefficients d and u were introduced in Sec. 12.3. The optimal mode that minimizes (12.36) is $|\vec{q}\,|^* = b^3/2a^{*2}$ [see Eq. (12.18)] and the critical temperature is renormalized upwards: $T_c = T_0 + b^6/4da^{*2}$. The elastic restoring force f_{elastic} calculated from (12.36) is

$$f_{\text{elastic}}(q) = E(q)\frac{\partial^2 u_q}{\partial x^2} \qquad (12.37)$$

with

$$E(q)/k_B T = d(T - T_0) + a^{*2}q^2 - b^3|q| \qquad (12.38)$$

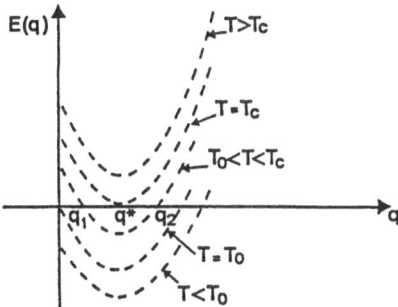

FIGURE 12.13. Elastic modulus E of a polar Langmuir monolayer as a function of the wavevector q for several temperatures. Due to the repulsive long-range dipolar interactions, the sign of E is reversed at a finite $q = q^*$ corresponding to the periodicity of the modulated phase, $2\pi/q^*$.

The membrane is unstable if the elastic restoring force is negative and $E(q) = 0$ defines the spinodal line. The dependence of $E(q)$ on q is shown in Fig. 12.13 for several temperatures. For $T > T_c$, E is always positive, i.e., there is no instability, whereas for $T = T_c$, the first unstable mode occurs at $q = q^*$. For $T < T_c$, the range of the unstable modes is $q_{min} < q^* < q_{max}$, and for even lower temperatures, $T < T_0$, the unstable q-modes start from $q_{min} = 0$ and terminate at q_{max}.

For a monolayer covering a thin water layer of thickness h and in the limit $qh \ll 1$, the lubrication approximation holds and any fluctuation in the monolayer induces a shear flow in the liquid. The viscous force on the monolayer is thus simply given by

$$f_{\text{visc}} = \frac{\eta}{h}\frac{\partial u_q}{\partial t} \tag{12.39}$$

where η is the water viscosity.

The balance between f_{elastic} (12.37) and f_{visc} (12.39) determines the equation of motion:

$$E(q)\frac{\partial^2 u_q}{\partial x^2} = \frac{\eta}{h}\frac{\partial u_q}{\partial t} \tag{12.40}$$

This leads to a diffusive mode

$$\frac{1}{\tau_q} = D_c q^2, \quad \text{with} \quad D_c = E\frac{h}{\eta} \tag{12.41}$$

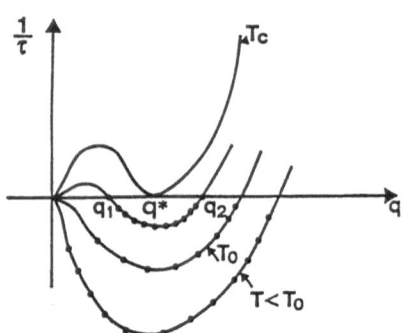

FIGURE 12.14. Relaxation time τ_q of the longitudinal mode of a planar Langmuir monolayer deposited on a thin liquid film. The first instability $(\tau_q^{-1} < 0)$ occurs at the same q-vector that determines the modulations periodicity.

For $E < 0$, $D_c < 0$ and the fluctuations are amplified. If $T \leq T_c$, the amplified modes are the ones close to q^*, and in the range $q_{min} < q < q_{max}$, as shown in Fig. 12.14. As T decreases, the lower threshold of the instability, q_{min}, goes to zero and the maximal wavevector q_{max} increases. q_{max} varies from being equal to q^* at T_c to a large value at low temperatures, $q_{max}^2 = d(T_0 - T)a^{*2}$. Note that in this limit q_{max} is independent of the dipole-dipole interactions.

The results described above can be extended to the case of a monolayer on a thick (bulk) water subphase. For a shallow quench to temperatures just below T_c, the characteristic time is still given by Eq. (12.41), with $h = q^{-1}$ and $D_c = E/q\eta$. For a deep temperature quench below T_c, $T \ll T_c$, the modes are propagative and the dipolar interactions are negligible.

12.5   Conclusions and Future Prospects

The growing body of experimental research on monolayers has resolved many long-standing controversies about the nature of monolayer phases but at the same time has posed many new questions. The nature of the transitions G → LE → LC has been clarified by experiments and much theoretical effort has gone into understanding the sequence of phases as a dilute monolayer is compressed. But the more recent experiments, which have confirmed the existence of many condensed phases, have in essence turned the problem about — how does one understand the sequence of phases that arise when a close-packed monolayer is expanded?

In part, the answer to this question can be found in theories that have been applied to liquid crystals, but the analogies between monolayers and smectics are not exact. The water surface and the bifunctional nature of the amphiphile break the symmetry and lead to new phenomena. In monolayers there is an intriguing coupling between microscopic properties — such as molecular structure, electrostatic interactions, and chirality — and the mesoscopic and macroscopic scales. The ease with which the density of a monolayer can be changed allows a multiplicity of phases to be explored for a single amphiphile and therefore makes it possible to examine in a systematic way how subtle changes in molecular interactions affect the phase behavior.

The line tension between monolayer phases is a key parameter in the theories discussed in this chapter. Measurements of the line tension at the LE-LC and the LE-G interfaces have very recently been undertaken, but the results are rather imprecise. Even less well determined is the anisotropy in the line tension, which manifests itself in the equilibrium shapes of domains and the formation of dendrites.

There is much to be learned about dynamic processes in monolayers. With the exception of studies of the evolution of monolayer foams, nothing is known about the kinetics of phase transitions and growth. While there have been light scattering from surface waves and classical viscometry measurements on monolayer systems, relatively little is known about transport processes in the films.

The interplay between the polar heads of the monolayer and solutes in the subphase is beginning to be examined. It is well known, for example, that divalent ions can induce profound changes in the molecular arrangement within the monolayer. Conversely, an ordered monolayer can be used as a convenient template to induce ordering of a layer of ions or molecules adsorbed from the subphase. This may lead the way to two-dimensional crystallization of low-molecular-weight compounds or even of proteins.

Another challenging future direction is to investigate the effect of mixing in Langmuir monolayers. For example: monolayers composed of two different amphiphiles, two surfactants or, even more generally, mixed monolayers of amphiphiles and polymers, or amphiphiles and proteins. The structure and phase behavior of the mixed monolayer depend both on the overall density (which can be controlled by the surface pressure) and the relative concentration of the two species. As in the case of DPPC + cholesterol, superstructures and spatial inhomogeneities can be strongly coupled to the relative composition.

Apart from their academic interest, mixed monolayers may have many practical applications. It may be possible, for example, to achieve nanolithography and further decrease the dimensions of electonic components by embedding proteins in a Langmuir monolayer and then transferring the monolayer to a solid support. The proteins can then be removed to

create holes of few tens of Angstroms in diameter. It may also be possible
to create ultra-thin membranes for use in separation processes by mixing
a polymerizable amphiphile with another non-reactive component. Once
crosslinking has been achieved, the unreacted species can be removed and
one is left with a two-dimensional mono-molecular membrane of controlled
pore size.

To conclude, Langmuir monolayers are easily accessible and highly versa-
tile two-dimensional systems. They also represent the simplest examples of
self assemblying systems and can be viewed as model systems for biological
cell membranes, vesicles, etc. Much remains to be done, but the activity in
the field of monolayer research remains high and the prognosis for major
advances is excellent.

Acknowledgements

A part of the theoretical results presented here have been obtained in collaboration with P. G. de Gennes and J. F. Joanny. This chapter was written
while one of us (DA) was a Rothschild fellow at the Institut Curie, Paris.
He would also like to acknowledge support from the U.S.-Israel Binational
Science Foundation under grant No. 87-00338, and the Israel Academy of
Sciences and Humanities. CMK acknowledges the support of the U.S. National Science Foundation under Grant CHE-9040092.

References

1. G. L. Gaines, Jr., *Insoluble Monolayers at Liquid-Gas Interfaces* (Interscience, New York, 1966); a) I. Langmuir, J. Am. Chem. Soc. **39**, 354 (1917); b) I. Langmuir, J. Chem. Phys. **1**, 756 (1933).

2. A. W. Adamson, *Physical Chemistry of Surfaces*, 4th Ed. (Wiley, New York, 1982), Chap. IV.

3. S. Ställberg-Stenhagen and E. Stenhagen, Nature **156**, 239 (1945).

4. E. Stenhagen in *Determination of Organic Structures by Physical Methods*, W. A. Braude and F. C. Nachod, eds. (Academic Press, New York, 1955).

5. M. Lundquist, Chem. Ser. **1**, 5 (1971).

6. M. Lundquist, Chem. Ser. **1**, 197 (1971).

7. C. M. Knobler, Adv. Chem. Phys. **77**, 397 (1990).

8. H. Möhwald, Ann. Rev. Phys. Chem. **41**, 441 (1990).

9. R. Peters and K. Beck, Proc. Natl. Acad. Sci. USA **80**, 7183 (1983).

10. V. von Tscharner and H. M. McConnell, Biophys. J. **36**, 409 (1981).

11. M. Lösche and H. Möhwald, Rev. Sci. Instrum. **55**, 1968 (1984).

12. B. G. Moore, C. M. Knobler, S. Akamatsu, and F. Rondelez, J. Phys. Chem. **94**, 4588, (1990).

13. V. T. Moy, D. J. Keller, H. E. Gaub, and H. M. McConnell, J. Phys. Chem. **90**, 3198 (1986).

14. X. Qui, J. Ruiz-Garcia, K. J. Stine, C. M. Knobler and J. V. Selinger, Phys. Rev. Lett. **67**, 703 (1991).

15. K. Kjaer, J. Als-Nielsen, S. A. Helm, P. Tippmann-Krayer and H. Möhwald, Thin Solid Films **159**, 17 (1988).

16. M. J. Grundy, R. M. Richardson, S. J. Roser, J. Penfold and R. C. Ward, Thin Solid Films **159**, 43 (1988).

17. One must be aware, however, that substitution of D for H in either the amphiphile or subphase may cause gross changes in the phase behavior. See, e.g., O. Bouloussa and M. Dupeyrat, Biochem. Biophys. Acta **896**, 239 (1987) and D. Vaknin, K. Kjaer, J. Als-Nielsen, and M. Lösche, Biophys. J. **59**, 1324 (1991).

18. H. Möhwald, Thin Solid Films **159**, 1 (1988).

19. S. Barton, A. Goudot, O. Bouloussa, F. Rondelez, B. Lin, F. Novak, A. Acero and S.A. Rice, J. Chem. Phys. **96**, 1343 (1992).

20. T. Rasing, Y. R. Shen, M. W. Kim, and S. Grubb, Phys. Rev. Lett. **55**, 2903 (1985).

21. X. Zhao, C. Goh, and K. B. Eisenthal, J. Phys. Chem. **94**, 2222 (1990).

22. P. Guyot-Sionnest, J. H. Hunt, and Y. R. Shen, Phys. Rev. Lett. **59**, 1597 (1987).

23. R. A. Dluhy, J. Phys. Chem. **90**, 1373 (1986).

24. S. Garoff, H. W. Deckman, J. H. Dunsmuir, M. S. Alvarez and J. M. Bloch, J.de Physique **49**, 701 (1986).

25. A. Fischer, M. Lösche, M. Möhwald and E. Sackmann, J. de Physique Lett. **45**, L-785 (1984).

26. J. K. H. Hörber, C. A. Lang, T. W. Hänsch, W. M. Heckl and H. Möhwald, Chem. Phys. Lett. **145**, 151 (1988).

27. E. Meyer, L. Howald, R. M. Overney, H. Heinzelmann, J. Frommer, H.-J. Güntherodt, T. Wagner, H. Schier, and S. Roth, Nature **349**, 398 (1991).

28. G. S. Patil, S. S. Katti, and A. B. Biswas, J. Colloid Interface Sci. **25**, 462 (1967).

29. B. M. Abraham, K. Miyano, J. B. Ketterson, and W. Q. Xu, Phys. Rev. Lett. **51**, 1975 (1983).

30. P. J. Winch and J. C. Earnshaw, J. Phys.: Condens. Matter **1**, 7187 (1989).

31. Th. Rasing, H. Hsiung, Y. R. Shen, and M. W. Kim, Phys. Rev. A **37**, 2732 (1988).

32. S. Hénon and J. Meunier, Rev. Sci. Instrum. **62**, 936 (1991).

33. D. Hönig and D. Möbius, J. Phys. Chem. **91**, 4590 (1991).

34. S. Akamatsu and F. Rondelez, J. de Physique **1**, 1309 (1991).

35. M. W. Kim and D. S. Cannell, Phys. Rev. A **13**, 411 (1976).

36. N. R. Pallas and B. A. Pethica, J. Chem. Soc. Faraday Trans. I **83**, 585 (1987).

37. A detailed discussion of this problem can be found in Ref. 7.

38. N. R. Pallas and B. A. Pethica, Langmuir 1, 509 (1985).

39. J. C. Earnshaw and P. J. Winch, J. Phys. : Condens. Matter 2, 8499 (1990).

40. A. M. Bibo and I. R. Peterson, Adv. Mater. 2, 151 (1990).

41. K. Kjaer, J. Als-Nielsen, C. A. Helm, L. A. Laxhuber and H. Möhwald, Phys. Rev. Lett. 58, 2224 (1987).

42. C. A. Helm, H. Möhwald, K. Kjaer and J. Als-Nielsen, Europhys. Lett. 4, 697 (1987).

43. H. Möhwald, R. M. Kenn, D. Degenhardt. K. Kjaer and J. Als-Nielsen, Physica A 168, 127 (1991).

44. B. Lin, M. C. Shih, T. M. Bohanon, G. E. Ice and P. Dutta, Phys. Rev. Lett. 65, 191 (1990).

45. K. Kjaer, J. Als-Nielsen, C. A. Helm, P. Tippmann-Krayer and H. Möhwald, J. Phys. Chem. 93, 3200 (1989).

46. R. M. Kenn, C. Böhm, A. M. Bibo, I. R. Peterson, H. Möhwald, J. Als-Nielsen and K. Kjaer, J. Phys. Chem. 95, 2092 (1991).

47. M. L. Schlossman, D. K. Schwartz, P. S. Pershan, E. H. Kawamoto, G. J. Kellog and S. Lee, Phys. Rev. Lett. 66, 1599 (1991).

48. S. W. Barton, B. N. Thomas, E. B. Flom, S. A. Rice, B. Lin, J. B. Peng, J. B. Ketterson and P. Dutta, J. Chem. Phys. 89, 5898 (1988).

49. B. Lin, J. B. Peng, J. B. Ketterson, P. Dutta, B. N. Thomas, J. Buontempo and S. A. Rice, J. Chem. Phys. 90, 2393 (1989).

50. C. M. Knobler, Science 249, 870 (1990).

51. H. Bercegol, F. Gallet, D. Langevin and J. Meunier, J. de Physique 50, 2277 (1989).

52. P. Muller and F. Gallet, Phys. Rev. Lett. 67, 1106 (1991); P. Muller and F. Gallet, J. Phys. Chem. 95, 3257 (1991).

53. M. L. Mitchell and R. A. Dluhy, J. Am. Chem. Soc. 210, 712 (1988).

54. A. Fisher and E. Sackmann, J. de Physique 45, 517 (1984).

55. A. M. Bibo, C. M. Knobler and I. R. Peterson, J. Phys. Chem. 95, 5591 (1991)

56. S. B. Dierker, R. Pindak, and R. B. Meyer, Phys. Rev. Lett. 56, 1819 (1986).

57. L. Onsager, Ann. N. Y. Acad. Sci. **51**, 625 (1949).

58. See, for example, M. A. Cotter, J. Chem. Phys. **66**, 4710 (1977).

59. A. Halperin, I. Schechter and S. Alexander, J. Chem. Phys. **86**, 6550 (1987).

60. Z.-Y. Chen, J. Talbot, W. M. Gelbart and A. Ben-Shaul, Phys. Rev. Lett. **61**, 1376 (1988).

61. D. Kramer, A. Ben-Shaul, Z.-Y. Chen and W. M. Gelbart, J. Chem. Phys. **96**, 2236 (1992).

62. C. M. Roland, M. J. Zuckermann and A. Georgallas, J. Chem. Phys. **86**, 5812 (1987).

63. A. Caille, D. Pink, F. de Verteuil and M. J. Zuckermann, Can. J. Phys. **58**, 581 (1980).

64. J. L. Viovy, W. M. Gelbart and A. Ben-Shaul, J. Chem. Phys. **87**, 4114 (1987).

65. See, for example, discussion in the review by G. M. Bell, L. L. Combs and L. J. Dunne, Chem. Rev. **81**, 15 (1981).

66. S. Shin, Z.-G. Wang and S. A. Rice, J. Chem. Phys. **92**, 1427 (1990).

67. P. J. Flory, *Principles of Polymer Chemistry* (Cornell, 1953).

68. T. L. Hill, *Introduction to Statistical Thermodynamics* (Addison-Wesley, 1960).

69. R. S. Cantor and P. M. McIlroy, J. Chem. Phys. **90**, 4423 and 4431 (1989).

70. K. To, A. Goudot, O. Bouloussa and F. Rondelez, unpublished.

71. J. Harris and S. A. Rice, J. Chem. Phys. **89**, 5898 (1988).

72. J. P. Bareman, G. Cordini and M. L. Klein, Phys. Rev. Lett. **60**, 2152 (1988).

73. See, for example, K. A. Motakabbir and M. Berkowitz, Chem. Phys. Lett. **176**, 61 (1991); M. Townsend, J. Gryko and S. A. Rice, J. Chem. Phys. **82**, 4391 (1985).

74. H. M. McConnell and V. T. Moy, J. Phys. Chem. **92**, 4520 (1988).

75. A. Miller and H. Möhwald, J. Chem. Phys. **86**, 4258 (1987).

76. H. M. McConnell, Ann. Rev. Phys. Chem. **42**, 171 (1991); J. Phys. Chem. **42**, 17, (1991).

77. T. K. Vanderlick and H. Möhwald, J. Phys. Chem. **94**, 886 (1990).

78. D. Andelman, F. Brochard and J. F. Joanny, J. Chem. Phys. **86**, 3673 (1987).

79. D. Andelman, F. Brochard, P. G. de Gennes and J. F. Joanny, C. R. Acad. Sci. (Paris) **301**, 675 (1985).

80. D. Andelman, Mat. Res. Soc. Symp. Proc. **177**, 337 (1990).

81. A. Miller and H. Möhwald, Europhys. Lett. **2**, 67 (1986).

82. M. Flörsheimer and H. J. Möhwald, Chem. Phys. Lipids **49**, 231 (1989).

83. S. Subramaniam and H. M. McConnell, J. Phys. Chem. **91**, 1715 (1987).

84. M. Seul and M. J. Sammon, Phys. Rev. Lett. **64**, 1903 (1991).

85. T. Garel and S. Doniach, Phys. Rev. B **26**, 325 (1982).

86. C. Kooy and U. Enz, Phillips Res. Rep. **15**, 7 (1960); M. Seul, unpublished.

87. R. E. Rosensweig, *Ferrohydrodynamics* (Cambridge, 1985).

88. C. A. Helm and H. Möhwald, J. Phys Chem. **92**, 1261 (1988).

89. M. Lösche, H.-P. Duwe and H. Möhwald, J. Colloid Interface Sci. **126**, 432 (1988).

90. K. To, S. Akamatsu and F. Rondelez, unpublished.

91. P. Pieranski, Phys. Rev. Lett. **45**, 569 (1980).

92. S. A. Brazovskii, Zh. Eksp. Teor. Fiz. **68**, 175 (1975) Sov. Phys. JETP **41**, 85 (1975)].

93. The exact summation of the inter-stripe electrostatic interactions is performed in Ref. 94 and is shown also in Ref. 80. In Ref. 78, the identical inter-stripe contribution is expressed as an infinite sum. Taking only the first few terms in the infinite sum gives qualitatively similar results.

94. D. J. Keller, H. M. McConnell and V. T. Moy, J. Phys. Chem. **90**, 2311 (1986).

95. P. G. de Gennes, C. R. Acad. Sci. (Paris) **290**, 119 (1980).

96. V. G. Levich, *Physicochemical Hydrodynamics* (Prentice Hall, New York, 1962).

97. P. G. de Gennes, in *Physics of Disordered Materials*, D. Adler, H. Fritzsche, and S. R. Ovshinsky, eds. (Plenum, New York, 1985), p. 227.

98. J. Crank, *The Mathematics of Diffusion* (Clarendon Press, Oxford, 1975).

99. J. D. Gunton, J. M. San Miguel and P. S. Sahni, in *Phase Transitions and Critical Phenomena*, Vol 8, C. Domb and J. Lebowitz, eds. (Academic Press, New York, 1983).

100. J. D. Gunton and M. Droz, *Introduction to the Theory of Metastable and Unstable States* (Springer, Heidelberg, 1983).

101. J. Lucassen, Trans. Faraday Soc. **64**, 2221 (1968).

102. L. Kramer, J. Chem. Phys. **55**, 2097 (1971).

103. F. Brochard, J. F. Joanny, D. Andelman, in *Physics of Amphiphilic Layers*, J. Meunier, D. Langevin and N. Boccara, eds. (Springer, Heidelberg, 1987).

Index